# QUANTUM NANOCHEMISTRY

*(A Five-Volume Set)*

## Volume II:
## Quantum Atoms and Periodicity

# QUANTUM NANOCHEMISTRY

*(A Five-Volume Set)*

## Volume II:
## Quantum Atoms and Periodicity

**Mihai V. Putz**

Assoc. Prof. Dr. Dr.-Habil. Acad. Math. Chem.
West University of Timişoara,
Laboratory of Structural and Computational Physical-Chemistry
for Nanosciences and QSAR, Department of Biology-Chemistry,
Faculty of Chemistry, Biology, Geography,
Str. Pestalozzi, No. 16, RO-300115, Timişoara, ROMANIA
Tel: +40-256-592638; Fax: +40-256-592620

&

Principal Investigator of First Rank, PI1/CS1
Institute of Research-Development for Electrochemistry
and Condensed Matter (INCEMC) Timisoara,
Str. Aurel Paunescu Podeanu No. 144,
RO-300569 Timişoara, ROMANIA
· Tel: +40-256-222-119; Fax: +40-256-201-382

E-mail: mv_putz@yahoo.com
URL: www.mvputz.iqstorm.ro

Apple Academic Press Inc.
3333 Mistwell Crescent
Oakville, ON L6L 0A2 Canada

Apple Academic Press Inc.
9 Spinnaker Way
Waretown, NJ 08758 USA

ISBN 13: 978-1-77463-100-3 (pbk)
ISBN 13: 978-1-77188-134-0 (hbk)

### Library and Archives Canada Cataloguing in Publication

Putz, Mihai V., author
Quantum nanochemistry / Mihai V. Putz (Assoc. Prof. Dr. Dr. Habil. Acad. Math. Chem.) West University of Timişoara, Laboratory of Structural and Computational Physical-Chemistry for Nanosciences and QSAR, Department of Biology-Chemistry, Faculty of Chemistry, Biology, Geography, Str. Pestalozzi, No. 16, RO-300115, Timişoara, ROMANIA, Tel: +40-256-592638; Fax: +40-256-592620, & Institute of Research-Development for Electrochemistry and Condensed Matter (INCEMC) Timişoara, Str. Aurel Paunescu Podeanu No. 144, RO-300569 Timişoara, ROMANIA Tel: +40-256-222-119; Fax: +40-256-201-382, E-mail: mv_putz@yahoo.com, URL: www.mvputz.iqstorm.ro.

Includes bibliographical references and index.
Contents: Volume I: Quantum theory and observability -- Volume II: Quantum atoms and periodicity -- Volume III: Quantum molecules and reactivity -- Volume IV: Quantum solids and orderability -- Volume V: Quantum structure–activity relationships (Qu-SAR).
Issued in print and electronic formats.
ISBN 978-1-77188-133-3 (volume 1 : hardcover).--ISBN 978-1-77188-134-0 (volume 2: hardcover).--ISBN 978-1-77188-135-7 (volume 3 : hardcover).-- ISBN 978-1-77188-136-4 (volume 4 : hardcover).--ISBN 978-1-77188-137-1 (volume 5 : hardcover).--ISBN 978-1-4987-2953-6 (volume 1 : pdf).--ISBN 978-1-4987-2954-3 (volume 2 : pdf).--ISBN 978-1-4987-2955-0 (volume 3 : pdf).--ISBN 978-1-4987-2956-7 (volume 4 : pdf).--ISBN 978-1-4987-2957-4 (volume 5 : pdf) 1. Quantum chemistry. 2. Nanochemistry. I. Title.
QD462.P88 2016          541'.28          C2015-908030-4          C2015-908031-2

### Library of Congress Cataloging-in-Publication Data

Names: Putz, Mihai V., author.
Title: Quantum nanochemistry / Mihai V. Putz.
Description: Oakville, ON, Canada ; Waretown, NJ, USA : Apple Academic Press, [2015-2016] | "2015 | Includes bibliographical references and indexes.
Identifiers: LCCN 2015047099| ISBN 9781771881388 (set) | ISBN 1771881380 (set) | ISBN 9781498729536 (set ; eBook) | ISBN 1498729533 (set ; eBook) | ISBN 9781771881333 (v. 1 ; hardcover) | ISBN 177188133X (v. 1 ; hardcover) | ISBN 9781498729536 (v. 1 ; eBook) | ISBN 1498729533 (v. 1 ; eBook) | ISBN 9781771881340 (v. 2 ; hardcover) | ISBN 1771881348 (v. 2 ; hardcover) | ISBN 9781498729543 (v. 2 ; eBook) | ISBN 1498729541 (v. 2 ; eBook) | ISBN 9781771881357 (v. 3 ; hardcover) | ISBN 1771881356 (v. 3 ; hardcover) | ISBN 9781498729550 (v. 3 ; eBook) | ISBN 149872955X (v. 3 ; eBook) | ISBN 9781771881364 (v. 4 ; hardcover) | ISBN 1771881364 (v. 4 ; hardcover) | ISBN 9781498729567 (v. 4 ; eBook) | ISBN 1498729568 (v. 4 ; eBook) | ISBN 9781771881371 (v. 5 ; hardcover) | ISBN 1771881372 (v. 5 ; hardcover) | ISBN 9781498729574 (v. 5 ; eBook) | ISBN 1498729576 (v. 5 ; eBook) Subjects: LCSH: Quantum chemistry. | Chemistry, Physical and theoretical. | Nanochemistry. | Quantum theory. | QSAR (Biochemistry)
Classification: LCC QD462 .P89 2016 | DDC 541/.28--dc23
LC record available at http://lccn.loc.gov/2015047099

*To solve the electronic frontier (valence or bonding) problems on the basis of the many-electronic ground state stands for quantum chemists almost as "to determine the weight of the captain of a large ship by weighing the ship when he is and when he is not on board"*
(Coulson, 1960)

**To XXI Scholars**

# CONTENTS

# LIST OF ABBREVIATIONS

| | |
|---|---|
| AIM | atoms-in-molecules |
| BE | Becke-Edgecombe |
| BP | boiling point |
| BS | basis set |
| CKS | Chapman-Komogorov-Smoluchowski |
| DFT | density functional theory |
| DOS | density of state |
| EA | electron affinity |
| ELF | electronic localization functions |
| EN | electronegativity |
| EXP | experimental |
| FD | finite-difference |
| FK | Feynman-Kleinert |
| FP | Fokker-Planck |
| GGA | general gradient approximation |
| HSAB | hard and soft acids and bases |
| IP | ionization potential |
| KM | Kramers-Moyal operator |
| KMS | Kubo-Martin-Schwinger |
| LCAO | linear combination of atomic orbitals |
| LDA | local density approximation |
| LSDA | local spin density approximation |
| MH | maximum hardness |
| MJ | Mulliken-Jaffe |
| MK | Moyal-Kramers series |
| MP | melting point |
| PI | path integral |
| QM | quantum mechanics |
| QME | quantum master equation |
| QS | quantum statistics |

| QSPR | quantitative structure-property relationship |
| RMS | root mean square |
| RS | stability ratio |
| SCF | self-consistent field |
| TF | Thomas-Fermi theory |
| TFDW | Thomas-Fermi-Dirac-von Weizsaecker theory |
| TRK | Thomas-Reiche-Kuhn |
| VSEPR | valence shell electron pair repulsion |
| WKB | Wentzel-Kramers-Brillouin |

# PREFACE TO FIVE-VOLUME SET

Dear Scholars (Student, Researcher, Colleague),

I am honored to introduce *Quantum Nanochemistry*, a handbook comprised of the following five volumes:

*Volume I: Quantum Theory and Observability*
*Volume II: Quantum Atoms and Periodicity*
*Volume III: Quantum Molecules and Reactivity*
*Volume IV: Quantum Solids and Orderability*
*Volume V: Quantum Structure–Activity Relationships (Qu-SAR)*

This treatise, a compilation of my lecture notes for graduates, post-graduates and doctoral students in physical and chemical sciences as well as my own post-doctoral research, will serve the scientific community seeking information in basic quantum chemistry environments: from the fundamental quantum theories to atoms, molecules, solids and cells (chemical–biological/ligand–substrate/ligand–receptor interactions); and will also creatively explain the quantum level concepts such as observability, periodicity, reactivity, orderability, and activity explicitly.

The book adopts a three-way approach to explain the main principles governing the electronic world:

- firstly, *the introductory principles* of quantumchemistry are stated;
- then, they are analyzed as *primary concepts* employed to understand the microscopic nature of objects;
- finally, they are explained through *basic analytical equations* controlling the observed or measured electronic object.

It explains the first principles of quantum chemistry, which includes quantum mechanics, quantum atom and periodicity, quantum molecule and reactivity, through two levels:

- *fundamental* (or *universal*) character of matter in isolated and interacting states; and
- the primary concepts elaborated for a beginner as well as an advanced researcher in quantum chemistry.

Each volume tells the "story of quantum chemical structures" from different viewpoints offering new insight to some current quantum paradoxes.

- The **first volume** covers the concepts of nuclear, atomic, molecular and solids on the basis of quantum principles—from Planck, Bohr, Einstein, Schrödinger, Hartree–Fock, up to Feynman Path Integral approaches;
- The **second volume** details an atom's quantum structure, its diverse analytical predictions through reviews and an in-depth analysis of atomic periodicities, atomic radii, ionization potential, electron affinity, electronegativity and chemical hardness. Additionally, it also discusses the assessment of electrophilicity and chemical action as the prime global reactivity indices while judging chemical reactivity through associated principles;
- The **third volume** highlights chemical reactivity through molecular structure, chemical bonding (introducing bondons as the quantum bosonic particles of the chemical field), localization from Hückel to Density Functional expositions, especially how chemical principles of electronegativity and chemical hardness decide the global chemical reactivity and interaction;
- The **fourth volume** addresses the electronic order problems in the solid state viewed as a huge molecule in special quantum states; and
- The **fifth volume** reveals the quantum implication to bio-organic and bio-inorganic systems, enzyme kinetics and to pharmacophore binding sites of chemical–biological interaction of molecules through cell membranes in targeting specific bindings modeled by celebrated QSARs (Quantitative Structure–Activity Relationships) renamed here as Qu–SAR (Quantum Structure–Activity Relationships).

Thus, the five-volume set attempts, for the first time ever, to unify the introductory principles, the primary concepts and the basic analytical equations against a background of quantum chemical bonds and interactions (short, medium and long), structures of matter and their properties: periodicity of atoms, reactivity of molecules, orderability of solids, and activity of cells (through an advanced multi-layered quantum structure–activity unifying concepts and algorithms), and observability measured throughout all the introduced and computed quantities (Figure 0.0).

It provides a fresh perspective to the "quantum story" of electronic matter, collecting and collating both research and theoretical exposition the "gold" knowledge of the quantum chemistry principles.

The book serves as an excellent reference to undergraduate, graduate (Masters and PhDs) and post-doctoral students of physical and chemical sciences; for it not only provides basics and essentials of applied quantum theory, but also leads to unexplored areas of quantum science for future research and development. Yet another novelty of the book set is the intelligent unification of the quantum principles of atoms, molecules, solids and cells through the qualitative–quantitative principles underlying the observed quantum phenomena. This is achieved through unitary analytical

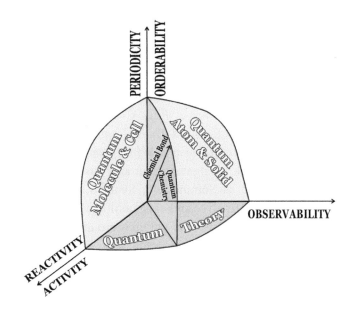

**FIGURE 0.0**    The featured concepts of the "First Principles of Quantum Chemistry" five-volume handbook as placed in the paradigmatic chemical orthogonal space of atoms and molecules.

exposition of the quantum principles ranging from quanta's nature (either as ondulatory and corpuscular manifestations) to wave function, path integral and electron density tools.

The modern quantum theories are reviewed mindful of their implications to quantum chemistry. Atomic, molecular, solid-state structures along cell/biological activity are analytically characterized. Major quantum aspects of the atomic, molecular, solid and cellular structure, properties/activity features, conceptual and quantitative correlations are unitarily reviewed at basic and advanced physical-chemistry levels of comprehension.

Unlike other available textbooks that are written as monographs displaying the chapters as themes of interests, this book narrates the "story of quantum chemistry" as *an extended review paper*, where theoretical and instructional concepts are appropriately combined with the relevant schemes of quantization of electronic structures, through path integrals, Bohmian, or chemical reactivity indices. The writing style is direct, concise and appealing; wherever appropriate physical, chemical and even philosophical insights are provided to explain quantum chemistry at large.

The author uses his rich university teaching experience of 15 years in physical chemistry at West University of Timisoara, Romania, along with his research expertise in treating chemical bond and bonding through conceptual and analytical quantum mechanical methods to explain the concepts. He has been a regular contributor to many physical-chemical international journals (*Phys Rev, J Phys Chem, Theor Acc Chem, Int J Quantum Chem, J Comp Chem, J Theor Comp Chem, Int J Mol Sci, Molecules, Struct Bond, Struct Chem, J Math Chem, MATCH*, etc.).

In a nutshell, the book amalgamates an analysis of the earlier works of great professors such as Sommerfeld, Slater, Landau and Feynman in a methodological, informative and epistemological way with practical and computational applications. The volumes are layered such that each can be used either individually or in combination with the other volumes. For instance, each volume reviews quantum chemistry from its level: as quantum formalisms in Volume I, as atomic structure and properties in Volume II, as detailed molecular bonding in Volume III, as crystal/solid state (electronic) in Volume IV, and as pharmacophore activity targeting specific bindings in Volume V.

To the best of my knowledge, such a collection does not exist currently in curricula and may not appear soon as many authors prefer to publish well-specialized monographs in their particular field of expertise. This multiple volumes' work, thus, assists academic and research community as a complete basic reference of conceptual and illustrative value.

I wish to acknowledge, with sincerity, the quantum flaws that myself and many researchers and professors make due to stressed delivery of papers using computational programs and software to report and interpret results based on inter-correlation. I feel, therefore, the need of a new comprehensive quantum chemistry reference approach and the present five-volume set fills the gap:

- *Undergraduate students* may use this work as an *introductory and training textbook* in the quantum structure of matter, for basic course(s) in physics and chemistry at college and university;
- *Graduate (Master and Doctoral) students* may use this work as the *recipe book* for analytical research on quantum assessments of electronic properties of matter in the view of chemical reactivity characterization and prediction;
- *University professors and tutors* may use this work as a *reference textbook* to plan their lectures and seminars in quantum chemistry at undergraduate or graduate level;
- *Research (Academic and Institutes) media* may use this work as a *reference monograph* for their results as it contains many tables and original results, published for the first time, on the atomic-molecular quantum energies, atomic radii and reactivity indices (e.g., electronegativity, chemical hardness, ionization and electron affinity results). It also has a collection of original, special and generally recommended literature, integrated results about quantum structure and properties.
- *Industry media* may use this work as a *working tool book* while assessing envisaged theoretical chemical structures or reactions (atoms-in-molecule, atoms-in-nanosystems), including molecular modeling for pharmaceutical purposes, following the presented examples, or simulating the physical–chemical properties before live production;

- *General media* may use this work as an *information book* to get acquainted with the main and actual quantum paradigms of matter's electronic structures and in understanding and predicting the chemical combinations (involving electrons, atoms and molecules) of Nature, because of its educative presentation.

I hope the academia shares the same enthusiasm for my work as the author while writing it and the professionalism and exquisite cooperation of the Apple Academic Press in publishing it.

Yours Sincerely,

*Mihai V. Putz,*
*Assoc. Prof. Dr. Dr.-Habil. Acad. Math. Chem.*
West University of Timişoara
& R&D National Institute for Electrochemistry and Condensed Matter Timişoara
(Romania)

# ABOUT THE AUTHOR

**Mihai V. PUTZ** is a laureate in physics (1997), with an MS degree in spectroscopy (1999), and PhD degree in chemistry (2002), with many post-doctorate stages: in chemistry (2002-2003) and in physics (2004, 2010, 2011) at the University of Calabria, Italy, and Free University of Berlin, Germany, respectively. He is currently Associate Professor of theoretical and computational physical chemistry at West University of Timisoara, Romania. He has made valuable contributions in computational, quantum, and physical chemistry through seminal works that appeared in many international journals. He is Editor-in-Chief of the *International Journal of Chemical Modeling* (at NOVA Science Inc.) and the *New Frontiers in Chemistry* (at West University of Timisoara). He is member of many professional societies and has received several national and international awards from the Romanian National Authority of Scientific Research (2008), the German Academic Exchange Service DAAD (2000, 2004, 2011), and the Center of International Cooperation of Free University Berlin (2010). He is the leader of the Laboratory of Computational and Structural Physical Chemistry for Nanosciences and QSAR at Biology-Chemistry Department of West University of Timisoara, Romania, where he conducts research in the fundamental and applicative fields of quantum physical-chemistry and QSAR. In 2010 Mihai V. Putz was declared through a national competition the Best Researcher of Romania, while in 2013 he was recognized among the first Dr.-Habil. in Chemistry in Romania. In 2013 he was appointed Scientific Director of the newly founded Laboratory of Structural and Computational Physical Chemistry for Nanosciences and QSAR in his alma mater of West University of Timisoara, while from 2014, he was recognized by the Romanian Ministry of Research as Principal Investigator of first rank/degree (PI1/CS1) at National Institute for Electrochemistry and Condensed Matter (INCEMC) Timisoara. He is also a full member of International Academy of Mathematical Chemistry.

# FOREWORD TO *VOLUME II: QUANTUM ATOMS AND PERIODICITY*

The frontiers of quantum physics and quantum chemistry are constantly changing with new discoveries and even bolder theoretical speculations. In the recent past, quantum mechanics have reached to such enormous heights that even seasoned theoreticians find it difficult to comprehend theories published in their own field of expertise. In my opinion understanding quantum mechanics is very much like climbing a steep hill and then moving on a smooth valley. Once one is able to overcome the barrier, one is then able to see clearly, the beautiful philosophy behind the abstract notions.

Quantum mechanics has come a long way from understanding atoms, invention of laser, quantum Hall effect, quantum nano-structures to quantum gases. The starting point of understanding advanced quantum mechanics is to first understand the more basic building blocks of nature, i.e., atoms and molecules and their properties. To this end, one then needs to appreciate atoms and molecules both from the perspective of a physicist as well as a chemist. Both these perspectives are essential to fully appreciate the interdisciplinary nature of science prevalent now days.

As varied as the interests of landmark entrepreneur and scientist Alfred Nobel, the present book *Quantum Atom and Periodicity* by Mihai V. Putz, explores the story of periodicity of chemical elements by incorporating topics spanning from history to topics as complex as Feynman-Kleinert variational theory, density functional theory as well as quantum theory of periodicity. The author has very nicely established a connection between the classical views and quantum theory about periodicity of elements. With its clear descriptions and explanations, readers from both physics as well as chemistry can appreciate and comprehend the book.

I recommend the book highly to students of both physics and chemistry who wish to understand the quantum theory of periodicity of elements.

***Aranya B. Bhattacherjee***
School of Physical Sciences
Jawaharlal Nehru University
New Delhi, India
*November 2015*

# PREFACE TO *VOLUME II:*
*QUANTUM ATOMS AND PERIODICITY*
*(UNIVERSE IN MACRO, AS IN*
*MICRO-SHELLS)*

## THE SCIENTIFIC PREMISES

Most of the Universe is made up of energy and substance. While energy is the ability to do mechanical work, movements and variations, the substance is characterized by mass and occupies a determined space. It is further divided into entities with distinct physical and chemical properties, the elements and the atoms. The elements contain a single type of isotopes' atoms (e.g., isotopes' atom of carbon – C), which can take different forms (for Carbon: diamond, graphite or fullerenes). Greek philosophers Leucippus and Democritus were the first to propose atomism – a theory that places atoms as the indivisible units of matter at the Universe's base. Scientific knowledge expanded later, starting from the works of Enrico Fermi (1901–1954), to prove that atom is further made up of *nucleus* (protons and neutrons) with *electrons* revolving around; division of nucleus (by fission in nuclear explosions, for example) generates enormous energy.[1]

Each atom has at least one *proton*, the positively charged particles (each with the charge $+1e$) present at the center of an atom (the nucleus). Elements differ from each other by the number of protons, for example Hydrogen has one proton, Helium has two protons, and so on. The *atomic number*, based on the number of protons in an atom, is a unique characteristic associated with each element. The atomic nucleus also contains neutral particles, *neutrons*, with electric charge but mass almost equal to the protons. Some researchers accept the neutrons as composite particles,

---

[1]Putz, M. V. (2006). The Structure of Quantum Nanosystems (in Romanian), West University of Timişoara Publishing House, Timişoara.

comprising one proton and particles same as electrons, thus accounting for the neutral charge and total mass, slightly above that of a proton (1 atomic mass unit, 1 AMU) characterizing neutrons.

Mass of protons and neutrons together constitute the atomic mass, also called the *atomic weight*. Atoms of elements with same number of protons but differing number neutrons are referred to as *isotopes*. The isotopes have many practical applications; they are used in archaeological dating, biochemical processes, determining the human diet from mummified tissues or bones and so on.

Additionally, some isotopes are *radioisotopes* – they generate radioactivity (electromagnetic waves with specific wavelengths) by transforming spontaneously into other elements (or their isotopes). Examples of radioactive isotope are Carbon-14 ($^{14}C$) or deuterium (hydrogen-2, $^{2}H$), while $^{12}C$ and $^{1}H$ are the stable isotopes of carbon and hydrogen, respectively. Radioisotopes are easy to identify due to the radiations they emit.

The electron is the smallest stable particle of an atom; about 1800 electrons equal the mass of a proton. It orbits around the nucleus in an atomic-orbital area at a speed nearly equal to the speed of light. Its negative charge ($-1e$) and its number equal to the number of protons in an atom provide the so-called *atomic neutrality*. However, many atoms exist in *ionized atomic states* also; when they have more or less electrons than the number of protons. For example, the iron atoms (Fe) have 26 protons in total; however, some Fe atoms have only 23 electrons (they are called $Fe^{3+}$ cations) and others have only 24 electrons ($Fe^{2+}$ cations). There are about 93 stable atoms occurring naturally. New elements with atomic numbers higher than 93 can be obtained under laboratory conditions (through nuclear reactions), but only a few of them occur naturally in the universe.

Approximately 15 billion years ago, the universe modeled from an infinitesimally small point of energy with an infinitely large gravity. The actual universe emerged from that "point zero" by an event called the Big Bang (Great Explosion). Initially, there were only hydrogen atoms, but with continuous expansion following the Great Explosion, the universe cooled and the hydrogen clouds condensed, by gravity, to become stars. Figure P.1 (left)

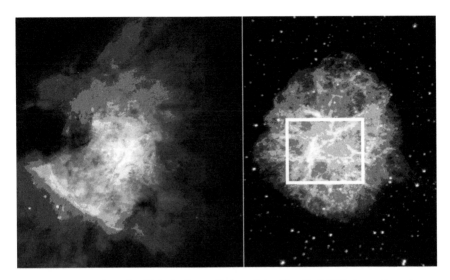

**FIGURE P.1**    Left: Orion Nebula Constellation, right: The Crab Nebula neutron star, as they were registered by Hubble Space Telescope (California) [Hubble Site (2014). http://hubblesite.org/newscenter/archive/releases/2005/37/].

shows the formation/aggregation of Orion Nebula, a constellation (a group of stars), taken by Hubble Space Telescope. The image shows formation of new stars by gravitational collapse of the interstellar dust. Electromagnetic waves emission due to electronic transitions between various atomic and molecular orbitals account for the observed color (from visible spectrum – radiation wavelengths corresponding to the red color until the one specific to purple color).

Figure P.1 (right) shows the remnant of a star that exploded as a supernova: the Crab Nebula, registered by Hubble Space Telescope. This event was first reported in Chinese and Amero-Indian annals around 1054. No European witness of this explosion was recorded till now. However, the so-called neutronic star is at the center of Crab Nebula, consisting of remnants neutrons after the total gravitational collapse; radiation emitted by the neutron star excites electrons from the surrounding matter, generating the visible effect of cosmic dust lighting. As stars grow due to gravitational attraction of surrounding cosmic dust, they become large enough to increase their internal pressure. This enormous pressure causes the fusion of hydrogen atoms. The most important reaction that occurs is the conversion of hydrogen to helium, in a non-nuclear elementary notation: H → He, as noted in Table P.1. This process continues till the formation of iron inside the stars; after which the process can no longer continue because the appearance of iron increases the gravitational attraction from inside and eventually leads to final collapse of the star as a supernova. Our Sun, for example, has not yet reached the age to produce Fe inside. Table P.1 summarizes the Sun mass ($M_S$) and temperature (in Kelvin units, [K]) required for a typical fusion reactions to occur.

The Solar Nebula System composition (containing elements heavier even than Fe) indicates that it is a solar system formed by "recycled material" from previous stars that exploded as supernovae.

Figure P.2 shows the relative abundance of the elements in the universe. It is seen that 99.9% of the elements in the universe are H and He; the abundance scale being a quasi-logarithmic one indicates H to be 10 times more abundant than He. Figure P.2 (right) shows planets' differentiation in our solar system in relation to abundance of atmosphere, silicates (products with Si) and other metals. Planets are made up of rare materials that include elements with affinity for oxygen (*litophile*) Si, Al, Ti, Cr, Mn, $Fe^{3+}$; alkaline elements; alkaline metals (crystals); rare elements; those with affinity for sulfur and oxygen (*calcophile*): Cu, Co, Ni, Zn, Pb, Sb, Mo, $Fe^{2+}$; and metal alloy of iron (*siderophile*): Fe, Ni, Pt, Ir, Os, Re, Au, Rh.

**TABLE P.1**   Nuclear Fusion Reactions Stages and the Elements' Transformation Inside Stars, According to Their Mass (Related to Our Sun $M_S$) and Absolute Temperature Inside Stars (in degrees Kelvin, [K]) (*)

| Stage | Mass | Temperature | Production of nuclei |
|-------|------|-------------|----------------------|
| 1 | $M_S$ | $10^7$ [K] | H → He |
| 2 | | $10^8$ [K] | He → C,O |
| 3 | | $5 \times 10^8$ [K] | C, O → Si |
| 4 | $>30 M_S$ | $5 \times 10^9$ [K] | Si → Fe |

*Mineralogy (2001). *Lectures' Notes of Mineralogy*, University of Bristol (Curator: Prof. D. M. Sherman).

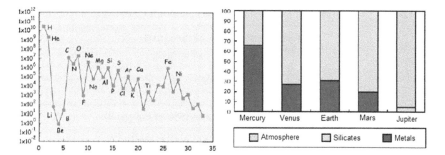

**FIGURE P.2**   Left: the (relative) abundance of elements in Universe; right: the planets differentiation in our solar system according to the (relative) abundance of atmosphere, silicates and metals; after [Mineralogy (2001). *Lectures' Notes of Mineralogy*, University of Bristol (Curator: Prof. D. M. Sherman)].

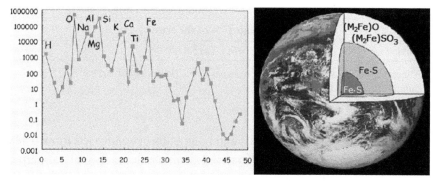

**FIGURE P.3**   Left: the (relative) abundance of Earth's Crust elements; right: the Earth differentiation in metal core and the oxide/silicate mantle; after [Mineralogy (2001). *Lectures' Notes of Mineralogy*, University of Bristol (Curator: Prof. D. M. Sherman)].

Earth's composition presents a high degree of chemical evolution compared to stars, especially considering the differentiation of H and He. Figure P.3 (left) shows the relative abundance of elements in Earth's crust: O and Si are the most abundant elements. Figure P.3 (right) shows sidero-phile elements in the metal core and litophile in the mantle and crust, i.e., the elements of K, Na, Al and B. At the crust level, elements were differentiating between the oceanic and the continental crust. In any case, a classification of silicates at the crust level can be made (Table P.2). It is important to understand that the Earth and planet formation involves

**TABLE P.2**    Classification of Silicate Compounds in Earth's Crust (*)

| Tectosilicates (63% of the crust) | |
|---|---|
| Plagio Class | $(Ca, Na)(Al, Si)AlSi_2O_8$ |
| Orto Class | $KAlSi_3O_8$ |
| Quartz | $SiO_2$ |
| **Phyllosilicates, planar silicates (5% of the crust)** | |
| Biotite | $K(Mg, Fe)_3(AlSi_3)O_{10}(OH, F)_2$ |
| Muscovite | $KAl_2(AlSi_3)O_{10}(OH)_2$ |
| Clorite | $(Mg, Fe)_5Al(AlSi_3)O_{10}(OH)_8$ |
| **Chain silicates (Pyroxene and Amfibole) (16% of the crust)** | |
| Orthopyroxene | $(Mg, Fe)SiO_3$ |
| Augite | $Ca(Mg, Fe)Si_2O_6$ |
| Hornblende | $NaCa_2(Mg,FeAl)_5[(Al, Si)_4O_{11}]_2(OH)_2$ |
| **Inosilicate (3% of the crust)** | |
| Olivine | $(Mg, Fe)_2SiO_4$ |

*Mineralogy (2001). *Lectures' Notes of Mineralogy*, University of Bristol (Curator: Prof. D. M. Sherman).

complex combinations of metals and nonmetals in condensed form. Their study helps in extraction of materials, which have special properties (mechanical, electrical, magnetic, optical, catalytic, etc.) and impacts the industrial and daily life. We present in the following paragraphs, along with examples, the reasons that make them useful.[2]

- *Mechanical properties* make them useful in metal alloys production: Ti for aircraft industry; for construction: cement $(Ca_3SiO_5)$; in ceramics industry: BN, SiC; as lubricants: graphite, $MoS_2$; as abrasives: diamond, quartz $(SiO_2)$.
- *Electrical properties* allow them to be used as metallic conductors – Cu, Ag; as semiconductors – Si, GaAs; as superconductors – $Nb_3Sn$, $YBa_2Cu_3O_7$; as electrolytes – Li in pacemaker batteries; as piezoelectric – α-quartz in watches.

---

[2]Inorganic Chemistry (1999). *Four Lectures in the 1st Year Inorganic Chemistry Course*, Oxford University (Curator: Dr. S. J. Heyes).

- *Magnetic properties* allow them to be used in audio and video technology: $CrO_2$, $Fe_3O_4$.

- *Optical properties* make them useful for producing the required pigments in paint: $TiO_2$; for color television: $Eu^{3+}$ in $Y_2O_3$ produces red color in TV; for laser effect: $Cr^{3+}$ in $Al_2O_3$ in ruby laser; for other physical effects: doubling the light frequency produced by the light passing through $LiNbO_3$.

- *Catalytic properties* find uses in the oil industry, Zeolite ZSM-5 (an aluminosilicate) in refining oil by the reaction: methanol→octane. Other functionalities: The oxygen as sensor: $ZrO_2/CaO$ in solid solution, and a lot more other applications.

The last few decades of research has focused on synergistic quantum approaches involving structure and properties of natural complex systems (polyatomic and biomolecular). A unified approach between physics and chemistry is also visible at the quantum level, mostly related to natural chemical bonds: ionic, covalent, metallic, hydrogenic and the van der Walls (as driven, induction and diffusion) ones. As chemical bonds coexist in various degrees and combinations in matter, a unitary quantum treatment, based on the first physical-chemical principles, is needed to produce an estimation of the structure–properties correlations across the complex natural nano-systems: metals, clusters, fullerenes, liquid crystals, polymers, ceramics, biomaterials, metaloenzymes.

Nanosystems represent a synergistic field of research:

- When the spatiality of the chemical bond is about. When atomic systems condense into smaller volume than the isolated components, the resulting composite nanosystems display exceptional properties of coherences which can be later used in processing, storing and communication of quantum information.[3]

- Alternatively, when dealing with chemical concentration of elements, it is already proved that the range of nano-Molls better

[3]Brinkmann, G., Fowler, P. W., Justus, C. (2003). A catalog of isomerization transformations of fullerene polyhedra. *J. Chem. Inf. Comput. Sci.* 43, 917–927; Liu, J., Shao, M., Chen, X., Yu, W., Liu, X., Qian, Y. (2003). Large-scale synthesis of carbon nanotubes by an ethanol thermal reduction process. *J. Am. Chem. Soc.* 125, 8088–8089; Kobayashi, S.-I., Mori, S., Iida, S., Ando, H., Takenobu, T., Taguchi, Y., Fujiwara, A., Taninaka, A., Shinohara, H., Iwasa, Y. (2003). Conductivity and field effect transistor of La2@C80 metallofullerene. *J. Am. Chem. Soc.* 125, 8116–8117

reflects the complex bio-organic and -inorganic combinations, especially considering the doze zones responses for an essential element, its role in selection or inhibition of a certain biological function in organisms, effect in growth and reproduction of cells and living organisms.[4]

Thus, a unitary approach linking the quantum mechanical formalisms at the chemical bonding level has been extensively researched.[5] It was recently established that for an adequate treatment in quantum space of the polyatomic combinations, the electronic density $\rho(r)$ and not the historical wave function $\psi(r_1,...,r_N)$ stays as the main variable for a system with $N$ electrons. It is so because quite contrary to the wave function, the electronic density is an experimentally detectable quantity defined in the real three-dimensional space, and not within a $3N$ Hilbert abstract space. It is also directly related with the total number of electrons in the concerned system through the functional relation: $\int \rho(\mathbf{r})d\tau = N$. Therefore, the electronic density receives the central role within the newest quantum paradigm of matter, the Density Functional Theory (Walter Kohn as its father, the Nobel laureate in Chemistry for this theory in 1998).

The new quantum view of matter on the nature of chemical bond offered both qualitatively and quantitatively schemes of structure analysis together with the chemical–physical transformations, accurately confirmed through computational and experimental expertise.[6] Consequently, many new chapters of the structural physical chemistry have appeared with *atoms in molecules, reactivity indices,* and *electronic localization* approaches. However, all of them are rooted in the *atomic properties caused by electronic structure*, especially at valence level for the quantum chemical realm. The present volume deals with such topics for the benefit of students and researches in academia and nanotechnology industry. The author is open and grateful to constructive observations and suggestions from the readers.

[4]Sato, K., Hosokawa, K., Maeda, M. (2003). Rapid aggregation of gold nanoparticles induced by non cross-linking DNA hybridization. *J. Am. Chem. Soc.* 125, 8102–8103.
[5]Bader, R. F. W. (2003). Quantum mechanics, or orbitals? *Int. J. Quantum Chem.* 94, 173–177.
[6]Leopoldini, M., Marino, T., Russo, N., Toscano, M. (2004). Density functional computations of the energetic and spectroscopic parameters of quercetin and its radicals in the gas phase and in solvent. *Theor. Chem. Acc.* 111, 210–216.

## VOLUME LAYOUT

The present volume is the *second* in the five-volume set on *Quantum Nanochemistry* listed as:

*Volume I: Quantum Theory and Observability*
**Volume II: Quantum Atoms and Periodicity**
*Volume III: Quantum Molecules and Reactivity*
*Volume IV: Quantum Solids and Orderability*
*Volume V: Quantum Structure–Activity Relationships (Qu-SAR)*

This book consists of the following chapters:

**Chapter 1 (*Historical Highlights on the Periodicity of the Chemical Elements*)**: This chapter introduces the principal objects of Chemistry, the chemical atoms and their elements against the background of the original thoughts of the creators of modern chemistry ranging from Dalton to Berzelius to Mendeleev, appropriately selected, tacitly adnotated and ordered, while preserving their "aetheral" philosophy of science; the route of modern chemistry starting from alchemical philosophy to the natural, physical-chemistry approach is well explained.

**Chapter 2 (*Quantum Assessment for Atomic Stability*)**: This chapter details the alternative quantum mechanical description of total energy explained through path integrals of Feynman–Kleinert formalism, the Bohr quantification of hydrogen atom, solution to the stability issue by existing quantum fluctuation, modeling by periodic paths and Matsubara frequencies, and treating both ground and excited states from quantum statistical perspective.

**Chapter 3 (*Periodicity by Quantum Propagators in Physical Atom*)**: the semiclassical path integral approach provides new definitions, atomic Scales of electronegativity and chemical hardness. The considered quantum probability amplitude up to the fourth-order expansion provides intrinsic electronegativity and chemical hardness analytical expressions in terms of principal quantum number of the concerned valence shell and of the effective atomic charge including screening effects. The present electronegativity scale strikes on different order of magnitudes down the groups of Periodic Table still satisfying the main required acceptability criteria respecting the finite difference based scale. The actual chemical hardness scale improves the trend across periods of Periodic System avoiding the usual irregularities within the old-fashioned energetic picture. The current quest introduces electronegativity of an element as the power with which the frontier electrons are attracted to the center of the atom being a stability measure of the atomic system as a whole. However, both electronegativity and chemical hardness are analyzed for their quantum nature in Fock spaces of electronic occupancies, while maintaining their dichotomy in observability.

**Chapter 4 (*Periodicity by Peripheral Electrons and Density in Chemical Atom*)**: This chapter aims to affirm specific physical-chemical quantities of electronegativity and hardness as the major electronic indicators of structure and reactivity. Their systematic definitions are presented and discussed for valence atomic region, Bohmian quantum mechanics and the associated density functionals, along their related reactivity index as electrophilicity, within conceptual density functional theory in general and for softness bilocal to global quantum observability in particular. It may serve for further analytical studies of periodicity for atomic properties (atomic radii, diamagnetic susceptibility, or polarizability), as well as for future understanding and chemical bonding, reactivity, aromaticity, the biological activity modeling of atoms in molecules and in nanostructures.

**Chapter 5 (*Quantum Algebraic and Stochastic Dynamics for Atomic Systems*)**: This chapter reviews and advances the two related quantum ways for quantum description of valence/interacting/exchanged electrons among atoms at their turn involved in binding or molecular systems (i) by abstract formalization within quantum algebra of open systems; (ii) by analytical formulation within stochastic/dissipative systems. This way one models the electronic distribution, exchange and localization in chemical (inherently open) system, dealing therefore with that chemistry is at its ultimate description: the science of moving electrons from one state to another (either by intra- or inter-atomic framework).

Accordingly, special features of the present volume are that it:

- Presents historical contributions of researchers, appropriately annotated and adapted for instructive and research purpose;

- Continues the quantum mechanical path integral modeling with the specialized Feynman–Kleinert variational formalism leading to a comprehensive understanding of atomic stability;
- Introduces the quantum statistical version of path integral approach of quantum mechanics, while employing the developed quantum propagators/amplitudes in the second- and fourth-order semiclassical expansion, with innovative applications on electronegativity and chemical hardness scales of Periodic Table;
- Characterizes electronegativity and chemical hardness by both quantum observability perspective as well as by Bohmian subquantum approach, under the most disputed and discussed parabolic energetic shape of valence electrons occupancy;
- Reviews the main pictures for electronegativity, from Pauling and Mulliken to the author's analytical density functionals within the softness density functional theory with proper illustration of atomic periodicity;
- Formulates, computes, and discusses the atomic periodic scale for electronegativity and chemical hardness related scales of atomic radii, diamagnetic susceptibility, and polarizability (with new formulation for the last quantity based on quantum Bethe rules);
- Formalizes the atomic structure and reactivity, i.e., the chemical atom, by the algebra of quantum states, eventually continued with Thomas-Fermi realization as the density functional theory precursor, along the modern approach of the electronic localization problem in terms of electronic density combinations;
- Presents the Fokker–Planck quantum description of open systems, by considering the drift and diffusion contributions in quantum evolution, as based on a non-equilibrium Lagrangian, lading with generating of the so called Markovian families of electronic localization functions, accommodating the Thom's catastrophe theory in an innovative manner, thus directly describing the atomic valence distribution as well as the molecular formation.

Kind thanks are expressed to individuals, universities, institutions, and publishers that inspired and supported the topics included in the present volume; a few of them include:

- *Supporting individuals*: Prof. Hagen Kleinert (Free University of Berlin); Priv. Doz. Dr. Axel Pelster (Free University of Berlin); Prof. Adrian Chiriac (West University of Timişoara); Prof. Nino Russo (University of Calabria); Prof. Pratim K. Chattaraj (Department of Chemistry and Center for Theoretical Studies, Indian Institute of Technology, Kharagpur); Prof. Dulal C. Ghosh (Department of Chemistry, University of Kalyani); Prof. Jan C.A. Boeyens (Center for Advancement of Scholarship, University of Pretoria, South Africa); Prof. Laszlo Szentpaly (Institute of Theoretical Chemistry, University of Stuttgart);

- *Supporting universities*: West University of Timişoara (Faculty of Chemistry, Biology, Geography/Biology-Chemistry Department/ Laboratory of Computational and Structural Physical Chemistry for Nanosciences and QSAR); Free University of Berlin (Physics Department/Institute for Theoretical Physics/Research Center for Einstein's Physics, Centre for International Cooperation); University of Calabria (Faculty of Mathematics and Natural Sciences/Chemistry Department);

- *Supporting institutions and grants*: DAAD (German Service for Academic Exchanges) by Grants: 322 A/17690/2004, 322 A/05356/2011; CNCSIS (Romanian National Council for Scientific Research in Higher Education) by Grant: AT54/2006-2007; CNCS-UEFISCDI (Romanian National Council for Scientific Research) by Grant: TE16/2010-2013;

- *Supporting publishers*: Wiley (New York); Springer (London, Berlin-Heidelberg); ACS – American Chemical Society (Washington); Nova Science, Inc. (New York); World Scientific (Singapore); Multidisciplinary Digital Publishing Institute – MDPI (Basel, Switzerland); University of Kragujevac (Serbia). Multidisciplinary Digital Publishing Institute – MDPI (Basel); Wiley (Hoboken, NJ), Springer Science (Berlin – Germany, London – UK, and New York – USA); World Scientific (Singapore); Chemistry Central (London – UK); American Chemical Society (Washington DC, USA); IGI Global (formerly Idea Group, Inc.), Hershey (Pasadena, USA).

The author wishes to express his gratitude to his family, especially his lovely little daughters *Katy & Ela*, for always energizing his atmosphere and creating the work-and-play atmosphere. He is hopeful of passing the enthusiasm to readers and students pursuing scientific knowledge about quantum *atoms* and of their *periodicity* description.

The author especially thanks the publisher, Apple Academic Press (AAP) team and in particular to Ashish (Ash) Kumar, the AAP President and Publisher, and Sandra (Sandy) Jones Sickels, Vice President, Editorial and Marketing, for professional and supervised production of the five volume series, *Quantum Nanochemistry*. He welcomes the *chemical theory* of atomic structure as well its importance in the years to come in science and technology

Keep close and think high!

Yours Sincerely,

**Mihai V. Putz,**
*Assoc. Prof. Dr. Dr.-Habil. Acad. Math. Chem.*
West University of Timişoara
& R&D National Institute for Electrochemistry and Condensed Matter Timişoara
(Romania)

**CHAPTER 1**

# HISTORICAL HIGHLIGHTS ON THE PERIODICITY OF THE CHEMICAL ELEMENTS

## CONTENTS

## ABSTRACT

This chapter is dedicated introducing the main objects of chemistry, namely the chemical atoms and their elements, mostly by reloading of the original thoughts of the creators of modern chemistry, from Dalton, to Berzelius, to Mendeleev, appropriately cited and ordered, for having a glimpse of the flavor in passing from alchemical philosophy to the natural, physical-chemistry approach in the forthcoming sections of this volume.

## 1.1   INTRODUCTION

The concept of chemical element has made a long and sinuous journey in history, while the knowledge of the structure of matter was developing. In ancient and alchemical period, by elements were understood more like "*personifications*" of some property classes, rather than specific forms of substance. In this manner it was defined a small number of essential elements, in general between one and five, which would be the basis of all material forms. As a single item was considered, in turn, *the water* (Thales of Miletus, 624–546 BC), *the air* (Anaximenes, 585–525 BC), and *the fire* (Heraclitus of Ephesus, 535–475 BC). All the substances were made—to the ancient Egyptians and Arab alchemists—from sulfur and mercury, while Paracelsus (1493–1541) proposed three elements: the sulfur, the mercury and the salt. Of course, these names were not related to the substances that we know today, they rather designated certain principles and mixed properties. For instance, the sulfur was related to fire, and the mercury to liquid property of bodies. Nevertheless, some Indian philosophers also believed in these three elements: the earth, the water and the air. Instead, the four elements of Empedocles (490–430 BC) as indestructible and immutable manifestations of Nature were the fire, the water, the air and the earth, all subjected to the two antagonist forces: the love and the hate; he also gave the "composition" of certain materials, e.g., the bones were made from ½ fire, ¼ earth, ¼ water, flesh and blood in equal quantities. Aristotle (384–322 BC) also added to the four elements of Empedocles, a fifth one, the "quintessence," a kind of ether: the first four items were ideated by him as being formed by combining two properties of four primary qualities: hot, cold, wet and dry; these items can transform themselves one to another. Instead, in other part of the world, according to some Chinese philosophers, the five essential elements would be: the water, the fire, the metal, the earth and wood (Horovitz et al., 2000).

The first definition of the chemical element, in the modern sense, is due to R. Boyle (1627–1629); in his book "Skeptical Chemist," he defined the chemical element as a substance which *cannot* be decomposed into other substances. So, the elements (simple bodies) are not composed of other substances, but they themselves are the products in which are decomposed, ultimately, all other substances. Boyle did not shown specifically what these items were about, for example, he could not decide if metals or their oxides had a character of an element. On the other hand, *the negation*, which stay on the basis of his statement, make it uncertain as a practical definition, because even not known the decomposition methods for a substance at a given time (historical epoch), this not necessarily means that such a decomposition is not possible in an arguable future. In any case, the use of the term *element*, in the sense of elemental substance, i.e., simple substance, was maintained until now, with the note that only its significance was made clear as time passed by. An important step was the transition from the meaning of the element notion as "principle," encompassing a collection of properties, to some specific to substance type. Noteworthy, Lavoisier, in his famous "Elementary Treatise of Chemistry (1789) included in the simple substances category approximate 40 substances, of which 25 were really elements (Horovitz et al., 2000).

In the following decades, the number of known elements grew fast, new and new names had appear, some of them disappearing very soon, as being prematurely assigned to "ghost" elements. Certain properties, experimentally observed, were attributed to some inexistent elements. Along with defining the elements as simple substances, it became predominant the idea of the impossibility for elements to transform themselves one to each other. But the faith in a "primordial matter," common to all elements, continued to be attractive to many researchers. In parallel with defining the element notion and with increasing the number of known elements, the atomistic theory was developed turning eventually into atomic structure of chemical matter (Horovitz et al., 2000).

Stacking to the atomistic theory of nature, the Greek philosopher Leucippus (in the first half of fifth century BC) claimed that atoms, indivisible because of their smallness, are in infinite number, are moving in vacuum, are immutable, but may be organized into different aggregates. Contemporarily, Democritus (about 460 BC), considered the atoms as being without qualities and indivisible because of their high hardness. Atoms would not be only constituents of substances, but also of light, so giving the reasoning/thoughts and to the soul a material substrate. Plato (427–247 BC) accepted the idea that the earth, water, air, and fire, are made by particles, but do not admitted the existence of the vacuum: the atoms are eternal, have different shapes, weights, properties, and bond with each other, through "hooks," giving incomplete structures for solids. It was only with the Dalton's ideas (1808) the atomic model of chemists replaced that one proposed by the philosophers: it assumes (and remains as such also nowadays) that each element corresponds to a certain type of atoms, having well definite mass, certain fixed (or at least discernable) dimensions and presenting chemical properties characteristic of that specific element (Horovitz et al., 2000).

After the discovering of the electrolysis laws by Faraday (1834) the concept of ion was introduced from where the idea of electrically charged particles was born, which will stay at the basis of atomic structure. Apparently the physicist Weber was the first who referred to "positive and negative electric atoms," with a negligible mass, in addition to regular neutral atoms, with "ponderable atoms." The discovery of electrically charged particles, precisely the electron, was primarily the result of studies on electrical discharges in gases, especially highlighting the cathodic rays. Remarkably, the existence of the electron (the mediator of negative charge) also involved the presence of certain positive charge constituents in atoms so anticipating the nuclei (with protons) existence (Horovitz et al., 2000).

Nevertheless, the discovery of radioactive processes gave the coup de grace to the conception of the atomic indivisibility. After the discovery of radioactivity, many products of radioactive decay were considered as new elements. Gradually, the concept of isotope was born, which means the possibility those atoms with different masses to belong to the same item/ element. Sometimes the term allotropes it is used also for different forms of an element, consisting of identical units, but differently arranged on a lattice: for example, the rhombic sulfur and the monocyclic one, which have different networks, yet both consisting of $S_8$ molecules. After the modern compiling the periodic table of chemical elements, the number indicating the position of an element in this system has become a new feature, more and more considered, to which, however, only the quantum mechanics succeeded to give a physical rationale (Horovitz et al., 2000).

However, the conceptual "battle" between the main characteristic of atoms of an element as atomic weight and atomic number lead with the Periodic System as one of the few universal symbols of nature instantly recognized worldwide, yet whose chemcial and physical roots are here to be restoried, with the hope of fresh impetus envisaging new atomic structural revelations.

## 1.2 DALTON'S LEGACY FOR MODERN CHEMISTRY

### 1.2.1 DALTON'S NEW PHILOSOPHIC SYSTEM FOR CHEMISTRY (1805–1827)

Dalton's cornerstone atomic theory of nature was laid on the next natural philosophy (Dalton, 1805, 1808, 1810, 1827):

*There are three kinds of bodies, or three states, which have particular required the attention of philosophical chemists, especially those marked by the terms as elastic fluids, liquids or solid; as an example, a body in water behavior, which in some circumstances, it is capable of assuming all three states. In steam we recognize the perfectly elastic fluid, in water the perfect liquid, while in ice a complete solid. These observations tacitly led to the conclusions that seem universal adopted: all relevant bodies, liquids or solids, are constituted of a large number of small particles or matter atoms, together bound by an attraction force, which is more or less strong, depending the circumstances, and which try to prevent their separation (in such case called the cohesion attraction). But once they collect themselves from a dissipated state, the chemical force it is called the aggregation attraction or simply, the affinity. Whatever name it may go by, it has the same power. The Berthollet's chemical affinity law states: the chemical affinity is proportional with the mass, and in all chemical unions there are undetectable gradations in the main constituent proportions.*

*From what there is known, there is no reason to apprehend the diversity in these particulars: if the chemical force does not exist in water, it has to be in equally measure in the elements which constitute the water, namely hydrogen and oxygen. If some of the water particles are heavier than others, there must be supposed that it affects the specific mass gravity, which is an unknown circumstance. Similar observations were made on other substance. Therefore, one can conclude that the ultimate particle of all homogeneous bodies is perfectly similar in weight, figure etc. In other words, every particle of water is the same as other particles of water. Each particle of hydrogen is the same as other particle of hydrogen. Except the attraction force, which in one character or another, universally belongs to the ponderable bodies, there was found another force (also universal) which act upon the matter, and comes as a following of the rendered consequently, namely, a repulsion force. This kind of constant fluid atmosphere surrounds the atoms of all bodies and prevents them to be drawn into a contact. The constitution of bulk can be reduced by abstraction of some of its own heat, but from what has been said in the last section, it seems that the enlargement or the reduction of a bulk depends perhaps more on the rearrangement than on the ultimate particle size, for example, the solid corps, like ice, containing a large part (maybe 4/5) of the heat that was found in the elastic state, as the steam one.*

*When any corps exists in elastic state, its last particles are separated from each other on a much grater distance than in any other state. Each particle occupies the center of a large sphere, and worthily resists by supporting the others, which by their gravity, or otherwise, are arranged in order to defy the possible limit, to a respectable distance. When is attempting to find the particles number in atmosphere, it is as attempting to find the number of stars from universe. But if the subject is limited, by taking a volume of any given gas, leaving the minute division, the number of particles has to be finite, similar with the fact that in a given space of Universe, the number of stars and planets cannot be infinite.*

*The chemical analysis and synthesis go no farther than the particles separation of one from another and by the involved assemble. There is no creation or destruction of matter in the action radius of the chemical agent. We can try to introduce a new planet in the solar system, or to annihilate an already existing one, just like a particle of hydrogen can be created or destroyed. All the changes are produced consistently in particles separation that were in a cohesion or combination state, and also associate with those which were at certain distance.*

*In the all chemical investigations a significant object has been considered that one of ascertain the relative weights of the simples which form a compound. But, unfortunately, everything has been stopped here. From the relative mass weights the relative weights of the last particles or the bodies' atoms can be deduced and from here the number and the weight in various compounds, in order to assist and guide further investigations, and also correct the results. The importance and the advantage of establishing the relative weights of the ultimate particles had been showed for simple and compound bodies, with emphasis on the number of simple elementary particles which constitute a compound and the number of less particles which enter into another compound by particle aggregation.*

*If there were two bodies, A and B, which are willing to combine eachother, the following scheme represent the order in which the combination will occur, starting with the simplest one, namely* (see Figure 1.1)*:*

- *1 atom of A + 1 atom of B = 1 atom of C, binary.*
- *1 atom of A + 2 atoms of B = 1 atom of D, ternary.*
- *2 atoms of A + 1 atom of B = 1 atom of E, ternary.*
- *1 atom of A + 3 atoms of B = 1 atom of F, quaternary.*
- *3 atoms of A + 1 atom of B = 1 atom of G, quaternary.*

*The following general rules were proposed, which may be adopted in all investigations which respect the chemical synthesis:*

- *When only one combination of two bodies can be obtained, it must be presumed to be a binary one, unless by some other cause it appears to be the contrary.*
- *When two combinations are observed, there must be assumed that it should be binary and ternary;*
- *When three combinations are observed, there must be assumed that one should be binary and the other two tertiary;*
- *When four combinations are observed, we can expect at one binary, two tertiary and one quaternary;*

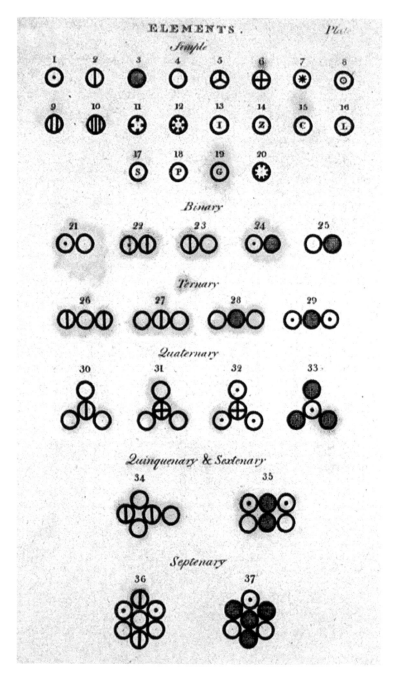

FIGURE 1.1    Combinations of (atomic) elements (Dalton, 1827).

- *A binary compound is always heavier than the mixture of its two ingredients;*
- *A ternary compound is heavier than a simple or binary mixture, which should be part of its constitution;*
- *The rules above, apply equally when the two bodies, such as C and D, D and E, are combined.*

*In order to apply these rules to the chemical phenomena which already exist, we deduce the following conclusions* (see Table 1.1):

- *The water is a binary compound of hydrogen and oxygen, and the relative weights of the two elementary atoms are in ratio 1:7;*

**TABLE 1.1** Dalton's Relative Atomic Weights (Dalton, 1827)

| No. | Symbol | Weight |
|---|---|---|
| *1.* | *Hydrogen* | 1 |
| *2.* | *Azoth* | 5 |
| *3.* | *Carbon* | 5 |
| *4.* | *Oxygen* | 7 |
| *5.* | *Phosphorus* | 9 |
| *6.* | *Sulfur* | 13 |
| *7.* | *Magnesia* | 20 |
| *8.* | *Calcium oxide* | 23 |
| *9.* | *Sodium carbonate* | 28 |
| *10.* | *Potassium* | 42 |
| *11.* | *Strontium oxide* | 46 |
| *12.* | *Barium oxide* | 68 |
| *13.* | *Iron* | 38 |
| *14.* | *Zinc* | 56 |
| *15.* | *Copper* | 56 |
| *16.* | *Lead* | 95 |
| *17.* | *Silver* | 100 |
| *18.* | *Platinum* | 100 |
| *19.* | *Gold* | 140 |
| *20.* | *Mercury* | 167 |

- *The ammonia is a binary compound of hydrogen and azoth, and the relative weights of the two atoms are 1:5;*
- *The nitrous gas is a binary compound of azoth and oxygen; the nitric acid is a binary or tertiary compound, and contain one atom of azoth and two of oxygen, together having 19 weight units; the nitrous oxide is a compound similar to the nitric acid, and contain one atom of oxygen and two of azoth, weighing 17 units;*
- *The carbonic oxide is a binary compound, consisting of one atom of charcoal and one of oxygen, together weighing 12 units; the carbonic acid is a tertiary compound, with two atoms of oxygen and two of charcoal, weighing 19 units;*

*In all these cases the weights are expressed in atoms of hydrogen, which represent the unit* (Dalton, 1808, 1810, 1827).

### 1.2.2   DALTON'S CONCEPTS ON ELEMENTARY MOLECULES, BY AVOGADRO (1811)

The Dalton's atomic theory was immediately faced with molecular samples as stated by Avogadro (1811):

*Gay-Lussac has shown that gases always unite in a very simple proportion by volume, and when the result of the union is a gas, its volume is finding in a simple relation with the compounds. But the quantitative proportion of substances in mixture seems to depend only by the relative number of molecules which are combining and by the number of composed molecules which result. One should admit that there is a very simple relation between the volume of gaseous substance and the number of simple or compound molecules which form them. The first hypothesis which presents this connection (apparently the only one permitted) is represented by the assumption that the number of integral molecules in any gas is always the same for equal volumes, or always proportional with the volumes. Indeed, if one assumes that the number of molecules contained in a given volume are different for different gases, it would be possible to conceive that the law which regulates the molecules distance could give, in all cases, simple relations as those in which the facts are analyzed in order to know the volume and the number of molecules.*

*On the other hand, there is more likely that the molecules of gases being at such a distance, their mutual attraction cannot be exercised, their attraction varying for caloric measure could be limited to the atmosphere condensation formed by this fluid having a bigger extension in a given case than in other and, in consequence, without the distance between the molecules to vary; or, in other words, without the number of molecules contained in a given volume to be different.*

*Dalton, proposed a direct hypothesis (opposed to this one): the caloric quantity is always the same for the molecules of all bodies, whatever the aggregation state is, and the more or less is the attraction for caloric sample it only leads to a greater or a smaller condensation production of this quantity around the molecules, thus varying the distance between the molecules themselves. There was no evidence helping to decide in favor of one of these theories, therefore, one has been inclined towards a neutral hypothesis, according which the distances between molecules and caloric quantities vary according to unknown laws, which is not the case of the presented hypothesis, i.e., based on the simplicity of the relationship between the gases volumes in combination, which otherwise seem inexplicable.*

*Starting from this assumption, there is obvious that such laws have the mean for easily determining the relative weights of the substances obtained in gaseous state and the relative number of these molecules in compounds. The molecular masses ratio is the same with the densities of different gases (on equal temperature and pressure) and the relative number of molecules in a compound is given by the ratio of the gases volumes that form it. For example, since the numbers 1.10359 and 0.07321 express the densities of two hydrogen and oxygen gases (compared with the air atmosphere as unit), the ratio of these two numbers therefore represent the ratio between masses, at equal volumes of these two gases; it will also represent (in the given hypothesis) the ratio of the masses of their molecules. So, the oxygen molecule mass will be 15 times higher than the hydrogen molecule, or more precisely, 15.074 to 1. Similarly, the mass of nitrogen molecule will be respecting the hydrogen one as 0.96913 is to 0.07321, so giving 13 or more specifically 13.238 to 1 proportion. On the other hand, as soon as the ratio of hydrogen and oxygen volumes in the water formation was known to be 2 to 1, the water is seen as the result of the union of each molecule of oxygen with two molecules of hydrogen. Similarly, according to the proportion of volumes, fixed by Gay-Lussac for the elements of ammonia, nitrogen oxide and nitrogen gas, the ammonia results from the union of one molecule of nitrogen and three molecules of hydrogen, the nitrogen gas lays on the combination of a nitrogen molecule with one of oxygen, and the nitric form is based on one molecule of nitrogen and two of oxygen.*

*There is a consideration which, at a first sight, seems in contradiction with the present hypothesis, referring to the composition of substances. It seemed that a molecule composed from two or more elementary molecules should have its mass equal with the sum of masses of these molecules. In particular, if entering a composition, a molecule of a substance is viewed by the union of two or more molecules of other substances, while the molecules number of the compound should remain the same as for the adduct substances. According with the present hypothesis, when a gas is combined with two or more times the volume of another gas, the resulted compound (if it is a gas) must had a volume equal to the first gas. For example, the volume of water in gaseous state was, according to Gay-Lussac, carries two times the volume of oxygen component (or what was entering at the same time) and one of the hydrogen component, instead of being equal to that of the oxygen only. It was assumed that the molecules constituents of a simple gas are not formed by solitary elements of a molecule, but these constituents were by a certain numbers of molecules of this type, united by attraction to form a sigle entity. And then, when the molecules of other substance are further added to the first one, in order to form a compound molecule, the integral molecule which should result, divided itself in two or more parts, made up of half, quarter, etc., the number of elementary molecules that formed the constituent molecule of the first, combined with half, the number of constituent molecules of the second substance which entered in combination with a constituent of the molecule belonging to the first substances, so that the entire number of the compound molecules becomes double etc.*

*Concerning the most known compound gases, there were found examples of duplication to the relative volume to the volume of one of the constituents that are combined with one or more volumes in the order. It was observed this phenomenon for water. Similarly, there is known that the volume of ammonia gas is twice of the nitrogen one which entire to it. Gay-Lussac also showed that the volume of nitrous oxide is equal with that of the nitrogen which forms it, and consequently, it is twice that of oxygen. Therefore, nitrous gas (containing equal volumes of nitrogen and oxygen) had a volume equal to the sum of the two gaseous constituents. In all these cases one had to imagine a split of molecules into two, being possible, in other situations, into four, eight, etc. The possibility to divide the compound molecules can be a priori considered; otherwise, the integral molecules of bodies composed by few substances with a large number of molecules should have an excessive weight, in comparison with the substance of simple molecules. It is therefore conceivable that the Nature has some means of bringing them to the state of the latter and these facts have proved the existence of such means. Then there is also a consideration that made us to admit, in some cases, the molecular division/combination hypothesis in question: the way in which one conceive of the reality of the combination between two gaseous substances united in equal volume without condensation, such as the formation of nitrous gas took place. Assuming that the molecules remain at such a distance in which the mutual attraction of each gas could not be expressed, one cannot imagine that a new attraction could occur between molecules of a gas with those of other* (Avogadro, 1811).

### 1.2.3   MORE ON DALTON'S ELEMENTARY MOLECULES, BY AVOGADRO (1811)

Avogadro (1811) further developed the (atomic) proportion theory into molecules (so anticipating a kind of atoms-in-molecules' theory) by the following considerations.

*Dalton had supposed that the water is formed by the union of hydrogen and oxygen, molecule to molecule. From this and from the mass ratio of the two components resulted in the oxygen one concluded that the molecule mass will be the same as the hydrogen as $7^{1/2}$ to 1 (or, after the evaluation of Dalton, as 6 to 1).*

*The present approach, in accordance with the molecular (weight) hypothesis gives, as it was said in previous section, twice bigger result so than as 15 to 1. For the water molecule, its mass seems to be express by $15 + 2 = 17$ mass units (taking hydrogen as unit), if it will not be a division of molecule in two; but, taking into account such division reduces the mass to half, $8^{1/2}$ (or more precisely 8.537), as one may find itself by dividing the active vapors densities 0.625 (Gay-Lussac) with the hydrogen density 0.0732; yet this mass number differs from 7 (the one given by Dalton) by the difference between values for water composition, so that, with respecting the results made by Dalton, it was correctly approximate the two compensatory errors in combination, the error in the oxygen molecular mass and by the molecular division omission.*

*Dalton assumed that in the nitrogen gas the combination of the nitrogen with the oxygen is not varying from molecule to molecule. Therefore, Dalton found the same molecular mass for nitrogen, always considering that the hydrogen is the unit; he did not employed another value for oxygen and took the same value for the quantities of elements in gas after weight. But, assuming that the oxygen molecule is smaller than half of what was found, he was forced to make the nitrogen equal with less than half of the assigned value, i.e., 5 instead of 13 units. Regarding the actual molecule, he neglected the molecular division, and made it as 6 + 5 = 11, roughly as (15 + 13)/2 = 14, or, precisely like (15.074+13.238)/2=14.156, as he divided 1.03636, the nitrous gas density after Gay-Lussac, with 0.07321. Dalton also fixed in the same manner the relative number of molecules in nitrous oxide and in nitric acid; in the first case as well, the same circumstances were considered as rectifying the result for the molecule magnitude. He made 6+2×5=16 which should have been (15.074+2×13.238)/2=20.775, a number which is obtained by dividing 1.52092 (Gay-Lussac value for nitrous oxide density) on the hydrogen density too.*

*In the ammonia case, Dalton supposition on the relative number of molecules in mixture was totally wrong. He assumed that the nitrogen and the hydrogen were united molecule to molecule, whereas it was observed that a nitrogen molecule unite three hydrogen molecules instead. Molecule division (which do not enter in Dalton calculations) partially correlate, in this case, the error which could result from (his) other suppositions.*

*All the discussed compounds were produced by the union of one molecule with one or more molecule from other. However, in the nitrous acid they had another compound from two substances already discussed, in which the ratio of terms between the molecules number differ from unity. From the Gay-Lussac experiments there seem that this acid is composed by a part of the oxygen volume and three parts of nitrogen gas, which is reduced to the same as three parts nitrogen and five parts oxygen; so, this molecule should be consisting from three nitrogen molecules and five oxygen molecules, leading to the possibility of eliminating the division. But this way of combining could make reference only on the simple forms by considering the result of the union of one oxygen molecule with three nitrogen gas, resulting in three molecules, each of them consisting in a half of an oxygen molecule and a half of a nitrogen molecule, which actually included the division of some oxygen molecules which enter in the nitrous acid composition. Assuming that there should not be a division, the mass of this molecule should have been 57.542 (with hydrogen as a unit) and the nitrous acid density would be 4.21267 (with air density as unit). But it is possible that here may exist another division in two, so, in consequence a density reduction in two.*

*Gay-Lussac showed that it was admitted that the dry sulfuric acid is made by 100 parts sulfur and 138 parts oxygen (after the weight) and that the sulfurous acid gas was 2.265 (referring to air as humidity) and if admitted (as a result of Gay-Lussac experiments) that the sulfuric acid is made by two parts of sulfurous acid, the volume is almost equal with the one of the oxygen which enter in the compound, and this equality will be exact if the method upon they are calculated will be the same. One founds in the sulfurous acid 100 parts sulfur taking 95.02 parts oxygen which enters in compound and in consequence in the sulfuric acid one has: 95.02+95.02/2=142.53 instead of 138. If by contrary one assumes as exact the analysis of the sulfuric acid, it results that the acid contains 92 oxygen for100 parts of sulfur and this is specific to the 2.30314 gravitation weights, instead of 2.265.*

*Once it appears the consideration for weight instead of the first guessing the sulfurous gas density, as it was confirmed or rectified by new experiments, meaning, it must have been an error in sulfurous acid composition determination, which made increasing the quantity of radical, or, which is the same thing, the diminution of the oxygen quantity. The determination was made from the dry sulfuric acid produced. Therefore, one can assume that the sulfurous acid should be composed by 95.02 parts oxygen to 100 parts sulfur (or, more likely, sulfuric radical) instead of 92.*

*In order to determine the sulfuric radical molecule mass it would have been necessary the proportion after volume of this radical in gaseous state incorporating the oxygen in sulfurous acid formation to be known.*

*The analogy with other combinations (where a volume doubling or a division in half of the molecule generally exists) was leading to the assumption that the same behavior occur in the other cases: the sulfur volume as gas is half of sulfurous acid and therefore the same of half of the oxygen with which is combined.*

*With this supposition, the gas sulfur density will be the one of the oxygen as 100 to 95.02/2, or 47.51, which gives 2.323 for the gaseous density of sulfur, taking the air as unit. The molecular masses being (in conformity with the given hypothesis) in the same ratio with the density which belongs*

*to the gases, i.e., the sulfuric radical mass will be the hydrogen one as
2.323 to 0.07321, or as 31.73 to 1.*

*One of these molecules, combined with two of oxygen will form the sulfu-
rous acid and combined with another oxygen molecule will form the sulfuric
acid. The sulfurous acid molecule (regarding the diffusion) will be equal with
(31.73+2x15.074)/2, or 30.94, as it should be obtained directly from dividing
the density 2.265 of the sulfurous acid gas through the hydrogen gas. For the
sulfuric acid molecule, the weight was impossible to determine because it
was not know if there is or there is not a molecular division in its formation.*

*The phosphor had a bigger analogy with the sulfur so that, apparently, it
could be supposed that the phosphoric acid is made by three oxygen molecules
to one radical and the phosphoric acid is structured only by two oxygen's
atoms and one radical. On this assumption we can calculate, approximately,
the phosphoric radical molecule mass. One may found (through a method
analog with the one used for the sulfuric acid, for example, by Rose) that
the phosphoric acid contains 115 parts oxygen to 100 weights of phosphor.
It should have been existed a bit more oxygen if one supposes that the phosphor contains hydrogen. As an approximation one can increase the figures in
the same proportion as in the sulfurous acid case, where in agreement with
the specific gravity of the sulfuric acid the oxygen quantity was increased to
120 units. With such hypothesis it was found that the mass of the phosphoric
radical is 38, with hydrogen as unit. Dalton adopted for the phosphor and
the phosphoric acid the analog hypothesis as made for sulfur and sulfuric
acid, but because he used different values for the elements of those acids
after weight, it came to determination of phosphor molecule which do not
fit in with the same determination ratio of sulfur molecule, between these
molecules: he determine that the phosphor have the mass 8, the hydrogen
being the unit.*

*He tries to determine the conjuncture which could be created by a mol-
ecule mass of a substance with a bigger role than sulfur or phosphor,
namely the carbon. The volume of the carbonic acid was equal with the
one of the oxygen which enter in the composition, then, if one should admit
that the carbon volume, supposedly gaseous, which forms the other element, is double by the division of molecules in two, same as this type of
combination, it might be necessary to assume that this volume is half of
the one of oxygen with which is combined, and therefore the carbonic acid
results from the union of one carbon molecule and two oxygen molecules,*

*and because of that, holds the analogy with the sulfuric and phosphoric acid, in concordance with the previous suppositions. In this case from the weight proportions between oxygen and carbon it was found that the density of the carbon as gas should be 0.832 respecting the air units, and the mass of this molecule should be 11.6 with respecting the hydrogen.*

*There is still a difficulty in this assumption: because one gives to the carbon molecule a mass smaller than the masses of nitrogen and oxygen, one is tempted to attribute the aggregation solidity at the highest temperature for a bigger molecular mass (same as it was observed in case of sulfuric and phosphoric radicals). So, there was assumed that in order to avoid this difficulty by assuming the molecular division in four, and even in eight, in forming the carbonic acid, the carbon molecule should have been twice or four times bigger than it was fixed to serve this purpose. But this type of composition will no longer be analog with the other acids and, in concordance with the known examples, the assuming or not assuming the gaseous state do not depend only by the molecule magnitude, but also by some substance properties, unknown yet.*

*The sulfuric acid is in gaseous form, on a usual temperature and pressure of atmosphere, but not the entire molecule, which is almost equal with the solid form of the sulfuric acid radical. The oxygenate acid of chlorine gas have a density, and therefore a molecular mass, much more considerable. Therefore, it was nothing that could prevent treating the carbonic acid in the manner previously presented, this way (analog with the nitric, sulfuric and phosphoric acids) the carbon molecule also results with a mass expressed by 11.36 units.*

*Assuming the indicated values for the carbon molecular masses and its gas density, the carbonic oxide will be consisting, in conformity with the Gay-Lussac experiments, from equal volume parts of carbon and oxygen, and its volume will be equal with the sum of their constituents' volumes. The resulted molecule will be formed from carbon and oxygen units, molecule to molecule, with an analog behavior to the nitrous gas. The mass of the carbonic acid molecule will be $(11.36+2\times15.074)/2=20.75=1.5196/0.07321$ and the one of the carbonic oxide will be given by $(11.36+15.074)/2=13.22=0.96782$.*

*In this context, Gay-Lussac assumed that the mercurous oxide, in whose formation 100 parts mercury absorb 4.16 parts oxygen is combined in half of the volume with the oxygen gas as 100 to 8.32, which will give 13.25 as density with unity air and the mercury molecule mass 181 with hydrogen as unity. In this case the mercury oxide which contains twice more oxygen should be formed from mercury and oxygen, molecule to molecule. But by some reasoning may found that the mercury oxide is actually mercurous oxide and that in the mercury oxide a mercury molecule is combining with two oxygen molecules. Then, the mercury gas density and its molecule mass should be double, starting from the hypothesis 261/2 for the first one and 362 for the second one (by analogy with other metals and in particular with the iron). However, from different chemical experiments (e.g. as analyzed by Hassenfrantz) it may result that two well known oxides of the iron (black and red oxide) are made from 31.8, respectively, 45 parts of weight oxygen to 100 iron. If one observes that the second from these quantities of oxygen was with almost half bigger than the first, there is natural to assume that in the first oxide an iron molecule is combining with two oxygen molecule, and in the second with three. This way and while admitting the proportion for the black oxide, the proportion for the red oxide should be 47.7 for 100 of iron, which is very close of the proportion found by Proust. The iron molecule mass should have been the oxygen molecule mass as 100 to 15.9, which gives approximately 94 with respecting the hydrogen as unit. From these considerations it seems that there should be another iron oxide which contains 15.9 oxygen to 100 iron, although the experiments made on this substance contain a big proportion of oxygen. Now, about the given two mercury oxides, one which contain twice more oxygen than the other, it should be apparently analogs with the last iron oxide and the blank oxide, that the red oxide not having analog in case of mercury.*

*Gay-Lussac suspected that the volume's equality between the gaseous alcanes and the acids (with which they unite and form a neutral salt) can be a general rule. The neutral salt consists in acid and alkali uniting molecule to molecule. Some consideration seems to be opposing to admitting this principle in a general acceptation. The acidity, alkalinity and neutrality ideas seem to be the most comfortable for this phenomenon. According with this approach, all substances form between each other though a series, in which they play either as acid, either as alkali, with respecting each others. Also, these series were the same as those depending on the negative or positive electrical properties which are developing on their mutual contact.*

*In conclusion, the Dalton system assumes that the compounds are made generally by fix proportions and this will show up in the experiments with respecting the compounds more stable and more interesting for chemists. It seems that only this type of combination occurs among the majority of gases, considering the enormous size of molecules which would result from ratios expressed through high numbers, despite the molecular division, which definitely occurs in small limits. However, there was perceived that the molecule packing in solids and liquids (which goes to the integral distinction between the same type of molecules respecting the elementary ones) can lead to even more complicated ratios, and even combinations in any kind of proportion; these compounds should have been a starting model and this distinction can serve to the Berthollet idea, according which the compounds occur because of their fixed (combined) properties* (Avogadro, 1811).

### 1.2.4   ANALOGIES IN THE STRUCTURE OF SUBSTANCES, BY HUMPHRY (1812)

David Humphry took a somehow bottom-down method and characterized the atomic behavior in molecules by observing and analyzing the molecular similarities, through the following sentences of his classical paper (Humphry, 1812):

*The substances which were not included in compounds, were analogs with each other and definitely were found among metals; one of them are so similar such that refined observations are necessary and sometimes (new) experiments in order to distinguish them. There is a gradual chain of resemblance which can be followed along the metallic corps series; in the same time there are several properties similar and characteristics which belong to meals with respecting to each other. Silver and palladium, antimony and tellurium are alike qualitative. Potassium and platinum, if one accepts the observations regarding polish, color and the possibility of conducing electricity, are extremely different elements. Still, by arranging the metals in the order of their natural properties which resemble between them, these last two substances can be made parts of a natural elemental chain: potassium, sodium and barium are most alike; barium is close to magnesium, zinc, iron and antimony. Platinum is analog to gold, silver and palladium. And palladium is in connection with zinc and iron. Arsenic and chromium, among the most different metals, are the same in the properties of the acid mater formation by combining with oxygen. Chlorine and oxygen are separated by the flammable elements by a number of distinctions. Still, sulfur is alike chlorine in forming the acid by combination with hydrogen and it have a weak attraction for chlorine and a powerful attraction for metallic substances.*

### 1.2.5  ON CHEMICAL PROPORTIONS, BY BERZELIUS (1813–1814)

With Berzelius' thoughts begins the quantification of chemical substances, in atomic proportions, while assessing to the atomic term a universal ontological signification (Berzelius, 1813–1814):

*The fact that elements are combining in different proportions when there is no force to be opposing to their reunion, added up to the observation that when two bodies A and B are combining in different proportions (the additional proportions of one of them are always multiple of integer number 1, 2, 3) lead to the conclusion that there is a cause which made other combinations impossible. Now, what was this cause? It was clear that the answer of this question should constitute the principal of the theoretical chemistry.*

*When reflecting on this cause, it seems obvious that there is a natural mechanism and which represent itself most probable and comfortable to the experience, employing the idea that the elements consist in atoms or molecules which are combining as: 1 to 1, 1 with 2 or 3 with 4. Even the chemical proportions' law seems to clearly result from this. An idea like this was very simple and most likely to not be adopted and even to not be proposed until our present time. As best as we known the English philosopher John Dalton (guided by the experiments of Bergman, Richter, Wenzel, Berthollet, Proust and others) was the first person which was ventured to establish such hypothesis.*

*On the other side, there seems necessary that when an atom of the element A is combined with one or more atoms of the element B (in order to form a new atomic compound) the atom A should touch every atoms of B. From this forward, the atomic compound was formed from the juxtaposing of several elementary atoms, same as an aggregate was formed from juxtaposing of different homogenous atoms. But the difference consist in the fact that, in the first case will occur an electric discharge of the polarity specific to heterogeneous atoms, which could not occur between the homogenous atoms.*

*An atomic compound, for obvious rationalizations, cannot be considered spherical. But, because it is made by atomic mechanical indivisible parts (or which cannot be separated by mechanical methods) the atomic compound is almost completely mechanically invisible as an elementary atom. Also, it was obvious that an atom made from A+3B should be bigger, and has a different figure respecting an atom made by A+B. The figure (structure) should have the form of an equilateral and triangular pyramid, while the last one should have a linear form.*

*The atoms were divided in two classes:*

1. *Elementary atoms;*
2. *Composed atoms.*

*The composed atoms were by three different species:*

1. *Atoms formed by two elementary substances united, called composed atom of first order (kind).*
2. *Atoms formed of more than two elementary substances (how many was only found in the organic elements or elements obtained by the organic matter destruction) called organic atoms.*
3. *Atoms formed by the union of two or more composed atoms (as example, the salts) called composed atoms of second order (kind).*

*The biggest number of spherical atoms of the same diameter, capable of touching only one atom of the same diameter was 12. From this there followed that A+12B contains the bigger number of atoms which can be contained in an composed atom of first kind. If, on the other hand, we are paying attention to the electric polarity of atoms, an atom A cannot combine with more than 9B atoms if the atom A+9B conserves any part of the original electrical polarity of A. For example, the oximuriatic acid which was a compound of an atom of muriatic radical and 8 oxygen atoms it conserves another part of the radical original polarity by meanings with which will react.*

*There appears as contrary to the sound logics the representation of a compound atom of first kind as being composed by two or more A atoms combined with two or more B atoms. For example, 2A+2B, 2A+3B, 7A+7B, etc. In this case there were no obstacles, nor mechanical nor chemical, in order to prevent such an atom to be divided, by pure mechanical means, in two or more atoms by a simple (de)composition. Then, such a composition almost destroys totally the chemical properties. From such a result, by declaring the result of this analysis comfortable on the corpuscular theory, one might have to consider some of the (decomposing resulted) constituents as unitary.*

*Another method for seeing the chemical properties was a method founded by the phenomena discovered by Gay-Lussac: the elements which are in gaseous state are uniting either in equal volume, or 1 volume from one is combined with 2 and 3 volumes from the other partner. This fact was verified but some distinguished chemists. By what it was known, with respecting the defined proportions, it follows that there will be kept all the elements on temperature and pressure which assume the gaseous form. But there is no difference between the atomic theory and the volume theory, unless that one represents the elements in solid form and the other in gaseous form, respectively. However, it was clear that what is called in a theory as an atom in the other one is called as volume. The volume theory had the advantage to be constituted on the well known (observable) phenomena, while the other one has as a fundament only a supposition. In the volume theory it could imagine a half volume, while in the atom theory a demy-atom seems absurd. On the other hand, the volume theory has the disadvantage, the one which the atom theory did not present it, precisely the existence of the composed compound (especially of organic nature) which cannot be assumed to exist in gaseous form.*

*In the volume theory it was not possible the assumption of combining 2 volume with 3, for this type of supposition there is no reason to explain the fact why 4 volumes should not combine with 5, 7, with 1000. In this case, there is no reason to believe in chemical proportions. Like in the atoms theory, it was absolutely necessary that in each compound one of the constituents should be considered a single volume.*

*It was obvious that if the weight of elementary element volumes should have been known (and express in numbers) they do not have anything else to do, in the analysis case, than to count the relative number of constituents parts volumes, regardless their aggregation form, but in order to obtain the relative weight of the elementary volumes expressed in numbers (meaning to be obtained the specific gravity in gas form) they must had a general measure to compare with. One could chose between the elementary elements, as ontological entities, when the volume weight will be the unit, same as the water was chosen as unit for determining the specific gravity of liquids and solids.*

*There were only two elements which posed the necessary qualities to serve as unit. They were the oxygen and the hydrogen. But the hydrogen has disadvantages which the oxygen did not present. The hydrogen volume weight was so small that if it is taken as unit, the number which represents the volume for some metals became very big. Then, the hydrogen enters in fewer compounds than the oxygen; and of course the number 100, when it was applied to hydrogen, does not facilitated the calculation as when it had been applied to oxygen. Among the elementary bodies, the oxygen constituted a particular class; it is the center around which chemistry turned on. There was a greater number of inorganic bodies and without exception in all the products of organic nature.*

*Firstly, it seems reasonable to assume that the bodies tend to combine in equal volumes in general; but by examining a great number of the elementary bodies' combinations, we found that those who distinguished (a strong affinity between the constituent parts, and through the force of the chemical affinity for other bodies) obviously contain more than a volume of one of their elements (as in water, carbonic acid, nitrous gas) and with few exceptions, was almost always the case the electro-negative element the volume of which was multiplied.*

*The experiment seemed to prove that if a combustible radical is preferably combined with two or three volumes of oxygen, it also combines with two or three volumes of sulfur. If this oxide was neutralized by an acid, it was supposed that the resulting neutral combination should have contained (for one volume of the oxide radical) as many volume of the acid radical as the oxide contains volume of oxygen; and therefore, that the number of times which the acid contains the oxygen of the oxide will be the number of the oxygen volumes combined with one volume of acid radical; for example, the sulfuric acid was considered as consisting of a one volume of radical and three volumes of oxygen, because it was very likely that the amount of sulfur and oxygen to be able to combine (at a high temperature) with a given portion of lead and having equal volumes. But if we want to know, by another method, how many volumes of oxygen are in the sulfuric acid?* (Wiki, 2013a)

*We have to examine the composition of some sulfate* (Wiki, 2013b)

*The black oxide of the iron contains one volume of metal and two volumes of oxygen. Considering this, there results that the iron black oxide should be neutralized by an amount of acid containing two volumes of sulfur for each volume of iron, so that the number of the sulfur volumes of acid and the oxygen of base to should be equal. Yet the acid contains three times more oxygen respecting a base; therefore, it was composed of three volumes of oxygen and a volume of sulfur. If, instead of the iron sulfate, the iron persulfate should be chosen, appears obvious that in this case the iron was combined with three volumes of sulfur, otherwise the result will be the same* (Berzelius, 1813–1814).

### 1.2.6 STRUCTURE ANALOGIES ON SUBSTANCES' GROUPING, BY DLIBEREINER (1829)

The first illustrative quantitative proportions of atoms in molecules as driven by similarity of atomic compositions/contribution to molecular samples were given by Döbereiner (1829) through the following considerations:

*The specific gravity and the atomic weight of the strontium are very close to the average of the specific gravities and the atomic weights of the calcium oxide and the barium oxide*

$$[356.019(= Ca) + 956.88(= Ba)]/2 = 656.449(= Sr) \qquad (1.1)$$

*which is very close to the atomic value of soda.*

*If the sulfur, selenium and tellurium belong to a group, that could be assumed (because the specific gravity of selenium was exactly the average of the specific gravities of the sulfur and tellurium) and all the three substances are combining with hydrogen to form characteristic hydrogen acids, then selenium forms the middle member*

$$[32.239(= S) + 129.243(= Te)]/2 = 80.741 \qquad (1.2)$$

*and the empirically value found for the selenium atom is 79.263.*

*Indeed, the fluorine belongs to the elements which forms salts, but certainly not to the chlorine, bromine and iodine group. Rather it belongs to another class of substances which can form salt that can relate with the first, the alkalis and alkaline earths. Because it has a very small value, apparently it stays as the first member of an assumed group; also in this case there are another two more members to discover, if the triads represent the law for all the chemical substances groups. This happens if the values of the atomic weights of the substances grouped together are compared with the intensity of the chemical affinity that these substances manifest; consequently one founds that the alkalis and the alkaline earths are directly proportional, but in formation the elementary salts are inversely proportional.*

*One may combine potassium, which has the largest value between the alkalis (was the strongest), with the lithium, which have the smallest value (was the weakest), and soda, which has the middle value between potassium and lithium, i.e., weaker than the potassium and stronger than lithium. The barium oxide, the calcium and the strontium oxides behave the same way. However, the chlorine which has the smallest value is the strongest and the iodine which presents the greatest value is the weakest in the formation of salt and the bromine stay between these two. If one expresses the intensity of the chemical affinity of the substances groups with the numbers 1, 2 and 3, these considerations were arranged as in* Table 1.2.

*The hydrogen, the oxygen, the nitrogen and the carbon seem to stay isolated from the substances which form bases, acids and salts. The fact that the average of atomic weight of the oxygen=16.026 and of the carbon=12.256 express the atomic weight of nitrogen=14.138, could not be considered here, as long as no analogy took place between these three substances.*

**TABLE 1.2** Combination of Elements Forming Salts, Acids, Along Alkaline and Earth Alkaline Ions Paralleling Their Chemical Affinity Changes (Döbereiner, 1829)

| Elements Which Form Salts and Their Acids | Intensity of Chemical Affinity |
|---|---|
| $221.325 = Cl$  $455.129 = HCl$  $942.650 = \overset{\cdots}{Cl}$ | 3 |
| $789.145 = I$  $1590.770 = HI$  $2078.29 = \overset{\cdots}{I}$ | 1 |
| $\dfrac{1010.460}{2} = Br$  $\dfrac{2045.056}{2} = HBr$  $\dfrac{3020.940}{2} = \overset{\cdots}{Br}$ | 2 |
| **Elements Which Form Acids and Their Acids** | |
| $201.165 = S$  $213.644 = HS$  $501.164 = \overset{\cdots}{S}$ | 3 |
| $806.452 = Te$  $831.412 = HTe$  $1106.452 = \overset{\cdots}{Te}?$ | 1 |
| $\dfrac{1007.617}{2} = Se$  $\dfrac{1045.056}{2} = HSe$  $\dfrac{1607.617}{2} = \overset{\cdots}{Se}$ | 2 |
| **Elements Which Form Alkali and Alkalis** | |
| $95.310 = L$  $195.310 = \overset{.}{L}$ | 1 |
| $489.916 = K$  $589.916 = \overset{.}{K}$ | 3 |
| $\dfrac{585.226}{2} = Na$  $\dfrac{785.226}{2} = \overset{.}{Na}$ | 2 |
| **Elements Which Form Alkaline Earths** | |
| $256.019 = Ca$  $356.019 = \overset{.}{Ca}$ | 1 |
| $856.880 = Ba$  $956.880 = \overset{.}{Ba}$ | 3 |
| $\dfrac{1112.899}{2} = Sr$  $\dfrac{1312.899}{2} = \overset{.}{Sr}$ | 2 |

*The earth metals and the earths themselves were in the same place, according to their similarities, but one had not successfully ordered them. It is true that they form boron and silicon, aluminum and beryllium, yttrium and cerium, which are special groups, but each of them lacked in the third member. The magnesium stayed alone and the zirconium belongs to the titanium and the tin.*

*The heavy metal group of the substances was fulfilled. Its factors are the iron oxide, the magnesium oxide and the chromium oxide; the last formed the middle member because we have*

$$(979.426 \text{ Fe} + 1011.574 \text{ Mn})/2 = 995.000 \text{ Cr} \qquad (1.3)$$

*According to Mitscherlich, Fe, Mn, Ni, Co, Zn and Cu are isomorphic with the magnesium. This was an interesting series of substances, because, firstly, they belonged to the magnetic metals and secondly, they are the best conductors of electricity. But how they have arranged themselves if the triad was the principle of grouping? In nature, Fe, Mn and Co are found in oxides which are frequently found together and the Ni, Zn, and Cu oxides are found together with minerals, from which by the Chinese people prepared copper, and to which the Germans call it Argentan. If this happens, then in the first group the manganese forms the third member, because:*

$$(439.213 \text{ Fe} + 468.991 \text{ Co})/2 = 454.102 \text{ Mn} \qquad (1.4)$$

*and in the second group the copper is occupying this position:*

$$(469.675 \text{ Ni} + 503.226 \text{ Zn})/2 = 486.450 \text{ Cu} \qquad (1.5)$$

*However, the copper weight was 495.695 and the specific gravity of copper was not the average of the specific gravity of nickel and zinc, therefore, these six oxides should be grouped differently. A rigorous experimental review of the specific gravity and the atomic weights could remove these doubts. The most interesting analogous metals series are those found in the minerals of platinum, where are found: the platinum, palladium, rhodium, iridium, osmium and pluranium. They exist in two groups, according to their atomic weights. The platinum, iridium and osmium belong to the first group, and the palladium, rhodium and pluranium to the second one. For the first group members, the atomic weights (according to the work of Berzelius) are as follows: for platinum 1233.60, for iridium 1233.260 and for osmium 1244.210. Because the specific gravity of the iridium was almost the average of the weight of platinum and of osmium, the iridium must be considered the middle member: (1233.260 + 1244.210)/2 = 1238.735. The atomic weights for the second group members (according to the same researchers) are: for palladium 665.840 and for rhodium 651.400. Therefore, we have for pluranium 636.960; if the atomic weight stays so near to the platinum and iridium, then the rhodium will be the middle member of this group.*

*The specific gravity and the atomic weight of the lead are close to the specific gravity average and of the atomic weights of silver and mercury; so, these three metals can be put together* (Döbereiner, 1829).

### 1.2.7 ELEMENTAL EQUIVALENTS AND THE OCTAVES' LAW, BY NEWLANDS (1863)

The atomic universality of atoms belonging to the same element was advanced by Newlands (1863) by a philosophical analogy with the music of spheres, through the octaves (so anticipating the octet rule), and consecrated as following:

*Many chemists (and M. Dumas in particular), on several occasions, pointed out the existence of some interesting relations between the equivalents of the bodies which belong to the same natural family or group.*

*The following relations are among the best; they are observed by comparing the analogous elements equivalents (in order to avoid the frequent repetition of the word "equivalent," it was generally used the name of various elements as representing their equivalent; so when the term zinc is used it is also the average of magnesium and cadmium, so trying to be induced that the equivalent of the zinc is the average of magnesium and cadmium).*

*Group I. The alkalis metals: Lithium 7; Sodium 23, Potassium 39, Rubidium 85, Cesium 123, Thallium 204. The relationship between the equivalents of this group can be presented as in Table 1.3.*

**TABLE 1.3**   Succesive Combinations of Elements of the First Group (Newlands, 1863)

| 1x | Li | +1x | K | = | 2xNa (sodium) |
|----|----|-----|---|---|---------------|
| 1x | Li | +2x | K | = | 1xRb (rubidium) |
| 1x | Li | +3x | K | = | 1xCs (cesium) |
| 1x | Li | +4x | K | = | 163 equivalent of an undiscovered element |
| 1x | Li | +5x | K | = | 1xTl (thallium) |

*Group II. The alkaline-earths metals: Magnesium 12; Calcium 20; Strontium 43.8; Barium 68.5. In this group, the strontium is the average between calcium and barium.*

*Group III. Earth metals: Beryllium 6.9; Aluminum 13.7; Zirconium 33.6; Cerium 47; Lanthanium 47; Didymium 48; Thorium 59.6. The aluminum is twice the beryllium, or the third part from the sum of beryllium with zirconium* (see Table 1.4).

*Group IV. The metals whose protoxides are isomorphic with magnesium: Magnesium 12; Chromium 26.7; Manganese 27.6; Iron 28; Cobalt 29.5; Nickel 29.5; Copper 31.7; Zinc 32.6; Cadmium 56. Between magnesium and cadmium (the extremities of this group) zinc is the average. The cobalt and nickel are identical. Between cobalt and zinc, copper is the average. The iron is half of the cadmium. Between the iron and chromium, the magnesium is the average.*

*Group V: Fluorine 19; Chlorine 35.5; Bromine 80; Iodine 127. In this group, the bromine is the average between chlorine and iodine.*

*Group VI: Oxygen 8; Sulphur 16; Selenium 39.5; Tellurium 64.2. In this group the selenium is the average between the sulfur and tellurium.*

*Group VII: Nitrogen 14; Phosphorus 31; Arsenic 75; Osmium 99.6; Antimony 120.3; Bismuth 213. In this group the arsenic is the average between phosphorus and antimony. The osmium is closer to the average between arsenic and antimony, and is almost exactly half of the difference between nitrogen and bismuth, the two extremities of this group,*

**TABLE 1.4**   Selective Combinations of Earth Metal Elements (the Lanthanium and the Didymium Were Identical With Cerium, or Close) (Newlands, 1863)

| 1x | Zr | +1x | Al | = | 1xCr (cerium) |
|----|----|-----|----|---|---------------|
| 1x | Zr | +2x | Al | = | 1xTh (thorium) |

*therefore 99.5. The bismuth is equal to 1 of antimony + 3 of phosphorus, thus: 120.3+93=213.3.*

*Group VIII: Carbon 6; Silicon 14.2; Titanium 25; Tin 58. In this group the difference between tin and titanium is almost three times higher than that of titanium and silicon.*

*Group IX: Molybdenum 46; Vanadium 68.6; Tungsten 92; Tantalium 184. In this group the vanadium is the average between molybdenum and tungsten. The tungsten is equal to 2 of molybdenum and the tantalium was 4 of molybdenum.*

*Group X: Rhodium 52.2; Ruthenium 52.2; Palladium 53.3; Platinum 98.7; Iridium 99. In this group the first three are identical (or closed) and there are more than half of the other two.*

*Group XI: Mercury 100; Lead 103.7; Silver 108. The lead is the average of the other two. If one deduces the group member with the lowest equivalent from the immediately one (from above it), it is frequently observed that the numbers thus obtained lead to a simple relation for everyone, as in the examples of* Table 1.5.

*Worth drawing the attention about the existence of a law with the following effect: "the atomic weights of the elementary bodies, with few exceptions, are exactly or very close multiple of eight."*

*Next, we present some explicative remarks for the various groups in the* Table 1.6.

*Group II: Boron is classed with gold, both elements being triatomic, although the last is sometimes monatomic.*

*Group III: Silicon and tin are placed next to each other, as the extremities of a triad. Titanium is placed next to them and occupies an intermediate position between silicon and the central term or the triad average,*

**TABLE 1.5**   Example of Quivalents and Their Atomic Weights' Differences (Newlands, 1863)

| Magnesium 12 | Calcium 20 | Δ | =8 |
|---|---|---|---|
| Oxygen 8 | Sulphur 16 | Δ | =8 |
| Carbon 6 | Silicon 14.2 | Δ | =8.2 |
| Lithium 7 | Sodium 23 | Δ | =16 |
| Fluorine 19 | Chlorine 35.5 | Δ | =16.5 |
| Nitrogen 14 | Phosphorus 31 | Δ | =17 |

**TABLE 1.6** The Equivalent Relationship Among Elements From Various Groups (Newlands, 1863)

| | | TRIAD | | | |
|---|---|---|---|---|---|
| Group | | Lowest term | Mean | Higest term | |
| I. | Li 7 | +17 = Mg 24 | Zn 65 | Cd 112 | |
| II. | B 11 | | | | Au 196 |
| III. | C 12 | +16 = Si 28 | | Sn 118 | |
| IV. | N 14 | +17 = P 31 | As 75 | Sb122 | +88 = Bi 210 |
| V. | O 16 | +16 = S 32 | Se 79.5 | Te 129 | +70 = Os 199 |
| VI. | F 19 | +16.5 = Cl 35.5 | Br 80 | I 127 | |
| VII. | +16 = Na 23 | +16 = K 39 | Rb 85 | Cs 133 | +70 = Tl 203 |
| VIII. | +17 = Mg 24 | +16 = Ca 40 | Sr 87.5 | Ba 137 | +70 = Pb 207 |
| IX. | | Mo 96 | V 137 | W 184 | |
| X. | | Pd 106.5 | | Pt 197 | |

*therefore (Si 28 + Sn 118)/2 = 73, (Si 28 + the triad average 73)/2 = 50.5, Ti=50.*

*Group IV: The equivalent of the antimony is almost the average of phosphorus and bismuth, and thus: (31+210)/2 = 120.5.*

*Group VII: The relations with which M. Dumas pointed out the differences between this group members were well known; a little change had to be made, the atomic weight of cesium has to be increased, as exposed in* Table 1.7.

**TABLE 1.7** The Elements' Combinations Through Their Atomic Wights Equivalents (Newlands, 1863)

| | | |
|---|---|---|
| Li + K = 2 Na | ≡ | 7 + 39 = 46 |
| Li + 2 K = Rb | ≡ | 7 + 78 = 85 |
| 2 Li + 3 K = Cs | ≡ | 14 + 117 = 131 |
| Li + 5 K = Tl | ≡ | 7 + 195 = 202 |
| 3 Li + 5 K =2 Ag | ≡ | 21 + 195 = 216 |
| Li + Ca = 2 Mg | ≡ | 7 + 40 = 47 |
| Li + 2 Ca = Sr | ≡ | 7 + 80 = 87 |
| 2 Li + 3 Ca = Ba | ≡ | 14 + 120 = 134 |
| Li + 5 Ca = Pb | ≡ | 7 + 200 = 207 |

*The silver equivalent is connected with the one of alkali metals; It could be seen as being made of sodium and rubidium equivalents, therefore: 23 + 85 = 108. Also, it is almost half of rubidium and cesium: (85 + 133)/2 = 109.*

*Group VIII: if lithium was considered as connected with this group as well as with the previous one, the same calculations as before are reported in the lower four positions of Table 1.7.*

*Again, there are two triads in the alkali metals group, from which one was known (lithium, sodium and potassium) and the other was observed by Mr.C.W. Quin (in the Chemical News, November 9, 1861) consisting of potassium, rubidium and cesium. Therefore, the potassium is the highest term for a triad and the lowest term for another triad.*

*Likewise, if lithium is included, we will have among the alkaline earths metals two triads: the first one including lithium, magnesium and calcium, and the second one with calcium, strontium and barium. Calcium is the highest term for a triad and the lowest for another.*

*The lead occupied a position (in relation with the metals from the alkaline earths) similar with that occupied by thallium in the group of alkali metals. Osmium seems to play a similar role in the sulfur group and the bismuth in the phosphorus group. The analogous term in the chlorine group was not known yet. Thallium, with its physical properties, had some similarities with the lead and frequently happened that the similar terms from different groups (such as oxygen or nitrogen, or sulfur and phosphorus) to bear more physical similarities with each other than with the members of the same group; from chemical reasons were required to assign them as such.*

*Worth being noted that the difference between the equivalents of tellurium and osmium, cesium and thallium, barium and lead, is the same in every case; the group X-Palladium and platinum seems to be the extremity of the triad, the average of something unknown* (Newlands, 1863).

John A.R. Newlands in the work "The Law of Octaves and the Causes of Numerical Relations among the Atomic Weights" claimed to have discovered a law according to which the elements with analogous properties have relations peculiar similar to those in music, between a note and its octave, see Table 1.8 (Newlands, 1863). Starting from the atomic weights of the Cannizzarro's system, the author arranges the known elements in

**TABLE 1.8**    Elements Arranged in Octaves (Newlands, 1863)

| "Do" | "Re" | "Mi" | "Fa" | "Sol" | "La" | "Si" | "Doo" |
|------|------|------|------|-------|------|------|-------|
| H 1 | F 8 | Cl 15 | Co & Ni 22 | Br 29 | Pd 36 | I 42 | Pt & Ir 50 |
| Li 2 | Na 9 | K 16 | Cu 23 | Rb 30 | Ag 37 | Cs 44 | Os 51 |
| G 3 | Mg10 | Ca 17 | Zn 24 | Sr 31 | Cd 38 | Ba & V 45 | Hg 52 |
| Bo 4 | Al 11 | Cr 19 | Y 25 | Ce & La 33 | U 40 | Ta 46 | Tl 53 |
| C 5 | Si 12 | Ti 18 | In 26 | Zr 32 | Sn 39 | W 47 | Pb 54 |
| N 6 | P 13 | Mn 20 | As 27 | Di & Mo 34 | Sb 41 | Nb 48 | Bi 55 |
| O 7 | S 14 | Fe 21 | Se 28 | Ro & Ru 35 | Te 43 | Au 49 | Th 56 |

order of succession, starting with the lowest weight (hydrogen) and ending with thorium, however placing nickel, cobalt, iridium, cerium and lanthanum in positions of absolute equality in the same line. The 56 items form eight octaves and the author found that chlorine, bromine, iodine and fluorine are brought on the same line, or occupy appropriate places in this scale. So, the doors of the chemical Copernican revolution were opened for the most-expected Periodic Systems of elements.

## 1.3  MENDELEEV'S LEGACY FOR MODERN CHEMISTRY

### 1.3.1  *FROM ALCHEMY TO CHEMISTRY, BY MENDELEEV (1869)*

Dalton revived the atomic matter theory, adding the key idea that different elements have different characteristics of atomic weights. In his paper, some of the earliest descriptions of molecular structure, in which the atoms from a molecule are arranged in space in a certain manner are also presented. The Dalton atomic theory was the first to give significance to the relative weights of the last (in elementary meaning) particles of the all known compounds and provided a quantitative explanation of the phenomena that accompany chemical reactions. Dalton believed that all matter was composed from indestructible and indivisible atoms, of different weights, each weight corresponding to a chemical element and that these atoms remain unchanged during chemical processes. The experiments with atoms relative weights, forced Dalton to construct the first periodic table of elements and to formulate laws concerning their combinations and to provide schematic representations of different combinations of atoms. His conception of "atom" and "chemical element" was of fundamental importance, because it provided to the chemists a new and enormous model of reality. Of the various representations Dalton ensured the symbol for atoms of different elements and their compounds. Based on these symbols and their accompanying theory, Dalton is known as the "father of modern chemistry.

Mendeleev's discovery (of the periodic law and the periodic table of elements) was for the first time announced to the European researchers in a short German article, published on *Zeitschrift für Chemie* in 1869. Mendeleev discovered the periodic law while he was preoccupied to writing the first edition of the chemistry textbook, Osnovy Khimii (see Figure 1.2). Mendeleev's approach was based on four aspects of matter, which reveal close relationships between certain chemical elements. These four aspects were: isomorphism, the specific volume for the similar compounds or elements, the composition of compound salts and the relationship between atomic weight of elements. Since the periodic law depended on the quantitative relations between the atomic weight (as an independent variable), the physical element and the chemical property, in 1869, Mendeleev took over the developing of a "whole natural system of elements." It unfolded a deduction in order to discover the boldest and the ultimate logical consequence of the law so that by checking it, to confirm the law itself. Among these consequences it is the prediction of chemical property of several unknown elements. Already in 1870 two elements were discovered, those we now know as gallium and germanium, and having almost exactly the properties predicted by Mendeleev.

### 1.3.2 ON ATOMIC WEIGHTS—ELEMENTAL PROPERTIES' RELATIONSHIPS, BY MENDELEEV (1869)

The Mendeleev's chemical revolution starts with its 1969 the landmark paper, shortly revived as following (see Figure 1.2):

*The elements arrangement (according to the increasing of the atomic weights) in vertical columns is made so that the horizontal rows contain analogous elements (also arranged by the increasing of atomic weight) is obtaining as in* Table 1.9 (Mendeleev, 1869a).

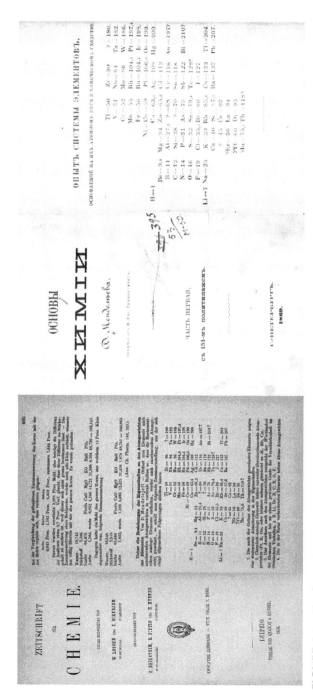

**FIGURE 1.2** The first rationalization of the Periodic Table jointly published by Mendeleev in 1969 in the German Journal of Chemistry (Zeitschrift fuer Chemie)—the second slide from left to right side, as well as in his Russian textbook on Principles of Chemistry (Osnovy Khimii)—in the second slide from right side (Mendeleev, 1869a,b).

**TABLE 1.9**  The Analogy Between Properties and Atomic Weight of Elements (Mendeleev, 1869a)

| | | | | | |
|---|---|---|---|---|---|
| | | | Ti=50 | Zr=90 | ?=180 |
| | | | V=51 | Nb=94 | Ta=182 |
| | | | Cr=52 | Mo=96 | W=186 |
| | | | Mn=55 | Rh=104,4 | Pt=197,4 |
| | | | Fe=56 | Ru=104,4 | Ir=198 |
| | | | Ni=Co=59 | Pd=106,6 | Os=199 |
| H=1 | | | Cu=63,4 | Ag=108 | Hg=200 |
| | Be=9,4 | Mg=24 | Zn=65,2 | Cd=112 | |
| | B=11 | Al=27,4 | ?=68 | Ur=116 | Au=197? |
| | C=12 | Si=28 | ?=70 | Sn=118 | |
| | N=14 | P=31 | As=75 | Sb=122 | Bi=210? |
| | O=16 | S=32 | Se=79,4 | Te=128? | |
| | F=19 | Cl=35,5 | Br=80 | J=127 | |
| Li=7 | Na=23 | K=39 | Rb=85,4 | Cs=133 | Tl=204 |
| | | Ca=40 | Sr=87,6 | Ba=137 | Pb=207 |
| | | ?=45 | Ce=92 | | |
| | | ?Er=56 | La=94 | | |
| | | ?Yt=60 | Di=95 | | |
| | | ?In=75,6 | Th=118? | | |

*Accordingly, one notes the fundamental features, namely* (Mendeleev, 1869a):

1. *The elements, if arranged according to their own atomic weights, present a periodicity of properties.*
2. *The analogous chemical elements have either similar atomic weights, or their weight increases with equal increments.*
3. *The arrangement according to the atomic weight corresponds to the element valence and until a specific extents which determines the chemical behavior, such as Li, Be, B, C, N, O, F.*
4. *The elements chaotically distributed in nature have smaller atomic weights and all these elements are distinguished by their behavior (they are representative elements), and therefore, the lightest element H is in right chosen as the most representative.*

5. *The magnitude of the atomic weight determines the properties of an element. Therefore, in the study of compounds, not just the quantities, properties of the elements and their reciprocally behavior should be considered, but also the atomic weight of elements. Therefore, the compounds of S and Te, Cl and J, have enough similarities, but also significant differences.*

6. *the discovery of some new elements can be predicted; for example the analogs of Si and Al with atomic weights of 65 to 75.*

7. *Some atomic weights probably require corrections, for example Te cannot have the atomic weight 128, but rather 123–126.*

### 1.3.3 ON ATOMIC WEIGHTS—ELEMENTAL PROPERTIES' RELATIONSHIPS, BY MEYER (1870)

Mendeleev's Keplerian's ideas of periodicity were nevertheless also advanced, by another contemporary scientist, Meyer (1870), yet lacking somehow of elemental prediction in terms of atomic weight and or atomic number position:

*The regularities that exist between the numerical values of the atomic weights are not observed by various authors (between different elements), yet are also very distinctly presented. Since we are not basing our considerations on the argument of the so-called "the equivalents" of Gmelin, but we use the atomic weights (determined by Avogadro and Dulong and Petit) the presentation of these regularities was considerably simplified. Even in 1864 there were found some regularities which lead to the family scheme of the chemical elements, considered until then as different. By the right determination of the various atomic weights, it was possible to arrange in the same scheme of the discovered elements. Mendeleev showed that such an arrangement is obtained if the atomic weight of all the elements (with no arbitrate selection) are merely arranged in series, according to the numerical values of sizes. These series are divied in sections and are together added in the succession.*

*The table contains all the elements (of those atomic weights established through the compounds gas density or through the heat capacity) arranged according to the increase of the atomic weight, with only one exception of the hydrogen (which seems to be anomalous) and also including Be and*

*In, of those atomic weight probably derive from the equivalent weights, in total 56 elements. Besides H, it is missing only Y, Eb, Ce, La, Di, Th, U, Jg, with no atomic weight which was known and even some of which equivalent was undiscovered. The* Table 1.10 *is actually identical to the one given by* Mendeleev (1869a).

*While the elements will follow the nine vertical rows (to the first until the last depending by their molecular weight), the horizontal rows include the natural families. In order to obtain this arrangement, certain elements (of which atomic weight was found almost equal and which probably was not very carefully determined) have to be rearranged: tellurium before iodine, osmium before iridium and platinum, and these before the gold.*

*If we assumed that the atoms are the aggregate of the same material and are different only in terms of masses, then we can consider the elements properties as depending on the atomic weighting size. They appear as a direct function of the atomic weight. The table gives to us the conception that the elements properties are, most of them, periodic function of the atomic weight. Same properties, or similar, reappear if the atomic weight has a certain size, after 16 units, then after 46, and finally 88 until 92 units.*

*For example, if we are starting with Li, we find that after an increase of 16 units, these properties are found in Na, and again after 16 units, in K. However, we find in the various series, the first elements Be, B, C, N, O, F and then Mg, Al, Si, P, Cl, apparently with no average of transition, but the atoms saturation ability increase and decrease in regular way and is equal in both intervals* (Meyer, 1870).

## 1.3.4 ON ELEMENTAL PERIODICITY, BY MENDELEEV (1889)

As later Gilbert Newton Lewis advanced the idea of atomic structure and chemical bonding, Mendeleev (1889) made the necessary passage from atomic periodicity feature to molecular composition by employing the atomic physicochemical features, eventually originating in atomic weights and number (latter with quantum mechanical relevance):

*The observations made with the spectroscope, permitted to analyze the chemical constitution by the distant lights, at the beginning, applicable to the challenge in determining the nature of the atoms themselves, but the laboratory work immediately demonstrated that the spectral*

**TABLE 1.10** The Arrangement of the Elements Discovered in 1870 (Meyer, 1870)

| I | II | III | IV | V | VI | VII | VIII | IX |
|---|---|---|---|---|---|---|---|---|
|  | B=11.0 | Al=27.3 | -- | -- | -- | ?In=113.4 | -- | Tl=202.7 |
|  | C=11.97 | Si =28 | Ti=48 | -- | Zr=89.7 | Sn=117.8 | -- | Pb=206.4 |
|  | N=14.01 | P=30.9 | V=51.2 | As=74.9 | Nb=93.7 | Sb=122.1 | Ta=182.2 | Bi=207.5 |
|  | O=15.96 | S=31.98 | Cr=52.4 | Se=78 | Mo=95.6 | Te=128? | -- | -- |
| -- | F=19.1 | Cl=35.38 | Mn=54.8 | Br=79.75 | Ru=103.5 | J=126.5 | W=183.5 | -- |
|  |  |  | Fe=55.9 |  | Rh=104.1 |  | Os=198.6? |  |
|  |  | Co = Ni = 58.6 |  |  | Pd=106.2 |  | Ir=196.7 |  |
|  |  |  |  |  |  |  | Pt=196.7 |  |
| Li=7.01 | Na=22.99 | K=39.04 |  | Rb=85.2 | Ag=107.66 | Cs=132.7 | Au=196.2 | -- |
| ?Be=9.8 | Mg=23.9 | Ca=39.9 | Zn=64.9 | Sr=87.0 | Cd=111.6 | Ba=136.8 | Hg=199.8 | -- |

*characteristics are determined not directly from atoms, but from molecules in which the atoms are packed. And therefore became obvious that more verifiable facts should be collected before being possible to formulate a new generalization, able to replace that ordinary based on the concept of the simple elements and atoms.*

*The Merignac's research (on the niobium) and those of Roscoe (on vanadium) were special moments. Sweeping analogies between vanadium and phosphorus on the one hand, and between vanadium and chromium on the other hand, which have become so apparent in the investigations connected with that element, have induced the possibility of comparison of V = 51 with Cr = 52, Nb = 94 with Mo = 96, and Ta = 192 with W = 194; on the other hand, P = 31 can be compared with S = 32, As = 75 with Se = 79, and Sb = 120 with Te = 125. From these approximations there was only a step until the discovering of the periodicity law.*

*The periodicity law was, therefore, a direct consequence of a package of generalities and of the establishment of the factors accumulated until the end of the decade 1860–1870. There was a combination of data in expressions more or less systematic. In this combination was the secret of the importance assessed to the periodicity law; it gave a new position to the generalizations, offering an unexpected aid of the chemist researchers, which promise to be more productive in the future.*

*A periodical function became obviously in the elements case, depending on the atomic mass. The primary concept of the elements masses or of the atomic masses belongs to a category that presents the state of a forbidden science, because of the fact that there was no mean to analyze and dissect such conception. All that was known about the masses dependent functions had origin to Galileo and Newton and indicated that this functions increase or decrease once with increasing the mass (as the attraction of the celestial bodies).*

*The numerical expression of the phenomenon was always found as proportional with the mass and in any case the increasing mass was not been followed by a recurrence of the properties as a periodical law of the elements. This was a novelty in the natural phenomena study; even if it was not appropriate to the purpose which define the truth about the mass concept, it was not indicated that the explanation of such a concept to be searched in the atomic masses. Moreover, as long as the masses are not something else than aggregations or additions of the chemical atoms, they can be better described as chemical individuals. It is remarked that by the way of seeing the "individual" mean a translation of the Greek word "atom"; the history had traced a distinguished line between the two words, and nowadays the chemical concept of the atoms is nearer to the one defined by the Latin "individual," rather than the Greek one, although the word has a special specification unknown to the classics. The periodic law had shown that the chemical individuals present a harmonic periodicity of the properties, according to their masses.*

*If we put on the abscissa the series of the lengths proportional with the angles and on the ordered the proportion with the trigonometrically functions, we obtain curves with harmonic character. It seemed at a first sight, that once with the increasing of the atomic weight, the elements propeties function should vary in the same harmonic way. But in this case, there were no such continuing changes like for the curves, because the periods do not contain an infinite number of points to form a curve, but only a finite number of points. To illustrate this observation one showcases next atomic weights:*

| Ag = 108 | Cd = 112 | In = 113 | Sn = 118 | Sb = 120 | Te = 125 | I = 127 |
|---|---|---|---|---|---|---|

*for which a steadily increase is observed, and this growth of them is accompanied by a modification of many properties which constitute the Periodic Law. Thus, for example, the density of the elements above decreases steadily, being respectively while their oxides contain a increasing amount of oxygen as ilustrated in the next series of compounds:*

| $Ag_2O$ | $Cd_2O_2$ | $In_2O_3$ | $Sn_2O_4$ | $Sb_2O_5$ | $Te_2O_6$ | $I_2O_7$ |
|---|---|---|---|---|---|---|
| 10.5 | 8.6 | 7.4 | 7.2 | 6.7 | 6.4 | 4.9 |

*However, by connecting through a curve on an ordinate axis the summation of these properties one would involve the rejection of the Dalton law, i.e., of the multiple proportions. There is not only the fact that there are no intermediate elements between silver (which gives AgCl) and cadmium (giving $CdCl_2$), but according to the essence of the periodic law there it could not exist any element. In fact, the uniform curve would be inapplicable in this case, because it leads to the waiting of some elements with* *special properties in any point of the curve. Therefore, the periods of the elements have a very different character from those which are so simple represented by geometers. They attach to each point (from numbers) the sudden changes of masses, without a continuing evolution. In these sudden changes the destitute of the steps or of the intermediate positions appears (in the absence of the intermediate elements between, let's say, silver and cadmium, or aluminum and silicon); we must admit that it is a problem in which no direct application of the analysis to the infinite small scale can be applied. Therefore, neither the trigonometrically functions proposed by Ridberg and Flavitzky, neighter the pendulum oscillations suggested by Crookes, or the cubic curves of Haughton, proposed for expressing the periodic law, cannot represent the chemical elements periods.*

*Another attempt was made in 1888, by B.N. Tchitcherin. This author has placed the periodicity issue in the first place, but had investigated only the alkaline metals. Firstly, he noted the simple relations existent between atomic volumes of the all alkaline metals; these relations can be expressed by the formula: $A(2-0.00535An)$, where A is the atomic weight, and n equals to 8 for lithium and sodium, 4 for potassium, 3 for rubidium and 2 for cesium.*

*If n remained equal to 8, while increasing A, the volume became zero at $A = 46^{2/3}$, and reached the maximum at $A = 23^{1/3}$. One noticed the close approximation of the minimum number to the difference between the atomic weights of the analogous (as Cs-Rb, I-Br) and of the maximum number of the weight of sodium. For this reason, what could be sure until then was that the attempts as the two above, have to be repeated and multiplied, because the periodic law clearly shown that the atoms masses increase abruptly, step by step, which are clearly connected in some way with the Dalton law (of the multiple proportions) and because of the elements periodicity which found the expression in the transition from RX to $RX_2$, $RX_3$, $RX_4$, and so on until the $RX_8$; in the point where the energy of the combinatorial forces was consumed, the series start again from the RX to $RX_2$, and so on.*

*Kant said that in the world exist "two things that will not stop to amaze us, or to born our admiration: the moral law within ourselves and the stellar sky above us." But when we focus our thoughts on the Nature's elements and the periodic law, we must admit a third topic, namely "the*

*elementary Nature's individuals who are found everywhere around us."*
*Without them, the stellar sky itself is incomprehensible; in the atoms we*
*see at once the peculiar individualities, the infinite multiplication of their*
*individualities, and by which they seem free to infer the general harmony*
*of the nature.*

*When we try to explain the origin of the unique matter idea, we easily*
*conclude that in the absence of some inductions from the experiments, we*
*use the philosophical scientific theories, to reveal a kind of unity in the*
*huge diversity of individualities that around us. In classical times, such*
*a tendency can be satisfied only by the conceptions about the immaterial*
*world. Regarding the material world, our ancestors were content to stick*
*to some assumptions and have adopted the idea of unity in a material*
*form, because they were not able to evolve the conception of another*
*possible unity in order to connect the multifarious relations of the matter.*
*Responding to the same legitimate scientific tendency, the natural sci-*
*ence has found through the Universe a unity plan, a unity force, a unity*
*matter and the convincing conclusions of the modern science that have*
*convinced anyone to approve such units. But as long as we admit unity*
*in many things, we need to explain the individuality and the apparent*
*diversity that we cannot rate to follow it everywhere. It was said of the*
*olds: "give me a fulcrum strong enough and I'll move the world." And*
*we must say, "Give any which is individualized and the apparent diversity*
*will be easily understood," otherwise how the unity could result in the*
*multitude?*

*Dr. Pelopidas in 1883, presented a scientific communication about*
*the periodicity of the hydrocarbon radicals, highlighting the parallelism*
*(which was noted) between the change of the hydrocarbon radicals' prop-*
*erties and the elements (which were classified into groups). In 1886, the*
*Professor Carnelley had developed an identical parallelism:*

| I | II | III | IV | V | VI | VII | VIII |
|---|---|---|---|---|---|---|---|
| $C_6H_{13}$ | $C_6H_{12}$ | $C_6H_{11}$ | $C_6H_{10}$ | $C_6H_9$ | $C_6H_8$ | $C_6H_7$ | $C_6H_6$ |

*Before the periodic law to be formulated, the atomic weights of the*
*elements were merely empirical numbers, so that the magnitude of the*
*equivalent and the atomicity, or the value in the substitution possessed*
*by an atom, could be tested by the critical examination of the determi-*
*nation methods, but never directly by considering the numerical values*

*themselves; in a few words, we were forced to work in dark, to replace the facts, instead of being theirs masters.*

*The trivalency of yttrium (which is now present in $Y_2O_3$ oxide instead of YO) was provided by the periodic law and now become very likely that Cleve and all the other investigators of the rare metals, not only have adopted, but have applied it without any new demonstration to the elements so imperfect (as the groups of cerite and gadolinite) especially since when Hildebrand had determined the specific heat of lanthanum and didymium and he had confirmed the expectations suggested by the periodic law.*

*Berzelius had determined the atomic weight of tellurium as being 128, while the periodic law required an atomic weight under iodine, which was fixed by Stas to 126.5, which certainly is not higher than 127. Brauner, had taken the investigations and had shown that the real atomic weight for tellurium is lower than the iodine one, going near to 125. For titanium Thorpe's extensive research had confirmed the atomic weight of Ti = 48, indicated by the law and already intuited by Rose, but contradicted by analysts of Pierre and other chemists.*

*The periodic law expectations were confirmed, firstly, by the new determinations of the platinum atomic weight (for Seubert, Dittmar and M'Arthur), which proved to be near to 196 (taking O = 16, as proposed by Marignac, Brauner and others); secondly, Seubert has proved that the atomic weight of osmium was really lower than that of platinum, almost 191. Thirdly, the investigation of Kruss and Thorpe proved that the atomic weight of gold overcome the one of platinum, being approximated to 197. These atomic weights required corrections; the periodic law had indicated as being affected by errors and therefore had proved that the law allows some testing means for the experimental results.*

*The indium indicate that the gradually increasing of the elements power to combine with the oxygen is accompanied by a corresponding decrease in their power to combine with hydrogen; the periodic law had shown that there is an oxidation limit, as well as the elements limit to combine with hydrogen. A single atom of an element combines with maximum four atoms, either hydrogen or oxygen; while $CH_4$ and $SiH_4$ represent the strongest hydrides, so $RuO_4$ and $OsO_4$ are the strongest oxides. Therefore, one came to know the types of oxides, as one came to know the types of hydrides. The oxides were obviously following the periodical* *law subject, both in terms of physical and chemical properties, especially if we consider the case of polymerization which was also observed when comparing $CO_2$ to $Si_nO_{2n}$. To prove this, it was compared the densities with and the specific volumes for the oxides of two periods. For an easier analysis, the oxides were represented as $R_2O_n$ (Table 1.11).*

*The size of the atomic weight, the matter essence itself, was a number which did not relate with the division state of a simply element, but related with the material part which was common to the simple element in all its compounds. The atomic weight does not belong to the coal or diamond, but to the carbon. The property that Gerhardt and Cannizzaro determined as the atomic weight of the elements is based on such a statement and on certain assumptions, that for the most objects (especially for the simple ones, whose heat capacity in the free state was determined) does not remain any doubt on the atomic weight (the atomic weights were often confused with the equivalents and were determined by various bases and often contradictory). This is the reason for which it was chosen the system based on the elements atomic weights.*

*The first attempt made in this regard was the follow: the bodies with the lowest atomic weight were selected, and then were arranged in order of their atomic weight increasing size. This showed that there is a period of the simple elements properties, even in terms of their atomicity, i.e., the elements follows one after another, in arithmetic order, successively to the atoms sizes. The elements arranged according to the size of the atomic weight clearly shown periodic properties:*

**TABLE 1.11**  The Oxides, Their Physical Properties and the Atomic Wights Differences ($\Delta$) Inside the Compounds, According With Mendeleev (1889)

| Compound | Density | Specific volume | $\Delta$ | Compound | Density | Specific volume | $\Delta$ |
|---|---|---|---|---|---|---|---|
| $Na_2O$ | 2.6 | 24 | −22 | $K_2O$ | 2.7 | 35 | −55 |
| $Mg_2O_2$ | 3.6 | 22 | -3 | $Ca_2O$ | 3.15 | 36 | −7 |
| $Al_2O_3$ | 4.0 | 26 | +1.3 | $Sc_2O_3$ | 3.86 | 35 | 0 |
| $Si_2O_4$ | 2.65 | 45 | 5.2 | $Li_2O_4$ | 4.2 | 38 | +5 |
| $P_2O_5$ | 2.39 | 59 | 6.2 | $V_2O_5$ | 3.49 | 52 | 6.7 |
| $S_2O_6$ | 1.96 | 82 | 8.7 | $Cr_2O_6$ | 2.74 | 73 | 9.5 |

1. *The elements which are similar in their chemical function have either atomic weights which are close (as Pt, Ir, Os) or show a uniform increase of the atomic weight (as K, Rb, Cs). In such comparisons there were not used the conclusions of Gerhardt, Regnault, Cannizzaro and other, who established the real value of the elements' atomic weights.*

2. *The comparison of the elements or their groups according to their atomic weight size, determine the so-called atomicity of them and by extension, their differences in the chemical nature (a fact clearly evident in the group of Li, Be, B, C, N, O, F and repeated in other groups too).*

3. *The simple elements (which are randomly distributed in nature) have small atomic weights and all the elements that have small atomic weights are characterized by their property specificity. They are, therefore, typical elements. Hydrogen, as the lightest element, is typical for such purpose.*

4. *The size of the atomic weight determines the element character, as the size of the molecule determines the properties of a complex element; when we are studying the compounds, one should consider not only the properties and the quantity of elements, not only the reaction, but also the atomic weights. For example, the compounds like S and Te, Cl, I etc., although present similarities, have clear differences as well.*

5. *We expect to discover many simple elements still unknown, for example, those similar to Al and Si, elements with atomic weight between 65 and 75.*

6. *Some analogies of the elements are discovered from the atomic weights size. Uranium is presented as analogous to aluminum, evidence which is proved when comparing their compounds* (Mendeleev, 1889).

## 1.3.5  DISCOVERING NEW ELEMENTS BY PERIODICITY LAW, BY MENDELEEV (1869, 1871, 1889)

The atomic-elemental system of Mendeleev leads with systematic prediction such that the Periodic Law assures the compactness of the

Periodic System itself, through the landmark analysis (Mendeleev, 1869, 1871, 1889):

*An established system is limited by the order of known and unknown elements. With the periodic and atomic relations (existent between all atoms and the properties of their elements) were observed the possibility to not only note the absence of some of them, but even to determine (with a high certitude) the properties of the elements still undiscovered; it was possible to predict their atomic weight, the density in the free state or as oxides, acids, bases, the oxidation degree, and the ability to be reduced and to form salts and to describe the metalloorganic properties of the compounds for a given element. It was also possible to describe in detail the properties of some compounds of these undiscovered elements.*

*Among the common elements, the lack of a number of boron and aluminum analogs was very striking (being in group III) and was sure that it was missing an element of this group immediately after the aluminum. It had to be found near, or in the second series, immediately after potassium and calcium. Their atomic weights being almost 40, and as in this line is following the element of the group IV, titanium, Ti = 50, there results that the atomic weight of that missing element would be about 45. As this element belongs to an equal (isoelectronic) series, it should have properties more basic than the lower elements of the group III (boron or aluminum); this means that $R_2O_3$ oxide should be a strong base. An indication of this behavior was the titanium oxide, $TiO_2$, with weak acid properties, and with many signs to be almost basic. Grounded on these properties, the metal oxide should be still weak, as the weak basic properties of the titanium dioxide. Compared with the aluminum, this oxide should have a stronger basic character and therefore, probably, would not decompose the water, and would combine with acids and alkalis with formation of simple salts. The ammonia will not dissolve, but maybe the hydrate will be weak dissolved in potassium hydroxide; this fact was unlikely because the element belonged to an equal series and to a group of elements of whose oxides contain a small quantity of oxygen* (see Figure 1.3).

*There appears the preliminary name of ekaboron, which is derived from the fact that it was the first neighboring element after the boron (the first element of a new group) and the eka syllable means "the first," Eb=45.*

*The ekaboron had to be a metal with the atomic volume of 15, because in these elements of the second series (and for all new series) the atomic*

| Series | First Group | Second Group | Third Group | Fourth Group | Fifth Group | Sixth Group | Seventh Group | Eighth Group |
|---|---|---|---|---|---|---|---|---|
| | $R_2O$ | $RO$ | $R_2O_3$ | $RH_4$ $RO_2$ | $RH_3$ $R_2O_5$ | $RH_2$ $RO_3$ | $RH$ $R_2O_7$ | $(R_2H)$ $(RO_4)$ |
| 1 / 2 | Li 7 | Be 9 | B 11 | C 12 | N 14 | O 16 | F 19 | |
| 3 / 4 | 23 Na / K 39 | 24 Mg / Ca 40 | 27 Al / ? 44 | 28 Si / Ti 48 | 31 P / V 51 | 32 S / Cr 52 | 35 Cl / Mn 55 | Fe 56, Co 59, Ni 59, Cu 63 |
| 5 / 6 | (63 Cu) / Rb 85 | 65 Zn / Sr 87 | 68 ? / Yt 88 | 72 ? / Zr 90 | 75 As / Nb 94 | 78 Se / Mo 96 | 80 Br / ? 100 | Ru 104, Rh 104, Pl 106, Ag 108 |
| 7 / 8 | (108 Ag) / Cs 133 | 112 Cd / Ba 137 | 113 In / ?Di 138 | 118 Sn / Ce 140 | 122 Sb / — | 125 Te / — | 127 I / — | Os 195, Ir 197, Pt 198, Au 199 |
| 9 / 10 | — | — | Er 178 | ?La 180 | Ta 182 | W 184 | ? 190 | — |
| 11 / 12 | (199 Au) | 200 Hg | 204 Tl | 207 Pb / Th 231 | 208 Bi | U 250 | | |

**FIGURE 1.3**   Mendeleev's Periodic Table (Mendeleev, 1869, 1871, 1889)

*volume decreases fast, starting from the first group to the next. The potassium volume was about 50, 25 for calcium, for titanium and vanadium almost 9, while for chromium, molybdenum and iron almost 7. Therefore, the specific gravity of the metal should be close to 3.0, because the atomic weight was 45. The metal will not be volatile, because it could hardly be discovered by the usual methods of the spectral analysis. It will not decompose the water at ordinary temperatures, but at certain temperatures it wills, as also many other metals from that series which form the basic oxides. Finally, it will dissolve in acids. Its chloride $EbCl_3$ (maybe $Eb_2Cl_6$), should be a volatile substance, but with a salt corresponding to the basic oxide. The water will act on it like the calcium and magnesium chloride, the ekaboron chloride will be hygroscopic and will be able to convert the hydrogen chloride without having a character of a hydrochloride. Since the calcium chloride volume is 49 and the titanium chloride is 109, the volume of ekaboron chloride should have been close to 78 and therefore the specific gravity would probably be 2.0. The ekaboron oxide, $Eb_2O_3$, should be a non-volatile substance, and probably will not merge. Will be insoluble in water, because even the calcium oxide is slightly soluble in water, but it will be probably dissolved in acid. Its specific volume will be approximately 39, because in series, the potassium oxide has a volume of 35, $CaO=18$, $TiO=20$ and $CrO_8=36$. This, by considering that we have only an atom of oxygen, the volume falls quickly to the right (in Periodic System), so that for potassium = 35, calcium = 18, titanium = 10,* *chromium = 12 and therefore the volume of ekaboron oxide with an atom of oxygen should be about 13, and the $Eb_2O_3$ formula would correspond to a volume of almost 39, and then anhydrous ekaboron would have a specific weight about 3.5. However, because it is a strong enough base, this oxide should show a small tendency to form alums, although it will give alum-compounds, double salt with potassium sulfate. Finally, the ekaboron will not form metalloorganic compounds, because is the first of the metals of the new series. Judging by the known data for the elements which accompanied cerium, none of them belong to the place designated to the ekaboron, so this metal is not a member of the cerium complex* (Mendeleev, 1869, 1871, 1889).

## 1.4   STANDARD FORM OF THE PERIODIC TABLE

The modern Periodic Table contains a huge amount of useful information for those interested in the structure of atomic or molecular phenomena and their combinations. In the eighteenth and nineteenth century, the chemistry had progressed a lot and, once with the discovery of many chemical elements, there appears the necessity of their classification. Resuming over the previous sections, the first chemist who proposed a classification model for elements was Johann Döbereiner. He formed several groups of three elements with similar properties such as the chlorine, the bromine and the iodine or the calcium, the strontium, the barium etc. Even though Döbereiner tried to extend this model of triads also to the other known elements, the model proved to be quite limited. The next attempt was made by the English chemist John Newlands, who in 1864, shows that the elements can be arranged in the increasing order of their atomic mass in octaves. This means that, some properties of elements are repeated for *every eighth element*, in a manner similar to the musical scale, which is repeated for every eighth tone. Even if this model lead only to a few groups of elements with similar properties, it was, generally, considered a success. The current form of periodic table has been regularly performed simultaneously by two chemists: the German Julius Lothar Meyer and the Russian Dmitri Ivanovich Mendeleev, based on the relationship between the atomic weight and the physical and chemical properties of elements. Greater credit was given to support the periodic table developed by Mendeleev, because he made the vanguard hypothesis for using the table towards the possibility of prediction of the existence and the properties of elements still unknown. For example, in 1872 when Mendeleev published his first table, the elements gallium, scandium, germanium were unknown, but he correctly predicted the existence and the properties of these elements of the unoccupied seats, based on the properties of other elements in neighboring groups. The chemical elements were arranged in the Mendeleev periodic table (Figure 3.1) based on the periodicity law, according to: *"physical and chemical properties of elements and their combinations are periodic functions of their atomic weight"* (Aldea et al., 2000).

The periodic system contained only 62 elements known at that time. The unknown elements arranged on the spots predicted by the peridicity law, were named using eka or dvi prefixes, for example Ekabe (Sc), eka-aluminum (Ga), eka silicon (Ge), eka manganese (Tc), dvimangan (Re). By using his table, Mendeleev corrected some atomic weights. For example, the initial weight of Indium was 76, based on the assumption that the indum oxide had the formula InO. This atomic weight places the Indium, which has metallic properties, among the nonmetals. Mendeleev assumed that the atomic weight is probably incorrect and suggests as a real formula of indium oxide the $In_2O_3$. Based on this formula, the indium has an atomic weight of approximately 113, as can be seen also in the following Table 1.12 with physical and chemical properties of the elements from the current periodic table of Aldea et al. (2000).

The confirmation of Mendeleev periodic table was made by G. Moseley, around 1913, which, by using the X-Ray spectra, established the serial number Z of the elements and therefore their right position in periodic table. He showed that the elements properties are periodic functions of atomic number Z and also proved that that most elements are composed of isotopes. Based on the discovery of new elements and the deepening the atom studies, Niels Bohr completed the short form of the periodic system developed by D. Mendeleev, with the rare-gases group (noted by zero) and the other newly discovered elements. This system consists in seven horizontal rows, called periods, each containing 2, 8, 8, 18, 18, 32 cells and 9 vertical columns, called groups (I to VIII plus the zero-th group). For the first seven groups, each contains two subgroups: a principal (A) and a secondary (B) one. The eighth group took a special place in the system:

**TABLE 1.12** Physical and Chemical Properties of the Periodic Table Elements (Aldea et al., 2000)

| Z | Name | Symbol | Weight | Atomic radius (pm) | Melting Point (K) | Boiling Point (K) | Atomic volume (cm³/mol) | Density (g/cm³) |
|---|------|--------|--------|--------------------|-------------------|-------------------|-------------------------|-----------------|
| 1. | *Hydrogen* | H | 1.008 | 37 | 14 | 20 | 14.2 | 0.07 |
| 2. | *Helium* | He | 4.003 | 128 | 1.0 | 4.1 | 27.2 | 0.15 |
| 3. | *Lithium* | Li | 6.941 | 156 | 454 | 1615 | 13.1 | 0.53 |
| 4. | *Beryllium* | Be | 9.012 | 113 | 1551 | 2745 | 4.87 | 1.848 |
| 5. | *Boron* | B | 10.81 | 97 | 2573 | 4273 | 4.62 | 2.34 |
| 6. | *Carbon* | C | 12.011 | 92 | 4100 | 5100 | 5.34 | 2.25 |
| 7. | *Nitrogen* | N | 14.007 | 55 | 63 | 77.4 | 17.3 | 0.88 |
| 8. | *Oxygen* | O | 15.999 | 60 | 54.8 | 90.2 | 14.0 | 1.15 |
| 9. | *Fluorine* | F | 18.998 | 71 | 59.5 | 85 | 17.1 | 1.51 |
| 10. | *Neon* | Ne | 20.179 | 65 | 24.5 | 27.1 | 16.7 | 1.20 |
| 11. | *Sodium* | Na | 22.990 | 186 | 370.9 | 1156 | 23.7 | 0.971 |
| 12. | *Magnesium* | Mg | 24.305 | 160 | 922 | 1380 | 14.1 | 1.738 |
| 13. | *Aluminium* | Al | 26.982 | 143 | 933 | 2792 | 10.0 | 2.702 |
| 14. | *Silicon* | Si | 28.086 | 117 | 1683 | 3538 | 12.1 | 2.33 |
| 15. | *Phosphorus* | P | 30.974 | 110 | 317 | 553 | 16.5 | 1.82 |
| 16. | *Sulfur* | S | 32.066 | 104 | 386 | 718 | 15.5 | 2.07 |
| 17. | *Chlorine* | Cl | 35.453 | 99 | 172.2 | 138.3 | 22.7 | 1.56 |
| 18. | *Argon* | Ar | 39.948 | 174 | 83.8 | 87.3 | 28.6 | 1.38 |
| 19. | *Potassium* | K | 39.098 | 231 | 337 | 1033 | 45.4 | 0.862 |
| 20. | *Calcium* | Ca | 40.078 | 197 | 1112 | 1713 | 25.8 | 1.55 |

**TABLE 1.12** Continued

| Z | Name | Symbol | Weight | Atomic radius (pm) | Melting Point (K) | Boiling Point (K) | Atomic volume (cm³/mol) | Density (g/cm³) |
|---|------|--------|--------|--------------------|-------------------|-------------------|--------------------------|------------------|
| 21. | *Scandium* | Sc | 44.956 | 161 | 1814 | 3103 | 15.0 | 2.99 |
| 22. | *Titanium* | Ti | 47.88 | 145 | 1941 | 3560 | 10.6 | 4.54 |
| 23. | *Vanadium* | V | 50.942 | 131 | 2173 | 3723 | 8.28 | 5.8 |
| 24. | *Chromium* | Cr | 51.996 | 125 | 2130 | 2938 | 7.2 | 7.19 |
| 25. | *Manganese* | Mn | 54.938 | 137 | 1518 | 2334 | 7.4 | 7.43 |
| 26. | *Iron* | Fe | 55.847 | 125 | 1808 | 3135 | 7.09 | 7.87 |
| 27. | *Cobalt* | Co | 58.933 | 125 | 1768 | 3200 | 6.6 | 8.90 |
| 28. | *Nickel* | Ni | 58.693 | 124 | 1726 | 3186 | 6.59 | 8.90 |
| 29. | *Copper* | Cu | 63.546 | 128 | 1356.4 | 2840 | 7.3 | 8.92 |
| 30. | *Zinc* | Zn | 65.39 | 133 | 693 | 1179 | 9.2 | 7.14 |
| 31. | *Gallium* | Ga | 69.723 | 122 | 302.9 | 2477 | 11.8 | 5.91 |
| 32. | *Germanium* | Ge | 72.61 | 123 | 1210.4 | 3103 | 13.6 | 5.32 |
| 33. | *Arsenic* | As | 74.922 | 125 | 1090 | 886 | 13.1 | 5.72 |
| 34. | *Selenium* | Se | 78.96 | 116 | 490 | 958.1 | 16.4 | 4.79 |
| 35. | *Bromine* | Br | 79.904 | 114 | 265.9 | 331.9 | 25.6 | 3.12 |
| 36. | *Krypton* | Kr | 83.80 | 110 | 116.6 | 121 | 38.9 | 2.6 |
| 37. | *Rubidium* | Rb | 85.468 | 243 | 312.2 | 961 | 55.8 | 1.53 |
| 38. | *Strontium* | Sr | 87.62 | 215 | 1041 | 1653 | 33.7 | 2.60 |
| 39. | *Yttrium* | Y | 88.906 | 180 | 1795 | 3610 | 19.9 | 4.47 |
| 40. | *Zirconium* | Zr | 91.22 | 161 | 2125 | 4650 | 14.0 | 6.49 |

**TABLE 1.12** Continued

| Z | Name | Symbol | Weight | Atomic radius (pm) | Melting Point (K) | Boiling Point (K) | Atomic volume (cm³/mol) | Density (g/cm³) |
|---|---|---|---|---|---|---|---|---|
| 41. | *Niobium* | Nb | 92.906 | 147 | 2741 | 5015 | 10.8 | 8.4 |
| 42. | *Molybdenum* | Mo | 95.94 | 136 | 2883 | 4885 | 9.39 | 10.2 |
| 43. | *Technetium* | Tc | 97.907 | 135 | 2445 | 5150 | 8.6 | 11.5 |
| 44. | *Ruthenium* | Ru | 101.07 | 132 | 2607 | 4423 | 8.15 | 12.3 |
| 45. | *Rhodium* | Rh | 102.906 | 134 | 2239 | 3970 | 8.29 | 12.4 |
| 46. | *Palladium* | Pd | 106.42 | 138 | 1825 | 3213 | 8.85 | 12.0 |
| 47. | *Silver* | Ag | 107.868 | 144 | 1235 | 2436 | 10.3 | 10.5 |
| 48. | *Cadmium* | Cd | 112.411 | 149 | 594.1 | 1038.6 | 13.0 | 8.65 |
| 49. | *Indium* | In | 114.18 | 163 | 429.3 | 2353 | 15.7 | 7.31 |
| 50. | *Tin* | Sn | 118.71 | 140 | 505 | 2875 | 16.3 | 7.28 |
| 51. | *Antimony* | Sb | 121.757 | 182 | 903.6 | 1860 | 18.2 | 6.68 |
| 52. | *Tellurium* | Te | 127.60 | 137 | 723 | 1263 | 20.5 | 6.24 |
| 53. | *Iodine* | I | 126.905 | 138 | 386.7 | 457 | 25.7 | 4.93 |
| 54. | *Xenon* | Xe | 131.29 | 218 | 161 | 165 | 37.3 | 2.94 |
| 55. | *Caesium* | Cs | 132.905 | 265 | 301.6 | 951.6 | 71.0 | 1.873 |
| 56. | *Barium* | Ba | 137.327 | 210 | 998 | 1913 | 39.2 | 3.51 |
| 57. | *Lanthanum* | La | 138.906 | 187 | 1193 | 3743 | 20.7 | 6.7 |
| 58. | *Cerium* | Ce | 140.115 | 183 | 1068 | 3741 | 21.1 | 6.77 |
| 59. | *Praseodymium* | Pr | 140.908 | 183 | 1208 | 3790 | 20.8 | 6.77 |
| 60. | *Neodymium* | Nd | 144.24 | 181 | 1297 | 3347 | 20.6 | 7.00 |

**TABLE 1.12** Continued

| Z | Name | Symbol | Weight | Atomic radius (pm) | Melting Point (K) | Boiling Point (K) | Atomic volume (cm³/mol) | Density (g/cm³) |
|---|------|--------|--------|--------------------|--------------------|--------------------|--------------------------|-----------------|
| 61. | *Promethium* | Pm | 144.913 | 183 | 1315 | 3300 | 22.3 | 6.475 |
| 62. | *Samarium* | Sm | 150.36 | 180 | 1345 | 2067 | 20.0 | 7.54 |
| 63. | *Europium* | Eu | 151.96 | 204 | 1095 | 1800 | 29.0 | 5.259 |
| 64. | *Gadolinium* | Gd | 157.25 | 179 | 1586 | 3546 | 19.9 | 7.90 |
| 65. | *Terbium* | Tb | 158.925 | 178 | 1629 | 3500 | 19.3 | 8.27 |
| 66. | *Dysprosium* | Dy | 162.50 | 177 | 1682 | 2840 | 19.0 | 8.54 |
| 67. | *Holmium* | Ho | 164.930 | 177 | 1747 | 2973 | 18.75 | 8.80 |
| 68. | *Erbium* | Er | 167.26 | 176 | 1802 | 3136 | 18.45 | 9.05 |
| 69. | *Thulium* | Tm | 168.934 | 175 | 1818 | 2220 | 18.1 | 9.33 |
| 70. | *Ytterbium* | Yb | 173.04 | 194 | 1097 | 1467 | 24.8 | 6.98 |
| 71. | *Lutetium* | Lu | 174.967 | 174 | 1936 | 3668 | 17.8 | 9.84 |
| 72. | *Hafnium* | Hf | 178.49 | 154 | 2495 | 4875 | 13.4 | 13.2 |
| 73. | *Tantalum* | Ta | 180.948 | 143 | 3269 | 5700 | 10.9 | 16.6 |
| 74. | *Tungsten* | W | 183.85 | 137 | 3683 | 5933 | 9.5 | 19.40 |
| 75. | *Rhenium* | Re | 186.207 | 138 | 3453 | 5900 | 8.85 | 21.03 |
| 76. | *Osmium* | Os | 190.23 | 134 | 3327 | 5300 | 8.49 | 22.40 |
| 77. | *Iridium* | Ir | 192.22 | 136 | 2637 | 4403 | 8.6 | 22.42 |
| 78. | *Platinum* | Pt | 195.08 | 139 | 2045 | 4100 | 9.1 | 21.45 |
| 79. | *Gold* | Au | 196.967 | 144 | 1337.6 | 3081 | 10.4 | 19.29 |
| 80. | *Mercury* | Hg | 200.59 | 147 | 234 | 630 | 14.8 | 13.55 |

**TABLE 1.12** Continued

| Z | Name | Symbol | Weight | Atomic radius (pm) | Melting Point (K) | Boiling Point (K) | Atomic volume (cm³/mol) | Density (g/cm³) |
|---|------|--------|--------|--------------------|-------------------|-------------------|--------------------------|------------------|
| 81. | *Tallium* | Tl | 204.383 | 170 | 576.7 | 1730 | 17.2 | 11.86 |
| 82. | *Lead* | Pb | 207.2 | 175 | 601 | 2013 | 18.3 | 11.34 |
| 83. | *Bismuth* | Bi | 208.980 | 155 | 544.5 | 1833 | 21.4 | 9.80 |
| 84. | *Polonium* | Po | 208.982 | 167 | 527 | 1235 | 22.2 | 9.40 |
| 85. | *Astatine* | At | 209.987 | - | 575 | 610 | - | 11.30 |
| 86. | *Radon* | Rn | 222.018 | 132 | 202 | 211 | 50.5 | 4.40 |
| 87. | *Francium* | Fr | 223.020 | 270 | 300 | 950 | 45.2 | - |
| 88. | *Radium* | Ra | 226.025 | 223 | 973 | 1413 | 22.6 | 5.0 |
| 89. | *Actinium* | Ac | 227.028 | 203 | 1323 | 4273 | 19.8 | 10.06 |
| 90. | *Thorium* | Th | 232.038 | 181 | 2023 | 4123 | 15.0 | 11.70 |
| 91. | *Protactinium* | Pa | 231.036 | 161 | 1873 | 4300 | 12.6 | 15.37 |
| 92. | *Uranium* | U | 238.029 | 138 | 1405 | 4091 | 11.6 | 19.05 |
| 93. | *Neptunium* | Np | 237.048 | 150 | 913 | 4193 | 12.3 | 20.05 |
| 94. | *Plutonium* | Pu | 244.064 | 151 | 914 | 3505 | 17.8 | 19.8 |
| 95. | *Americium* | Am | 243.061 | 182 | 1267 | 2880 | 18.3 | 13.7 |
| 96. | *Curium* | Cu | 247.070 | 174 | 1613 | - | 18.6 | 13.5 |
| 97. | *Berkelium* | Bk | 247.070 | - | 1259 | - | 16.6 | 13.3 |
| 98. | *Californium* | Cf | 251.080 | 186 | 1173 | - | - | 15.1 |
| 99. | *Einsteinium* | Es | 252.083 | 186 | - | - | - | - |
| 100. | *Fermium* | Fm | 257.095 | - | - | - | - | - |

**TABLE 1.12** Continued

| Z | Name | Symbol | Weight | Atomic radius (pm) | Melting Point (K) | Boiling Point (K) | Atomic volume (cm$^3$/mol) | Density (g/cm$^3$) |
|---|------|--------|--------|--------------------|-------------------|-------------------|----------------------------|--------------------|
| 101. | *Mendeleevium* | Md | 258.09 | - | - | - | - | - |
| 102. | *Nobelium* | No | 259.101 | - | - | - | - | - |
| 103. | *Lawrencium* | Lr | 262.110 | - | - | - | - | - |
| 104. | *Rutherfordium* | Rf | 261.109 | - | - | - | - | - |
| 105. | *Dubnium* | Db | 262.114 | - | - | - | - | - |
| 106. | *Seaborgium* | Sg | 263.119 | - | - | - | - | - |
| 107. | *Bohrium* | Bh | 262.123 | - | - | - | - | - |
| 108. | *Hassium* | Hs | 265.131 | - | - | - | - | - |
| 109. | *Meitnerium* | Mt | 266.138 | - | - | - | - | - |
| 110. | *Ununnilium* | Uun | 269 | - | - | - | - | - |
| 111. | *Roentgenium* | Rg | 272 | - | - | - | - | - |
| 112. | *Ununbium* | Uub | 277 | - | - | - | - | - |
| 116. | *Ununhexium* | Uuh | - | - | - | - | - | - |
| 118. | *Ununoctium* | Uuo | - | - | - | - | - | - |

*Note that for element Z=111 (A=272), the current name is Roentgenium replacing the temporary IUPAC's Unununium assingnement (and as such accepted by IUPAC on November 1, 2004) (Wiki, 2013c).

it contains 9 elements grouped three by three. The elements with the serial numbers between 58 and 71, called lanthanides, were located in the same cell of the periodic system with lanthanum (Z = 57), and the elements with the serial numbers between 90 and 103 (actinides), were in the actinium (Z = 89) box. Subsequently, A. Werner has developed the long-form of the periodic system, where the secondary groups elements were arranged separately and inserted between the elements from groups IIA and III A, as a series of ten elements, while the lanthanide and the actinides were listed in the bottom of the periodic table, in two horizontal rows, 14 items each (Aldea et al., 2000).

The properties of the elements which depend on the atomic nucleus, present a linear variation, meaning that they are nonperiodic. These nonperiodic properties are:

- the atomic number, which increase from Z=1 to Z=118;
- the atomic mass, which increases from 1.008 to 266;
- X-Ray spectra, where the square root of the radiation frequency is a linear function of the number Z.

The properties which depend on the external electronic shell structure vary periodically with the Z number. The most important periodic properties are: the atomic radius, the atomic volume, the ionic radius, the ionic volume, the melting point, the boiling point, the ionization energy, the electron affinity, the electronegativity, the valence, the acid-base character etc. (Aldea et al., 2000).

The remaining of this Volume will widely discuss about the electronegativity and related chemical periodic indices, as a starting point for (in principle) all other periodic properties of elements from the Periodic Table.

## 1.5   CONCLUSION

The periodic law is not of the physical but the chemical kind, although not often properly realized. It resides in the fact of attributing four quantum numbers (principal, orbital, magnetic, and spin) to each electron in arranging them in the so-called configuration by the *aufbau principle* according which the Periodic Table is constructed. Certainly, such quantum labeling obviously departs the physical principle according which electrons are indistinguishable particles—therefore, they cannot be identified and classified as 1s, 2s, 2p, 3s, 3p, 3d, etc. This is a Chemical Idea! This is not a Physical principle! This is a Chemical Principle! Perhaps the biggest one! Nobody can explain it in terms of physical rationale of elementary particle properties for many-electronic atoms likewise the atomic partial charge in molecule seems to be the rule not the exception in chemical bonding! However, the aufbau principle has proved itself as a reasonable reality (when assumed as a working tool) in explaining the vast variety of chemical compounds and combination through the resulting concepts of valence, frontier orbitals, electronegativity, chemical hardness. In this respect, Chemistry appears well equipped with special Principles and concepts that work despite their apparent refutation by the physical principles (Putz, 2011).

Periodic law belongs indeed to Chemistry alone although also being contributed to by remarkable physicists, for example, Bohr, Pauli, Curie, Langmuir, Hartree, Fock, Hoyle, Fowler, Bethe, just to name a few.

Criteria such as atomic weight, the [*principal (n) + orbital (l) quantum numbers*] or Madelung rule, and the simple atomic number (Z) ordering allow Scerri to recognize among three variants of regarding chemical elements: as neutral elements by Weinhold and Bent (2007), as bonded atoms by Schwarz (2007) or as macroscopic elements characterized by atomic number Z by Mendeleev (1891), Paneth (1962), and Scerri (2004a, 2004b, 2007), respectively. It so happens that the last idea is embraced by the present author as well, while being recently justified throughout a quantitative structure-property relationship (QSPR) analysis showing that among physical + chemical properties and chemical concepts, those closely related with Chemistry (such are the electronegativity and chemical hardness) better resemble the prediction of the atomic number in the Periodic System (see next chapters of the present volume).

The idea of attributing Z alone (at least for some level of chemical comprehension) was shown by Scerri to be fruitful in deciding upon the best form of the Periodic Table. In this respect, although he previously proposed the option for the left-step long form, as based on Madelung rule, he recently has revived the *atomic triads idea*, i.e., among thee elements placed one in top of other in neighborhood periods (therefore, somehow having $n+l$ rule included), the middle one has its atomic number expressed as semi-sum of the extreme ones. This allows him to reclaim, for instance the element He at the top of the inert gases group since the sequence of Z numbers: 2, 10, 18 resonates with triadic rule; the same with H at the top of the halogen group since the sequence of Z numbers: 1, 9, 17 also accords with the triadic rule, Figure 1.4. The same holds, for instance for the transitional metal triads Ru (44), Os (76), Hs (108), or from the alkali metals triad Li(3), Na(11), K(19)—as holds exactly for about 50% of all triads of Periodic Table (Scerri, 2009; Putz, 2011).

Nevertheless, the other way around, it is worth recognizing that Z alone is not enough for providing further insight into the atomic chemical properties such as atomic radii, electronegativity and chemical hardness, when further account of the principal, orbital and core shielding effects count as well (see next chapters of the present volume).

However, the crusade in discovering Chemical Principles complementing those of Physics in governing the chemical structure and reactivity is continuously open as the synthesis of newly designed chemicals targeting optimization in chemical medicine, technology, and environment demand. Equally, validating the Periodic Law by different approaches, i.e., by computing reliable atomic radii, electronegativity or chemical hardness—eventually within the celebrated DFT, see next chapters of the present volume, would serve to better identify the right concepts and tools that are then be used to fulfill chemical intuition by quantifying the molecular chemical realm (Hefferlin & Kuhlman, 1980; Hefferlin, 1989).

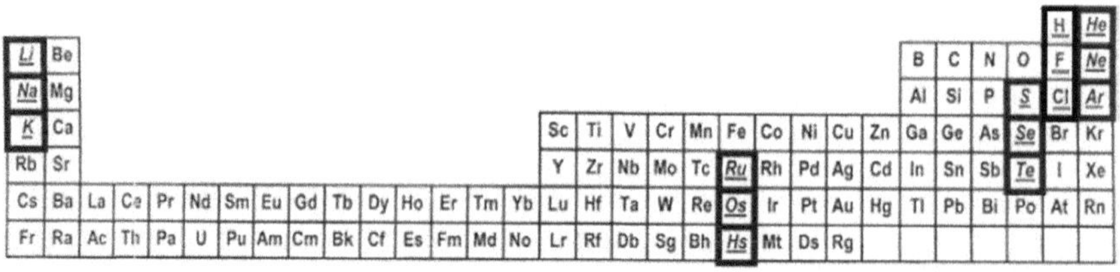

**FIGURE 1.4** Scerri's long-form proposal for the Periodic Table, while emphasizing on some triads of elements with the middle one having its atomic number as the semi-sum of the external ones (Scerri, 2009; Putz, 2011).

## KEYWORDS

- Avogadro's Law
- Berzelius Chemistry
- Dalton's Chemistry
- Döbereiner analogies
- Humphry's Theory
- Mendeleev's Law
- Meyer's relationships
- Newlands law
- Periodic table

## REFERENCES

### *AUTHOR'S MAIN REFERENCE*

Putz, M. V. (2011). Big chemical ideas in context: the periodic law and Scerri's periodic table. *International Journal of Chemical Modeling*, 3(1–2), 15–22.

### *SPECIFIC REFERENCES*

Aldea, V., Chivaroşi, V. (2000). *Anorganic Chemistry-Fundamental Principles (in Romanian)*, Medical Publishing House, Bucharest, 81–106.

Avogadro, A. (1811). Essay on a manner of determining the relative masses of the elementary molecules of bodies, and the proportions in which they enter into these compounds. *Journal de Physique* 73, 58–76.

Berzelius, J. J. (1813–1814). Essay on the cause of chemical proportions, and on some circumstances relating to them: together with a short and easy method of expressing them. *Annals of Philosophy* 2, 3, 51–52, 93–106, 244–255, 353–364, 443–454 [reprinted by Knight, D. M., ed., *Classical Scientific Papers* (New York: American Elsevier, 1968)].

Dalton, J. (1805). Experimental enquiry into the proportion of the several gases or elastic fluids, constituting the atmosphere. *Memoirs of the Literary and Philosophical Society of Manchester* 1, 244–258, 1, 271–287.

Dalton, J. (1808). *A New System of Chemical Philosophy*—Part I, Manchester.

Dalton, J. (1810). *A New System of Chemical Philosophy*—Part II, Manchester.

Dalton, J. (1827). *A New System of Chemical Philosophy*—Part III, Manchester.

Döbereiner, J. W. (1829). An attempt to group elementary substances according to their analogies. Poggendorf's *Annalen der Physik und Chemie* 15, 301–307 [reprinted by Leicester, H. M. & Klickstein, H. S., eds., *A Source Book in Chemistry, 1400–1900* (Cambridge, MA: Harvard, 1952)].

Hefferlin, R. (1989). *Periodic System of Molecules and their Relation to the Systematic Analysis of Molecular Data*, Edwin Mellin Press, Lewiston, New York.

Hefferlin, R., Kuhlman, H. (1980). The periodic system for free diatomic molecules III. *J. Quant. Spectro. Rad. Transf.* 24, 379–383.

Horovitz, O., Sârbu, C., Pop, H. (2000). *Rational Classification of the Chemical Elements (in Romanian)*, Dacia Publishing House, Cluj-Napoca, pp.11–17, 221–247.

Humphry, D. (1812). Of the analogies between the undecompounded substances: ideas respecting their nature. In: *Elements of Chemical Philosophy*, Vol. 1, Part 1, pp. 478–479, London [reprinted by Crosland, M., ed., *The Science of Matter: a Historical Survey* (Harmondsworth, UK: Penguin, 1971)].

Mendeleev, D. I. (1869a). On the relationship of the properties of the elements to their atomic weights (firstly appeared in Russian). *Journal of the Russian Chemical Society* 1, 60–77; then published as a short abstract in *Zeitschrift für Chemie* 12, 405–406 [translated from German by Carmen Giunta and published by David, M. Knight, ed., In: *Classical Scientific Papers-Chemistry, Second Series*, (New York: American Elsevier, 1970)].

Mendeleev, D. I. (1869b). *Osnovy Khimii. St.Petersburg* [with German edition: Mendelejeff, D. (1891). *Grundlagen der Chemie,* Ricker, St. Petersburg; and with English edition: Mendeleyev, D. I. (1897). *The Principles of Chemistry.* Longmans, Green and Co, London].

Mendeleev, D. I. (1871). A natural system of the elements and its use in predicting the properties of undiscovered elements. *Journal of the Russian Chemical Society* 3, 25–56.

Mendeleev, D. I. (1889). The periodic law of the chemical elements [Faraday Lecture of Mendeleev]. *Journal of the Chemical Society* 55, 634–656.

Meyer, J. M. (1870). The nature of the chemical elements as a function of their atomic weights. *Annalen der Chemie, Supplementband* 7, 354–364.

Newlands, J. A. R. (1863). On relations among the equivalents. *Chemical News* 7, 70–72.

Paneth, F. A. (1962). The epistemological status of the chemical concept of element. *Brit. J. Philos. Sci.* 13, 1–14 (Part I) and 144–160 (Part II) [reprinted in 2003 on *Found. Chem.* 5, 113–145].

Scerri, E. R. (2004a) Just how ab initio is ab initio quantum chemistry? *Foundations Chem.* 6, 93–116.

Scerri, E. R. (2004b) The role of the $n+l$ rule and the concept of an element as a basic substance. In: Rouvrary, D. and King, R. B. (eds.), *The Periodic Table: Into the twenty-first century*, Science Studies Press, Bristol, pp.143–160.

Scerri, E. R. (2007). *The Periodic Table. Its Story and Its Significance.* Oxford University Press, Oxford.

Scerri, E. R. (2009). The dual sense of the term "element," attempts to derive the madelung rule, and the optimal form of the periodic table, if any. *Int. J. Quantum Chem.* 109, 959–971.

Schwarz, W. H. E. (2007). Recommended questions on the road towards a scientific explanation of the periodic system of chemical elements with the help of the concepts of quantum physics. *Found. Chem.* 9, 139–188.
Weinhold, F., Bent, H. A. (2007). News from the periodic table: an introduction to "periodicity symbols, tables, and models for higher-order valency and donor–acceptor kinships." *J. Chem. Educ.* 84, 1145–1146.
Wikipedia (2013a). http://ro.wikipedia.org/wiki/Acid_sulfuric
Wikipedia (2013b). http://en.wikipedia.org/wiki/Sulfate
Wikipedia (2013c). http://en.wikipedia.org/wiki/Roentgenium

# CHAPTER 2

# QUANTUM ASSESSMENT FOR ATOMIC STABILITY

## CONTENTS

## ABSTRACT

The alternative quantum mechanical description of total energy given by path integrals specialized to Feynman-Kleinert formalism, while recovering the Bohr quantification of Hydrogen atom in a great extend, yet within a more general framework in which the stability issue is solved by existing quantum fluctuation, here explicitly modeled by periodic paths and Matsubara frequencies; both ground and exited states are treated in quantum statistical perspective.

## 2.1  INTRODUCTION

Being about to recently celebrating the International Year of Chemistry (2011) worth rethinking about what Chemistry indeed added as plus-value to the humankind knowledge, and what was not accomplished by other disciplines, Physics for instance—often seen (and rightly in general or quantitative sense) as the foreground bath of natural laws, if not the depositary of all Natural laws (and eventually the philosophy) as we earn from the Great Brit Newton. However, worth noting that in the times after the Middle Age recrudescence against any natural laws that were presumably against the theological doctrine of "believing without researching" the Chemistry was in fact the single discipline emerged as independent field of knowledge. This happened perhaps because it was regarded as containing enough mystery such that the crude minds will be never corrupted from seeing the Nature through the veil of initiation. In fact, the Alchemy, was the only pseudo-science present in those days, to which Newton itself, among others preeminent figures, were adhered in addition of being proud to be considered as belonging to. Just remember his vision about light as being composed from elementary particles "sympathetic one each other," i.e., manifesting a sort of chemistry among them, or describing the body interactions by engaged "sympathies" among corps, what were duly later recognized as corresponding nonetheless than to the "second quantization" of Physics (fields to elementary/quantable corpuscles) (Putz, 2011).

Therefore, one may say that Chemistry is a special way to see the physical phenomena, there where the mass or number of particles does not count in macroscopic way. And this is the first big leap of Chemistry: it seems to deal with macroscopic things when in fact it deals with observable quantities—that is different approach of reality; this, because observability is not a given fact but an emerging effect whose cause lays in microscopic reality of quantum fields and particles.

Moreover, Chemistry deals also with the non-observable quantities that Physics ignore somehow: take the *radius of atoms*—the reality is obvious as any substance may be seen as atoms-in-molecule compounds, yet there is no clear frontier of individual atoms neither in isolated or in interacted state; beside this there is no quantum mechanical operator of atomic radii—it eventually resulted through a variational process of the total energy.

Then, take the *electronic charge*—physically (in macroscopic or observable sense) indivisible below the elementary charge carry—however, there is rarely integer amounts of elementary charges are attributed to certain atoms in a molecule, according with quantum mechanical calculation at any level of approximation for a given many-electronic many-nuclear bonded structure.

However, having these physic al concepts particularly defined in Chemistry, allows therefore asserting against the first impression reductionism of physical laws to those of Chemistry. Beyond these there are few but astonishing important laws of nature that seems to especially belong to Chemistry despite their apparent refutation by ordinary physical thinking. One of these, perhaps the greatest one, is in the following discussed (Putz, 2011).

Recently, the crucial problem regarding whether chemical phenomena are reducible to physical ones has had an increasingly strong impact on the current course of conceptual and theoretical chemistry. For instance, the fact that elements arrange themselves in atomic number (Z) triads in approximately 50% of the periodic system seems to escape custom ordering quantifications (see also the Conclusion of the precedent Chapter) (Scerri, 2007; Putz, 2011). The same applies to the following: the fascinating golden ratio ($\tau$) limit for the periodicity of nuclei beyond any physical first-principle constants, which provides specific periodic laws for the chemical realm (see the Chapter 4 of the present volume) (Boeyens, 2005, 2008, 2011; Boeyens & Levendis, 2008); the fact that atoms have no definite atomic radii in the sense of a quantum operator, and even the Aufbau principle, which, although chemically workable, seems to violate the Pauli Exclusion Principle (Kaplan, 2002); at the molecular level, the well-celebrated reaction coordinate, which, although formally defined in the projective energy space, does not constitute a variable to drive optimization in the course of chemical reactions, appearing merely as a consequence of such reactions (Scerri, 2004); the problem of atoms in molecules (Bader, 1990), i.e., how much of the free atoms enter molecules and how much independency the atoms preserve in bonding; and chemical bonding itself, which ultimately appears to be reinterpreted as a special case of bosonic condensation with the aid of author's bondons—the quantum bosons of chemical bonding, which, without being elementary, imbue chemical compounds with a specific reality, see Volume III of the present five-volume book (Putz, 2016a).

However, in all cases the stability of atomic systems make possible all molecular and upwards structural constitution of matter; there is therefore essentially to be properly understand on quantum bases, this way also better distinguishing between the physical and chemical nature of the atom itself: paradoxically (or not) this will be firstly achieved through the celebrated Heisenberg principle by means of involving or not the "magic" measure of matter by means of the golden ratio first, then imbalance, then driving periodicity (see next chapters of the present volume). Beside the golden ratio, the *path integral* formalism, see Volume I of the present five-volume book (Putz, 2016b), will be here reloaded and extended under the Feynman-Kleinert formalism allowing for properly describing the atomic stability by combining the periodic paths with wave packets inside of atom so that having the quantum fluctuation at the origin of the atomic stability itself.

## 2.2   ATOMIC VALENCE STABILITY BY GOLDEN RATIO IMBALANCE

Atomic stability and periodicity remain major issues in the structural theories of matter; fortunately, they both have been largely solved by wave-particle (W/P) complementarily quantum behavior; phenomenologically, such relationship can be expressed as *"WAVE ⊗ PARTICLE = constant,"* while it may be quantized (by Planck's constant $h$) in the light of Heisenberg principle as (Putz, 2010, 2012)

$$WAVE \otimes PARTICLE = n_{W/P}h \qquad (2.1)$$

Remarkably, when fixing the particle's observable property, say $O$, while letting wave information to vary, say $\Delta O$, Eq. (2.1) takes the workable form

$$\Delta O \times O = n_O h \qquad (2.2)$$

having as the preeminent realization the Bohr-de Broglie formulation, leading with the first rationalization of the atomic periodicity (Pauling & Wilson, 1985): for circular orbits, the lowest ones in each atomic shells— including the valence ones, one has $\Delta O = \Delta r = 2\pi r$, with $r$ the orbital radii thereof, while $O = p$ is the fixed particle's momentum on that orbit; therefore, when combined into Eq. (2.2) they provide the celebrated Bohr-de Broglie relationship $rp = n\hbar$ solving the atomic spectra of Hydrogen atom in principal quantum numbers ($n$).

However, when about the atomic chemical reactivity a similar analysis may be provided in terms of the number of electrons to atomic number ratio ($N/Z$): one may fix the observable ("particle") character of the reactive atomic system by the ratio itself

$$O = \frac{N}{Z} \tag{2.3}$$

while modeling its evolving ("wave") character by the natural variation of the previous ratio in terms of exchanged electrons respecting the neutral state:

$$\Delta O = \frac{\Delta N}{Z} = \frac{N - Z}{Z} \tag{2.4}$$

When combining Eqs. (2.3) and (2.4) into Eq. (2.2) on the lowest quantized state (X. $n_O = 1$, the "ground state" of atomic reactivity that is the atom in its *valence* state so to speak) and within atomic units' formulation (i.e., by putting $h = 1$, since the actual reactivity quantification involves only numbers with no dimension), one has the so-called *Heisenberg imbalance equation* for *valence atoms*

$$\frac{N - Z}{Z} \times \frac{N}{Z} = 1 \tag{2.5}$$

that can be rewritten as

$$Z^2 + NZ - N^2 = 0 \tag{2.6}$$

Equation (2.6) has the elementary acceptable solution

$$Z = \frac{-N + \sqrt{N^2 + 4N^2}}{2} = N\tau \tag{2.7}$$

which establishes, the direct "chemical" connection between the number of electrons and the atomic charge by means of the golden ratio

$$\tau = \frac{-1+\sqrt{5}}{2} = 0.6180 \qquad (2.8)$$

generalizing the "physical" connection between nuclear (cosmic) synthesis at high pressure and atomic stability in the gas phase ($Z = N$); one has therefore the actual physical-to-chemical *electronic charge—atomic number* relationships (Putz, 2012)

$$\frac{Z}{N} = \begin{cases} 1...STABLE\,(PHYSICAL)\,ATOM \\ \tau...REACTIVE\,(CHEMICAL)\,ATOM \end{cases} \qquad (2.9)$$

Worth remarking the results of type (2.7) and (2.9), here based on chemical reactivity specialization of Heisenberg type equations (2.1) and/or (2.2), were previously obtained at the level of neutron-protonic imbalance, inside the atomic nuclei, based on well-founded empirical observations (Boeyens & Levendis, 2008). The present golden ratio appearance is ultimately sustained also by the deviation from the $N=Z$ condition for so-called "quark atoms" (as another way in considering the atoms in a quantum valence state), earlier identified as true matter's entities responsible for matter's reactivity at the atomic level (Lackner & Zweig, 1983).

Therefore, the atomic structure branching (2.9) can be regarded as the present *golden ratio* extension *to valence atom* and as such employed; actually, its consequences regarding the characterization of the quantum valence states of atoms within the Bohmian quantum potential are the main aims of the present endeavor and will be discussed next.

## 2.3   RELOADING QUANTUM PATH INTEGRAL FORMALISM FOR CHEMISTRY

Since the recent most celebrated quantum theory of Chemistry—the Density Functional Theory (DFT) is mainly based on density functionals, which relay on their turn on the many-body densities, the seek for electronic density both as computational (here understood as analytical) and conceptual assignments stays as a crucial endeavor in quantum chemistry comparable with the landmark theoretical predictions of spectra in the early age of quantum physics (Putz, 2009).

Yet, for achieving such challenging task the complex mathematical-informatics and mathematical-physics seems to be at the foremost background for computational and conceptual Density Functional Chemistry, respectively. The present review was dedicated to the later goal that is to present the analytical framework in which the many-electronic systems may be described by the associate densities at various levels of conceptualization, approximation, and applications.

As such, through presenting the basics of density matrix theory, the precursor of DFT, the path integral concept appears as the natural solution for expressing the time-space electronic density. Indeed, the Feynman path integral formulation has been revealed as the natural generalization of the Schrödinger equation, being in close relation with the propagators and Green function of a given quantum system, either at equilibrium or coupled with a temperature bath or particle environment.

Nevertheless, the density matrix—path integral description allows the general formulation for the many-electronic density through the so-called canonical density algorithm; it prescribes that the system is firstly solved for the single electron evolution under the concerned potential for which the time-space density matrix is analytically formulated, in an evolution manner, as the propagator $(x_b, t_b; x_a, t_a)$; then, the partition function is computed by closing the paths such that the spatially end-points to coincide,

$$Z(t_b;t_a) = \int dx(x,t_b;x,t_a) \qquad (2.10)$$

this step assures nevertheless that all possible energetic or eigen-configurations are accounted, thus including all the virtual single-eigen-states to be occupied when the systems will be eventually filled with electrons; moreover, it allows for the final writing of the $N$-electronic density formulation simply as

$$\rho_N(x;t_b - t_a) = Z^{-1}(t_b;t_a)[N \times (x,t_b;x,t_a)] \qquad (2.11)$$

Remarkably, the partition function involvement in this density algorithm was widely and most extensive used by the Feynman-Kleinert approach which was proved to furnish meaningful approximations either for the ground state (as was the case for atomic Hydrogen and the Bohr's orbitalic proofed stability) as well for the higher temperature or excited or the valence states (that resembles the semiclassical approximation).

Regarding the realization forms of path integral approaches of a quantum problem/system there were individuated three major pictures: the quantum mechanical (QM), the quantum statistical (QS) and the Fokker-Planck (FP) ones; the passage among them as well as their inter-conversion and equivalence being realized through the time transformations presented in the Table 2.1. It, nevertheless, leads both with philosophical and practical consequences: epistemologically, there seems that the time itself may suffer transformations being of quantum order ($\sim\hbar$) and correlated with inverse of the thermic energy ($\sim\beta$), while passing from quantum to statistical description of Nature; as well, the mass in equilibrium states plays the role in open systems of inverse of diffusion (naturally, since presenting inertia) but also with a quantum manifestly nature, while the ordinary (QM or QS) harmonic oscillations' frequency becomes friction in FP description of non-equilibrium systems.

**TABLE 2.1**   The Parametric Correspondence Between the Quantum Mechanics (QM), Quantum Statistics (QS) and Fokker-Planck (FP) Path Integral Representations; $m$ Stays for the Particle's Mass, $\omega$ for the Harmonic Frequency (of Paths' Fluctuation, Eventually), $\beta$ for the Inverse of the Thermic Energy $k_B T$, $D$ for the Diffusion Constant, $\gamma$ for the Friction Constant, while $t_a$ & $t_b$ are the End-Point Times for the Observed Evolution (Putz, 2009)

| QM | QS | FP |
|---|---|---|
| $m$ | $M$ | $\dfrac{h}{2D}$ |
| $\omega$ | $\omega$ | $\gamma$ |
| $\dfrac{1}{i}(t_b - t_a)$ | $\hbar\beta$ | $t_b - t_a$ |
| $\sin[\omega(t_b - t_a)]$ | $\dfrac{1}{i}\sinh(\omega\hbar\beta)$ | $\dfrac{1}{i}\sinh[\gamma(t_b - t_a)]$ |
| $\cos[\omega(t_b - t_a)]$ | $\cosh(\omega\hbar\beta)$ | $\cosh[\gamma(t_b - t_a)]$ |

Moreover, practically, such inter-conversion table allows for immediately transferring of one result obtained within a quantum picture into another without the need of entirely problem reformulation. This procedure was largely considered and applied throughout the present review; it nevertheless leads with another epistemological conclusion, namely that the QM oscillatory description is equivalently converted into the hyperbolic function for statistical and Markovian (or FP) frameworks, which further means the quantum modeling by the Gaussian wave-packet and Green functions. At this point, worth being mentioned the explicit proof for the de Broglie equivalence with the Gaussian wave function by the smearing out procedure of the fluctuation of the closed paths with the effective partition function approach; even more, such equivalence is justified by the very roots of quantum theory since the Born normalization of the de Broglie wave-packet is finely satisfied by the Gaussian form of its Fourier coefficients (amplitude), see Volume I of the present five-volume book.

From the chemically point of view, the valence states are those situated in the "chemical zone"-and they are the main concern for the chemical reactivity by employing the frontier or the outer electrons; consequently, the semiclassical approximation that models the excited states was expressly presented either as an extension of the quantum Feynman path integral or as a specialization of the Feynman-Kleinert formalism for higher temperature treatment of quantum systems (see Section 2.5). However, due to the correspondences of Table 2.1 one may systematically characterize the semiclassical (or quantum chemical) approaches as one of the limiting situations (Putz, 2009):

- $\hbar \to 0$: the quantum semiclassical limit;
- $\hbar\beta \to 0$: the QS short-time limit;
- $T \to \infty$: the high-temperature limit;
- $\omega \to 0$: the flat potential, or the quasi-homogeneous (Thomas-Fermi) limit; yet, this may be easier visualized by noting the discrete-to-quasi continuum transformation of eigen-levels intervals in the exited zones of quantum systems (atoms, molecules), i.e., where the approximation $\omega\hbar\beta = \omega\hbar/(k_B T) \ll 1$ holds; moreover, this limit nicely overlaps with the "free harmonic approximation" used in this work, when the interplay between the free and harmonic motion helped in elucidating and solving (by integrating out) the quantum fluctuations along the classical paths;

- $N \rightarrow \infty$ or $Z \rightarrow \infty$: the bosonic limit due to the scaled equivalence $T \sim N^{1/3}$ or $T \sim Z^{1/3}$ when the system is thermally expanded, being it related with the Thomas-Fermi theory, in Chapter 5 of the present volume exposed.

Yet, the use of path integral formalism for electronic density prescription presents several advantages: assures the inner QM description of the system by parameterized paths; averages the quantum fluctuations; behaves as the propagator for time-space evolution of quantum information; resembles Schrödinger equation; allows QS description of the system through partition function computing. In this framework, four levels of path integral formalism were presented: the Feynman quantum mechanical, the semiclassical, the Feynman-Kleinert effective classical, and the FP non-equilibrium (see Chapter 5 of the present volume) ones. In each case the density matrix or/and the canonical density are rigorously defined and presented. The practical specializations for quantum free and harmonic motions, for statistical high and low temperature limits, the smearing justification for the Bohr's quantum stability postulate with the paradigmatic Hydrogen atomic excursion, along the quantum chemical calculation of semiclassical electronegativity and hardness, of chemical action and Mulliken electronegativity, as well as by the Markovian generalizations of Becke-Edgecombe electronic focalization functions—all advocate for the reliability of assuming path integral formalism of quantum mechanics as a versatile one, suited for analytically and/or computationally modeling of a variety of fundamental physical and chemical reactivity concepts characterizing the (density driving) many-electronic systems.

## 2.4 PERIODIC PATH INTEGRALS

### 2.4.1 SURVEY ON MATSUBARA FREQUENCIES AND THE QUANTUM PERIODIC PATHS

Since the actual path picture uses the periodic paths, they will be seen as the Fourier series (Feynman & Hibbs, 19665; Feynman, 1972; Schulman, 1981; Wiegel, 1986; Kleinert, 2004; Putz, 2009)

$$x(\tau) = \sum_{m=-\infty}^{+\infty} x_m \exp(i\omega_m \tau) \tag{2.12}$$

in terms of the so-called Matsubara frequencies $\omega_m$ that are explicitly found through imposing on Eq. (2.12) the actual statistical closed path constraint

$$x_a = x(0) = x(\hbar\beta) = x_b \tag{2.13}$$

resulting in the equality

$$1 = \exp(i\omega_m \hbar\beta) \tag{2.14}$$

with the solution

$$\omega_m = \frac{2\pi}{\hbar\beta} m \ , \ m \in \mathbf{Z} \tag{2.15}$$

which certifies the quantization of paths (2.12).

Moreover, under the condition the quantum paths (2.12) are real,

$$x^*(\tau) = x(\tau) \tag{2.16}$$

its equivalent expanded form with the conjugated path

$$x^*(\tau) = \sum_{m=-\infty}^{+\infty} x_m^* \exp(-i\omega_m\tau) = \sum_{m=-\infty}^{+\infty} x_m \exp(i\omega_m\tau) \tag{2.17}$$

yields for the coefficients of the periodical paths the relationship:

$$x_m = x_m^* = x_{-m} \tag{2.18}$$

With this, the quantified form of periodic path frequencies, Eq. (2.15), allows rewriting of the paths (2.12) under a separated form into constant and complex conjugated oscillating contributions

$$x(\tau) = x_0 + \sum_{m=1}^{+\infty} x_m \exp(i\omega_m\tau) + c.c. \tag{2.19}$$

with the 0th terms viewed more than the "zero-oscillator" or free motion path but the thermal averaged path over entire quantum paths (2.12), see Volume I/Chapter 4 of this five-volume book:

$$\frac{1}{\hbar\beta} \int_0^{\hbar\beta} x(\tau)d\tau = x_0 \tag{2.20}$$

thus resulting in the Feynman centroid formula.

However, beside revealing the Feynman variable integration of the classical partition function

$$Z = \oint \mathscr{D}x(\tau) \exp\left\{ -\frac{1}{\hbar} \int_0^{\hbar\beta} d\tau \left[ m \frac{\dot{x}^2(\tau)}{2} + V(x(\tau)) \right] \right\} \quad (2.21)$$

as being of path integral (average) nature, the result enlighten on the actual periodic path decomposition (2.19) as furnishing another level for parameterize quantum paths, which goes beyond characterizing them as quantum fluctuations around classical motion; they are here constructed as periodic oscillations (back and forth—see the complex conjugation, in analogy with conjugated plane waves traveling in opposite directions) around the averaged path value (interpreted as thermic average, or, more plastic, as centroid of the quantum fluctuations themselves).

The symbol $\oint \mathscr{D}x(\tau)$ in Eq. (2.21) denotes the fact that the path integral is performed over all paths that fulfill the periodicity $x(0) = x(\hbar\beta)$ of Eq. (2.13). Is now clear that we prefer to deal with such integrals because their completeness respecting with all possible (statistically closed) paths between two quantum events. Therefore, just for this path parameterization perspective the present level seems involving quite complex quantum phenomenology; this will be further enriched in the sections to follow.

### 2.4.2  MATSUBARA HARMONIC PARTITION FUNCTION

With the quantum path decomposition (2.19) the Feynman path integral measure in Eq. (2.21) factorizes accordingly (Putz, 2009)

$$\int_{x(0)=x(\hbar\beta)} \mathscr{D}''x(\tau) \equiv \left( C_0 \int_{-\infty}^{+\infty} dx_0 \right) \left[ \prod_{m=1}^{\infty} C_m \int_{-\infty}^{+\infty} d\operatorname{Re}(x_m) \int_{-\infty}^{+\infty} d\operatorname{Im}(x_m) \right] \quad (2.22)$$

with the integration constants $C_0$ and $C_m$ to be determined from identifying the known partition function of the harmonic oscillator, see Volume I/ Chapter 4 of this five-volume work, as the result for the path integral representation of the same partition function with the measures (2.22) and paths (2.19):

$$Z_\Omega = \int\limits_{x(0)=x(\hbar\beta)} \mathcal{D}''x(\tau)\exp\left\{-\frac{1}{\hbar}\int\limits_0^{\hbar\beta}d\tau\left[\frac{m}{2}\dot{x}^2(\tau)+\frac{m}{2}\Omega^2x^2(\tau)\right]\right\}$$

$$=\left(C_0\int\limits_{-\infty}^{+\infty}dx_0\right)\left[\prod_{m=1}^\infty C_m\int\limits_{-\infty}^{+\infty}d\operatorname{Re}(x_m)\int\limits_{-\infty}^{+\infty}d\operatorname{Im}(x_m)\right]$$

$$\exp\left\{-\frac{1}{\hbar}\int\limits_0^{\hbar\beta}d\tau\left[\frac{m}{2}\dot{x}^2(\tau)+\frac{m}{2}\Omega^2x^2(\tau)\right]\right\}$$

$$\overset{!}{=}\frac{1}{2\sinh(\hbar\beta\Omega/2)} \tag{2.23}$$

Firstly, let's separately compute the kinetic and harmonic quantum path terms appearing under the integral (2.23). For kinetic term we successively get:

$$\frac{m}{2}\int\limits_0^{\hbar\beta}d\tau\dot{x}^2(\tau)=\frac{m}{2}\int\limits_0^{\hbar\beta}d\tau\left[\sum_{m=-\infty}^{+\infty}ix_m\omega_m\exp(i\omega_m\tau)\right]$$

$$\times\left[\sum_{m'=-\infty}^{+\infty}ix_{m'}\omega_{m'}\exp(i\omega_{m'}\tau)\right]$$

$$=-\frac{m}{2}\sum_{m=-\infty}^{+\infty}\sum_{m'=-\infty}^{+\infty}x_mx_{m'}\omega_m\omega_{m'}$$

$$\times\int\limits_0^{\hbar\beta}d\tau\exp\left[i(\omega_m+\omega_{m'})\tau\right]$$

$$=-\frac{m}{2}\hbar\beta\sum_{m=-\infty}^{+\infty}\sum_{m'=-\infty}^{+\infty}x_mx_{m'}\omega_m\omega_{m'}\delta_{m+m',0}$$

$$=-\frac{m}{2}\hbar\beta\sum_{m=-\infty}^{+\infty}x_mx_{-m}\omega_m\omega_{-m}$$

$$\overset{\omega_{-m}=-\omega_m}{=}\frac{m}{2}\hbar\beta\sum_{m=-\infty}^{+\infty}\omega_m^2|x_m|^2$$

$$=m\hbar\beta\sum_{m=1}^{+\infty}\omega_m^2\left[(\operatorname{Re}x_m)^2+(\operatorname{Im}x_m)^2\right] \tag{2.24}$$

while the harmonic contribution casts as:

$$\frac{m}{2}\Omega^2 \int_0^{\hbar\beta} d\tau x^2(\tau) = \frac{m}{2}\Omega^2 \int_0^{\hbar\beta} d\tau \left[ \sum_{m=-\infty}^{+\infty} x_m \exp(i\omega_m \tau) \right]$$

$$\times \left[ \sum_{m'=-\infty}^{+\infty} x_{m'} \exp(i\omega_{m'} \tau) \right]$$

$$= \frac{m}{2}\Omega^2 \sum_{m=-\infty}^{+\infty} \sum_{m'=-\infty}^{+\infty} x_m x_{m'}$$

$$\times \int_0^{\hbar\beta} d\tau \exp[i(\omega_m + \omega_{m'})\tau]$$

$$= \frac{m}{2}\Omega^2 \sum_{m=-\infty}^{+\infty} \sum_{m'=-\infty}^{+\infty} x_m x_{m'} \hbar\beta\delta_{m+m',0}$$

$$= \frac{m}{2}\Omega^2 \hbar\beta \sum_{m=-\infty}^{+\infty} x_m x_{-m}$$

$$= \frac{m}{2}\Omega^2 \hbar\beta \sum_{m=-\infty}^{+\infty} x_m x_m^*$$

$$= \frac{m}{2}\Omega^2 \hbar\beta \left\{ x_0^2 + 2\sum_{m=1}^{+\infty} \left[ (\operatorname{Re} x_m)^2 + (\operatorname{Im} x_m)^2 \right] \right\} \quad (2.25)$$

and together combined in the partition function (2.23):

$$Z_\Omega = \left( C_0 \int_{-\infty}^{+\infty} dx_0 \right) \left[ \prod_{m=1}^{\infty} C_m \int_{-\infty}^{+\infty} d\operatorname{Re}(x_m) \int_{-\infty}^{+\infty} d\operatorname{Im}(x_m) \right]$$

$$\times \exp\left\{ -m\beta \sum_{m=1}^{+\infty} \omega_m^2 \left[ (\operatorname{Re} x_m)^2 + (\operatorname{Im} x_m)^2 \right] \right.$$

$$\left. -\frac{m}{2}\Omega^2 \beta x_0^2 - m\Omega^2 \beta \sum_{m=1}^{+\infty} \left[ (\operatorname{Re} x_m)^2 + (\operatorname{Im} x_m)^2 \right] \right\}$$

$$= \left[ C_0 \int_{-\infty}^{+\infty} dx_0 \exp\left( -\frac{m}{2} \Omega^2 \beta x_0^2 \right) \right.$$

$$\left. \times \prod_{m=1}^{\infty} C_m \left\{ \int_{-\infty}^{+\infty} d\operatorname{Re}(x_m) \exp\left[ -m\beta \sum_{m=1}^{+\infty} \left( \omega_m^2 + \Omega^2 \right) \left( \operatorname{Re} x_m \right)^2 \right] \right\} \right]^2$$

$$= C_0 \sqrt{\frac{2\pi}{m\beta\Omega^2}} \prod_{m=1}^{\infty} C_m \frac{\pi}{m\beta\left( \omega_m^2 + \Omega^2 \right)} \tag{2.26}$$

Within the frequency choice

$$C_m = C\omega_m^2 \tag{2.27}$$

the partition function (2.26) further resumes as:

$$Z_\Omega = C_0 \frac{1}{\Omega} \sqrt{\frac{2\pi}{m\beta}} \frac{\pi}{m\beta} Cf(\Omega) \tag{2.28}$$

with the newly introduced function:

$$f(\Omega) = \prod_{m=1}^{\infty} \frac{\omega_m^2}{\omega_m^2 + \Omega^2} \tag{2.29}$$

like a series of Matsubara frequencies. There is clear that in order the Matsubara partition function (2.28) be solved the product series (2.29) has to be evaluated. This is done through three more transformations, namely by rewriting it as

$$f(\Omega) = \exp\left[ g(\Omega) \right] \tag{2.30}$$

with

$$g(\Omega) = \sum_{m=1}^{\infty} \ln \frac{\omega_m^2}{\omega_m^2 + \Omega^2} \tag{2.31}$$

followed by its derivative:

$$g'(\Omega) = \sum_{m=1}^{\infty} \frac{\omega_m^2 + \Omega^2}{\omega_m^2} \frac{-\omega_m^2}{\left(\omega_m^2 + \Omega^2\right)^2} 2\Omega = -2\Omega R(\Omega) \qquad (2.32)$$

that is recognized as being related with he Riemann generalized series

$$R(\Omega) = \sum_{m=1}^{\infty} \frac{1}{\Omega^2 + \omega_m^2} \qquad (2.33)$$

Once calculated, the Riemann series (2.33) is replaced in Eq. (2.32) which, at its turn, is employed in the integral manner

$$g(\Omega) = g(0) + \int_{0}^{\Omega} d\tilde{\Omega} g'(\tilde{\Omega}) \qquad (2.34)$$

for finally providing the searched function (2.29).

The calculus of the generalized Riemann series (2.33) will be in next section exposed.

### 2.4.3   THE GENERALIZED RIEMANN' SERIES

Computation of the generalized Riemann series (2.33) requires few intermediate steps (Putz, 2009):

- Writing it under the form

$$R(\Omega) = \frac{1}{2}\left[ F(\Omega) - \frac{1}{\Omega^2} \right] \qquad (2.35)$$

in terms of the extended series:

$$F(\Omega) = \sum_{m=-\infty}^{\infty} \frac{1}{\Omega^2 + \omega_m^2} = \frac{\hbar^2 \beta^2}{4\pi^2} \sum_{m=-\infty}^{\infty} \frac{1}{\alpha^2 + m^2} \ , \alpha \equiv \frac{\hbar^2 \beta^2 \Omega^2}{4\pi^2} \qquad (2.36)$$

- Applying the Poisson (comb function) formula for series, see Volume I/Appendix A.1.2 of this five-volume work (Putz, 2016b), on series (2.36)

$$F(\Omega) = \frac{h^2 \beta^2}{4\pi^2} \sum_{n=-\infty}^{\infty} \left[ \int_{-\infty}^{+\infty} dq \frac{1}{q^2 + \alpha^2} \exp(-i2\pi qn) \right] \quad (2.37)$$

- Computing the integral under the sum of Eq. (2.37) by complex integration, according with the contours of integration identified in Figure 2.1, around the poles $q = \pm i\alpha$, throughout applying the residues' theorem:

$$\oint_{C(z_0)} f(z)dz = 2\pi i \operatorname*{\mathbf{Res}}_{z=z_0} f(z) = 2\pi i \lim_{z \to z_0} (z - z_0) f(z) \quad (2.38)$$

while summing the convergent cases:

$$\begin{aligned} n \geq 0 &\Rightarrow \operatorname{Im} q < 0 \Rightarrow \textit{outline II} \\ n < 0 &\Rightarrow \operatorname{Im} q > 0 \Rightarrow \textit{outline I} \end{aligned} \quad (2.39)$$

arisen from the observation that

$$\left| \exp\left[ -i2\pi(\operatorname{Re} q + i \operatorname{Im} q)n \right] \right| = \exp(2\pi n \operatorname{Im} q) \quad (2.40)$$

Note that the outline ($I$) is considered as being the one circulated with trigonometric direction, while for the ($II$) outline the anti-trigonometric sense resulted, being is equivalent with the ($-$) sign in front of its integral, which, for instance explicitly gives:

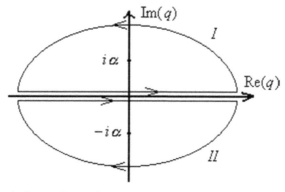

**FIGURE 2.1**   The integration outlines around the poles $q = \pm i\alpha$.

$$\int_{-\infty(II)}^{+\infty} dq \frac{1}{q^2+\alpha^2}\exp(-i2\pi qn)$$

$$=(-1)\oint_{C(II)} dq \frac{1}{q^2+\alpha^2}\exp(-i2\pi qn)=-2\pi i \underset{q=-i\alpha}{\text{Res}} \frac{\exp(-i2\pi qn)}{q^2+\alpha^2}$$

$$=-2\pi i \lim_{\substack{q=-i\alpha \\ \alpha=\frac{\hbar\beta\Omega}{2\pi}}} (q+i\alpha)\frac{\exp(-i2\pi qn)}{(q+i\alpha)(q-i\alpha)}=\frac{2\pi^2}{\hbar\beta\Omega}\exp(-\hbar\beta\Omega n) \qquad (2.41)$$

while the other contour integration leads with similar result.

- Combining the integrations' result sin the expression (2.37) while attributing to each contour integration result the series summing range in accordance with the constraints resulted in Eq. (2.39):

$$F(\Omega)=\frac{\hbar^2\beta^2}{4\pi^2}\left[\sum_{n=1}^{\infty}\int_{-\infty(I)}^{+\infty} dq \frac{\exp(i2\pi qn)}{q^2+\alpha^2}+\sum_{n=0}^{\infty}\int_{-\infty(II)}^{+\infty} dq \frac{\exp(-i2\pi qn)}{q^2+\alpha^2}\right]$$

$$=\frac{\hbar^2\beta^2}{4\pi^2}\frac{2\pi^2}{\hbar\beta\Omega}\left[\sum_{n=1}^{\infty}\exp(-\hbar\beta\Omega n)+\sum_{n=0}^{\infty}\exp(-\hbar\beta\Omega n)\right]$$

$$=\frac{\hbar\beta}{2\Omega}\left\{2\sum_{n=0}^{\infty}[\exp(-\hbar\beta\Omega)]^n -1\right\}$$

$$=\frac{\hbar\beta}{2\Omega}\coth\left(\frac{\hbar\beta\Omega}{2}\right) \qquad (2.42)$$

- With evaluation (2.42) for the series (2.36) the Riemann series (2.35) results as:

$$R(\Omega)=\sum_{m=1}^{\infty}\frac{1}{\Omega^2+\omega_m^2}=\frac{\hbar^2\beta^2}{4\pi^2}\sum_{m=1}^{\infty}\frac{1}{(\hbar\beta\Omega/2)^2+m^2}$$

$$=\frac{1}{4}\frac{\hbar\beta}{\Omega}\left[\coth\left(\frac{\hbar\beta\Omega}{2}\right)-\frac{2}{\hbar\beta\Omega}\right] \qquad (2.43)$$

- The check with the usual Riemann series by means of turning the harmonic to free motion picture, as already done within path integral evaluations based on free motion- harmonic motion interplay; In fact having to evaluate the limit

$$\lim_{\Omega \to 0} R(\Omega) = \frac{\hbar^2 \beta^2}{4\pi^2} \sum_{m=1}^{\infty} \frac{1}{m^2} = \lim_{\Omega \to 0} \frac{1}{4} \frac{\hbar \beta}{\Omega} \left[ \coth\left(\frac{\hbar \beta \Omega}{2}\right) - \frac{2}{\hbar \beta \Omega} \right] \quad (2.44)$$

it is practically (with mathematically sufficiency) applied only to the hyperbolic cotangent function

$$[\coth(\bullet)]_{\bullet \to 0} \cong \frac{1}{\bullet} + \frac{\bullet}{3} \quad (2.45)$$

since resulting in the identity:

$$\frac{\hbar^2 \beta^2}{4\pi^2} \sum_{m=1}^{\infty} \frac{1}{m^2} = \frac{1}{4} \frac{\hbar^2 \beta^2}{6} \quad (2.46)$$

from where springs out the Riemann series custom limit:

$$\sum_{m=1}^{\infty} \frac{1}{m^2} = \frac{\pi^2}{6} \quad (2.47)$$

We have now all prerequisites to compute the Matsubara harmonic partition function (2.28) aiming to find out the Matsubara normalization of periodic path integrals. This will be addressed in the sequel.

### 2.4.4  PERIODIC PATH INTEGRAL MEASURE

Turning to the Matsubara harmonic partition function algorithm (2.28)–(2.34) one successively has (Putz, 2009):

- Computing the function (2.32) where inserting the above Riemann generalized series (2.43):

$$g'(\Omega) = -2\Omega R(\Omega) = -\frac{\hbar \beta}{2} \left[ \coth\left(\frac{\hbar \beta \Omega}{2}\right) - \frac{2}{\hbar \beta \Omega} \right] \quad (2.48)$$

- Compute the function (2.31) by the aid of Eq. (2.34) rule through considering the variable change $z = \hbar \beta \Omega / 2$ in Eq. (2.48):

$$g(\Omega) = -\int_0^{\hbar\beta\Omega/2} dz\left(\coth z - \frac{1}{z}\right) = -\left[\ln(\sinh z) - \ln z\right]_0^{\hbar\beta\Omega/2}$$

$$= -\left[\ln\left(\frac{\sinh z}{z}\right)\right]_0^{\hbar\beta\Omega/2}$$

$$= \ln\left[\frac{\hbar\beta\Omega}{2\sinh(\hbar\beta\Omega/2)}\right] \tag{2.49}$$

- Computing the function (2.29) with the help of Eqs. (2.30) and (2.49):

$$f(\Omega) = \prod_{m=1}^{\infty} \frac{\omega_m^2}{\omega_m^2 + \Omega^2} = \exp[g(\Omega)] = \frac{\hbar\beta\Omega}{2\sinh(\hbar\beta\Omega/2)} \tag{2.50}$$

- Releasing the Matsubara partition function for the harmonic motion by replacing function (2.50) into expression (2.28):

$$Z_\Omega = C_0 C\sqrt{\frac{2\pi}{m\beta}}\frac{\pi\hbar}{m}\frac{1}{2\sinh(\hbar\beta\Omega/2)} \tag{2.51}$$

- Comparing the form (2.51) with the consecrated result for harmonic oscilators' partition function (see Volume I/Chapter 4 of the present five-volume book) one remains with the condition:

$$1 = C_0 C\sqrt{\frac{2\pi}{m\beta}}\frac{\pi\hbar}{m} \tag{2.52}$$

- Choosing for the Feynman centroid integral the normalization factor that regains the inverse of the thermal length (280):

$$C_0 = \frac{1}{\lambda_{th}} = \frac{1}{\sqrt{2\pi\hbar^2\beta/m}} \tag{2.53}$$

- Plugging expression (2.53) in Eq. (2.52) there result the constant

$$C = \frac{\beta m}{\pi} \tag{2.54}$$

and then through the relation (2.27) also the Matsubara constants

$$C_m = C\omega_m^2 = \frac{\beta m_0 \omega_m^2}{\pi} \tag{2.55}$$

- Once determined the constants (2.53) and (2.55) are replaced in Eq. (2.22) to yield the normalized measure of the periodic integrals in terms of the Matsubara quantum frequencies (2.15)

$$\int_{x(0)=x(\hbar\beta)} \mathcal{D}'' x(\tau) \equiv \left( \int_{-\infty}^{+\infty} \frac{dx_0}{\sqrt{2\pi\hbar^2\beta/m}} \right) \left[ \prod_{m=1}^{\infty} \int_{-\infty}^{+\infty}\int_{-\infty}^{+\infty} \frac{d\operatorname{Re}x_m d\operatorname{Im}x_m}{\pi/(m\beta\omega_m^2)} \right] \tag{2.56}$$

Note that the measure given in Eq. (2.56) is rather universal for periodic paths, while the involvement of the harmonic oscillator was only a tool (and always an inspiring exercise) for determining it through the complete quantum and statistical solution at hand for harmonic motion and its versatile properties respecting the perturbation or limiting the free motion as well as for modeling the quantum fluctuations (by quantifying the displacements away from classical equilibrium or path).

## 2.5 FEYNMAN-KLEINERT VARIATIONAL FORMALISM

### 2.5.1 FEYNMAN-KLEINERT PARTITION FUNCTION

Being equipped with the periodic path integral technique we can present one of the most efficient ways for approximate the effective-classical partition function (2.21); it firstly unfolds for a general external potential like the *exact* integral (Putz, 2009):

$$Z = \int_{x(0)=x(\hbar\beta)} \mathcal{D}'' x(\tau) \exp\left\{ -\frac{1}{\hbar} S^+[x,\dot{x},\tau] \right\}$$

$$= \int_{x(0)=x(\hbar\beta)} \mathcal{D}'' x(\tau) \exp\left\{ -\frac{1}{\hbar} \int_0^{\hbar\beta} d\tau \left[ m\frac{\dot{x}^2(\tau)}{2} + V(x(\tau)) \right] \right\}$$

$$= \left( \int_{-\infty}^{+\infty} \frac{dx_0}{\sqrt{2\pi\hbar^2\beta/m}} \right) \left[ \prod_{m=1}^{\infty} \int_{-\infty}^{+\infty}\int_{-\infty}^{+\infty} \frac{d\operatorname{Re}x_m d\operatorname{Im}x_m}{\pi/(m\beta\omega_m^2)} \right]$$

$$\times \exp\left[ -\beta m \sum_{m=1}^{\infty} \omega_m^2 |x_m|^2 - \frac{1}{\hbar}\int_0^{\hbar\beta} d\tau V(x(\tau)) \right] \tag{2.57}$$

Since the exact solution for expression (2.57) is hard to formulate for an unspecified potential form, it may be eventually reformulated in a workable way by involving another partition function, the so-called *Feynman-Kleinert* (FK) *partition function* $Z_{FK}$, and its special average recipe, respectively as

$$Z = \int_{x(0)=x(\hbar\beta)} \mathscr{D}'x(\tau)\exp\left\{-\frac{1}{\hbar}S_{FK}^+[x,\dot{x},\tau]\right\}$$

$$\times \exp\left\{-\frac{1}{\hbar}\left(S^+[x,\dot{x},\tau]-S_{FK}^+[x,\dot{x},\tau]\right)\right\}$$

$$= Z_{FK}\left\langle\exp\left\{-\frac{1}{\hbar}\left(S^+[x,\dot{x},\tau]-S_{FK}^+[x,\dot{x},\tau]\right)\right\}\right\rangle_{FK} \qquad (2.58)$$

and

$$\langle O[x]\rangle_{FK} = Z_{FK}^{-1}\int_{x(0)=x(\hbar\beta)} \mathscr{D}'x(\tau)O[x]\exp\left\{-\frac{1}{\hbar}S_{FK}^+[x,\dot{x},\tau]\right\} \qquad (2.59)$$

In Eqs. (2.58) and (2.59) the Feynman-Kleinert partition function takes the general form

$$Z_{FK} = \int_{x(0)=x(\hbar\beta)} \mathscr{D}'x(\tau)\exp\left\{-\frac{1}{\hbar}S_{FK}^+[x,\dot{x},\tau]\right\} \qquad (2.60)$$

wile the working ansatz looks like

$$S_{FK}^+[x,\dot{x},\tau]=\int_0^{\hbar\beta}d\tau\left[m\frac{\dot{x}^2(\tau)}{2}+\frac{m}{2}\Omega^2(x_0)(x(\tau)-x_0)^2+L_{FK}(x_0)\right] \qquad (2.61)$$

thus being constructed as such, unlike the general partition function (2.57), to explicitly account for the path fluctuations around the Feynman centroid (2.20), the only integration variable in effective-partition function (2.21), through the term $(x(\tau)-x_0)^2$, driven harmonically by the frequency $\Omega^2(x^0)$, with a role in optimizing the quantum fluctuations in order state equilibrium be achieved, while the supplementary of Feynman-Kleinert perturbation function $L_{FK}(x^0)$ assures the global optimization for the action, and implicitly for the Feynman-Kleinert partition function, so approaching the best the exact partition function (2.57) or its associate total ground state energy of the system given by the free energy:

$$F = -\beta^{-1} \ln Z \qquad (2.62)$$

In fact, the Feynman-Kleinert action (2.61) is being to be involved in two-fold optimization algorithm in providing the best approximation of the partition function (2.57). This will favor a close analogy with the double search for electronic density, in density functional theory (DFT), as will be latter discussed.

Yet, the Feynman-Kleinert partition function is to be unfolded within the actual periodic path integral representation, with the help of Eqs. (2.28) and (2.50)

$$Z_{FK} = \int\limits_{x(0)=x(\hbar\beta)} \mathscr{D}'' x(\tau) \exp\left\{-\frac{1}{\hbar}\int_0^{\hbar\beta} d\tau \left[m\frac{\dot{x}^2(\tau)}{2} + \frac{m}{2}\Omega^2(x_0)(x(\tau)-x_0)^2 + L_{FK}(x_0)\right]\right\}$$

$$= \left(\int_{-\infty}^{+\infty} \frac{dx_0 \exp\left[-\beta L_{FK}(x_0)\right]}{\sqrt{2\pi\hbar^2\beta/m}}\right)$$

$$\times \underbrace{\left[\prod_{m=1}^{\infty} \int_{-\infty}^{+\infty}\int_{-\infty}^{+\infty} \frac{d\,\mathrm{Re}\,x_m\, d\,\mathrm{Im}\,x_m}{\pi/(m\beta\omega_m^2)}\right] \exp\left\{-m\beta\sum_{m=1}^{\infty}\left[\omega_m^2 + \Omega^2(x_0)\right]|x_m|^2\right\}}_{f(\Omega)\ see\ eqs.\ (292)\&(311)}$$

$$= \int_{-\infty}^{+\infty} \frac{dx_0}{\sqrt{2\pi\hbar^2\beta/m}} \frac{\hbar\beta\Omega(x_0)/2}{\sinh(\hbar\beta\Omega(x_0)/2)} \exp\left[-\beta L_{FK}(x_0)\right]$$

$$\qquad (2.63)$$

while being formally expressed under the effective-classical form (2.21)

$$Z_{FK} = \int_{-\infty}^{+\infty} \frac{dx_0}{\sqrt{2\pi\hbar^2\beta/m}} \exp\left[-\beta W_{FK}(x_0)\right] \qquad (2.64)$$

it provides the Feynman-Kleinert potential (Feynman & Kleinert, 1986)

$$W_{FK}(x_0) = \frac{1}{\beta}\ln\left[\frac{\sinh(\hbar\beta\Omega(x_0)/2)}{\hbar\beta\Omega(x_0)/2}\right] + L_{FK}(x_0) \qquad (2.65)$$

to be optimized in respect with its harmonic frequency (equilibrium optimization) and for perturbation (ground state optimization) in what follows.

### 2.5.2 FEYNMAN-KLEINERT OPTIMUM POTENTIAL

The optimization of the Feynman-Kleinert partition function (2.64) is performed employing the Jensen-Peierls inequality (Putz, 2009),

$$\langle \exp[O] \rangle \geq \exp[\langle O \rangle] \tag{2.66}$$

whose the phenomenological proof is given in the Figure 2.2, on the partition function relationship (2.58) leading to the lower bounded partition function

$$Z \geq Z_{FK} \exp\left\{-\frac{1}{\hbar}\left\langle S^{+}[x,\dot{x},\tau] - S^{+}_{FK}[x,\dot{x},\tau]\right\rangle_{FK}\right\} \tag{2.67}$$

or, by calling the Eq. (2.62), to the higher bounded free energy

$$F \leq F_{FK} + \frac{1}{\hbar\beta}\left\langle S^{+}[x,\dot{x},\tau] - S^{+}_{FK}[x,\dot{x},\tau]\right\rangle_{FK} \tag{2.68}$$

Yet, the last inequality, rewritten with the help of Euclidian actions for general partition function and the Feynman-Kleinert specialization, Eqs. (2.57) and (319), respectively, one notes the disappearing of the kinetic (free motion) terms, while the resulting expression

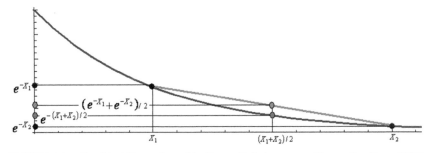

**FIGURE 2.2**   Graphical illustration of the Jensen-Peierls averages' inequality (Putz, 2009).

$$F \le F_{FK} + \frac{1}{\hbar\beta} \int_0^{\hbar\beta} d\tau \left\langle V(x(\tau)) - \frac{m}{2}\Omega^2(x_0)(x(\tau) - x_0)^2 - L_{FK}(x_0) \right\rangle_{FK} \quad (2.69)$$

provides the searched variational architecture:

$$\delta F = 0 \Rightarrow \left\langle V(x(\tau)) - \frac{m}{2}\Omega^2(x_0)(x(\tau) - x_0)^2 - L_{FK}(x_0) \right\rangle_{FK} = 0 \quad (2.70)$$

where all involved terms combines external, perturbation and quantum fluctuation influences.

Being the variational problem formulated it remains to individually compute the terms appearing in Feynman-Kleinert average (2.70), by using the definition (2.59) with the action (2.61).

For the external potential average we have in the first instance the periodic path integral representation, with the help of Eqs. (2.28) and (2.50)

$$\langle V(x(\tau)) \rangle_{FK} = Z_{FK}^{-1} \int_{x(0)=x(\hbar\beta)} \mathcal{D}' x(\tau) V(x(\tau)) \exp\left\{ -\frac{1}{\hbar} S_{FK}^+[x, \dot{x}, \tau] \right\}$$

$$= Z_{FK}^{-1} \left( \int_{-\infty}^{+\infty} \frac{dx_0 \exp[-\beta L_{FK}(x_0)]}{\sqrt{2\pi\hbar^2 \beta / m}} \right) \left[ \prod_{m=1}^{\infty} \int_{-\infty}^{+\infty}\int_{-\infty}^{+\infty} \frac{d\operatorname{Re}x_m \, d\operatorname{Im}x_m}{\pi/(m\beta\omega_m^2)} \right]$$

$$\times \exp\left\{ -m\beta \sum_{m=1}^{\infty} [\omega_m^2 + \Omega^2(x_0)]|x_m|^2 \right\}$$

$$\times \int_{-\infty}^{+\infty} \frac{dk}{2\pi} V(k) \exp\left\{ ik\left[ x_0 + \sum_{n=1}^{\infty} x_n \exp(-i\omega_n\tau) + c.c. \right] \right\}$$

$$= Z_{FK}^{-1} \int_{-\infty}^{+\infty} \frac{dx_0 \exp[-\beta L_{FK}(x_0)]}{\sqrt{2\pi\hbar^2 \beta / m}} \int_{-\infty}^{+\infty} \frac{dk}{2\pi} V(k) \exp(ikx_0)$$

$$\times \left[ \prod_{m=1}^{\infty} \int_{-\infty}^{+\infty}\int_{-\infty}^{+\infty} \frac{d\operatorname{Re}x_m \, d\operatorname{Im}x_m}{\pi/(m\beta\omega_m^2)} \right] \exp\left\{ -m\beta \sum_{m-1}^{\infty} [\omega_m^2 + \Omega^2(x_0)]|x_m|^2 \right.$$

$$\left. + ik\left[ \sum_{n=1}^{\infty} x_n \exp(-i\omega_n\tau) + c.c. \right] \right\}$$

$$= Z_{FK}^{-1} \int_{-\infty}^{+\infty} \frac{dx_0 \exp\left[-\beta L_{FK}(x_0)\right]}{\sqrt{2\pi\hbar^2\beta/m}} \int_{-\infty}^{+\infty} \frac{dk}{2\pi} V(k) \exp\left[ikx_0 - \frac{1}{2}a^2(x_0)k^2\right]$$

$$\times \left[\prod_{m=1}^{\infty} \int_{-\infty}^{+\infty}\int_{-\infty}^{+\infty} \frac{d\,\mathrm{Re}\,x_m d\,\mathrm{Im}\,x_m}{\pi/\left(m\beta\omega_m^2\right)}\right]$$

$$\times \exp\left\{-m\beta\sum_{m=1}^{\infty}\left[\omega_m^2 + \Omega^2(x_0)\right]\left[\left(\mathrm{Re}\,x_m - ik\frac{1/(m\beta)}{\omega_m^2 + \Omega^2(x_0)}\cos\omega_m\tau\right)^2 \right.\right.$$
$$\left.\left. + \left(\mathrm{Im}\,x_m - ik\frac{1/(m\beta)}{\omega_m^2 + \Omega^2(x_0)}\sin\omega_m\tau\right)^2\right]\right\}$$

$$\underbrace{\phantom{xxxxxxxxxxxxxxxxxxxxxxxxxxxxxxxxxxxxxxxxxxxxxxxxxxxxxxxxxxxxxxxxxxxxxxxxxxxx}}_{f(\Omega) \ see \ eqs. \ (2.28)\&(2.50)}$$

$$= Z_{FK}^{-1} \int_{-\infty}^{+\infty} \frac{dx_0 \exp\left[-\beta L_{FK}(x_0)\right]}{\sqrt{2\pi\hbar^2\beta/m}} \frac{\hbar\beta\Omega(x_0)/2}{\sinh(\hbar\beta\Omega(x_0)/2)}$$

$$\times \int_{-\infty}^{+\infty} \frac{dk}{2\pi} V(k) \exp\left[ikx_0 - \frac{1}{2}a^2(x_0)k^2\right]$$

$$(2.71)$$

where the Fourier $k$-(wave vector) representation was implemented for external potential so that the quantum path to explicitly appear in evaluation, followed by quadratic completion of paths in the view of harmonic-like integration of type (2.26) with the result (2.50); nevertheless, in course of these operations the new quantity was introduced, namely

$$a^2(x_0) = 2\frac{1}{m\beta}\sum_{m=1}^{\infty}\frac{1}{\omega_m^2 + \Omega^2(x_0)} \qquad (2.72)$$

which can be immediately analytically evaluated since recognized as directly related with generalized Riemann series (2.43) with the form (Feynman & Kleinert, 1986)

$$a^2(x_0) = 2\frac{1}{m\beta}R(\Omega) = \frac{1}{m\beta\Omega^2}\left[\frac{\hbar\beta\Omega}{2}\coth\left(\frac{\hbar\beta\Omega}{2}\right) - 1\right] \quad (2.73)$$

Expression (2.71) can be even more simplified when solving out the k-integral by considering the back Fourier transformation for the potential and then proceeding with the quadratic completion toward the Poisson standard integration

$$V_{a^2(x_0)}(x_0) = \int_{-\infty}^{+\infty}\frac{dk}{2\pi}V(k)\exp\left[ikx_0 - \frac{1}{2}a^2(x_0)k^2\right]$$

$$= \int_{-\infty}^{+\infty}\frac{dk}{2\pi}\int_{-\infty}^{+\infty}dxV(x)\exp(-ikx)\exp\left[ikx_0 - \frac{1}{2}a^2(x_0)k^2\right]$$

$$= \frac{1}{2\pi}\int_{-\infty}^{+\infty}dxV(x)\exp\left[-\frac{(x-x_0)^2}{2a^2(x_0)}\right]$$

$$\times \int_{-\infty}^{+\infty}dk\exp\left[\left(\frac{x-x_0}{\sqrt{2a(x_0)}} - i\frac{a(x_0)}{\sqrt{2}}k\right)^2\right]$$

$$= \frac{1}{2\pi}\int_{-\infty}^{+\infty}dxV(x)\exp\left[-\frac{(x-x_0)^2}{2a^2(x_0)}\right]$$

$$\times \int_{-\infty}^{+\infty}dk'\exp\left[-\frac{a^2(x_0)}{2}k'^2\right]$$

$$= \frac{1}{\sqrt{2\pi a^2(x_0)}}\int_{-\infty}^{+\infty}dxV(x)\exp\left[-\frac{(x-x_0)^2}{2a^2(x_0)}\right]$$

$$\equiv \langle V(x)\rangle_{a^2(x_0)} \quad (2.74)$$

The potential (2.74) is known as the smeared out potential and has a major role in explaining the quantum stabilization of matter, as will be largely discussed in the next section. For the moment it is regarded jus as the integral transformation of the original applied potential through convoluted with a Gaussian packet with the width $a^2(x_0)$ accounting for the existing quantum fluctuation in the system.

Nevertheless, with Eq. (2.74) back in Eq. (2.71) we have for the Feynman-Kleinert average of external potential the result (Feynman & Kleinert, 1986):

$$\langle V(x(\tau))\rangle_{FK} = Z_{FK}^{-1} \int_{-\infty}^{+\infty} \frac{dx_0}{\sqrt{2\pi\hbar^2 \beta / m}} \langle V(x)\rangle_{a^2(x_0)} \frac{\hbar\beta\Omega(x_0)/2}{\sinh(\hbar\beta\Omega(x_0)/2)} \exp\left[-\beta L_{FK}(x_0)\right]$$

(2.75)

Next, for the rest of the averaged terms in Eq. (2.70) things are considerably more easy since or each of them we have firstly to compute their smeared out version (2.74), with the respective results

$$\left\langle \frac{m}{2}\Omega^2(x_0)(x-x_0)^2 \right\rangle_{a^2(x_0)} = \frac{m}{2}\Omega^2(x_0)\frac{1}{\sqrt{2\pi a^2(x_0)}} \int_{-\infty}^{+\infty} dx(x-x_0)^2$$

$$\times \exp\left[-\frac{(x-x_0)^2}{2a^2(x_0)}\right] = \frac{m}{2}\Omega^2(x_0)a^2(x_0) \quad (2.76)$$

$$\langle L_{FK}(x_0)\rangle_{a^2(x_0)} = L_{FK}(x_0)\frac{1}{\sqrt{2\pi a^2(x_0)}}$$

$$\times \int_{-\infty}^{+\infty} dx \exp\left[-\frac{(x-x_0)^2}{2a^2(x_0)}\right] = L_{FK}(x_0) \quad (2.77)$$

When replaced in average form (2.75) the forms (2.76) and (2.77) cumulate with the smeared out potential (2.74) in the final Feynman-Kleinert average equation (2.70), now featuring the form:

$$0 = \int_{-\infty}^{+\infty} \frac{dx_0}{\sqrt{2\pi\hbar^2\beta/m}} \left\{ V_{a^2(x_0)}(x_0) - \frac{m}{2}\Omega^2(x_0)a^2(x_0) - L_{FK}(x_0) \right\}$$

$$\times \frac{\hbar\beta\Omega(x_0)/2}{\sinh(\hbar\beta\Omega(x_0)/2)} \exp\left[-\beta L_{FK}(x_0)\right]$$

(2.78)

from where the first stage of variational algorithm is fulfilled by the obvious choice

$$L_{FK}(x_0) = V_{a^2(x_0)}(x_0) - \frac{m}{2}\Omega^2(x_0)a^2(x_0) \qquad (2.79)$$

With Eq. (2.79) the Feynman-Kleinert potential (2.65) now displays as (Feynman & Kleinert, 1986)

$$W_{FK}(x_0) = \frac{1}{\beta}\ln\left[\frac{\sinh(\hbar\beta\Omega(x_0)/2)}{\hbar\beta\Omega(x_0)/2}\right]$$

$$+ V_{a^2(x_0)}(x_0) - \frac{m}{2}\Omega^2(x_0)a^2(x_0) \qquad (2.80)$$

There remains only to finally optimizing the explicit potential (2.80) for the harmonic (trial) frequency assuring therefore the equilibrium of the gained lowest approximation of the ground state for the concerned system. This is simply achieved through the chain derivative:

$$0 = \frac{dW_{FK}(x_0)}{d\Omega^2(x_0)} = \frac{\partial W_{FK}(x_0)}{\partial\Omega^2(x_0)} + \frac{\partial W_{FK}(x_0)}{\partial a^2(x_0)}\frac{\partial a^2(x_0)}{\partial\Omega^2(x_0)} \qquad (2.81)$$

seeing that also the fluctuation width (2.73) depends on harmonic frequency. Moreover, due to the derivative equivalence $2(\partial/\partial\Omega^2) = (1/\Omega)(\partial/\partial\Omega) \bullet$ the first in Eq. (2.81) is arranged to observe its vanishing nature when recalling the fluctuation width (2.73),

$$\frac{\partial W_{FK}(x_0)}{\partial\Omega^2(x_0)} = \frac{m}{2}\left\{\frac{1}{m\beta\Omega^2}\left[\frac{\hbar\beta\Omega}{2}\coth\left(\frac{\hbar\beta\Omega}{2}\right) - 1\right] - a^2(x_0)\right\} = 0 \quad (2.82)$$

This way, Eq. (2.81) remains only the simple condition

$$\frac{\partial W_{FK}(x_0)}{\partial a^2(x_0)} = 0 \qquad (2.83)$$

that provides from Eq. (2.80) the optimum (stabilization) frequency for quantum fluctuation

$$\Omega^2(x_0) = \frac{2}{m}\frac{\partial V_{a^2(x_0)}(x_0)}{\partial a^2(x_0)} \qquad (2.84)$$

Nevertheless, through observing the huge role both the smeared out potential (2.74) and the fluctuation width (2.73) play in deriving the approximated equilibrium ground state they deserve be further analyzed and commented in relation with matter stability.

### 2.5.3  QUANTUM SMEARED EFFECTS AND THE STABILITY OF MATTER

The intriguing role the smeared potential in special and the smearing effect in general play in optimization of the total energy and partition function of a quantum system opens the possibility analyzing the "smearing" phenomenon of the quantum fluctuation in a more fundamental way (Putz, 2009).

1. Firstly, there was noted that the smearing potential (2.74) appears as a Gaussian convolution of the applied potential, although modeling the evolution of a wave-packet under that potential; in other terms, there appears the fundamental question whether the Gaussian and wave function "kernels" behave in similar way throughout the smearing effect of quantum fluctuations; analytically, one likes to see whether there holds the smearing average equality:

$$\left\langle \exp\left(-ikx\right)\right\rangle_{a^2(x_0)} \overset{?}{=} \left\langle \exp\left(-k^2x^2\right)\right\rangle_{a^2(x_0)} \tag{2.85}$$

In order to check Eq. (2.85) one separately computes each of its side separately by the aid of $k$-form of (2.74) and gets successively the smearing average for wave-function:

$$\left\langle \exp\left(-ikx\right)\right\rangle_{a^2(x_0)} = \int_{-\infty}^{+\infty} \frac{dk}{2\pi} \exp\left[-ikx + ikx_0 - \frac{1}{2}a^2(x_0)k^2\right]$$

$$= \int_{-\infty}^{+\infty} \frac{dk}{2\pi} \exp\left[-ik(x-x_0) - \frac{1}{2}a^2(x_0)k^2\right]$$

$$= \exp\left[-\frac{(x-x_0)^2}{2a^2(x_0)}\right] \int_{-\infty}^{+\infty} \frac{dk}{2\pi} \exp\left\{-\frac{a^2(x_0)}{2}\left[k + i\frac{x-x_0}{a^2(x_0)}\right]^2\right\}$$

$$= \frac{1}{2\pi} \exp\left[-\frac{(x-x_0)^2}{2a^2(x_0)}\right] \int_{-\infty}^{+\infty} dk' \exp\left\{-\frac{a^2(x_0)}{2}k'^2\right\}$$

$$= \frac{1}{\sqrt{2\pi a^2(x_0)}} \exp\left[-\frac{(x-x_0)^2}{2a^2(x_0)}\right] \tag{2.86}$$

and respectively for the Gaussian packet:

$$\left\langle \exp\left(-k^2 x^2\right)\right\rangle_{a^2(x_0)} = \int_{-\infty}^{+\infty} \frac{dk}{2\pi} \exp\left[-k^2 x^2 + ikx_0 - \frac{1}{2}a^2(x_0)k^2\right]$$

$$= \int_{-\infty}^{+\infty} \frac{dk}{2\pi} \exp\left[-k^2\left(x^2 + \frac{a^2(x_0)}{2}\right) + ikx_0\right]$$

$$= \exp\left[-\frac{x_0^2}{4(x^2 + a^2(x_0)/2)}\right] \int_{-\infty}^{+\infty} \frac{dk}{2\pi}$$

$$\times \exp\left\{-(x^2 + a^2(x_0)/2)\left[k - i\frac{x_0}{2(x^2 + a^2(x_0)/2)}\right]^2\right\}$$

$$= \frac{1}{2\pi} \exp\left[-\frac{x_0^2}{4(x^2 + a^2(x_0)/2)}\right] \int_{-\infty}^{+\infty} dk' \exp\left\{-(x^2 + a^2(x_0)/2)k'^2\right\}$$

$$= \frac{1}{\sqrt{2\pi[2x^2 + a^2(x_0)]}} \exp\left[-\frac{x_0^2}{2(2x^2 + a^2(x_0))}\right] \tag{2.87}$$

Now for closely compare the expressions (2.86) and (2.87) the most elegant way is to make once more recourse to the smearing procedure, this time referring both to the entire paths and Feynman centroid; to this end, the previous result (2.76) is here used explicitly as:

$$\left\langle (x-x_0)^2 \right\rangle_{a^2} = a^2 \tag{2.88}$$

It allows the additional similar relationships:

$$\left\langle x^2 \right\rangle_{a^2} = \left\langle x_0^2 \right\rangle_{a^2} = \frac{a^2}{2} \tag{2.89}$$

Based on the fact that the first two terms of Eq. (2.89) are equal due to the symmetry of the smearing average formula (2.74) at the interchange $x \leftrightarrow x_0$, while the mixed term of Eq. (2.88) expansion vanishes, $\langle xx_0 \rangle_{a^2(x_0)} = 0$, in any path representation. With these, practically we can reconsider Eqs. (2.86) and (243b) by performing the formal equivalences

$$(x - x_0)^2 \approx a^2, \; x^2 \approx \frac{a^2}{2}, \; x_0^2 \approx \frac{a^2}{2} \qquad (2.90)$$

yielding with:

$$\langle \exp(-ikx) \rangle_{a^2} \approx \frac{1}{\sqrt{2\pi a^2}} \exp\left[-\frac{1}{2}\right] \qquad (2.91)$$

$$\langle \exp(-k^2 x^2) \rangle_{a^2} \approx \frac{1}{\sqrt{4\pi a^2}} \exp\left[-\frac{1}{8}\right] = \langle \exp(-ikx) \rangle_{a^2} \frac{\exp(3/8)}{\sqrt{2}} \qquad (2.92)$$

Since the difference between these expressions is numerically proportionally with the factor

$$\frac{\exp(3/8)}{\sqrt{2}} \cong 1.029 \qquad (2.93)$$

they can be considered as identical in quantum smearing effects and Eq. (2.85) as valid.

Yet, the quantum identity between the plane-wave and Gaussian packet has profound quantum implication, while revealing for instance the de Broglie—Born identity in Gaussian normalization of the de Broglie moving wave-packet. It may express as well the observational Gaussian character of the wave-function evolution in Hilbert space. Finally, and very important, it leads with *explanation* of the Bohr first postulate, i.e., is able to explain the stationary wave on orbits under singular (Coulombic) potential thus explaining the matter stabilization on rigorous quantum base, rather than to admit it by the power of a postulate. This is to be in next proofed (Feynman & Kleinert, 1986; Kleinert, 2004; Putz, 2009).

2. Let's consider a quantum system evolving under the influence of the Yukawa potential, as a generalization of the Coulomb interaction, available also on the sub-nuclear world:

$$V_{Yuk}(r) = \frac{A}{r}\exp(-\alpha r) \ , \ r = x - x_0 \tag{2.94}$$

which goes to the celebrated Hydrogen Coulomb central potential in the limit:

$$\lim_{\substack{\alpha \to 0 \\ A = -e_0^2}} V_{Yuk}(r) = V_H(r) = -\frac{e_0^2}{r} \ , \ e_0^2 = -\frac{e^2}{4\pi\varepsilon_0} \tag{2.95}$$

Now, we like to investigate the smeared version of the Yukawa potential (2.94). In 3D towards radial formulation the general definition (2.74) specializes as:

$$\langle V_{Yuk}(r = x - x_0) \rangle_{a^2} = A4\pi \int_0^{+\infty} \frac{r^2 dr}{\left(\sqrt{2\pi a^2}\right)^3} \frac{e^{-\alpha r}}{r} \exp\left[-\frac{(x-x_0)^2}{2a^2}\right]$$

$$= A2\pi e^{\alpha x_0} \int_0^{+\infty} \frac{d(r^2)}{\left(\sqrt{2\pi a^2}\right)^3} e^{-\alpha x} \exp\left[-\frac{r^2}{2a^2}\right] \tag{2.96}$$

In the last expression one can recognize the squared integration variable, of the same nature as fluctuation width, see Eq. (2.90) with $r = x - x_0$, so that the passage to integration upon the variable $a^2$ seems natural, yet meaning that the path dependent terms becomes smeared respecting the fluctuations, and the integration (lower) limit changes accordingly:

$$\langle V_{Yuk}(r_0) \rangle_{a^2} = A2\pi \langle e^{\alpha x_0} \rangle_{a^2} \int_{a^2}^{+\infty} \frac{d(\tilde{a}^2)}{\left(\sqrt{2\pi\tilde{a}^2}\right)^3} \langle e^{-\alpha x} \rangle_{\tilde{a}^2} e^{-\frac{r_0^2}{2\tilde{a}^2}} \tag{2.97}$$

In this new integral form only one smeared term is truly of the compulsory form (2.86), namely

$$\langle e^{-\alpha x_0} \rangle_{a^2} = \langle \exp[-i(i\alpha)x_0] \rangle_{a^2} \cong \langle \exp[-(i\alpha)^2 x_0^2] \rangle_{a^2} \tag{2.98}$$

where also the proofed identity (2.85) was considered upon it. Yet, the form (2.98) may be further approximated by the application of the Jensen-Peierls equality limit of the Eq. (2.66), to yield

$$\left\langle e^{-\alpha x_0}\right\rangle_{a^2} \cong \left\langle \exp\left[\alpha^2 x_0^2\right]\right\rangle_{a^2} \approx \exp\left[\alpha^2 \left\langle x_0^2\right\rangle_{a^2}\right] = \exp\left(\alpha^2 a^2 / 2\right) \quad (2.99)$$

when the smeared rules (2.89) was counted as well. The other similar term in Eq. (2.97) is however evaluated by the approximated inverse identity:

$$\left\langle e^{\alpha x}\right\rangle_{\tilde{a}^2} \cong \frac{1}{\left\langle e^{-\alpha x}\right\rangle_{\tilde{a}^2}} \cong \exp\left(-\alpha^2 \tilde{a}^2 / 2\right) \quad (2.100)$$

however, based on the unconnected version of the second order Wick cumulant

$$\left\langle e^{\alpha x}\right\rangle_{\tilde{a}^2}\left\langle e^{-\alpha x}\right\rangle_{\tilde{a}^2} \cong \left\langle e^{\alpha x} e^{-\alpha x}\right\rangle_{\tilde{a}^2} = \left\langle 1\right\rangle_{\tilde{a}^2} = 1 \quad (2.101)$$

With expressions (2.99) and (2.100) back into the smeared Yukawa potential (2.97) it becomes:

$$\left\langle V_{Yuk}(r_0)\right\rangle_{a^2} = A2\pi e^{\frac{\alpha^2 a^2}{2} + \infty} \int_{a^2}^{\infty} \frac{d\left(\tilde{a}^2\right)}{\left(\sqrt{2\pi\tilde{a}^2}\right)^3} e^{-\frac{\alpha^2 \tilde{a}^2}{2} - \frac{r_0^2}{2\tilde{a}^2}} \quad (2.102)$$

Now, through considering the variable exchange under the integral:

$$\zeta = \frac{r_0}{\sqrt{2\tilde{a}^2}} \quad (2.103)$$

there result the following transformations:

$$\tilde{a}^2 = \frac{r_0^2}{2\zeta^2}; \quad \frac{d\left(\tilde{a}^2\right)}{\left(\sqrt{\tilde{a}^2}\right)^3} = -\frac{2\sqrt{2}}{r_0} d\zeta \quad (2.104)$$

so that the smeared potential (2.102) finally cast as:

$$\left\langle V_{Yuk}(r_0)\right\rangle_{a^2} = A \frac{\exp\left(\alpha^2 a^2 / 2\right)}{r_0} \frac{2}{\sqrt{\pi}} \int_0^{r_0/\sqrt{2\tilde{a}^2}} d\zeta \exp\left[-\left(\zeta^2 + \frac{\alpha^2 r_0^2}{4\zeta^2}\right)\right] \quad (2.105)$$

which has no longer singularity at origin, since the integral in Eq. (2.105) is behaving like its integration interval for the limit $r_0 \to 0$, which gives:

$$\langle V_{Yuk}(0)\rangle_{a^2} = A\frac{2\exp\left(\alpha^2 a^2 / 2\right)}{\sqrt{2\pi a^2}} \tag{2.106}$$

Now, there is clear that under the Coulombic limit (2.95) the resulting atomic (say for Hydrogen case) smeared effect leaves it with the form:

$$\langle V_H(r_0)\rangle_{a^2} = -\frac{e_0^2}{r_0}\frac{2}{\sqrt{\pi}}\int_0^{r_0/\sqrt{2a^2}} d\zeta \exp\left(-\zeta^2\right) = -\frac{e_0^2}{r_0}\operatorname{erf}\left(r_0/\sqrt{2a^2}\right) \tag{2.107}$$

while its value on origin is of finite value:

$$\langle V_H(0)\rangle_{a^2} = -\frac{2e_0^2}{\sqrt{2\pi a^2}} \tag{2.108}$$

thus assuring (and explaining) why the atomic electron(s) do not fall onto nucleus.

Therefore, the smearing procedure plays a kind of renormalization role in transforming singular potential in finite interactions by means of quantum fluctuation effects. Such picture strongly advocates for the powerful path integral formalism in general and of that of Feynman-Kleinert in special since explicitly accounting for the fluctuation width in optimizing the quantum equilibrium states. Nevertheless, worth particularizing the Feynman-Kleinert formalism to the ground and excited states cases for better capture its realization and limits.

### 2.5.4   GROUND STATE ($\beta \to \infty$, $T \to 0K$) CASE

The basic ground state conditions in terms of thermodynamic factor ($\beta$) or the temperature ($T$),

$$\beta \to \infty \Leftrightarrow T \to 0 \tag{2.109}$$

aims to bring the Feynman-Kleinert formalism, through its working potential (2.80), to the variational ground state as usually provided by the consecrated quantum variational principle. For this purpose it will be firstly
specialized within the general limit (2.109) and then tested for the paradigmatic Hydrogen ground state case for investigating upon the accuracy
of the formalism itself (Putz, 2009).

As such, the components of the Feynman-Kleinert potential (2.80)
have the ground state limits:

$$\lim_{\beta \to \infty}\left\{\frac{1}{\beta}\ln\left[\frac{\sinh(\hbar\beta\Omega(x_0)/2)}{\hbar\beta\Omega(x_0)/2}\right]\right\}$$
$$=\frac{\hbar\Omega}{2}\lim_{\beta \to \infty}\frac{\cosh(\hbar\beta\Omega(x_0)/2)}{\sinh(\hbar\beta\Omega(x_0)/2)}-\lim_{\beta \to \infty}\frac{1}{\beta}=\frac{\hbar\Omega}{2} \qquad (2.110)$$

which recognizes the ground state of harmonic motion of trial fluctuations, while the ground state of the fluctuation width (2.73) reads as

$$\lim_{\beta \to \infty}a^2(x_0)=\lim_{\beta \to \infty}\left\{\frac{1}{m\beta\Omega^2}\left[\frac{\hbar\beta\Omega}{2}\coth\left(\frac{\hbar\beta\Omega}{2}\right)-1\right]\right\}$$
$$=\frac{\hbar}{2m\Omega}\lim_{\beta \to \infty}\frac{\cosh(\hbar\beta\Omega(x_0)/2)}{\sinh(\hbar\beta\Omega(x_0)/2)}-\lim_{\beta \to \infty}\frac{1}{m\beta\Omega^2}$$
$$=\frac{\hbar}{2m\Omega} \qquad (2.111)$$

from where also the trial fluctuations frequency springs as:

$$\lim_{\beta \to \infty}\Omega=\frac{\hbar}{2ma^2(x_0)} \qquad (2.112)$$

Through considering the relations (2.111) and (112) yields for the working general effective-classical approximation potential (2.80) the general
ground state limit:

$$W_{FK}^{T\to 0}(x_0)=\lim_{\beta \to \infty}\left\{\frac{1}{\beta}\ln\left[\frac{\sinh(\hbar\beta\Omega(x_0)/2)}{\hbar\beta\Omega(x_0)/2}\right]-\frac{m}{2}\Omega^2(x_0)a^2(x_0)\right\}+V_{a^2(x_0)}^{T\to 0}(x_0)$$

$$= \frac{\hbar\Omega}{4} + V^{T\to0}_{a^2(x_0)}(x_0)$$

$$= \frac{\hbar^2}{8ma^2} + V^{T\to0}_{a^2(x_0)}(x_0) \tag{2.113}$$

with the ground state smeared out potential remaining for individuation for a given problem.

Very interesting, the expression (2.113) entirely corresponds to the smeared out effect applied on the ordinary quantum Hamiltonian:

$$\hat{H} = -\frac{\hbar^2}{2m}\partial^2_x + V(x) \tag{2.114}$$

as one can immediately check out though applying the general smearing averaging definition (2.74) on it:

$$\left\langle \hat{H} \right\rangle_{a^2(x_0)} = \frac{1}{\sqrt{2\pi a^2(x_0)}} \int_{-\infty}^{+\infty} dx \left[ -\frac{\hbar^2}{2m}\partial^2_x + V(x) \right] \exp\left[ -\frac{(x-x_0)^2}{2a^2(x_0)} \right]$$

$$= -\frac{\hbar^2}{2m} \frac{1}{\sqrt{2\pi a^2(x_0)}} \int_{-\infty}^{+\infty} dx \left\{ \frac{\partial^2}{\partial x^2} \exp\left[ -\frac{(x-x_0)^2}{2a^2(x_0)} \right] \right\} + V^{T\to0}_{a^2(x_0)}(x_0)$$

$$= -\frac{\hbar^2}{2m} \frac{1}{\sqrt{2\pi a^2(x_0)}} \int_{-\infty}^{+\infty} dx \left\{ \left[ \frac{(x-x_0)^2}{2a^2(x_0)} -1 \right] \exp\left[ -\frac{(x-x_0)^2}{2a^2(x_0)} \right] \right\}$$

$$+ V^{T\to0}_{a^2(x_0)}(x_0)$$

$$= \frac{\hbar^2}{8ma^2} + V^{T\to0}_{a^2(x_0)}(x_0) \tag{2.115}$$

The identity between expressions (2.113) and (2.115) leaves with the important idea that the smearing operation produces in fact the average of quantum fluctuation for the ground state equilibrium. For the Coulomb interaction, say on the Hydrogen (H), either expression produces the working form

$$W^{T\to0}_{FK-H}(x_0) = \left\langle \hat{H}_H \right\rangle_{a^2(x_0)} = \frac{3\hbar^2}{8ma^2} - \frac{2e_0^2}{\sqrt{2\pi a^2}} \tag{2.116}$$

where the 3D version of the kinetic term of (2.115) was here considered aside the smearing out potential in the origin (2.108) to produce the form ready for ordinary minimization respecting the fluctuation width:

$$\frac{\partial}{\partial a^2(x_0)}\left\langle \hat{H}_H \right\rangle_{a^2(x_0)} = 0 \qquad (2.117)$$

The solution of the Eq. (2.117) with the form (2.116) produces the optimum width for quantum fluctuations:

$$a_{FK}^{opt} = \frac{3\hbar^2\sqrt{2\pi}}{8me_0^2} \qquad (2.118)$$

which, in terms of the standard first Bohr radius

$$a_0 = \frac{\hbar^2}{me_0^2} \qquad (2.119)$$

reads as

$$a_{FK}^{opt} = \sqrt{\frac{9\pi}{32}}a_0 \cong 0.94a_0 \qquad (2.120)$$

thus producing only a 6% error in predicting the localization for the stabilization of electronic ground state orbit closer to the nucleus respecting the exact Bohr-Schrödinger solution. However, for predicted approximated ground state energy error is a bit higher due to the energy dependency

$$E_{FK-H}^{min} = \left\langle \hat{H}_H \right\rangle_{a^2(x_0)}\left(a_{FK}^{opt}\right) = -\frac{e_0^2}{\sqrt{2\pi}a_{FK}^{opt}}$$

$$= -\frac{8}{3\pi}\left(\frac{e_0^2}{2a_0}\right) = \frac{8}{3\pi}E_0^H \cong 0.84E_0^H \qquad (2.121)$$

this way laying about 16% higher than the exact ground state of Hydrogen atom.

Nevertheless, besides the approximated character of the formalism, the Feynman-Kleinert approach adapts very well to the singular potential, having the advantage of being compatible with a wide class of electronic potentials in atoms and molecules; moreover, it can be particularized to the ground state in the same degree it accounts for higher temperature cases, the other extreme of thermodynamic limit—see the next section, spanning this way I principle the entire range of statistical systems at equilibrium. Such "universal" QS picture of equilibrium is hard to found in the quantum theory, at the same level of elegance, analyticity and complex ideas (Dirac, 1944; Duru & Kleinert, 1979, 1982; Blinder, 1993; Kleinert, 1996).

### 2.5.5 EXCITED STATE ($\beta\rightarrow0$, $T\rightarrow\infty$) CASE: WIGNER EXPANSION

As before the components of the Feynman-Kleinert potential (2.80) are to be now evaluated in the limit

$$\beta \rightarrow 0 \Leftrightarrow T \rightarrow \infty \qquad (2.122)$$

The analytical terms are computed through reducing them as such to contain the hyperbolic functions and then applying the approximations of type (2.45). With this recipe we firstly evaluate for the fluctuation width the expansion of the term (Putz, 2009):

$$\left[\frac{\hbar\beta\Omega}{2}\coth\left(\frac{\hbar\beta\Omega}{2}\right)-1\right]_{\beta\rightarrow0} \cong \frac{\hbar^2\beta^2\Omega^2}{12} \qquad (2.123)$$

contribution to the fluctuation width higher temperature approximation

$$a_{\beta\rightarrow0}^2(x_0) \cong \frac{1}{m\beta\Omega^2}\frac{\hbar^2\beta^2\Omega^2}{12} = \frac{\hbar^2\beta}{12m} \qquad (2.124)$$

while observing the limitation to the first order in $\beta$ expansion.

The same limitation applies also for the harmonic fluctuation term

$$\Lambda = \ln\left[\frac{\sinh(\hbar\beta\Omega(x_0)/2)}{\hbar\beta\Omega(x_0)/2}\right] \qquad (2.125)$$

that collects all first order in $\beta$ expansion from the associate McLaurin expansion, so up to the second order truncation:

$$\Lambda_{\beta\rightarrow0} \cong \lim_{\beta\rightarrow0}\Lambda + \beta\lim_{\beta\rightarrow0}\left(\frac{\partial\Lambda}{\partial\beta}\right) + \frac{1}{2}\beta^2\lim_{\beta\rightarrow0}\left(\frac{\partial^2\Lambda}{\partial\beta^2}\right) \qquad (2.126)$$

with the components:

$$\lim_{\beta \to 0} \Lambda = \ln\left[\lim_{\beta \to 0} \frac{\sinh(\hbar\beta\Omega(x_0)/2)}{\hbar\beta\Omega(x_0)/2}\right] = 0 \qquad (2.127)$$

$$\lim_{\beta \to 0}\left(\frac{\partial\Lambda}{\partial\beta}\right) = \lim_{\beta \to 0}\left[\frac{\hbar\Omega}{2}\coth\left(\frac{\hbar\beta\Omega}{2}\right) - \frac{1}{\beta}\right] = \lim_{\beta \to 0}\left(\frac{\hbar\beta\Omega}{6}\right) = 0 \qquad (2.128)$$

$$\lim_{\beta \to 0}\left(\frac{\partial^2\Lambda}{\partial\beta^2}\right) = \lim_{\beta \to 0}\left\{\frac{1}{\beta^2} - \frac{\hbar^2\Omega^2}{4}\left[\operatorname{csch}\left(\frac{\hbar\beta\Omega}{2}\right)\right]^2\right\}$$

$$= \lim_{\beta \to 0}\left\{\frac{1}{\beta^2} - \frac{\hbar^2\Omega^2}{4}\left[\frac{2}{\hbar\beta\Omega} - \frac{\hbar\beta\Omega}{12}\right]^2\right\} = \lim_{\beta \to 0}\left\{\frac{\hbar^2\Omega^2}{12} - \frac{\hbar^4\beta^2\Omega^4}{4\times144}\right\} = \frac{\hbar^2\Omega^2}{12}$$

$$(2.129)$$

where, in last expression, the hyperbolic cosecant was approximated, in the spirit of Eq. (2.45), with the form:

$$[\operatorname{csch}(\bullet)]_{\bullet \to 0} \cong \frac{1}{\bullet} - \frac{\bullet}{6} \qquad (2.130)$$

With the partial limits (2.127)–(2.129) one constructs the approximation of term (2.125) by replacing them in the expansion (2.126) so becoming:

$$\Lambda_{\beta \to 0} \cong \beta^2 \frac{\hbar^2\Omega^2}{24} \qquad (2.131)$$

helping on its turn in the harmonic approximation

$$\left\{\frac{1}{\beta}\ln\left[\frac{\sinh(\hbar\beta\Omega(x_0)/2)}{\hbar\beta\Omega(x_0)/2}\right]\right\}_{\beta \to 0} \cong \beta\frac{\hbar^2\Omega^2}{24} \qquad (2.132)$$

being thus of the same $\beta$-order as the limit (2.124) of the fluctuation width.

The remaining term for approximating in higher temperature limit is the smeared potential (2.74); this will be done by considered the change of variable in the way that

$$z(x) = \frac{x - x_0}{\sqrt{2a^2(x_0)}}, \quad dz(x) = \frac{dx}{\sqrt{2a^2(x_0)}} \qquad (2.133)$$

allowing for the successively formulation of it within the so-called Wigner expansion:

$$V_{a^2(x_0)}^{\beta \to 0}(x_0) = \frac{1}{\sqrt{2\pi a^2(x_0)}} \sqrt{2a^2(x_0)} \int_{-\infty}^{+\infty} dz\, V\left(x_0 + \sqrt{2a^2(x_0)}z\right) \exp\left[-z^2\right]$$

$$\cong \frac{1}{\sqrt{\pi}} \int_{-\infty}^{+\infty} dz \left\{ V(x_0) + \left[\sqrt{2a^2(x_0)}z\right]V'(x_0) + \frac{1}{2}\left[\sqrt{2a^2(x_0)}z\right]^2 V''(x_0) \right\} \exp\left[-z^2\right]$$

$$= \frac{1}{\sqrt{\pi}} \int_{-\infty}^{+\infty} dz \left\{ V(x_0) + a^2(x_0)z^2 V''(x_0) \right\} \exp\left[-z^2\right]$$

$$= V(x_0) + \frac{1}{2}a^2(x_0)V''(x_0)$$

$$(2.134)$$

Expression (2.134) was obtained due to the small fluctuation width at higher temperatures, see the limit (2.124); it features, nevertheless, the small perturbation in terms of fluctuation width around the applied potential "centered" on Feynman centroid, therefore behaving as a sort of semi-classical expansion. Indeed, recalling the trial fluctuation frequency optimal definition (2.84) it specializes in the high temperature limit potential (2.134) to the working expression:

$$\Omega_{\beta \to 0}^2(x_0) = \frac{2}{m} \frac{\partial V_{a^2(x_0)}^{\beta \to 0}(x_0)}{\partial a^2(x_0)} = \frac{1}{m} V''(x_0) \qquad (2.135)$$

Finally, by plugging the optimum frequency (2.135) into the limit (2.132), and together with the limits (2.124) and (2.134), back in Feynman-Kleinert potential (2.80) it acquires the high temperature form:

$$W_{FK}^{T \to \infty}(x_0) = V(x_0) + \hbar^2 \frac{\beta}{24m} V''(x_0) \qquad (2.136)$$

which is nothing than the semiclassical potential appearing in the second order partition function (2.64), thus providing the identical Feynman-Kleinert partition function:

$$Z_{FK}^{T \to \infty} = \int_{-\infty}^{+\infty} \frac{dx_0}{\sqrt{2\pi\hbar^2 \beta/m}} \exp\left[-\beta V(x_0) - \frac{\hbar^2 \beta^2}{24m} V''(x_0)\right] \qquad (2.137)$$

From such identity one there is re-affirmed that important conceptual achievement according which the Feynman centroid (2.20) corresponds in periodic path approach with the end-point coordinate average (2.19) in semiclassical expansion. Moreover, there was inferred that the higher temperature limit of the Feynman-Kleinert periodic path integral approach regains the semiclassical expansion of the non periodic paths of a quantum particle (which becomes nevertheless periodic at higher temperatures due to higher oscillation about the equilibrium while possessing not sufficient kinetic energy to break that equilibrium by traveling too far away).

We are now fully convinced that Feynman-Kleinert path integral formulation works fine either at low and higher temperature limits, while recovering both the (Hydrogen) ground state and the semiclassical expansion with high fidelity, respectively. There nevertheless remains to stress on its further connection with the electronic density and consequently with the DFT towards the quantum chemical properties' computation. These issues will be in addressed next.

## 2.6   CONCLUSION

Overall, there seems the *chemical atom* may be viewed as the set of properties describing its quantum structure, say (*shielding principal + orbital quantum + atomic numbers*), so acquainting for both inner (shielding + valence) and global characterization of the many-body system, while for the *chemical molecule* one may consider the (*softness + electronegativity + chemical hardness density functional theory*) space for describing bonding and reactivity, again accounting for local (frontier) and global chemical behavior of atoms-in-molecules (Putz, 2008). The problem of the *chemical elemental* definition, as all pure things, probably has no definite (genuine, i.e., from first principles) formulation, being let for the chemical history and philosophy for deciding and properly advising—yet it may happened to be one day accepting that the chemical element is a purely chemical philosophical concept, belonging to the Plato's world of ideas, true in Nature but inaccessible in practice; it will be nevertheless a big philosophical concept coming from Chemistry side, whatsoever, arguing this way against any kind of physical-to-chemical reductionism. Anyway, the chemical law of periodicity seems to refer more to the chemical atoms rather than to the elements counterpart—and this should suffice for further conceptualizing and interpreting of the vast chemical manifestation of Nature. Other specific chemical laws (that somehow departs from physical laws and sometimes complement them) such are those of reactivity will be reviewed in the Volume III of the present five-volume book set.

## KEYWORDS

- **Feynman-Kleinert formalism**
- **Golden ratio**
- **Matsubara frequencies**
- **Matter's stability**
- **path integrals for chemistry**
- **Riemann's series**
- **Wiegner expansion**

## REFERENCES

### *AUTHOR'S MAIN REFERENCES*

Putz, M. V. (2016a). *Quantum Nanochemistry. A Fully Integrated Approach: Vol. III. Quantum Molecules and Reactivity.* Apple Academic Press & CRC Press, Toronto-New Jersey, Canada-USA.

Putz, M. V. (2016b). *Quantum Nanochemistry. A Fully Integrated Approach: Vol. I. Quantum Theory and Observability.* Apple Academic Press & CRC Press, Toronto-New Jersey, Canada-USA.

Putz, M. V. (2012). Valence atom with Bohmian quantum potential: the golden ratio approach. *Chemistry Central Journal*, 6, 135/16 pp. (DOI: 10.1186/1752-153X-6-135).

Putz, M. V. (2011). Big chemical ideas in context: the periodic law and Scerri's periodic table. *International Journal of Chemical Modeling*, 3(1–2), 15–22.

Putz, M. V. (2010). On Heisenberg uncertainty relationship, its extension, and the quantum issue of wave-particle duality. *International Journal of Molecular Sciences* 11, 4124–4139 (DOI: 10.3390/ijms11104124).

Putz, M. V. (2009). Path integrals for electronic densities, reactivity indices, and localization functions in quantum systems. *International Journal of Molecular Sciences* 10(11): 4816–4940 (DOI: 10.3390/ijms10114816).

Putz, M. V. (2008). *Absolute and Chemical Electronegativity and Hardness*, Nova Publishers Inc., New York.

## SPECIFIC REFERENCES

Bader, R. F. W. (1990). *Atoms in Molecules—A Quantum Theory*, Oxford University Press, Oxford.

Blinder, S. M. (1993). Analytic form for the nonrelativistic Coulomb propagator. *Phys. Rev. A* 43, 13–16.

Boeyens, J. C. A. (2005). *New Theories for Chemistry*, Elsevier, Amsterdam.

Boeyens, J. C. A. (2008). *Chemistry from First Principles*, Springer, Heidelberg-Berlin.

Boeyens, J. C. A. (2011). Emergent properties in bohmian chemistry. In: *Quantum Frontiers of Atoms and Molecules*, Putz, M. V. (Ed.), Nova Publishers Inc., New York, pp. 191–215.

Boeyens, J. C. A., Levendis, D. C. (2008). *Number Theory and the Periodicity of Matter*, Springer, Heidelberg-Berlin.

Dirac, P. A. M. (1944). *The Principles of Quantum Mechanics*, Oxford University Press, Oxford.

Duru, I. H., Kleinert, H. (1979). Solution of the path integral for the H-atom, *Phys. Lett. B* 84, 185–188.

Duru, I. H., Kleinert, H. (1982). Quantum mechanics of H-atom from path integrals. *Fortschr. Physik.* 30, 401–435.

Feynman, R. P. (1972). *Statistical Mechanics*, Benjamin, Reading.

Feynman, R. P., Hibbs, A. R. (1965). *Quantum Mechanics and Path Integrals*, McGrow Hill,New York.

Feynman, R. P., Kleinert, H. (1986). Effective classical partition function. *Phys. Rev. A* 34, 5080–5084.

Kaplan, I. G. (2002). Is the Pauli exclusive principle an independent quantum mechanical postulate? *Int. J. Quantum Chem.* 89, 268–276.

Kleinert, H. (1996). Path integral for a relativistic spinless coulomb system. *Phys. Lett. A* 212, 15–21.

Kleinert, H. (2004). *Path Integrals in Quantum Mechanics, Statistics, Polymer Physics, and Financial Markets*, 3rd Ed., World Scientific, Singapore.

Lackner, K. S., Zweig, G. (1983): Introduction to the chemistry of fractionally charged atoms: electronegativity. *Phys. Rev. D* 28, 1671–1691.

Pauling, L., Wilson, E. B. (1985). *Introduction to Quantum Mechanics with Applications to Chemistry*, Dover Publications, New York.

Scerri, E. R. (2004). Just how ab initio is ab initio quantum chemistry? *Found. Chem.* 6, 93–116.

Schulman, L. S. (1981). *Techniques and Applications of Path Integration*, Wiley, New York.

Wiegel, F. W. (1986). *Introduction to Path-Integral Methods in Physics and Polymer Science*, World Scientific, Singapore.

# CHAPTER 3

# PERIODICITY BY QUANTUM PROPAGATORS IN PHYSICAL ATOM

## CONTENTS

## ABSTRACT

The semiclassical path integral approach is undertaken to develop new definitions and atomic scales of electronegativity and chemical hardness. The considered quantum probability amplitude up to the fourth order expansion provides intrinsic electronegativity and chemical hardness analytical expressions in terms of principal quantum number of the concerned valence shell and of the effective atomic charge including screening effects. The present electronegativity scale strikes on different order of magnitudes down groups of Periodic Table still satisfying the main required acceptability criteria respecting the finite difference based scale. The actual chemical hardness scale improves the trend across periods of Periodic System avoiding the usual irregularities within the old-fashioned energetic picture. The current quest introduces the electronegativity of an element as the power with which the frontier electrons are attracted to the center of the atom being a stability measure of the atomic system as a whole. However, both electronegativity and chemical hardness are analyzed for their quantum nature in Fock spaces of electronic occupancies, while maintaining their dichotomy in observability.

## 3.1  INTRODUCTION

From the birthday of the modern chemistry, i.e., since Boyle had used for the first time a coherent atomic theory in science in his 1661 famous book *The Skeptical Chymist*, the fundamental principles and concepts of matter structure were constantly shared by the physics and chemistry. However, the divorce of chemistry from physics would have to come with many occasions by means of classical chemical concepts, e.g., valence, chemical bond, and electronegativity. The "mystery" by which the atoms are kept together and still preserving their intimate properties was searched by great minds, from Newton to Lewis, being the key furnished only with the advent of quantum theory. Within this new paradigm of matter there is the feeling that the physics and chemistry are united under the actual common view "that the whole chemistry is a huge manifestation of quantum phenomena" (Ballhausen, 1979).

Nevertheless, despite almost all the concepts of chemistry have been reviewed according with quantum principles, the electronegativity notion still resists both as a proper qualitative meaning and a suitable quantification scheme. The Pauling inspired 1932 insight according with the chemical reactivity and bonding can be qualitatively measured through bond energy difference leads with the statement that a third dimension of the Periodic Table (Allen, 1989) has to be introduced as "the power of an atom in a molecule to attract electrons to it" so introducing the modern concept of electronegativity (EN or $\chi$) (Pauling, 1932). Unfortunately, this is an *in situ* definition and experimentally searches for this "power" evidence had remained unclear over nearly 60 years. On the other way, the observed striking dependence on electronegativity of the superconducting transition temperature of the nano-materials (Askamani & Mahjula, 1989; Ichikawa, 1989) emphasizes on the importance in having a clear quantum picture of the intrinsic atomic electronegativity concept.

However, due to the fact that the principles of the quantum mechanics do not suggest any operator whose eigen-value to be electronegativity, years after Pauling many definitions and interpretations of electronegativity have been formulated (Mulliken, 1934; Gordy, 1946; Iczkowski & Margrave, 1961; Hinze & Jaffe, 1962; Klopman, 1965; Parr et al., 1978; Bartolotti et al., 1980; Parr & Bartolotti, 1983; Sen & Jørgensen, 1987; Sanderson, 1988). One of most preeminent was given by Mulliken in 1934 as the average of the ionization potential (*IP*) and electron affinity (*EA*) for the valence state of an atom (Mulliken, 1934). This empirical spectroscopic definition dominated chemistry almost half century until its quantitative counterpart was introduced by the works of Parr (Parr et al., 1978; Bartolotti et al., 1980; Parr & Bartolotti, 1983), as the minus chemical potential of a multi-electronic system. The link between these two definitions is acquired if the finite difference approximation is performed on the ground state energy, $E_N$, around the referential integer total number of electrons $N_0$ (Parr, 1985; Parr & Yang, 1989; Kohn et al., 1996):

$$\chi \equiv -\left(\frac{\partial E_N}{\partial N}\right)_{V(r)} \tag{3.1}$$

Yet, the problem with the equivalent EN forms in Eq. (3.1) refers to the mixed potential conditions that they imply. As such, since the EN definition in left side of the chain equations (3.1) involves ground state energy in a non-zero constant potential $V(r)$ it assumes almost vertical values of the energies when electrons are changed with environment. On contrary, at the right extreme of Eq. (3.1) the so-called finite-difference EN corresponds to the valence state and is thus characterized by the almost adiabatic case $V(r) = 0$, as no further electrons are attached to the system.

The compromise between these two limits was recently approached by appealing to the systematic energetic expansions respecting charge and potential variations within density functional softness theory, see Chapter 4, and Garza and Robles (1993). Still, because the electronegativity—unlike $IP$ or $EA$—is not a direct measurable quantity of an isolated atom worth questing for another structural quantum mechanically way of introducing EN.

In this respect, here we assume EN as the convolution of the imaginary time conditional probability $(r, \tau|0, 0)$ with the valence shell potential $V(r)$ (Putz, 2007),

$$\chi = \int V(r)(r, \tau|0,0)dr , \tau = \mathrm{Im}(it) \qquad (3.2)$$

so representing the power of entire atom (nucleus + core + valence shell) to attract electrons of the outer shell (fixed by radius $r$) to its center ($r = 0$). This way, the current EN definition may in both qualitatively and quantitatively manners to account for the whole stability of the atom with its electronic and nuclear subsystems.

Worth noting, that the imaginary time in Eq. (3.2) formally comes out from the analytic continuation procedure known in the path integral theory as the Wick rotation (Kleinert, 1995). It is based on equivalencies between the quantum amplitudes $\exp(-iHt/\hbar)$ and $\exp(-\beta H)$ of quantum-mechanics and quantum-statistics representation of the quantum theory, respectively. For this reason, in practical applications the working time in Eq. (3.2) has to be implemented as real one, i.e., taking the component of the imaginary axis, $\tau = \mathrm{Im}(it)$, since $\tau = it = \hbar\beta$ form above correspondence.

Having a viable EN quantum formulation, its natural companion named chemical hardness, $\eta$, can be immediately introduced (Parr & Pearson, 1983)

$$\eta = -\frac{1}{2}\left(\frac{\partial \chi}{\partial N}\right)_{V(r)} = \frac{1}{2}\left(\frac{\partial^2 E}{\partial N^2}\right)_{V(r)} \qquad (3.3)$$

with a major role in establishing the main chemical principles of reactivity: the hard-and-soft-acids-and-bases (HSAB) and the maximum hardness (MH),

Volume III of the present five-volume set and Refs. (Robles & Bartolotti, 1984; Gazquez & Ortiz, 1984; Sen & Mingos, 1993; Pearson, 1997).

Therefore, finding the analytically expression of the conditional probability EN from definition (3.2) stands as the main goal of the present chapter. It is accomplished by the semiclassical expansion up to the fourth order of the probability amplitude $(r, \tau|0, 0)$ to provide the proper relation with the intrinsic quantum characteristics of the atom.

The issue of quantum observability of electronegativity and chemical hardness are firstly treated such that to distinguishing among the possible occupancies they circumvent for a given quantum state under a parabolic Hamiltonian.

Then, the associated atomic electronegativity and chemical hardness scales will be computed under general path integral quantum statistic framework and their periodic characteristics discussed respecting the general guidelines of the acceptability criteria and the finite-difference counterparts.

## 3.2 ON QUANTUM NATURE OF ELECTRONEGATIVITY AND CHEMICAL HARDNESS

One starts by considering the fermionic Fock space built on the creation and annihilation particle operators (Putz, 2009a)

$$\hat{a}^+ = |1\rangle\langle 0| \tag{3.4}$$

$$\hat{a} = |0\rangle\langle 1| \tag{3.5}$$

so that the vacuum and uni-particle sectors complete the entire particle projection space:

$$\hat{1} = |0\rangle\langle 0| + |1\rangle\langle 1| = \hat{a}\hat{a}^+ + \hat{a}^+\hat{a} = \{\hat{a}, \hat{a}^+\} \tag{3.6}$$

thus fulfilling the dot product rules

$$\langle 0|1\rangle = \langle 1|0\rangle = 0 \tag{3.7}$$

$$\langle 0|0\rangle = \langle 1|1\rangle = 1 \tag{3.8}$$

and the operatorial actions,

$$\hat{a}^+|0\rangle = |1\rangle\langle 0|0\rangle = |1\rangle \tag{3.9}$$

$$\hat{a}|1\rangle = |0\rangle\langle 1|1\rangle = |0\rangle \tag{3.10}$$

They also allow the equivalent density normalization relationships:

$$
\begin{aligned}
1 &= \langle\psi_0|\psi_0\rangle = \langle\psi_0|\hat{1}|\psi_0\rangle \\
&= \langle\psi_0|\left(\hat{a}\hat{a}^+ + \hat{a}^+\hat{a}\right)|\psi_0\rangle = \langle\psi_0|\hat{a}\hat{a}^+|\psi_0\rangle + \langle\psi_0|\hat{a}^+\hat{a}|\psi_0\rangle \\
&= \left|\langle 0|\psi_0\rangle\right|^2 + \left|\langle 1|\psi_0\rangle\right|^2 = (1-\rho_0) + \rho_0 , \ \rho_0 \in [0,1]
\end{aligned} \tag{3.11}
$$

for unperturbed state $|\psi_0\rangle$ with associated eigen-energy $E_0$ for a given valence system throughout the conventional eigen-equation

$$\hat{H}|\psi_0\rangle = E_0|\psi_0\rangle \tag{3.12}$$

in an atom or molecule.

In these conditions, the chemical processes through electronic exchanges (releasing by ionization or accepting through affinity) are modeled by the associate ionization and affinity through second quantized wave-functions (Putz, 2009a)

$$\left|\psi_\lambda^I\right\rangle = \left(1 + \lambda\hat{a}\hat{a}^+\right)|\psi_0\rangle = |\psi_0\rangle + \lambda|0\rangle\langle 1|1\rangle\langle 0|\psi_0\rangle = |\psi_0\rangle + \lambda\sqrt{1-\rho_0}|0\rangle \tag{3.13}$$

$$\left|\psi_\lambda^A\right\rangle = \left(1 + \lambda\hat{a}^+\hat{a}\right)|\psi_0\rangle = |\psi_0\rangle + \lambda|1\rangle\langle 0|0\rangle\langle 1|\psi_0\rangle = |\psi_0\rangle + \lambda\sqrt{\rho_0}|1\rangle \tag{3.14}$$

by means of the perturbation factor $\lambda$; they will help calculating the perturbed energy

$$\left\langle E_{\lambda\in\Re}^{I\leftrightarrow A}\right\rangle := \frac{\left\langle\psi_\lambda^I\left|\hat{H}\right|\psi_\lambda^A\right\rangle}{\left\langle\psi_\lambda^I\middle|\psi_\lambda^A\right\rangle} \tag{3.15}$$

and electronic density

$$\rho_{\lambda \in \mathfrak{R}}^{I \leftrightarrow A} := \frac{\left\langle \psi_\lambda^I \middle| \hat{a}^+ \hat{a} \middle| \psi_\lambda^A \right\rangle}{\left\langle \psi_\lambda^I \middle| \psi_\lambda^A \right\rangle} \tag{3.16}$$

However, when electronegativity field of the system-environment (bath) complex is acting on this valence state it has to be calculated throughout the perturbation factor $\lambda$ as

$$\chi_\lambda = -\frac{\partial \langle E_\lambda \rangle}{\partial \rho_\lambda} = -\frac{\partial \langle E_\lambda \rangle}{\partial \lambda} \frac{\partial \lambda}{\partial \rho_\lambda} \tag{3.17}$$

while from the chemical hardness definition one has (Putz, 2010)

$$\begin{aligned}
\eta_\lambda &= \frac{1}{2} \frac{\partial^2 \langle E_\lambda \rangle}{\partial \rho_\lambda^2} \\
&= \frac{1}{2} \left\{ \left[ \frac{\partial}{\partial \lambda} \left( \frac{\partial \langle E_\lambda \rangle}{\partial \lambda} \right) \right] \frac{\partial \lambda}{\partial \rho_\lambda} + \frac{\partial \langle E_\lambda \rangle}{\partial \lambda} \left[ \frac{\partial}{\partial \lambda} \left( \frac{\partial \lambda}{\partial \rho_\lambda} \right) \right] \right\} \frac{\partial \lambda}{\partial \rho_\lambda}
\end{aligned} \tag{3.18}$$

when employing the chain derivation rule

$$\frac{\partial \bullet}{\partial \rho_\lambda} = \frac{\partial \bullet}{\partial \lambda} \cdot \frac{\partial \lambda}{\partial \rho_\lambda} \tag{3.19}$$

Starting with computing the perturbed occupancy, while using the above rules, we firstly get

$$\left\langle \psi_0 \middle| \hat{a}\hat{a}^+ \hat{a}^+ \hat{a} \middle| \psi_0 \right\rangle = \left\langle \psi_0 \middle| 0 \right\rangle \langle 1|1\rangle \langle 0|1\rangle \langle 0|0\rangle \langle 1 | \psi_0 \rangle = 0 \tag{3.20}$$

$$\left\langle \psi_0 \middle| \hat{a}^+ \hat{a}\hat{a}^+ \hat{a} \middle| \psi_0 \right\rangle = \left\langle \psi_0 \middle| 1 \right\rangle \langle 0|0\rangle \langle 1|1\rangle \langle 0|0\rangle \langle 1 | \psi_0 \rangle = \rho_0 \tag{3.21}$$

$$\left\langle \psi_0 \middle| \hat{a}\hat{a}^+ \hat{a}^+ \hat{a}\hat{a}^+ \hat{a} \middle| \psi_0 \right\rangle = \left\langle \psi_0 \middle| 0 \right\rangle \langle 1|1\rangle \langle 0|1\rangle \langle 0|0\rangle \langle 1|1\rangle \langle 0|0\rangle \langle 1 | \psi_0 \rangle = 0 \tag{3.22}$$

and then for the valence density (Putz, 2009a)

$$\rho_{\lambda \in \mathfrak{R}}^{I \leftrightarrow A} = \frac{\left\langle \psi_\lambda^I \middle| \hat{a}^+ \hat{a} \middle| \psi_\lambda^A \right\rangle}{\left\langle \psi_\lambda^I \middle| \psi_\lambda^A \right\rangle_{0 < < \rho_0 \leq 1}}$$

$$= \frac{\left\langle \psi_0 \left| \left(1 + \lambda \hat{a}\hat{a}^+\right)\hat{a}^+ \hat{a}\left(1 + \lambda \hat{a}^+ \hat{a}\right)\right| \psi_0 \right\rangle}{\left\langle \psi_0 \left| \left(1 + \lambda \hat{a}\hat{a}^+\right)\left(1 + \lambda \hat{a}^+ \hat{a}\right)\right| \psi_0 \right\rangle_{0 << \rho_0 \leq 1}}$$

$$= \frac{\left\langle \psi_0 \left| \left(\hat{a}^+ \hat{a} + \lambda \hat{a}^+ \hat{a}\hat{a}^+ \hat{a} + \lambda \hat{a}\hat{a}^+ \hat{a}^+ \hat{a} + \lambda^2 \hat{a}\hat{a}^+ \hat{a}^+ \hat{a}\hat{a}^+ \hat{a}\right)\right| \psi_0 \right\rangle}{\left\langle \psi_0 \left| \left(1 + \lambda \hat{a}^+ \hat{a} + \lambda \hat{a}\hat{a}^+ + \lambda^2 \hat{a}\hat{a}^+ \hat{a}^+ \hat{a}\right)\right| \psi_0 \right\rangle_{0 << \rho_0 \leq 1}}$$

$$= \rho_0 \frac{1 + \lambda}{1 + \lambda \rho_0} \tag{3.23}$$

yielding the needed expressions

$$\frac{\partial \lambda}{\partial \rho_\lambda} = \frac{\left(1 + \lambda \rho_0\right)^2}{\rho_0 \left(1 - \rho_0\right)} \tag{3.24}$$

$$\frac{\partial}{\partial \lambda}\left(\frac{\partial \lambda}{\partial \rho_\lambda}\right) = 2\frac{1 + \lambda \rho_0}{1 - \rho_0} \tag{3.25}$$

Similarly, since the eigen-equation of the non-perturbed valence state by means of its energy is accommodated with the creation-annihilation quantum rules:

$$\left\langle \psi_0 \left| \hat{H}\hat{a}^+ \hat{a}\right| \psi_0 \right\rangle = \left\langle \psi_0 \left| \hat{H}\right|1\right\rangle\left\langle 0\left|0\right\rangle\left\langle 1\right| \psi_0 \right\rangle$$
$$= E_0 \left\langle \psi_0 \left|1\right\rangle\left\langle 1\right| \psi_0 \right\rangle = E_0 \rho_0 \tag{3.26}$$

$$\left\langle \psi_0 \left| \hat{a}\hat{a}^+ \hat{H}\right| \psi_0 \right\rangle = \left\langle \psi_0 \left|0\right\rangle\left\langle 1\left|1\right\rangle\left\langle 0\left| \hat{H}\right| \psi_0 \right\rangle$$
$$= \left\langle \psi_0 \left|0\right\rangle\left\langle 0\right| \psi_0 \right\rangle E_0 = E_0\left(1 - \rho_0\right) \tag{3.27}$$

while the term

$$\left\langle \psi_0 \left| \hat{a}\hat{a}^+ \hat{H}\hat{a}^+ \hat{a}\right| \psi_0 \right\rangle = \left\langle \psi_0 \left|0\right\rangle\left\langle 1\left|1\right\rangle\left\langle 0\left| \hat{H}\right|1\right\rangle\left\langle 0\left|0\right\rangle\left\langle 1\right| \psi_0 \right\rangle \tag{3.28}$$

is set to zero based on the usual second quantization form of the Hamiltonian (Surján, 1989),

$$\hat{H} = \sum_{pq} h_{pq}\hat{a}^+_p \hat{a}_q + \frac{1}{2}\sum_{pqts} g_{pq,ts}\hat{a}^+_p \hat{a}^+_t \hat{a}_q \hat{a}_s \tag{3.29}$$

Note that Eq. (3.29) through its one and two particle terms (with the corresponding integrals $h_{pq}$ and $g_{pqts}$ over the $p$, $q$, $t$, and $s$ orbitals) will produce the zero giving quantum operation

$$\langle 0|\hat{H}|1\rangle \sim \langle 0|\hat{a}_p^+...|1\rangle = \langle 0|1\rangle\langle 0|...|1\rangle = 0 \tag{3.30}$$

one has for the average energy (Putz, 2009)

$$\langle E_{\lambda\in\Re}^{I\leftrightarrow A}\rangle = \frac{\langle \psi_\lambda^I|\hat{H}|\psi_\lambda^A\rangle}{\langle \psi_\lambda^I|\psi_\lambda^A\rangle_{0<<\rho_0\leq 1}}$$

$$= \frac{\langle \psi_0|(1+\lambda\hat{a}\hat{a}^+)\hat{H}(1+\lambda\hat{a}^+\hat{a})\psi_0\rangle}{\langle \psi_0|(1+\lambda\hat{a}\hat{a}^+)(1+\lambda\hat{a}^+\hat{a})\psi_0\rangle_{0<<\rho_0\leq 1}}$$

$$= \frac{\langle \psi_0|\hat{H}|\psi_0\rangle + \lambda\langle \psi_0|\hat{H}\hat{a}^+\hat{a}|\psi_0\rangle + \lambda\langle \psi_0|\hat{a}\hat{a}^+\hat{H}|\psi_0\rangle + \lambda^2\langle \psi_0|\hat{a}\hat{a}^+\hat{H}\hat{a}^+\hat{a}|\psi_0\rangle}{1+\lambda\rho_0}$$

$$= E_0\frac{1+\lambda}{1+\lambda\rho_0}$$

$$\tag{3.31}$$

It leaves the needed terms

$$\frac{\partial\langle E_\lambda\rangle}{\partial\lambda} = E_0\frac{1-\rho_0}{(1+\lambda\rho_0)^2} \tag{3.32}$$

$$\frac{\partial}{\partial\lambda}\left(\frac{\partial\langle E_\lambda\rangle}{\partial\lambda}\right) = -2E_0\rho_0\frac{1-\rho_0}{(1+\lambda\rho_0)^3} \tag{3.33}$$

Now, for electronegativity definition (3.17) we have the result (Putz, 2009a)

$$\chi_\lambda = -\frac{E_0}{\rho_0} = -\mu_0 = \begin{cases} \infty & , \rho_0 \to 0(E_0 < 0) \\ -E_0 = -\langle \psi_0|H|\psi_0\rangle, & \rho_0 \to 1 \end{cases} \tag{3.34}$$

while for the chemical hardness (3.18) the limits are inferred (Putz, 2010)

$$\eta_\lambda = 0\cdot E_0\frac{1+\lambda\rho_0}{\rho_0(1-\rho_0)} = \begin{cases} 0, & \rho_0 \in (0,1) \\ 0\cdot\infty = ?, & \rho_0 \to 0 \\ 0\cdot\infty = ?, & \rho_0 \to 1 \end{cases} \tag{3.35}$$

This way, it is obvious that at the conceptual (quantum fundamental) level one obtained that (Putz, 2011):

- electronegativity is revealed as a quantum (chemical potential) observable, i.e., electronegativity is spanned between the *pure electronic attraction* and *either electronic attraction or release* characters for the states described by the upper and lower branches of (3.34), respectively; note that albeit the electronegativity can have the value of the electronic energy this has to be consider for the global system's full occupancy density, since the second quantization approach specific to many body-systems, and not iteratively for atoms building molecule for which would appear that the more atoms, the more electronegative the system is.

- chemical hardness is not a quantum observable, having neither a *nonzero* nor a *definite* value for any electronic density realization or limit, respectively. However, the result (3.35) does not exclude "real" even as "hidden" or dispersed values of chemical hardness, as it is often associated with open states of chemical bonding for the fractional occupied states—the first upper branch of Eq. (3.35), or with not definite in eigen-value sense of the empty or fully occupied states that may equally be fully engaged or inert respecting chemical bonding; note that although seems a controversial result the chemical hardness as the general second derivative of the total energy in Eq. (3.3)—as it was considered as well in this section, and not as the finite difference 2C scheme where it acquires the clear meaning of energetic gap between frontier LUMO and HOMO orbitals—see Eq. (3.3), has indeed little observational meaning: this was revealed also by scarce fulfillment of HSAB principle for a series of simple molecules considered in aqueous solvent environment, see Volume III of the present five-volume work, as well as by earlier gas-phase works of Drago (Drago & Kabler, 1972; Pearson, 1972; Drago et al., 1987); another argument may be found also in the parallelism between chemical hardness and aromaticity, see above, that indicates that the chemical hardness has virtually the same observable character as aromaticity—a concept that is still in debate respecting its experimental counterpart or indexing because of its inherently relative scale—a behavior affirmed by Katritzky et al. as a "multidimensional characteristic" (Katritzky et al., 1998; Cyrański et al., 2002; Schleyer, 2005); on the other side, the "metallic" behavior of chemical hardness for the fractional states on the top branch of Eq. (3.35) clearly indicates that the electrons of the system can freely move under the electronegativity influence of Eq. (3.34), i.e., the system is not in the phase of second order adjustments of the total energy, thus the chemical hardness role is absent on this electronic density range or occupancy.

Overall, it follows that there are two different kinds of quantum manifestations (and the corresponding indices) to characterize the chemical structure in balance between accepting or engaging electrons in boding.

## 3.3 SEMICLASSICAL ANALYTICS OF QUANTUM EVOLUTION AMPLITUDE

### 3.3.1 PATH INTEGRAL SEMICLASSICAL EXPANSION

Semiclassical derivation of the evolution amplitude employs some of the previously Feynman path integral ideas refined due to the works of Kleinert and collaborators (Feynman & Hibbs, 1965; Kleinert, 2004; Dachen et al., 1974; Grosche, 1993; Manning & Ezra, 1994). They can be synthesized as (Putz, 2009b):

- The real time dependency is said to be (Wick) "rotated" into the imaginary time,

$$\tau = it = \hbar\beta \qquad (3.36)$$

and it is detailed as:

$$\tau_b - \tau_a = \hbar\beta = i(t_b - t_a) \qquad (3.37)$$

- The quantum paths of Eq. (2.19) are re-parameterized as

$$x(\tau) = \bar{x} + \eta(\tau) \qquad (3.38)$$

where the classical path of Eq. (2.19) is replaced by the fixed (non time-dependent) average:

$$x_{cl}(t) \rightarrow \bar{x} = \frac{x_a + x_b}{2} \qquad (3.39)$$

while the fluctuation path $\eta(\tau)$ remains to carry the whole path integral information, while being changed at the end of integration frontier from previously Dirichlet boundary conditions (2.13) to the actual different endpoint values:

$$\eta_a = \eta(\tau_a) = -\frac{\Delta x}{2} \tag{3.40}$$

$$\eta_b = \eta(\tau_b) = +\frac{\Delta x}{2} \tag{3.41}$$

in terms of the length of the "traveled" space:

$$\Delta x = x_b - x_a \tag{3.42}$$

In these conditions the quantum statistical path integral representation of quantum propagator, see Eq. (2.10) with Eq. (2.21), and Volume I/ Chapter 4 of the present five-volume set, and takes the form

$$\left(x_b \tau_b; x_a \tau_a\right)_{QS} = \int_{\eta(\tau_a)=-\Delta x/2}^{\eta(\tau_b)=+\Delta x/2} \wp \eta(\tau) \exp\left\{-\frac{1}{\hbar}\int_{\tau_a}^{\tau_b}\left[\frac{m}{2}\dot{\eta}^2(\tau) + V(\bar{x} + \eta(\tau))\right]d\tau\right\} \tag{3.43}$$

since we immediately noted the transformations

$$\wp' x(\tau) \rightarrow \wp \eta(\tau) \tag{3.44}$$

$$\dot{x}^2(\tau) \rightarrow \dot{\eta}^2(\tau) \tag{3.45}$$

$$V(x(\tau)) \rightarrow V(\bar{x} + \eta(\tau)) \tag{3.46}$$

based on the above Eqs. (3.38)–(3.42) parameterization.

There should be pointed out that the used re-parameterization is not modifying the value of the path integral but is intended to better visualizing of its properties, towards solving it. As such, from expression (3.43) now appears clearer than before that for the systems governed by smooth potentials, the series expansion may now be applied respecting the path fluctuation, here in the second order truncation:

$$V(\bar{x} + \eta(\tau)) \cong V(\bar{x}) + \partial_i V(\bar{x})\eta_i(\tau) + \frac{1}{2}\partial_i\partial_j V(\bar{x})\eta_i(\tau)\eta_j(\tau) \tag{3.47}$$

where the covariant notation for function products was assumed for maintaining the generality of the approach, in $D$-dimensions for instance. This way there is at once obtained a (truncated) series of path integral evolution amplitude of Eq. (3.43), and of any propagator in principle (Putz, 2009b):

$$
\begin{aligned}
(x_b\tau_b; x_a\tau_a)_{QS} &= \int\limits_{\eta(\tau_a)=-\Delta x/2}^{\eta(\tau_b)=+\Delta x/2} \mathscr{D}\eta(\tau) e^{-\frac{1}{\hbar}\int\limits_{\tau_a}^{\tau_b}\left[\frac{m}{2}\dot{\eta}^2(\tau)\right]d\tau} \; e^{-\frac{1}{\hbar}\int\limits_{\tau_a}^{\tau_b}V(\bar{x})d\tau} \\
&\quad \times e^{-\frac{1}{\hbar}\int\limits_{\tau_a}^{\tau_b}\left[\partial_i V(\bar{x})\eta_i(\tau)+\frac{1}{2}\partial_i\partial_j V(\bar{x})\eta_i(\tau)\eta_j(\tau)+...\right]d\tau} \\
&= \exp\left[-\frac{\tau_b-\tau_a}{\hbar}V(\bar{x})\right]\int\limits_{\eta(\tau_a)}^{\eta(\tau_b)}\mathscr{D}\eta(\tau)\exp\left\{-\frac{1}{\hbar}\int\limits_{\tau_a}^{\tau_b}\left[\frac{m}{2}\dot{\eta}^2(\tau)\right]d\tau\right\} \\
&\quad \times \left\{\begin{aligned}
&\left[1-\frac{1}{\hbar}\int\limits_{\tau_a}^{\tau_b}d\tau\left[\partial_i V(\bar{x})\eta_i(\tau)+\frac{1}{2}\partial_i\partial_j V(\bar{x})\eta_i(\tau)\eta_j(\tau)+...\right]\right. \\
&\left. +\frac{1}{2\hbar^2}\int\limits_{\tau_a}^{\tau_b}d\tau\int\limits_{\tau_a}^{\tau_b}d\tau'\left[\partial_i V(\bar{x})\partial_j V(\bar{x})\eta_i(\tau)\eta_j(\tau')+...\right]+...\right.
\end{aligned}\right\}
\end{aligned}
$$

(3.48)

as being driven by the quantum fluctuation' various orders contribution, here restrained to the second order. This is a natural approach since at the end the quantum nature of the path integral is given by the quantum fluctuations themselves, from where the present focus of path integrals over the quantum fluctuations. The series is known as the semiclassical expansion since is formally done in the "powers of $\hbar$-Planck."

Now, looking on Eq. (3.48) as compared with the previously used quantum mechanical form, see (2.10) with (2.21), and Volume I/Chapter 4 of the present five-volume work, the present propagator representation would be resumed as:

$$
(x_b\tau_b; x_a\tau_a)_{QS} = \underbrace{\exp\left[-\beta V(\bar{x})\right]}_{\substack{\text{CLASSICAL}\\\text{FACTOR}}}\underbrace{\int\limits_{\eta(\tau_a)}^{\eta(\tau_b)}\mathscr{D}\eta(\tau)F_{SC}[\eta]\exp\left\{-\frac{1}{\hbar}\int\limits_{\tau_a}^{\tau_b}\left[\frac{m}{2}\dot{\eta}^2(\tau)\right]d\tau\right\}}_{\text{SEMI-CLASSICAL CONTRIBUTION}}
$$

(3.49)

where we have used Eq. (3.37) as we will do any time this will help better visualizing the result, and where we have identified the series of Eq. (3.48) with the semiclassical factor $F_{SC}[\eta]$. Now, expression (3.49) may be further formally written as

$$
\left(x_b\tau_b;x_a\tau_a\right)_{QS} = \exp\left[-\beta V(\bar{x})\right]\left(\frac{\Delta x}{2}\tau_b;-\frac{\Delta x}{2}\tau_a\right)_{(0)}\langle F_{SC}[\eta]\rangle \quad (3.50)
$$

by introducing the so-called *free* imaginary time amplitude

$$
\left(\frac{\Delta x}{2}\tau_b;-\frac{\Delta x}{2}\tau_a\right)_{(0)} = \int_{\eta(\tau_a)=-\Delta x/2}^{\eta(\tau_b)=+\Delta x/2}\mathcal{D}\eta(\tau)\exp\left\{-\frac{1}{\hbar}\int_{\tau_a}^{\tau_b}\left[\frac{m}{2}\dot{\eta}^2(\tau)\right]d\tau\right\} \quad (3.51)
$$

readily given by accommodating the free-propagator solution to the present statistical and boundary transformation (3.37), see also Volume I/ Chapter 4 of the present five-volume work, namely

$$
\left(\frac{\Delta x}{2}\tau_b;-\frac{\Delta x}{2}\tau_a\right)_{(0)} = \sqrt{\frac{m}{2\pi\hbar^2\beta}}\exp\left\{-\frac{m}{2\hbar^2\beta}(\Delta x)^2\right\} \quad (3.52)
$$

having the normalization role for the semiclassical factor averages' contributions:

$$
\langle F_{SC}[\eta]\rangle = \frac{\displaystyle\int_{\eta(\tau_a)=-\Delta x/2}^{\eta(\tau_b)=+\Delta x/2}\mathcal{D}\eta(\tau)F_{SC}[\eta]\exp\left\{-\frac{1}{\hbar}\int_{\tau_a}^{\tau_b}\left[\frac{m}{2}\dot{\eta}^2(\tau)\right]d\tau\right\}}{\left(\dfrac{\Delta x}{2}\tau_b;-\dfrac{\Delta x}{2}\tau_a\right)_{(0)}} \quad (3.53)
$$

Therefore, the new form of path integral representation of evolution amplitude, given either by Eq. (3.48) or (3.51) looks like (Putz, 2009b):

$$
\left(x_b\tau_b;x_a\tau_a\right)_{QS} = \sqrt{\frac{m}{2\pi\hbar^2\beta}}\exp\left\{-\frac{m}{2\hbar^2\beta}(\Delta x)^2 - \beta V(\bar{x})\right\}
$$

$$
\times\left\{
\begin{aligned}
&\left[1-\frac{1}{\hbar}\partial_i V(\bar{x})\int_{\tau_a}^{\tau_b}d\tau\langle\eta_i(\tau)\rangle - \frac{1}{2\hbar}\partial_i\partial_j V(\bar{x})\int_{\tau_a}^{\tau_b}d\tau\langle\eta_i(\tau)\eta_j(\tau)\rangle\right.\\
&\left.+\frac{1}{2\hbar^2}\partial_i V(\bar{x})\partial_j V(\bar{x})\int_{\tau_a}^{\tau_b}d\tau\int_{\tau_a}^{\tau_b}d\tau'\langle\eta_i(\tau)\eta_j(\tau')\rangle + ...\right]
\end{aligned}
\right\} \quad (3.54)
$$

being now the problem of expressing the averaged values of the fluctuation paths in single or multiple time connection: $\langle \eta_i(\tau) \rangle$, $\langle \eta_i(\tau)\eta_j(\tau) \rangle$, $\langle \eta_i(\tau)\eta_j(\tau') \rangle$, etc.

From heuristically point of view there is normal to arrive at this form telling us that once the quantum fluctuation are averaged along the quantum evolution an then integrated in time the evolution amplitude to be determined. Observe also that the present semiclassical approach is not using the previous employed properties of classical action, and being somehow limited by its derivative behavior at edge of the space domain of integration, while having now the limitation in what respect the quantum fluctuation power. There is also useful remarking that the present semiclassical approach may use the *interplay* between the previous solved *free-and-harmonic quantum motions*, since the path integral (3.51) may equally be regarded as the free motion of the quantum fluctuation (naturally since they are not known *apriori* or with some possibility of instantaneously observation); at the same time, if one formally counts the kinetic term as the perturbative (aka fluctuation) oscillatory motion,

$$\frac{m}{2}\dot{\eta}^2(\tau) = \frac{m}{2}\left(\frac{d}{dt}\eta\right)^2 \approx \frac{m}{2}\underset{\partial_t}{\omega^2}\eta^2 \tag{3.55}$$

a more complex picture of quantum fluctuation is obtained; in conclusion, quantum fluctuated paths may be (or should be) treated as being a kind of *harmonically free* motion: harmonic since as fluctuations may be expanded in Fourier series (as originally perceived by Feynman) but also free since their unknown of instantaneous feature. Therefore, an appropriate use of both these feature will conduct to the reliable path integral representation. As we already have used the free-motion character of fluctuation paths, the harmonic one is entering the analysis in the next section.

### 3.3.2 CONNECTED CORRELATION FUNCTIONS

For calculating the quantum fluctuation paths' averages one has to understand their inner nature: in order reconciliation of free and harmonic features be achieved the so-called *quantum current* $j(\tau)$ is introduced (and presumed to appear in reality too as causing/driving the quantum fluctuations), so that the current dependent propagator, known as *the generating functional*, is formed (Kleinert, 1986, 2004; Putz, 2009b):

$$
\left(\frac{\Delta x}{2}\tau_b;-\frac{\Delta x}{2}\tau_a\right)[j(\tau)] = \int_{\eta(\tau_a)=-\Delta x/2}^{\eta(\tau_b)=+\Delta x/2}\wp\eta(\tau)\exp\left\{-\frac{1}{\hbar}\int_{\tau_a}^{\tau_b}\left[\frac{m}{2}\dot{\eta}^2(\tau)-j(\tau)\eta(x)\right]d\tau\right\}
$$

$$(3.56)$$

with which help one can recognized the equivalence:

$$
\langle\eta(\tau)\rangle = \frac{\hbar\dfrac{\delta}{\delta j(\tau)}\left(\dfrac{\Delta x}{2}\tau_b;-\dfrac{\Delta x}{2}\tau_a\right)[j(\tau)]\Big|_{j(\tau)=0}}{\left(\dfrac{\Delta x}{2}\tau_b;-\dfrac{\Delta x}{2}\tau_a\right)_{(0)}}
$$

$$
= \frac{\displaystyle\int_{\eta(\tau_a)=-\Delta x/2}^{\eta(\tau_b)=+\Delta x/2}\wp\eta(\tau)[\eta(x)]\exp\left\{-\frac{1}{\hbar}\int_{\tau_a}^{\tau_b}\left[\frac{m}{2}\dot{\eta}^2(\tau)-j(\tau)\eta(x)\right]d\tau\right\}}{\left(\dfrac{\Delta x}{2}\tau_b;-\dfrac{\Delta x}{2}\tau_a\right)_{(0)}}
$$

$$(3.57)$$

in accordance with general definition (3.53). One can nevertheless see that the quantum current appearance in Eq. (3.57) is under the perturbation form, so that it readily accounts for the deviation from the free fluctuation motion towards the (ordered) harmonically one. Therefore, although general correlation definition may be advanced by the rule

$$
\langle\eta(\tau_1)\cdots\eta(\tau_n)\rangle = \frac{\hbar\dfrac{\delta}{\delta j(\tau_1)}\cdots\hbar\dfrac{\delta}{\delta j(\tau_n)}\left(\dfrac{\Delta x}{2}\tau_b;-\dfrac{\Delta x}{2}\tau_a\right)[j(\tau_1),\cdots,j(\tau_n)]\Big|_{\substack{j(\tau_1)=0,\\ \cdots\\ j(\tau_n)=0}}}{\left(\dfrac{\Delta x}{2}\tau_b;-\dfrac{\Delta x}{2}\tau_a\right)_{(0)}}
$$

$$(3.58)$$

the problem of practically evaluation still remains. Aiming for solving it one observes the form (3.58) analogous with the partition function based electronic density, see Eq. (2.11) for instance; the alternative formulation thus looks at the canonical ($N=1$, mono-particle) level like:

$$\langle \eta(\tau_1)\cdots\eta(\tau_n) \rangle = \left\{ Z^{-1}[j]\hbar\frac{\delta}{\delta j(\tau_1)}\cdots\hbar\frac{\delta}{\delta j(\tau_n)}Z[j] \right\}_{j=0} \qquad (3.59)$$

where now the quantity $Z[j]$ plays the role of the generating functional of the quantum fluctuation correlation (or connection) average. Yet the writing Eq. (3.59) may suffer from disconnecting character due to the presence of simple $Z[j]$; this may be better view from the further equivalent writing of Eq. (3.59) under the so-called *n-point (correlation) functions*:

$$\langle \eta(\tau_1)\cdots\eta(\tau_n) \rangle = Z^{-1}\int \mathcal{D}\eta(\tau)[\eta(\tau_1)\cdots\eta(\tau_n)]\exp\left\{ -\frac{1}{\hbar}S_+ \right\} \qquad (3.60)$$

$$= \prod_{i=1}^{n+1}\left[ \int_{-\infty}^{+\infty}d\eta(\tau_i) \right](\eta_b\tau_b;\eta_n\tau_n)\cdot\eta(\tau_n)\cdot(\eta_n\tau_n;\eta_{n-1}\tau_{n-1})$$

$$\cdot\eta(\tau_{n-1})\cdots\eta(\tau_1)(\eta_1\tau_1;\eta_a\tau_a) \qquad (3.61)$$

with $S_+$ being the Euclidian action, while the space and time were considered through slicing intervals for the respective events as custom for path-integral approaches.

The disconnected character of correlations (3.59)–(214) may be surpassed remembering that when under logarithm the partition function provides the thermodynamic free energy, here under canonical ($N=1$) form,

$$\ln Z[j] = -\beta F[j] \qquad (3.62)$$

which, being measurable-observable energy, is compulsory containing the connected parts of $Z[j]$ (i.e., energy's pieces combines towards the total energy). Therefore, this leaves with the idea that through introducing another generating functional as:

$$W[j] = \ln Z[j] \qquad (3.63)$$

it leads with simple rewriting of Eq. (3.59), however producing the connected part of correlation $\langle \eta(\tau_1)\cdots\eta(\tau_n) \rangle$ that is naturally identified with noting else than with a sort of generalized n-points (events) Green function:

$$G_{con}^{(n)}(\tau_1,\cdots,\tau_n)=\left\{\frac{\delta}{\delta j(\tau_1)}\cdots\frac{\delta}{\delta j(\tau_n)}W[j]\right\}_{j=0} \tag{3.64}$$

For having a better "feeling" how the connected and disconnected correlation (fluctuation) functions (3.64) and (3.59) are linked, let's start evaluating some orders of such correlations.

As such, absorbing the constants in the involved functionals, for the first order of correlation of Eq. (3.59) we successively have:

$$\langle\eta(\tau)\rangle=Z^{-1}[j]\frac{\delta}{\delta j(\tau)}Z[j]=\frac{\delta}{\delta j(\tau)}W[j]=G_{con}^{(1)}(\tau)=\eta_{cl}(\tau) \tag{3.65}$$

until the single connected path (since fluctuation was already averaged out) that is nothing else that the classical path connecting the ending points of the quantum evolution.

Now, going to the second order of correlation of Eq. (3.59) one has (Putz, 2009b):

$$\langle\eta(\tau_1)\eta(\tau_2)\rangle=Z^{-1}[j]\frac{\delta}{\delta j(\tau_1)}\frac{\delta}{\delta j(\tau_2)}Z[j]$$

$$=Z^{-1}[j]\frac{\delta}{\delta j(\tau_1)}\left\{Z[j]\left(\frac{\delta}{\delta j(\tau_2)}W[j]\right)\right\}$$

$$=\frac{\delta}{\delta j(\tau_1)}\frac{\delta}{\delta j(\tau_2)}W[j]+Z^{-1}[j]\frac{\delta Z[j]}{\delta j(\tau_1)}\frac{\delta W[j]}{\delta j(\tau_2)}$$

$$=\frac{\delta}{\delta j(\tau_1)}\frac{\delta}{\delta j(\tau_2)}W[j]+\frac{\delta W[j]}{\delta j(\tau_1)}\frac{\delta W[j]}{\delta j(\tau_2)} \tag{3.66}$$

a result that can be wisely rearranged as:

$$\langle\eta(\tau_1)\eta(\tau_2)\rangle=G_{con}^{(2)}(\tau_1,\tau_2)+G_{con}^{(1)}(\tau_1)G_{con}^{(1)}(\tau_2)$$

$$=\underbrace{G_{con}^{(2)}(\tau_1,\tau_2)}_{\substack{connected\\events}}+\underbrace{\langle\eta(\tau_1)\rangle\langle\eta(\tau_2)\rangle}_{\substack{dis-connected\\events}} \tag{3.67}$$

or, even more practically for our purpose:

$$\langle \eta(\tau_1)\eta(\tau_2)\rangle = G_{con}^{(2)}(\tau_1,\tau_2)+\eta_{cl}(\tau_1)\eta_{cl}(\tau_2) \qquad (3.68)$$

In similar manner, while applying a sort of recursive rule, sometimes called also as *cluster decomposition* or *cumulant expansion*,

$$\langle \eta(\tau_1)\cdots\eta(\tau_n)\rangle = G_{con}^{(n)}(\tau_1,\cdots,\tau_n)+\langle \eta(\tau_2)\cdots\eta(\tau_n)\rangle G_{con}^{(1)}(\tau_1), n\geq 2 \quad (3.69)$$

with the pair-wise (Wick) decomposition of the *n*-points correlated function:

$$G_{con}^{(n)}(\tau_1,\cdots,\tau_n)=\sum_{j=2}^{n}G_{con}^{(2)}(\tau_1,\tau_j)\underbrace{\langle \eta(\tau_k)\cdots\eta(\tau_l)\rangle}_{(n-2)order\ function\ \substack{k\neq 1,j\\l\neq 1,j,k}} \qquad (3.70)$$

one can easily obtain the higher orders of correlations, while still observing that all connected orders of events reduces to the combinations of *pair-connected events*. For instance we get for the third order fluctuations the average contributions (Putz, 2009b)

$$\langle \eta(\tau_1)\eta(\tau_2)\eta(\tau_3)\rangle = \langle \eta(\tau_1)\rangle\langle \eta(\tau_2)\eta(\tau_3)\rangle + G_{con}^{(2)}(\tau_1,\tau_2)\langle \eta(\tau_3)\rangle$$
$$+ G_{con}^{(2)}(\tau_1,\tau_3)\langle \eta(\tau_2)\rangle$$
$$= \eta_{cl}(\tau_1)\left[\eta_{cl}(\tau_2)\eta_{cl}(\tau_3)+ G_{con}^{(2)}(\tau_2,\tau_3)\right]$$
$$+ G_{con}^{(2)}(\tau_1,\tau_2)\eta_{cl}(\tau_3)+ G_{con}^{(2)}(\tau_1,\tau_3)\eta_{cl}(\tau_2)$$
$$= \eta_{cl}(\tau_1)\eta_{cl}(\tau_2)\eta_{cl}(\tau_3)+\eta_{cl}(\tau_1)G_{con}^{(2)}(\tau_2,\tau_3)$$
$$+ \eta_{cl}(\tau_2)G_{con}^{(2)}(\tau_1,\tau_3)+\eta_{cl}(\tau_3)G_{con}^{(2)}(\tau_1,\tau_2) \qquad (3.71)$$

while for the fourth order more terms are involved:

$$\langle \eta(\tau_1)\eta(\tau_2)\eta(\tau_3)\eta(\tau_4)\rangle$$
$$= \langle \eta(\tau_1)\rangle\langle \eta(\tau_2)\eta(\tau_3)\eta(\tau_3)\rangle + G_{con}^{(2)}(\tau_1,\tau_2)\langle \eta(\tau_3)\eta(\tau_4)\rangle$$
$$+ G_{con}^{(2)}(\tau_1,\tau_3)\langle \eta(\tau_2)\eta(\tau_4)\rangle + G_{con}^{(2)}(\tau_1,\tau_4)\langle \eta(\tau_2)\eta(\tau_3)\rangle$$

$$= \eta_{cl}(\tau_1) \left[ \begin{array}{l} \eta_{cl}(\tau_1)\eta_{cl}(\tau_2)\eta_{cl}(\tau_3) + \eta_{cl}(\tau_1)G^{(2)}_{con}(\tau_2,\tau_3) \\ + \eta_{cl}(\tau_2)G^{(2)}_{con}(\tau_1,\tau_3) + \eta_{cl}(\tau_3)G^{(2)}_{con}(\tau_1,\tau_2) \end{array} \right]$$

$$+ G^{(2)}_{con}(\tau_1,\tau_2)\left[ G^{(2)}_{con}(\tau_3,\tau_4) + \eta_{cl}(\tau_3)\eta_{cl}(\tau_4) \right]$$

$$+ G^{(2)}_{con}(\tau_1,\tau_3)\left[ G^{(2)}_{con}(\tau_2,\tau_4) + \eta_{cl}(\tau_2)\eta_{cl}(\tau_4) \right]$$

$$+ G^{(2)}_{con}(\tau_1,\tau_4)\left[ G^{(2)}_{con}(\tau_2,\tau_3) + \eta_{cl}(\tau_2)\eta_{cl}(\tau_3) \right]$$

$$= \eta_{cl}(\tau_1)\eta_{cl}(\tau_2)\eta_{cl}(\tau_3)\eta_{cl}(\tau_4)$$

$$+ \eta_{cl}(\tau_1)\eta_{cl}(\tau_2)G^{(2)}_{con}(\tau_3,\tau_4) + \eta_{cl}(\tau_1)\eta_{cl}(\tau_3)G^{(2)}_{con}(\tau_2,\tau_4)$$

$$+ \eta_{cl}(\tau_1)\eta_{cl}(\tau_4)G^{(2)}_{con}(\tau_2,\tau_3)$$

$$+ \eta_{cl}(\tau_2)\eta_{cl}(\tau_3)G^{(2)}_{con}(\tau_1,\tau_4) + \eta_{cl}(\tau_2)\eta_{cl}(\tau_4)G^{(2)}_{con}(\tau_1,\tau_3)$$

$$+ \eta_{cl}(\tau_3)\eta_{cl}(\tau_4)G^{(2)}_{con}(\tau_1,\tau_2)$$

$$+ G^{(2)}_{con}(\tau_1,\tau_2)G^{(2)}_{con}(\tau_3,\tau_4) + G^{(2)}_{con}(\tau_1,\tau_3)G^{(2)}_{con}(\tau_2,\tau_4)$$

$$+ G^{(2)}_{con}(\tau_1,\tau_4)G^{(2)}_{con}(\tau_2,\tau_3) \tag{3.72}$$

Next, having these examples in hand, one tries to check them by deriving by an appropriate generating functional (3.56) working with the connected function definition (3.58). At this moment one uses the previously reasoned dual nature of fluctuation paths, as "free-harmonic motion"—see Eq. (3.55), to reconsider the free imaginary time amplitude (3.51) for free + harmonic fluctuation contribution (Putz, 2009b)

$$\left( \frac{\Delta x}{2}\tau_b; -\frac{\Delta x}{2}\tau_a \right)_{(\omega)} = \int_{\eta(\tau_a)=-\Delta x/2}^{\eta(\tau_b)=+\Delta x/2} \mathcal{D}\eta(\tau)\exp\left\{ -\frac{1}{\hbar}S^+_\omega(\eta,\dot\eta,\tau) \right\} \tag{3.73}$$

with harmonic Euclidian action of fluctuations:

$$S^+_\omega(\eta,\dot\eta,\tau) = \int_{\tau_a}^{\tau_b}\left[ \frac{m}{2}\dot\eta^2(\tau) + \frac{m}{2}\omega^2\eta^2(\tau) \right]d\tau \tag{3.74}$$

while recovering at the end of calculation the condition

$$\lim_{\omega \to 0}\left(\frac{\Delta x}{2}\tau_b;-\frac{\Delta x}{2}\tau_a\right)_{(\omega)} = \left(\frac{\Delta x}{2}\tau_b;-\frac{\Delta x}{2}\tau_a\right)[j] \qquad (3.75)$$

We like to rearrange the action (3.74) so that the quantum current contribution to appear (from harmonic presence); in achieving this one firstly rewrites it by performing the integration by parts:

$$S_\omega^+(\eta,\dot{\eta},\tau) = \int_{\tau_a}^{\tau_b}\left\{\frac{m}{2}\left[\frac{d}{d\tau}\left(\eta\frac{d\eta}{d\tau}\right)-\eta\frac{d^2\eta}{d\tau^2}\right]+\frac{m}{2}\omega^2\eta^2(\tau)\right\}d\tau$$

$$= \underbrace{\frac{m}{2}(\eta\dot{\eta})_{\tau_a}^{\tau_b}}_{0} + \int_{\tau_a}^{\tau_b}\left\{\frac{m}{2}\eta(\tau)\left[-\partial_\tau^2+\omega^2\right]\eta(\tau)\right\}d\tau$$

$$= \int_{\tau_a}^{\tau_b}\left\{\frac{m}{2}\eta(\tau)D_\omega(\tau)\eta(\tau)\right\}d\tau \qquad (3.76)$$

where we have recognized the appearance of the harmonic differential operator:

$$D_\omega(\tau) = -\partial_\tau^2 + \omega^2 \qquad (3.77)$$

The form (3.76) is very useful through employing the Green equation for harmonic motion

$$D_\omega(\tau)G_\omega(\tau,\tau') = \delta(\tau-\tau') \qquad (3.78)$$

and of its integral property

$$\int D_\omega(\tau)G_\omega(\tau,\tau')f(\tau,\tau')d\tau = \int f(\tau,\tau')\delta(\tau-\tau')d\tau = f(\tau',\tau') \quad (3.79)$$

to perform the path shifting of fluctuations by transformation

$$\eta(\tau) \to \eta'(\tau) = \eta(\tau)+\frac{1}{m}G_\omega(\tau,\tau')j(\tau) \qquad (3.80)$$

on its integrand, with the result:

$$\frac{m}{2}\eta(\tau)D_\omega(\tau)\eta(\tau) \to \frac{m}{2}\eta'(\tau)D_\omega(\tau)\eta'(\tau)$$

$$-\left[\int \eta'(\tau)D_\omega(\tau)G_\omega(\tau,\tau')j(\tau')\right]d\tau'$$

$$+\frac{1}{2m}\iint \left[j(\tau)D_\omega(\tau)G_\omega(\tau,\tau')G_\omega(\tau',\tau'')j(\tau'')\right]d\tau'd\tau'' \qquad (3.81)$$

so that by the prescribed integration in Eq. (3.76) the actual action transforms to:

$$S_\omega^+(\eta,\dot\eta,\tau)\rightarrow \int_{\tau_a}^{\tau_b}\left\{\frac{m}{2}\eta'(\tau)D_\omega(\tau)\eta'(\tau)\right\}d\tau$$

$$-\int_{\tau_a}^{\tau_b}\eta'(\tau')j(\tau')d\tau'+\frac{1}{2m}\int_{\tau_a}^{\tau_b}\int_{\tau_a}^{\tau_b} j(\tau')G_\omega(\tau',\tau'')j(\tau'')\frac{d\tau'}{i}\frac{d\tau''}{i} \qquad (3.82)$$

under the assumption the *physical* integration interval $(\tau_a, \tau_b)$ as assimilating the entirely evolution universe of the concerned problem, being thus assimilated with the *mathematical* interval $(-\infty, +\infty)$, so that the delta-Dirac integration property (3.79) is consistently applied. Also note that in expression (3.82) since the statistical Green function comes from its associate real time quantum mechanically problem, see bellow, it tracks also the time Wick "rotation $t = \tau/i$," in the integration measure, explaining the complex factors in the last term of (3.82).

With these, the harmonic fluctuation action of Eq. (3.82) may be reconsidered with the working form (Putz, 2009b):

$$\widetilde{S}_\omega^+(\eta,\dot\eta,\tau)=\int_{\tau_a}^{\tau_b}\left\{\frac{m}{2}\eta(\tau)\left[-\partial_\tau^2+\omega^2\right]\eta(\tau)\right\}d\tau$$

$$-\int_{\tau_a}^{\tau_b}\eta(\tau)j(\tau)d\tau-\frac{1}{2m}\int_{\tau_a}^{\tau_b}\int_{\tau_a}^{\tau_b}G_\omega(\tau,\tau')j(\tau)j(\tau')d\tau d\tau' \qquad (3.83)$$

which under the condition $\omega \rightarrow 0$, as prescribed by Eq. (3.76), it can be further rearranged so that the free terms action to appear distinctively:

$$\widetilde{S}_{\omega\rightarrow 0}^+(\eta,\dot\eta,\tau)=S_0^+(\eta,\dot\eta,\tau)-\int_{\tau_a}^{\tau_b}\eta(\tau)j(\tau)d\tau$$

$$-\frac{1}{2m}\int_{\tau_a}^{\tau_b}\int_{\tau_a}^{\tau_b}G_{\omega\rightarrow 0}(\tau,\tau')j(\tau)j(\tau')d\tau d\tau' \qquad (3.84)$$

with the free Euclidian action

$$S_0^+(\eta,\dot\eta,\tau)=\int_{\tau_a}^{\tau_b}\left[\frac{m}{2}\dot\eta^2(\tau)\right]d\tau \tag{3.85}$$

was recovered though considering on its expression from Eq. (3.83) the reverse integration by parts procedure unfolded in Eq. (3.76).

Finally, with the identification:

$$\eta(\tau)\to\eta_{cl}(\tau) \tag{3.86}$$

$$\frac{\hbar}{m}G_{\omega\to0}(\tau,\tau')\to G_{con}^{(2)}(\tau,\tau') \tag{3.87}$$

in action (230), there follows that now its last two terms are free of quantum fluctuation integration since considered under averaged forms, see Eqs. (3.65) and (3.68), respectively, releasing for the searched current-dependent amplitude of (3.75) the actual solution

$$\left(\frac{\Delta x}{2}\tau_b;-\frac{\Delta x}{2}\tau_a\right)[j]=\lim_{\omega\to0}\int_{\eta(\tau_a)=-\Delta x/2}^{\eta(\tau_b)=+\Delta x/2}\wp\eta(\tau)\exp\left\{-\frac{1}{\hbar}S_\omega^+(\eta,\dot\eta,\tau)\right\}$$

$$=\int_{\eta(\tau_a)=-\Delta x/2}^{\eta(\tau_b)=+\Delta x/2}\wp\eta(\tau)\exp\left\{-\frac{1}{\hbar}\tilde S_{\omega\to0}^+(\eta,\dot\eta,\tau)\right\}$$

$$=\left(\int_{\eta(\tau_a)}^{\eta(\tau_b)}\wp\eta(\tau)\exp\left\{-\frac{1}{\hbar}S_0^+(\eta,\dot\eta,\tau)\right\}\right)$$

$$\times\exp\left(\frac{1}{\hbar}\int_{\tau_a}^{\tau_b}\eta_{cl}(\tau)j(\tau)d\tau+\frac{1}{2\hbar^2}\int_{\tau_a}^{\tau_b}\int_{\tau_a}^{\tau_b}G_{con}^{(2)}(\tau,\tau')j(\tau)j(\tau')d\tau d\tau'\right) \tag{3.88}$$

which ultimately simply re-writes the current dependent propagator amplitude

$$\left(\frac{\Delta x}{2}\tau_b;-\frac{\Delta x}{2}\tau_a\right)[j]$$

$$= \left( \frac{\Delta x}{2} \tau_b; -\frac{\Delta x}{2} \tau_a \right)_{(0)}$$

$$\exp\left( \frac{1}{\hbar} \int_{\tau_a}^{\tau_b} \eta_{cl}(\tau) j(\tau) d\tau + \frac{1}{2\hbar^2} \int_{\tau_a}^{\tau_b} \int_{\tau_a}^{\tau_b} G_{con}^{(2)}(\tau, \tau') j(\tau) j(\tau') d\tau d\tau' \right) \quad (3.89)$$

as a wave-perturbative form of the free fluctuation amplitude (3.51), being intermediated by harmonic towards free limiting motion of quantum fluctuations. Let's further comment that the actual form (3.87) generalizes the previously "guessed" form (3.56) that provided the first order fluctuation correlation, having in addition to it the power to recover all other superior orders of correlation, for instance those given by Eqs. (3.68) and (3.69), by successive application of the formula (3.58).

With these the connected correlation function algorithm was provided and checked in details, being at disposition to be implemented in whatever order of semiclassical expansion of the path integral evolution amplitude (3.54); For exemplification, the next section will expose the analytic solution for the second order case.

### 3.3.3  CLASSICAL FLUCTUATION PATH AND CONNECTED GREEN FUNCTION

We already seen that aiming to evaluate any of the above-connected correlation functions one imperatively needs the working analytical forms of classical fluctuation path

$$\langle \eta_i(\tau) \rangle = \eta_{(cl)i}(\tau) \quad (3.90)$$

as well as the connected Green function identified from (3.87)

$$G_{con}^{(2)}(\tau, \tau') = \frac{\hbar}{m} G_{\omega \to 0}(\tau, \tau') \quad (3.91)$$

computed from the knowledge of the Green function of the harmonic oscillator problem (3.78) with (3.77) employed for the "free harmonic" limit $\omega \to 0$.

Therefore, having evaluated the quantities (3.90) and (3.91) any semiclassical problem can be solved out analytical. Yet, for both quantities in question the procedure of computing consists of three major stages (Putz, 2009b):

(i)   solving the associate real time harmonic problem;
(ii)  rotating the solution into imaginary time picture;
(iii) taking the "free harmonic limit" $\omega \to 0$.

### 3.3.3.1   Calculation of Classical Fluctuation Path

As elsewhere discussed, see Volume I/Chapter 4 of the present five-volume work (Putz, 2016), the classical path for quantum fluctuation will not be written directly from the ordinary path free motion but using the similar result for harmonic motion upon which the free-harmonic condition $\omega \to 0$ will be imposed; actually, the procedure is unfolded as follows.

The time classical path is firstly considered for harmonic motion, see Volume I/Section 4.3.3 of the present five-volume work (Putz 2016):

$$x_{(\omega)cl}(t) = \frac{1}{\omega} \dot{x}_a \sin[\omega(t - t_a)] + x_a \cos[\omega(t - t_a)]$$

$$= \frac{1}{\omega} \frac{\omega}{\sin[\omega(t_b - t_a)]} \{x_b - x_a \cos[\omega(t_b - t_a)]\} \sin[\omega(t - t_a)] + x_a \cos[\omega(t - t_a)]$$

$$= \frac{\begin{aligned}&x_b \sin[\omega(t - t_a)] \\ &+ x_a \{\sin[\omega(t_b - t_a)]\cos[\omega(t - t_a)] - \cos[\omega(t_b - t_a)]\sin[\omega(t - t_a)]\}\end{aligned}}{\sin[\omega(t_b - t_a)]}$$

$$= \frac{x_b \sin[\omega(t - t_a)] + x_a \sin[\omega(t_b - t)]}{\sin[\omega(t_b - t_a)]}$$

$$(3.92)$$

The real to imaginary time rotation is performed on the result (3.92) according with the Wick rule prescription of (3.36), being this equivalently of directly rewriting of expression (3.92) replacing the trigonometric functions by their hyperbolic counterparts, according with the previously explained conversion, see Table 2.1/Section 2.3:

$$x_{cl}(\tau) = \frac{x_b \sinh[\omega(\tau - \tau_a)] + x_a \sinh[\omega(\tau_b - \tau)]}{\sinh[\omega(\tau_b - \tau_a)]} \qquad (3.93)$$

The "free-harmonic" ($\omega \to 0$) limit is performed upon the expression (3.93) through employing the ordinary hyperbolic limit:

$$\lim_{\omega \to 0} \sinh[\omega \bullet] = \omega \bullet \qquad (3.94)$$

This gives:

$$x_{(\omega \to 0)cl}(\tau) = \frac{x_b(\tau - \tau_a) + x_a(\tau_b - \tau)}{\tau_b - \tau_a} \qquad (3.95)$$

which evidently does the same job as the classical free-motion result, see Volume I/Chapter 4 of the present five volume work, although not identical, since derived from a generalized perspective here.

The result (3.95) is implemented in the formula (3.38) to finally produce the classical fluctuation path:

$$\langle \eta(\tau) \rangle = \eta_{cl}(\tau) \equiv \eta_{(\omega \to 0)cl}(\tau) = x_{(\omega \to 0)cl}(\tau) - \bar{x}$$

$$= \frac{x_b(\tau - \tau_a) + x_a(\tau_b - \tau)}{\tau_b - \tau_a} - \frac{x_a + x_b}{2}$$

$$= \left[ \frac{\tau}{\tau_b - \tau_a} - \frac{1}{2} \frac{\tau_b + \tau_a}{\tau_b - \tau_a} \right] \Delta x \qquad (3.96)$$

which takes even the simpler form:

$$\langle \eta(\tau) \rangle = \left( \frac{\tau}{\hbar\beta} - \frac{1}{2} \right) \Delta x \qquad (3.97)$$

while rewritten within the thermodynamic picture:

$$\tau_a = 0 , \; \tau_b = \hbar\beta \qquad (3.98)$$

### 3.3.3.2   Calculation of Connected Green Function

Now, going to the evaluation of the expression (3.91) we se that beyond of terms of the single correlation nature given by Eq. (3.96) we need the Green function of the harmonic oscillator, whose equation is of (3.78) type; written in real time picture (Putz, 2009b):

$$\left[-\partial_t^2 + \omega^2\right] G_\omega(t,t') = \delta(t-t') \ , \ t > t' \in (t_a, t_b) \tag{3.99}$$

has the advantage of having the frontier values fixed by the Dirichlet boundary conditions:

$$G_\omega(t,t') = 0 \begin{cases} t = t_b \ , \ \forall t' \\ \forall t \quad , \ t' = t_a \end{cases} \tag{3.100}$$

in the same manner as the fluctuation paths in real time are set to cancels at the endpoint frontier, see Eq. (2.13). Such double boundary condition fixes the type of solution as being of the double trigonometric form, looking like

$$G_\omega(t,t') = C \sin[\omega(t_b - t)] \sin[\omega(t'-t_a)] \tag{3.101}$$

while the time ordering problem

$$\left[-\partial_{t'}^2 + \omega^2\right] G_\omega(t',t) = \delta(t'-t) \ , \ t' > t \in (t_a, t_b) \tag{3.102}$$

with the Dirichlet boundary conditions:

$$G_\omega(t',t) = 0... \begin{cases} t' = t_b \ , \ \forall t \\ \forall t' \quad , \ t = t_a \end{cases} \tag{3.103}$$

produces the variant Green function:

$$G_\omega(t',t) = C \sin[\omega(t_b - t')] \sin[\omega(t - t_a)] \tag{3.104}$$

yet with the same constant as for solution (3.101) since recognizing it belongs to the formally the same homogeneous equation of type:

$$\left[-\partial_{t_>}^2 + \omega^2\right] G_\omega(t_>, t_<) = 0 \ , \ t_> > t_< \in (t_a, t_b) \tag{3.105}$$

and going to be determined from the appropriate identification in the inho-mogeneous equation:

$$-\partial_t^2 G_\omega(t,t') = \delta(t-t') + \omega^2 G_\omega(t,t') \tag{3.106}$$

Now, the left side of Eq. (3.106) is formed from the difference of the first derivatives of the solutions (3.101) and (3.104) approaching each other the event times (Putz, 2009b):

$$-\partial_t^2 G_\omega(t,t') = -\lim_{t \to t'} \frac{\partial_t G_\omega(t,t') - \partial_t G_\omega(t',t)}{t-t'}$$

$$= C\omega \lim_{t \to t'} \frac{1}{t-t'} \{\cos[\omega(t_b-t)]\sin[\omega(t'-t_a)] + \sin[\omega(t_b-t')]\cos[\omega(t-t_a)]\}$$

$$= C\omega\{\cos[\omega(t_b-t')]\sin[\omega(t'-t_a)] + \sin[\omega(t_b-t')]\cos[\omega(t'-t_a)]\}\lim_{t \to t'}\frac{1}{t-t'}$$

$$= C\omega\sin[\omega(t_b-t_a)]\delta(t-t') \tag{3.107}$$

which, through compared with the right side first term of Eq. (3.106) gives the searched constant:

$$C = \frac{1}{\omega\sin[\omega(t_b-t_a)]} \tag{3.108}$$

leaving with the real time Green function solution of the harmonic oscillator

$$G_\omega(t,t') = \frac{\left\{\begin{array}{l}\Theta(t-t')\sin[\omega(t_b-t)]\sin[\omega(t'-t_a)]\\+\Theta(t'-t)\sin[\omega(t_b-t')]\sin[\omega(t-t_a)]\end{array}\right\}}{\omega\sin[\omega(t_b-t_a)]} \tag{3.109}$$

which combines both above solution with the help of Heaviside step-function:

$$\Theta(t-t') = \begin{cases} 1, & t > t' \\ \dfrac{1}{2}, & t = t' \\ 0, & t < t' \end{cases} \tag{3.110}$$

Next, as previously done with the fluctuation paths, the change to the imaginary time picture is done automatically through trigonometric-to-hyperbolic recipe Table 2.1/Section 2.3 to give

$$G_\omega(\tau,\tau') = \frac{\left\{\begin{array}{l}\Theta(\tau-\tau')\sinh[\omega(\tau_b-\tau)]\sinh[\omega(\tau'-\tau_a)] \\ +\Theta(\tau'-\tau)\sinh[\omega(\tau_b-\tau')]\sinh[\omega(\tau-\tau_a)]\end{array}\right\}}{\omega\sinh[\omega(\tau_b-\tau_a)]} \quad (3.111)$$

noting that in the course of transformation the factor

$$i\left(i\frac{1}{i\cdot i}\right) = 1 \quad (3.112)$$

was tacitly absorbed with the parenthesis complex indices coming from the trigonometric to hyperbolic rotation exposed in the Table 2.1/Section 2.3, while the outside index assures the equivalence of Green function contribution in the canonic-to-Euclidian path integrals action exponents:

$$\underbrace{\exp\left(\frac{i}{\hbar}[\cdots G_\omega(t,t')\cdots]\right)}_{CANONIC}$$

$$\rightarrow \underbrace{\exp\left(\frac{i}{\hbar}[\cdots iG_\omega(t\to\tau,t'\to\tau')\cdots]\right) = \exp\left(-\frac{1}{\hbar}[\cdots G_\omega(\tau,\tau')\cdots]\right)}_{EUCLIDIAN} \quad (3.113)$$

Expression (3.111) in then employed to the "free harmonic" limit (3.94) providing the result:

$$G_\omega(\tau,\tau') = \frac{\Theta(\tau-\tau')(\tau_b-\tau)(\tau'-\tau_a)+\Theta(\tau'-\tau)(\tau_b-\tau')(\tau-\tau_a)}{\tau_b-\tau_a} \quad (3.114)$$

which being free of harmonic influence it remains identically also from the quantity $G_{\omega\to0}(\tau,\tau')$. Still, it has to be converted into the searched connected Green function (3.91), leaving with the time imaginary form:

$$G_{con}^{(2)}(\tau,\tau') = \frac{\hbar}{m}\frac{\Theta(\tau-\tau')(\tau_b-\tau)(\tau'-\tau_a)+\Theta(\tau'-\tau)(\tau_b-\tau')(\tau-\tau_a)}{\tau_b-\tau_a} \quad (3.115)$$

or with its equivalent statistical one:

$$G_{con}^{(2)}(\tau,\tau') = \frac{\Theta(\tau-\tau')(\hbar\beta-\tau)\tau'+\Theta(\tau'-\tau)(\hbar\beta-\tau')\tau}{m\beta} \qquad (3.116)$$

when the thermodynamic picture (3.98) is considered.

### 3.3.4  SECOND ORDER FOR SEMICLASSICAL PROPAGATOR, PARTITION FUNCTION AND DENSITY

Aiming to evaluate the second order truncated expansion (3.54) one needs the evaluation of the quantities $\langle\eta_i(\tau)\rangle$, $\langle\eta_i(\tau)\eta_j(\tau)\rangle$, $\langle\eta_i(\tau)\eta_j(\tau')\rangle$ and of their integration. Given the previous discussions one immediately has, see for instance the Eqs. (3.65) and (3.68):

$$\langle\eta_i(\tau)\rangle = \eta_{(cl)i}(\tau) \qquad (3.117)$$

$$\langle\eta_i(\tau)\eta_j(\tau')\rangle = \eta_{(cl)i}(\tau)\eta_{(cl)j}(\tau') + G_{con}^{(2)}(\tau,\tau')\delta_{ij} \qquad (3.118)$$

With the help of expression (3.96) one can immediately compute the associate integral appearing on (3.54) to be:

$$\int_{\tau_a}^{\tau_b}\langle\eta(\tau)\rangle d\tau = \int_{\tau_a}^{\tau_b}\eta_{cl}(\tau)d\tau = 0 \qquad (3.119)$$

Going now to the double connected correlation functions, one has the working analytical expression:

$$\langle\eta_i(\tau)\eta_j(\tau')\rangle = \left(\frac{\tau}{\tau_b-\tau_a} - \frac{1}{2}\frac{\tau_b+\tau_a}{\tau_b-\tau_a}\right)\left(\frac{\tau'}{\tau_b-\tau_a} - \frac{1}{2}\frac{\tau_b+\tau_a}{\tau_b-\tau_a}\right)\Delta x_i\Delta x_j$$

$$+\frac{\hbar}{m}\frac{\Theta(\tau-\tau')(\tau_b-\tau)(\tau'-\tau_a)+\Theta(\tau'-\tau)(\tau_b-\tau')(\tau-\tau_a)}{\tau_b-\tau_a}\delta_{ij}$$

$$(3.120)$$

through replacing into the expression (3.118) the classical fluctuation paths and connected Green function components, see Eqs. (3.96) and (3.115), respectively.

Now there is immediate to compute the second order involved integrals (Putz, 2009b).

At coincident times we have:

$$\int_{\tau_a}^{\tau_b}\langle\eta_i(\tau)\eta_j(\tau)\rangle d\tau = \delta_{ij}\frac{\hbar}{m(\tau_b-\tau_a)}\int_{\tau_a}^{\tau_b}(\tau_b-\tau)(\tau-\tau_a)d\tau$$

$$+\Delta x_i\Delta x_j\int_{\tau_a}^{\tau_b}\left(\frac{\tau}{\tau_b-\tau_a}-\frac{1}{2}\frac{\tau_b+\tau_a}{\tau_b-\tau_a}\right)^2 d\tau$$

$$=\delta_{ij}\frac{\hbar}{m}\frac{(\tau_b-\tau_a)^2}{6}+\Delta x_i\Delta x_j\frac{\tau_b-\tau_a}{12} \tag{3.121}$$

or as:

$$\int_0^{\hbar\beta}\langle\eta_i(\tau)\eta_j(\tau)\rangle d\tau = \delta_{ij}\frac{\hbar^3\beta^2}{6m}+\Delta x_i\Delta x_j\frac{\hbar\beta}{12} \tag{3.122}$$

in thermodynamic environment given by Eq. (3.98).

At different times we get:

$$\int_{\tau_a}^{\tau_b}d\tau\int_{\tau_a}^{\tau_b}d\tau'\langle\eta_i(\tau)\eta_j(\tau')\rangle$$

$$=\delta_{ij}\frac{\hbar}{m(\tau_b-\tau_a)}\left[\int_{\tau_a}^{\tau_b}d\tau\int_{\tau_a}^{\tau}d\tau'(\tau_b-\tau)(\tau'-\tau_a)+\int_{\tau_a}^{\tau_b}d\tau'\int_{\tau_a}^{\tau'}d\tau(\tau_b-\tau')(\tau-\tau_a)\right]$$

$$+\left[\int_{\tau_a}^{\tau_b}\left(\frac{\tau}{\tau_b-\tau_a}-\frac{1}{2}\frac{\tau_b+\tau_a}{\tau_b-\tau_a}\right)d\tau\right]\left[\int_{\tau_a}^{\tau_b}\left(\frac{\tau'}{\tau_b-\tau_a}-\frac{1}{2}\frac{\tau_b+\tau_a}{\tau_b-\tau_a}\right)d\tau'\right]\Delta x_i\Delta x_j$$

$$=\delta_{ij}\frac{\hbar}{m}\frac{(\tau_b-\tau_a)^3}{12}$$

$$\tag{3.123}$$

with its quantum thermodynamic counterpart:

$$\int_{\tau_a}^{\tau_b}d\tau\int_{\tau_a}^{\tau_b}d\tau'\langle\eta_i(\tau)\eta_j(\tau')\rangle = \delta_{ij}\frac{\hbar^4\beta^3}{12m} \tag{3.124}$$

Now, replacing the values of Eqs. (3.119), (3.122), and (3.124) in the second order truncated semiclassical expression of imaginary time amplitude (3.54) one finally gets (Putz, 2009b):

$$
\left(x_b,\hbar\beta;x_a,0\right)_{QS}^{[II]} = \sqrt{\frac{m}{2\pi\hbar^2\beta}}\exp\left\{-\frac{m}{2\hbar^2\beta}(\Delta x)^2 - \beta V(\bar{x})\right\}
$$

$$
\times\left\{
\begin{array}{l}
1-\dfrac{1}{\hbar}\partial_i V(\bar{x})\displaystyle\int_0^{\hbar\beta} d\tau\langle\eta_i(\tau)\rangle-\dfrac{1}{2\hbar}\partial_i\partial_j V(\bar{x})\displaystyle\int_0^{\hbar\beta} d\tau\langle\eta_i(\tau)\eta_j(\tau)\rangle \\[3mm]
+\dfrac{1}{2\hbar^2}\partial_i V(\bar{x})\partial_j V(\bar{x})\displaystyle\int_0^{\hbar\beta} d\tau\int_0^{\hbar\beta} d\tau'\langle\eta_i(\tau)\eta_j(\tau')\rangle
\end{array}
\right\}
$$

$$
= \sqrt{\frac{m}{2\pi\hbar^2\beta}}\exp\left\{-\frac{m}{2\hbar^2\beta}(\Delta x)^2 - \beta V(\bar{x})\right\}
$$

$$
\times\left\{1-\frac{1}{2\hbar}\partial_i\partial_j V(\bar{x})\left[\delta_{ij}\frac{\hbar^3\beta^2}{6m}+\Delta x_i\Delta x_j\frac{\hbar\beta}{12}\right]+\frac{1}{2\hbar^2}\partial_i V(\bar{x})\partial_j V(\bar{x})\delta_{ij}\frac{\hbar^4\beta^3}{12m}\right\};
$$

$$
\left(x_b,\hbar\beta;x_a,0\right)_{QS}^{[II]} = \sqrt{\frac{m}{2\pi\hbar^2\beta}}\exp\left\{-\frac{m}{2\hbar^2\beta}(\Delta x)^2 - \beta V(\bar{x})\right\}
$$

$$
\times\left\{1-\frac{\hbar^2\beta^2}{12m}\nabla^2 V(\bar{x})-\frac{\beta}{24}(\Delta x\nabla)^2 V(\bar{x})+\frac{\hbar^2\beta^3}{24m}[\nabla V(\bar{x})]^2\right\}
$$

$$\tag{3.125}$$

Note that the expression (3.125) may provide the semiclassical canonical density in path integral based density functional theory (DFT) algorithm given by Eqs. (2.10) and (2.11), see also Volume I/Chapter 4 of the present five-volume set (Putz, 2016):

$$
\rho_{\otimes QS}^{[II]}(\bar{x},\beta)=\left(x_b,\hbar\beta;x_a,0\right)_{QS}^{[II]}\Big|_{x_a=x_b}
$$

$$
= \sqrt{\frac{m}{2\pi\hbar^2\beta}}\left\{1-\frac{\hbar^2\beta^2}{12m}\nabla^2 V(\bar{x})+\frac{\hbar^2\beta^3}{24m}[\nabla V(\bar{x})]^2\right\}\exp\left\{-\beta V(\bar{x})\right\} \tag{3.126}
$$

to be used in construction of N-body density at thermodynamic equilibrium

$$\rho_{N-QS}^{[II]}(\bar{x},\beta) = \frac{N}{Z^{[II]}(\beta)}\rho_{\otimes QS}^{[II]}(\bar{x},\beta) \qquad (3.127)$$

by means of partition function

$$Z^{[II]}(\beta) = \int \rho_{\otimes QS}^{[II]}(\bar{x},\beta)d\bar{x}$$

$$= \sqrt{\frac{m}{2\pi\hbar^2\beta}}\int\left\{1 - \frac{\hbar^2\beta^2}{12m}\nabla^2 V(\bar{x}) + \frac{\hbar^2\beta^3}{24m}[\nabla V(\bar{x})]^2\right\}\exp\{-\beta V(\bar{x})\}d\bar{x}$$

$$(3.128)$$

At this point expression (3.128) may be elegantly transformed through considering the Gauss theorem of canceling integrated divergence, in a general $D$-dimensional case:

$$\int\vec{\nabla}\{[\vec{\nabla}V(\bar{x})]\exp[-\beta V(\bar{x})]\}d^D\bar{x} = 0 \qquad (3.129)$$

which leaves with the useful differential relationship:

$$\int[\nabla V(\bar{x})]^2\exp[-\beta V(\bar{x})]d^D\bar{x} = \frac{1}{\beta}\int\nabla^2 V(\bar{x})\exp[-\beta V(\bar{x})]d^D\bar{x} \quad (3.130)$$

that helps in rewriting partition function (3.128) firstly as:

$$Z^{[II]}(\beta) = \sqrt{\frac{m}{2\pi\hbar^2\beta}}\int\left[1 - \frac{\hbar^2\beta^2}{24m}\nabla^2 V(\bar{x})\right]\exp\{-\beta V(\bar{x})\}d\bar{x} \quad (3.131)$$

and finally, after exponentially resuming, as:

$$Z^{[II]}(\beta) = \sqrt{\frac{m}{2\pi\hbar^2\beta}}\int\exp\left[-\beta V(\bar{x}) - \frac{\hbar^2\beta^2}{24m}\nabla^2 V(\bar{x})\right]d\bar{x} \quad (3.132)$$

In the same manner can be constructed also the higher orders of semiclassical expansion of density matrix (3.48) or (3.54), following the cumulant expansion (3.118), its fluctuation path and connected Green function components, as given by Eqs. (3.96) and (251), respectively, towards constructing the analytical canonical density, partition function and finally the many-body density to be used in DFT and of its (chemical) applications.

Such an application is to be in next presented for electronegativity and chemical hardness indices' computations.

## 3.4 SEMICLASSICAL PERIODICITY OF ELECTRONEGATIVITY AND CHEMICAL HARDNESS

### 3.4.1 FOURTH ORDER FOR SEMICLASSICAL ELECTRONEGATIVITY AND CHEMICAL HARDNESS

The present picture is based on specialization of the density matrix amplitude $(x_b, \tau_b | x_a, \tau_a)$ such as to become uniformly in the valence shell properties, such as the space-time Bohr-Slater quantification on orbits (in atomic units $m = \hbar = e^2/4\pi\varepsilon_0 = 1$), see (Bohr, 1921; White, 1934)

$$x_a = 0, \; x_b = x = \frac{n^2}{Z_{eff}} \tag{3.133}$$

$$\tau_a = 0, \; \tau_b = \tau = 2\pi\frac{n^3}{Z_{eff}^2} \tag{3.134}$$

under the atomic central field *absolute* representation

$$V(x) = \frac{Z_{eff}}{x} \tag{3.135}$$

where $Z_{eff}$ stands for the Slater effective atomic number specific for the multi-electronic atoms, being derived from the standard atomic number $Z$ by subtracting the shielding effects of the inner electrons (Slater, 1930), see at the end of this section the shielding Slater rules. Note that the use of the absolute potential (3.135) has two fundamental reasons:

(i)  we retain the positive values of electronegativity in Eq. (3.34), in accordance with the definition (3.1) as well—because EN is evaluated as a stability measure of such nuclear-electronic system;

(ii)  the counterattractive sign in Eq. (3.135) is in accordance with the electric field orientation that drives the sense of the electronic conditional probability of the imaginary evolution amplitude evaluated from the center of atom to the current valence shell radius. Therefore, the actual electronegativity can also be seen as *power of holding electrons in the valence shell opposing to that exercised upon them from the center of atom*. This way, the present EN definition and equation stand for the reconciliation of the two opposite phenomena acting upon the valence electrons: attraction to nucleus and repulsion from the other atomic inner electrons, this way accounting for the earlier EA and IP influences, respectively. Moreover, in this picture, since no IP and electronic affinity is present the chemical hardness definition can be relaxed by its ordinary factor (1/2) to be simple taken as

$$\eta_{[IV]}^{SC}\left(n,Z_{eff}\right)=\left[\frac{\partial}{\partial Z}\chi^{SC}\left(n,Z\right)\right]_{Z\to Z_{eff}} \tag{3.136}$$

Note that for the semiclassical chemical hardness (3.136) the basic definition (3.3) was employed taking account that for neutral atoms we have $N=Z$, and where the minus sign was as well reconsidered according with the potential (3.135), while the ½ prefactor was formally abolished since at present semiclassical level an integer quantum "leap" LUMO-HOMO is considered to be in agreement with the integer fluctuation domain of quantum propagation, see below Eqs. (3.38)–(7.43), see Eq. (4.158) of Section 4.2.3.3 as well as the discussion of Eq. (4.252) in Section 4.5 of the present volume.

Going to evaluating the quantum propagator $(x_b, \tau_b | x_a, \tau_a)$ this can be done through considering the parameterized quantum paths as in Eq. (3.38), see (Feynman & Hibbs, 1965), where the classical path is replaced by the fixed (non time-dependent) average, according with (3.39) while the fluctuation path $\eta(\tau)$ remains and accounts for the whole path integral information to the actual different endpoint values of Eqs. (3.40) and (3.41) not to be confounded with the chemical hardness since they always indicate the imaginary time dependency they carry; the working quantum statistical path integral representation of the time evolution amplitude has, therefore, the input form of Eq. (3.43), see (Kleinert, 2004); it turns into its semiclassical form upon the potential expansion respecting the path fluctuations, see Eq. (3.47), here up to the fourth order contributions

$$V\left(\bar{x}+\eta(\tau)\right)=V\left(\bar{x}\right)+\partial_i V\left(\bar{x}\right)\eta_i(\tau)+\frac{1}{2}\partial_i\partial_j V\left(\bar{x}\right)\eta_i(\tau)\eta_j(\tau)$$

$$+\frac{1}{6}\partial_i\partial_j\partial_k V\left(\bar{x}\right)\eta_i(\tau)\eta_j(\tau)\eta_k(\tau)$$

$$+\frac{1}{24}\partial_i\partial_j\partial_k\partial_l V\left(\bar{x}\right)\eta_i(\tau)\eta_j(\tau)\eta_k(\tau)\eta_l(\tau)+... \quad (3.137)$$

Note that we have maintained here the physical constants appearances so that the semiclassical expansion procedure is clearly understood in orders of Planck's orders, see below. As such, the path integral propagator representation can be resumed as in Eqs. (3.49) and (3.50) by introducing the so-called *free* imaginary time amplitude (3.51) readily given by the free-propagator solution (3.52), it has the normalization role for averaging the semiclassical factor contribution as in Eq. (3.53). All in all, the semiclassical form of path integral representation of evolution amplitude looks in the fourth order of Planck constant's expansion (Putz, 2011)

$$\left(x_b\tau_b;x_a\tau_a\right)_{QS}=\sqrt{\frac{m}{2\pi\hbar^2\beta}}\exp\left\{-\frac{m}{2\hbar^2\beta}(\Delta x)^2-\beta V(\bar{x})\right\}$$

$$\times\left\{1-\frac{1}{\hbar}\left[\partial_i V\left(\bar{x}\right)\int_0^\tau d\tau_1\langle\eta_i(\tau_1)\rangle+\frac{1}{2}\partial_i\partial_j V\left(\bar{x}\right)\int_0^\tau d\tau_1\langle\eta_i(\tau_1)\eta_j(\tau_1)\rangle\right.\right.$$

$$+\frac{1}{6}\partial_i\partial_j\partial_k V\left(\bar{x}\right)\int_0^\tau d\tau_1\langle\eta_i(\tau_1)\eta_j(\tau_1)\eta_k(\tau_1)\rangle$$

$$+\frac{1}{24}\partial_i\partial_j\partial_k\partial_l V\left(\bar{x}\right)\int_0^\tau d\tau_1\langle\eta_i(\tau_1)\eta_j(\tau_1)\eta_k(\tau_1)\eta_l(\tau_1)\rangle+...\right]$$

$$+\frac{1}{2\hbar^2}\left[\partial_i V\left(\bar{x}\right)\partial_j V\left(\bar{x}\right)\int_0^\tau d\tau_1\int_0^\tau d\tau_2\langle\eta_i(\tau_1)\eta_j(\tau_1)\rangle\right.$$

$$+\partial_k\partial_i V\left(\bar{x}\right)\partial_j V\left(\bar{x}\right)\int_0^\tau d\tau_1\int_0^\tau d\tau_2\langle\eta_k(\tau_1)\eta_i(\tau_1)\eta_j(\tau_2)\rangle$$

$$+\frac{1}{3}\partial_l\partial_k\partial_iV(\bar{x})\partial_jV(\bar{x})\int_0^\tau d\tau_1\int_0^\tau d\tau_2\langle\eta_l(\tau_1)\eta_k(\tau_1)\eta_i(\tau_1)\eta_j(\tau_2)\rangle$$

$$+\frac{1}{4}\partial_k\partial_iV(\bar{x})\partial_l\partial_jV(\bar{x})\int_0^\tau d\tau_1\int_0^\tau d\tau_2\langle\eta_k(\tau_1)\eta_i(\tau_1)\eta_j(\tau_2)\eta_l(\tau_2)\rangle+...\Bigg]$$

$$-\frac{1}{6\hbar^3}\Bigg[\partial_iV(\bar{x})\partial_jV(\bar{x})\partial_kV(\bar{x})\int_0^\tau d\tau_1\int_0^\tau d\tau_2\int_0^\tau d\tau_3\langle\eta_i(\tau_1)\eta_j(\tau_2)\eta_k(\tau_3)\rangle$$

$$+\frac{3}{2}\partial_l\partial_iV(\bar{x})\partial_jV(\bar{x})\partial_kV(\bar{x})\int_0^\tau d\tau_1\int_0^\tau d\tau_2\int_0^\tau d\tau_3\langle\eta_l(\tau_1)\eta_i(\tau_1)\eta_j(\tau_2)\eta_k(\tau_3)\rangle+...\Bigg]$$

$$+\frac{1}{24\hbar^4}\partial_iV(\bar{x})\partial_jV(\bar{x})\partial_kV(\bar{x})\partial_lV(\bar{x})\int_0^\tau d\tau_1\int_0^\tau d\tau_2\int_0^\tau d\tau_3$$

$$\int_0^\tau d\tau_4\langle\eta_i(\tau_1)\eta_j(\tau_2)\eta_k(\tau_3)\eta_l(\tau_4)\rangle+...\Bigg\}$$

$$(3.138)$$

Now, we are faced with expressing the averaged values of the fluctuation paths in single or multiple time connection, i.e., $\langle\eta_i(\tau)\rangle$, $\langle\eta_i(\tau)\eta_j(\tau)\rangle$, $\langle\eta_i(\tau)\eta_j(\tau')\rangle$, etc. For that, it can be readily shown that the first order of the averaged fluctuation path resembles the classical (observed) path, see also Eq. (3.117)

$$\langle\eta_i(\tau_1)\rangle=G_{con}^{(1)}(\tau_1)=\eta_i^{cl}(\tau_1)\qquad(3.139)$$

while the higher orders can be unfolded in connection with the Green functions/propagators according to the so-called *cluster decomposition* or *cumulant expansion*, see also Eq. (3.118)

$$\langle\eta(\tau_1)\cdots\eta(\tau_n)\rangle=G_{con}^{(n)}(\tau_1,\cdots,\tau_n)+\langle\eta(\tau_2)\cdots\eta(\tau_n)\rangle G_{con}^{(1)}(\tau_1),n\geq2$$

$$(3.140)$$

However, by involving the pair-wise (Wick) decomposition of the *n*-points correlated function

$$G_{con}^{(n)}(\tau_1,\cdots,\tau_n)=\sum_{j=2}^{n}G_{con}^{(2)}(\tau_1,\tau_j)\underbrace{\langle\eta(\tau_k)\cdots\eta(\tau_l)\rangle}_{(n-2)order\ function}{}^{k\neq1,j}_{l\neq1,j,k}\qquad(3.141)$$

one can easily obtain the higher orders of correlations, however, observing that all connected orders of events are reduced to the combinations of *pair-connected events*. For instance, we get for the second, third and fourth order fluctuations the respective average contributions, in the same manner as in Section 3.3.4

$$\langle\eta_i(\tau_1)\eta_j(\tau_2)\rangle=\eta_i^{cl}(\tau_1)\eta_j^{cl}(\tau_2)+G(\tau_1,\tau_2)\delta_{ij}\qquad(3.142)$$

$$\langle\eta_i(\tau_1)\eta_j(\tau_2)\eta_k(\tau_3)\rangle=\eta_i^{cl}(\tau_1)\eta_j^{cl}(\tau_2)\eta_k^{cl}(\tau_3)+G(\tau_1,\tau_2)\delta_{ij}\eta_k^{cl}(\tau_3)$$

$$+G(\tau_1,\tau_3)\delta_{ik}\eta_j^{cl}(\tau_2)+G(\tau_2,\tau_3)\delta_{jk}\eta_i^{cl}(\tau_1)$$
$$(3.143)$$

$$\langle\eta_i(\tau_1)\eta_j(\tau_2)\eta_k(\tau_3)\eta_l(\tau_4)\rangle=\eta_i^{cl}(\tau_1)\eta_j^{cl}(\tau_2)\eta_k^{cl}(\tau_3)\eta_l^{cl}(\tau_4)$$
$$+G(\tau_1,\tau_2)\delta_{ij}\eta_k^{cl}(\tau_3)\eta_l^{cl}(\tau_4)$$
$$+G(\tau_1,\tau_3)\delta_{ik}\eta_j^{cl}(\tau_2)\eta_l^{cl}(\tau_4)+G(\tau_1,\tau_4)\delta_{il}\eta_j^{cl}(\tau_2)\eta_k^{cl}(\tau_3)$$
$$+G(\tau_2,\tau_3)\delta_{jk}\eta_i^{cl}(\tau_1)\eta_l^{cl}(\tau_4)+G(\tau_2,\tau_4)\delta_{jl}\eta_i^{cl}(\tau_1)\eta_k^{cl}(\tau_3)$$
$$+G(\tau_3,\tau_4)\delta_{kl}\eta_i^{cl}(\tau_1)\eta_j^{cl}(\tau_2)+G(\tau_1,\tau_2)G(\tau_3,\tau_4)\delta_{ij}\delta_{kl}$$
$$+G(\tau_1,\tau_3)G(\tau_2,\tau_4)\delta_{ik}\delta_{jl}+G(\tau_1,\tau_4)G(\tau_2,\tau_3)\delta_{il}\delta_{jk}\qquad(3.144)$$

with $\delta_{ij}$ type being the delta-Kronecker tensor. Next, for the quantum objects in question, i.e., for the classical path and connected Green function, the computing procedure consists of three major stages:

(i)   Considering and solving the associate real time harmonic problem;
(ii)  Rotating the solution into imaginary time picture;
(iii) Taking back the "free harmonic limit," this way providing the respective results for the classical fluctuation path as in

Eq. (3.97) and for connected Green function as in Eq. (3.116), respectively.

With expressions (3.97) and (3.116) back in Eqs. (3.142)–(3.144) and then in Eq. (3.138) some imaginary-time integrals vanish, see also Section 3.3.4, namely (Putz, 2011):

$$\int_0^\tau d\tau_1 \langle \eta_i(\tau_1) \rangle = \int_0^\tau d\tau_1 \langle \eta_i(\tau_1) \eta_j(\tau_1) \eta_k(\tau_1) \rangle$$

$$= \int_0^\tau d\tau_1 \int_0^\tau d\tau_2 \langle \eta_k(\tau_1) \eta_i(\tau_1) \eta_j(\tau_2) \rangle$$

$$= \int_0^\tau d\tau_1 \int_0^\tau d\tau_2 \int_0^\tau d\tau_3 \langle \eta_i(\tau_1) \eta_j(\tau_2) \eta_k(\tau_3) \rangle = 0 \quad (3.145)$$

while for the non-vanishing imaginary-time integrals appearing in Eq. (3.138), one yields

$$\int_0^\tau d\tau_1 \langle \eta_i(\tau_1) \eta_j(\tau_1) \rangle = \frac{\tau}{12} \Delta x_i \Delta x_j + \frac{\hbar}{m} \frac{\tau^2}{6} \delta_{ij} \quad (3.146)$$

$$\int_0^\tau d\tau_1 \langle \eta_i(\tau_1) \eta_j(\tau_1) \eta_k(\tau_1) \eta_l(\tau_1) \rangle = \frac{\tau}{80} \Delta x_i \Delta x_j \Delta x_k \Delta x_l$$

$$+ \frac{\hbar}{m} \frac{\tau^2}{120} \left( \Delta x_i \Delta x_j \delta_{kl} + \Delta x_i \Delta x_k \delta_{jl} + \Delta x_i \Delta x_l \delta_{jk} \right.$$

$$\left. + \Delta x_k \Delta x_l \delta_{ij} + \Delta x_j \Delta x_l \delta_{ik} + \Delta x_j \Delta x_k \delta_{il} \right)$$

$$+ \frac{\hbar^2}{m^2} \frac{\tau^3}{30} \left( \delta_{ij} \delta_{kl} + \delta_{ik} \delta_{jl} + \delta_{il} \delta_{ik} \right) \quad (3.147)$$

$$\int_0^\tau d\tau_1 \int_0^\tau d\tau_2 \langle \eta_i(\tau_1) \eta_j(\tau_2) \rangle = \frac{\hbar}{m} \frac{\tau^3}{12} \delta_{ij} \quad (3.148)$$

$$\int_0^\tau d\tau_1 \int_0^\tau d\tau_2 \langle \eta_k(\tau_1) \eta_i(\tau_1) \eta_j(\tau_2) \eta_l(\tau_2) \rangle$$

$$= \frac{\tau^2}{144} \Delta x_i \Delta x_j \Delta x_l \Delta x_k + \frac{\hbar}{m} \frac{\tau^3}{72} \left( \Delta x_k \Delta x_i \delta_{jl} + \Delta x_j \Delta x_l \delta_{ki} \right)$$

$$+\frac{\hbar}{m}\frac{\tau^3}{720}\left(\Delta x_k \Delta x_j \delta_{il} + \Delta x_k \Delta x_i \delta_{ij} + \Delta x_i \Delta x_i \delta_{kj} + \Delta x_i \Delta x_j \delta_{kl}\right)$$

$$+\frac{\hbar^2}{m^2}\frac{\tau^4}{36}\delta_{ki}\delta_{jl} + \frac{\hbar^2}{m^2}\frac{\tau^4}{90}\left(\delta_{kj}\delta_{il} + \delta_{kl}\delta_{ij}\right) \tag{3.149}$$

$$\int_0^\tau d\tau_1 \int_0^\tau d\tau_2 \left\langle \eta_l(\tau_1)\eta_k(\tau_1)\eta_i(\tau_1)\eta_j(\tau_2)\right\rangle$$

$$=\frac{\hbar}{m}\frac{\tau^3}{240}\left(\Delta x_l \Delta x_k \delta_{ij} + \Delta x_l \Delta x_i \delta_{kj} + \Delta x_k \Delta x_i \delta_{lj}\right)$$

$$+\frac{\hbar^2}{m^2}\frac{\tau^4}{60}\left(\delta_{lk}\delta_{ij} + \delta_{li}\delta_{kj} + \delta_{lj}\delta_{ki}\right) \tag{3.150}$$

$$\int_0^\tau d\tau_1 \int_0^\tau d\tau_2 \int_0^\tau d\tau_3 \left\langle \eta_l(\tau_1)\eta_i(\tau_1)\eta_j(\tau_2)\eta_k(\tau_3)\right\rangle$$

$$=\frac{\hbar}{m}\frac{\tau^4}{144}\Delta x_l \Delta x_i \delta_{jk} + \frac{\hbar^2}{m^2}\frac{\tau^5}{72}\delta_{li}\delta_{jk}$$

$$+\frac{\hbar^2}{m^2}\frac{\tau^5}{120}\left(\delta_{lj}\delta_{ik} + \delta_{lk}\delta_{ij}\right) \tag{3.151}$$

$$\int_0^\tau d\tau_1 \int_0^\tau d\tau_2 \int_0^\tau d\tau_3 \int_0^\tau d\tau_4 \left\langle \eta_i(\tau_1)\eta_j(\tau_2)\eta_k(\tau_3)\eta_l(\tau_4)\right\rangle$$

$$=\frac{\hbar^2}{m^2}\frac{\tau^6}{144}\left(\delta_{ij}\delta_{kl} + \delta_{ik}\delta_{jl} + \delta_{il}\delta_{jk}\right) \tag{3.152}$$

With these, the earlier form (3.138) takes the particular expression for the propagator needed in electronegativity formulation (3.2) (Putz, 2007)

$$(x,\tau;0,0)=\sqrt{\frac{m}{2\pi\hbar\tau}}\exp\left\{-\frac{m}{2\hbar\tau}(\Delta x)^2 - \frac{\tau}{\hbar}V(\bar{x})\right\}$$

$$\times \left\{ 1 - \frac{1}{\hbar} \left[ \frac{1}{2} V''(\bar{x}) \left( \frac{\tau}{12} \Delta x + \frac{\hbar}{m} \frac{\tau^2}{6} \right) \right. \right.$$

$$\left. + \frac{1}{24} V''''(\bar{x}) \left( \frac{\tau}{80} (\Delta x)^4 + \frac{\hbar}{m} \frac{\tau^2}{20} (\Delta x)^2 + \frac{\hbar^2}{m^2} \frac{\tau^3}{10} \right) \right]$$

$$+ \frac{1}{2\hbar^2} \left[ \frac{\hbar}{m} V'(\bar{x})^2 \frac{\tau^3}{12} + \frac{1}{4} V''(\bar{x})^2 \left( \frac{\tau^2}{144} (\Delta x)^4 + \frac{\hbar}{m} \frac{\tau^3}{30} (\Delta x)^2 + \frac{\hbar^2}{m^2} \frac{\tau^4}{20} \right) \right.$$

$$\left. + V'''(\bar{x}) V'(\bar{x}) \left( \frac{\hbar}{m} \frac{\tau^3}{240} (\Delta x)^2 + \frac{\hbar^2}{m^2} \frac{\tau^4}{60} \right) \right]$$

$$- \frac{1}{6\hbar^3} \left[ \frac{3}{2} V''(\bar{x}) V'(\bar{x})^2 \left( \frac{\hbar}{m} \frac{\tau^4}{144} (\Delta x)^2 + \frac{11}{360} \frac{\hbar^2}{m^2} \tau^5 \right) \right]$$

$$\left. + \frac{1}{1152\hbar^4} V'(\bar{x})^4 \frac{\hbar^2}{m^2} \tau^6 \right\}$$

$$(3.153)$$

Finally, the electronegativity integral (3.2) is solved analytically by applying the saddle-point recipe (see Appendix of this volume)

$$I = \int g(r_b) \exp[f(r_b)] dr_b \cong g(r_b^0) \exp[f(r_b^0)] \sqrt{-\frac{2\pi}{f''(r_b^0)}} \quad (3.154)$$

very well accommodated for the present semiclassical context, where

$$r_b^0 = \frac{2\pi^{2/3} n^2}{Z_{eff}} \quad (3.155)$$

corresponds to the valence shell saddle radius expressed by the optimization condition $f'(r_b^0) = 0$. Then, the fourth order semiclassical expansion for electronegativity is obtained (Putz, 2007, 2011)

$$\chi_{[IV]}^{SC}(n, Z_{eff}) \cong \chi_0^{SC} + \chi_1^{SC} + \chi_2^{SC} + \chi_3^{SC} + \chi_4^{SC} \qquad (3.156)$$

with the components

$$\chi_0^{SC} = \frac{Z_{eff}^2 \exp\left(-3n\pi^{1/3}\right)}{2\sqrt{3}\pi^{2/3}n^2} \qquad (3.157)$$

$$\chi_1^{SC} = -\frac{Z_{eff}^2 \left[12 + 22\pi^{1/3}n + 6\pi^{2/3}n^2 + 5Z_{eff}\right]\exp\left(-3n\pi^{1/3}\right)}{30\sqrt{3}\pi n^3} \qquad (3.158)$$

$$\chi_2^{SC} = \frac{Z_{eff}^2 \left(9 + 2\pi^{1/3}n\right)\left(6 + 5\pi^{1/3}n\right)\exp\left(-3n\pi^{1/3}\right)}{90\sqrt{3}\pi^{2/3}n^2} \qquad (3.159)$$

$$\chi_3^{SC} = -\frac{Z_{eff}^2 \left(11 + 5\pi^{1/3}n\right)\exp\left(-3n\pi^{1/3}\right)}{45\sqrt{3}\pi^{1/3}n} \qquad (3.160)$$

$$\chi_4^{SC} = \frac{Z_{eff}^2 \exp\left(-3n\pi^{1/3}\right)}{36\sqrt{3}} \qquad (3.161)$$

while, for chemical hardness, the working form (3.136) applied on the formulas (3.156)–(3.161) yields the companion expansion (Putz, 2007, 2011)

$$\eta_{[IV]}^{SC}(n, Z_{eff}) \cong \eta_0^{SC} + \eta_1^{SC} + \eta_2^{SC} + \eta_3^{SC} + \eta_4^{SC} \qquad (3.162)$$

with the components

$$\eta_0^{SC} = \frac{Z_{eff} \exp\left(-3n\pi^{1/3}\right)}{\sqrt{3}\pi^{2/3}n^2} \qquad (3.163)$$

$$\eta_1^{SC} = -\frac{Z_{eff}\left[24 + 44\pi^{1/3}n + 12\pi^{2/3}n^2 + 15Z_{eff}\right]\exp\left(-3n\pi^{1/3}\right)}{30\sqrt{3}\pi n^3} \qquad (3.164)$$

$$\eta_2^{SC} = \frac{Z_{eff}\left(9 + 2\pi^{1/3}n\right)\left(6 + 5\pi^{1/3}n\right)\exp\left(-3n\pi^{1/3}\right)}{45\sqrt{3}\pi^{2/3}n^2} \quad (3.165)$$

$$\eta_3^{SC} = -\frac{Z_{eff}\,2\left(11 + 5\pi^{1/3}n\right)\exp\left(-3n\pi^{1/3}\right)}{45\sqrt{3}\pi^{1/3}n} \quad (3.166)$$

$$\eta_4^{SC} = \frac{Z_{eff}\,\exp\left(-3n\pi^{1/3}\right)}{18\sqrt{3}} \quad (3.167)$$

These results widely depend on the atomic effective numer $Z_{eff}$ depending on specific electronic shielding of the valence electrons, computed according to the *Slater's rules*, namely.

  (i)   the electrons found in the exterior groups of the considered electron, do not contribute to shielding ($s = 0$);
 (ii)   each electron in the same group with the one considered contributes with 0.33 to shielding, except the 1 s electron, whose contribution is 0.30;
(iii)   if the calculation electron is *ns or np* then each electron belonging to the layer with principal quantum number $(n-1)$ shields the nuclear charge by 0.85, and each electron with principal quantum number smaller than $(n-1)$ shielded with 1;
 (iv)   for the discussed electrons, of type *d* or *f*, each electron belonging to any internal group, contributes to the shielding with 1.

The Slater resulted values for atomic valence states, are in Table 3.1 presented, along the electronegativity and chemical hardness computed scales, respectively, in the next section.

### 3.4.2 FOURTH ORDER FOR SEMICLASSICAL ELECTRONEGATIVITY AND CHEMICAL HARDNESS

The associate atomic scales are obtained through implementing the intrinsic atomic parameters, $n$ and $Z_{eff}$, from Table 3.1 into the terms of

**TABLE 3.1** Synopsis of the Periodic Principal Quantum Number $n$, of the Effective Charge $Z_{eff}$ Calculated on the Slater Method (Slater, 1930), and of the Associated Electronegativity $\chi^{FD}$ and Chemical Hardness $\eta^{FD}$, Calculated on the Finite Difference (Lackner & Zweig, 1983), Respectively, for Ordinary Elements (Putz, 2007)

*Legend:* **Symbol of the Element**
Valence principal quantum number: $n$
Slater effective charge: $Z_{eff}$
Finite difference electronegativity: $\chi^{FD\,*}$
Finite difference chemical hardness: $\eta^{FD\,*}$

| Element | $n$ | $Z_{eff}$ | $\chi^{FD}$ | $\eta^{FD}$ |
|---|---|---|---|---|
| H | 1 | 1 | 7.18 | 6.45 |
| He | 1 | 1.7 | 12.27 | 12.48 |
| Li | 2 | 1.30 | 3.02 | 4.39 |
| Be | 2 | 1.95 | 3.43 | 5.93 |
| B | 2 | 2.60 | 4.29 | 4.06 |
| C | 2 | 3.25 | 6.24 | 5.00 |
| N | 2 | 3.90 | 7.23 | 7.30 |
| O | 2 | 4.55 | 7.59 | 6.14 |
| F | 2 | 5.2 | 10.4 | 7.07 |
| Ne | 2 | 5.85 | 10.71 | 10.92 |
| Na | 3 | 2.20 | 2.80 | 2.89 |
| Mg | 3 | 2.85 | 2.6 | 4.99 |
| Al | 3 | 3.50 | 3.22 | 2.81 |
| Si | 3 | 4.15 | 4.68 | 3.43 |
| P | 3 | 4.80 | 5.62 | 4.89 |
| S | 3 | 5.45 | 6.24 | 4.16 |
| Cl | 3 | 6.10 | 8.32 | 4.68 |
| Ar | 3 | 6.75 | 7.7 | 8.11 |
| K | 4 | 2.20 | 2.39 | 1.98 |
| Ca | 4 | 2.85 | 2.29 | 3.85 |
| Sc | 4 | 3.00 | 3.43 | 3.22 |
| Ti | 4 | 3.15 | 3.64 | 3.22 |
| V | 4 | 3.30 | 3.85 | 2.91 |
| Cr | 4 | 3.45 | 3.74 | 3.12 |
| Mn | 4 | 3.60 | 3.85 | 3.64 |
| Fe | 4 | 3.75 | 4.26 | 3.64 |
| Co | 4 | 3.90 | 4.37 | 3.43 |
| Ni | 4 | 4.05 | 4.37 | 3.22 |
| Cu | 4 | 4.20 | 4.47 | 3.22 |
| Zn | 4 | 4.35 | 4.26 | 5.2 |
| Ga | 4 | 5.00 | 3.22 | 2.81 |
| Ge | 4 | 5.65 | 4.58 | 3.33 |
| As | 4 | 6.30 | 5.3 | 4.47 |
| Se | 4 | 6.95 | 5.93 | 3.85 |
| Br | 4 | 7.60 | 7.59 | 4.26 |
| Kr | 4 | 8.25 | 6.86 | 7.28 |

**TABLE 3.1** Continued

| Rb | Sr | Y | Zr | Nb | Mo | Tc | Ru | Rh | Pd | Ag | Cd | In | Sn | Sb | Te | I | Xe |
|---|---|---|---|---|---|---|---|---|---|---|---|---|---|---|---|---|---|
| 5 | 5 | 5 | 5 | 5 | 5 | 5 | 5 | 5 | 5 | 5 | 5 | 5 | 5 | 5 | 5 | 5 | 5 |
| 2.20 | 2.85 | 3.00 | 3.15 | 3.30 | 3.45 | 3.60 | 3.75 | 3.90 | 4.05 | 4.20 | 4.35 | 5.00 | 5.65 | 6.30 | 6.95 | 7.60 | 8.25 |
| 2.29 | 1.98 | 3.43 | 3.85 | 4.06 | 4.06 | 3.64 | 4.06 | 4.26 | 4.78 | 4.47 | 4.16 | 3.12 | 4.26 | 4.89 | 5.51 | 6.76 | 5.82 |
| 1.87 | 3.74 | 2.91 | 3.02 | 2.91 | 3.12 | 3.64 | 3.43 | 3.22 | 3.64 | 3.12 | 4.78 | 2.70 | 3.02 | 3.85 | 3.54 | 3.74 | 6.34 |

* In units of eV (electron-volts).

Eqs. (3.156)–(3.167). However, one should use the calibrated expressions for the electronegativity and chemical hardness for H atom, with the obtained pre-factors are summarized in the Table 3.2. With these the results for the elements of the first five periods of Periodic System are listed in Table 3.3.

The main characteristic of the actual atomic scales of electronegativity and chemical hardness is that a striking difference in terms of orders of magnitudes is observed between elements down groups.

However, this is not surprising because the actual definition of electronegativity and chemical hardness reflects the holding power with which the whole atom attracts valence electrons to its center. There is therefore natural that as the atom is richer in core electrons down groups lesser is the attractive force on the outer electrons from the center of the atom. In this respect, the actual scales mirror at the best the atomic stability at the valence shell.

Regarding the different orders of semiclassical influence on the electronegativity and chemical hardness values there is clear form Table 3.3 that at least fourth order expansion is necessary to achieve convergence. For this reason, in what follow only the electronegativity and chemical hardness scales based on the combinations (3.156) and (3.162) will be discussed and compared with those obtained from the finite-difference approach displayed in Table 3.1.

To make the discussion more transparent $\chi_{[IV]}^{SC}$ and $\eta_{[IV]}^{SC}$, the data of Table 3.3 are supplied with their scaled graphical representations in Figure 3.1 respecting those of $\chi^{FD}$ and $\eta^{FD}$ from Table 3.1.

**TABLE 3.2**  Calibration Coefficients of the Electronegativity $\chi^{SC}$ and Chemical Hardness $\eta^{SC}$ for the Considered Orders of Semiclassical Expansions (3.156)–(3.167) Such Way Their Values for Atomic H to Recover the Respective Finite Difference Ones (as given in Table 3.1) (Putz, 2007)

|        | $\chi^{SC}$ | $\eta^{SC}$ |
|--------|--------------------------|--------------------------|
| [0]    | $27.21 \times 158.713$   | $27.21 \times 71.2882$   |
| [I]    | $27.21 \times (-86.9029)$| $27.21 \times (-36.7442)$|
| [II]   | $27.21 \times 93.0596$   | $27.21 \times 44.7874$   |
| [III]  | $27.21 \times 309.502$   | $27.21 \times 178.665$   |
| [IV]   | $27.21 \times 251.14$    | $27.21 \times 137.576$   |

**TABLE 3.3** The Electronegativity $\chi^{SC}$ and Chemical Hardness $\eta^{SC}$ Values of Ordinary Elements Through Employing the Components of Table 3.1 in Eqs. (3.156)–(3.167) with the Periodic Inputs ($n$, and $Z_{eff}$) of Table 3.1 and the Energetic Calibrated Pre-Factors of Table 3.2

| Atom | $\chi^{SC}_{[0]}$ | $\chi^{SC}_{[I]}$ | $\chi^{SC}_{[II]}$ | $\chi^{SC}_{[III]}$ | $\chi^{SC}_{[IV]}$ | $\eta^{SC}_{[0]}$ | $\eta^{SC}_{[I]}$ | $\eta^{SC}_{[II]}$ | $\eta^{SC}_{[III]}$ | $\eta^{SC}_{[IV]}$ |
|---|---|---|---|---|---|---|---|---|---|---|
| H | 7.18 | 7.18 | 7.18 | 7.18 | 7.18 | 6.45 | 6.45 | 6.45 | 6.45 | 6.45 |
| He | 20.75 | 22.56 | 18.81 | 14.30 | 15.52 | 10.96 | 12.32 | 9.32 | 4.4 | 5.91 |
| Li | $0.37\times10^{-1}$ | $0.42\times10^{-1}$ | $1.05\times10^{-1}$ | $1.04\times10^{-1}$ | $1.12\times10^{-1}$ | $0.26\times10^{-1}$ | $0.28\times10^{-1}$ | $0.76\times10^{-1}$ | $0.87\times10^{-1}$ | $0.91\times10^{-1}$ |
| Be | $0.84\times10^{-1}$ | $0.98\times10^{-1}$ | $2.32\times10^{-1}$ | $2.21\times10^{-1}$ | $2.43\times10^{-1}$ | $0.39\times10^{-1}$ | $0.45\times10^{-1}$ | $1.12\times10^{-1}$ | $1.2\times10^{-1}$ | $1.28\times10^{-1}$ |
| B | $1.5\times10^{-1}$ | $1.81\times10^{-1}$ | $4.06\times10^{-1}$ | $3.72\times10^{-1}$ | $4.15\times10^{-1}$ | $0.52\times10^{-1}$ | $0.63\times10^{-1}$ | $1.45\times10^{-1}$ | $1.46\times10^{-1}$ | $1.6\times10^{-1}$ |
| C | $2.34\times10^{-1}$ | $2.93\times10^{-1}$ | $6.23\times10^{-1}$ | $5.47\times10^{-1}$ | $6.2\times10^{-1}$ | $0.65\times10^{-1}$ | $0.82\times10^{-1}$ | $1.77\times10^{-1}$ | $1.64\times10^{-1}$ | $1.86\times10^{-1}$ |
| N | $3.37\times10^{-1}$ | $4.35\times10^{-1}$ | $8.82\times10^{-1}$ | $7.39\times10^{-1}$ | $8.54\times10^{-1}$ | $0.78\times10^{-1}$ | $1.03\times10^{-1}$ | $2.07\times10^{-1}$ | $1.75\times10^{-1}$ | $2.07\times10^{-1}$ |
| O | $4.59\times10^{-1}$ | $6.11\times10^{-1}$ | $11.81\times10^{-1}$ | $9.39\times10^{-1}$ | $11.08\times10^{-1}$ | $0.91\times10^{-1}$ | $1.26\times10^{-1}$ | $2.35\times10^{-1}$ | $1.79\times10^{-1}$ | $2.22\times10^{-1}$ |
| F | $6.\times10^{-1}$ | $8.22\times10^{-1}$ | $15.17\times10^{-1}$ | $11.4\times10^{-1}$ | $13.77\times10^{-1}$ | $1.04\times10^{-1}$ | $1.5\times10^{-1}$ | $2.61\times10^{-1}$ | $1.76\times10^{-1}$ | $2.31\times10^{-1}$ |
| Ne | $7.59\times10^{-1}$ | $10.71\times10^{-1}$ | $18.87\times10^{-1}$ | $13.33\times10^{-1}$ | $16.54\times10^{-1}$ | $1.17\times10^{-1}$ | $1.75\times10^{-1}$ | $2.86\times10^{-1}$ | $1.66\times10^{-1}$ | $2.35\times10^{-1}$ |
| Na | $0.06\times10^{-2}$ | $0.08\times10^{-2}$ | $0.3\times10^{-2}$ | $0.2\times10^{-2}$ | $0.3\times10^{-2}$ | $0.2\times10^{-3}$ | $0.3\times10^{-3}$ | $1.3\times10^{-3}$ | $1.2\times10^{-3}$ | $1.4\times10^{-3}$ |
| Mg | $0.1\times10^{-2}$ | $0.14\times10^{-2}$ | $0.49\times10^{-2}$ | $0.38\times10^{-2}$ | $0.48\times10^{-2}$ | $0.3\times10^{-3}$ | $0.4\times10^{-3}$ | $1.6\times10^{-3}$ | $1.5\times10^{-3}$ | $1.8\times10^{-3}$ |
| Al | $0.15\times10^{-2}$ | $0.22\times10^{-2}$ | $0.73\times10^{-2}$ | $0.57\times10^{-2}$ | $0.71\times10^{-2}$ | $0.4\times10^{-3}$ | $0.6\times10^{-3}$ | $2.\times10^{-3}$ | $1.7\times10^{-3}$ | $2.1\times10^{-3}$ |
| Si | $0.21\times10^{-2}$ | $0.31\times10^{-2}$ | $1.02\times10^{-2}$ | $0.78\times10^{-2}$ | $0.99\times10^{-2}$ | $0.45\times10^{-3}$ | $0.67\times10^{-3}$ | $2.3\times10^{-3}$ | $2.\times10^{-3}$ | $2.5\times10^{-3}$ |
| P | $0.28\times10^{-2}$ | $0.43\times10^{-2}$ | $1.36\times10^{-2}$ | $1.01\times10^{-2}$ | $1.3\times10^{-2}$ | $0.53\times10^{-3}$ | $0.8\times10^{-3}$ | $2.7\times10^{-3}$ | $2.2\times10^{-3}$ | $2.8\times10^{-3}$ |
| S | $0.36\times10^{-2}$ | $0.56\times10^{-2}$ | $1.75\times10^{-2}$ | $1.27\times10^{-2}$ | $1.64\times10^{-2}$ | $0.6\times10^{-3}$ | $0.93\times10^{-3}$ | $3.\times10^{-3}$ | $2.38\times10^{-3}$ | $3.06\times10^{-3}$ |
| Cl | $0.45\times10^{-2}$ | $0.71\times10^{-2}$ | $2.17\times10^{-2}$ | $1.54\times10^{-2}$ | $2.02\times10^{-2}$ | $0.7\times10^{-3}$ | $1.06\times10^{-3}$ | $3.33\times10^{-3}$ | $2.54\times10^{-3}$ | $3.33\times10^{-3}$ |
| Ar | $0.56\times10^{-2}$ | $0.87\times10^{-2}$ | $2.6\times10^{-2}$ | $1.8\times10^{-2}$ | $2.4\times10^{-2}$ | $0.74\times10^{-3}$ | $1.2\times10^{-3}$ | $3.66\times10^{-3}$ | $2.67\times10^{-3}$ | $3.58\times10^{-3}$ |
| K | $0.04\times10^{-4}$ | $0.07\times10^{-4}$ | $0.3\times10^{-4}$ | $0.2\times10^{-4}$ | $0.3\times10^{-4}$ | $0.17\times10^{-5}$ | $0.27\times10^{-5}$ | $1.38\times10^{-5}$ | $1.1\times10^{-5}$ | $1.46\times10^{-5}$ |

**TABLE 3.3** Continued

| Atom | $\chi^{SC}_{[0]}$ | $\chi^{SC}_{[1]}$ | $\chi^{SC}_{[11]}$ | $\chi^{SC}_{[111]}$ | $\chi^{SC}_{[1111]}$ | $\eta^{SC}_{[0]}$ | $\eta^{SC}_{[1]}$ | $\eta^{SC}_{[11]}$ | $\eta^{SC}_{[111]}$ | $\eta^{SC}_{[1111]}$ |
|---|---|---|---|---|---|---|---|---|---|---|
| Ca | $0.07\times10^{-4}$ | $0.12\times10^{-4}$ | $0.53\times10^{-4}$ | $0.36\times10^{-4}$ | $0.5\times10^{-4}$ | $0.22\times10^{-5}$ | $0.36\times10^{-5}$ | $1.78\times10^{-5}$ | $1.39\times10^{-5}$ | $1.87\times10^{-5}$ |
| Sc | $0.08\times10^{-4}$ | $0.13\times10^{-4}$ | $0.59\times10^{-4}$ | $0.39\times10^{-4}$ | $0.55\times10^{-4}$ | $0.23\times10^{-5}$ | $0.38\times10^{-5}$ | $1.87\times10^{-5}$ | $1.46\times10^{-5}$ | $1.96\times10^{-5}$ |
| Ti | $0.08\times10^{-4}$ | $0.14\times10^{-4}$ | $0.65\times10^{-4}$ | $0.43\times10^{-4}$ | $0.6\times10^{-4}$ | $0.24\times10^{-5}$ | $0.4\times10^{-5}$ | $1.96\times10^{-5}$ | $1.52\times10^{-5}$ | $2.05\times10^{-5}$ |
| V | $0.09\times10^{-4}$ | $0.16\times10^{-4}$ | $0.71\times10^{-4}$ | $0.47\times10^{-4}$ | $0.66\times10^{-4}$ | $0.25\times10^{-5}$ | $0.42\times10^{-5}$ | $2.05\times10^{-5}$ | $1.59\times10^{-5}$ | $2.15\times10^{-5}$ |
| Cr | $0.1\times10^{-4}$ | $0.17\times10^{-4}$ | $0.77\times10^{-4}$ | $0.51\times10^{-4}$ | $0.72\times10^{-4}$ | $0.26\times10^{-5}$ | $0.44\times10^{-5}$ | $2.14\times10^{-5}$ | $1.65\times10^{-5}$ | $2.23\times10^{-5}$ |
| Mn | $0.11\times10^{-4}$ | $0.19\times10^{-4}$ | $0.84\times10^{-4}$ | $0.56\times10^{-4}$ | $0.78\times10^{-4}$ | $0.27\times10^{-5}$ | $0.46\times10^{-5}$ | $2.23\times10^{-5}$ | $1.72\times10^{-5}$ | $2.33\times10^{-5}$ |
| Fe | $0.12\times10^{-4}$ | $0.21\times10^{-4}$ | $0.91\times10^{-4}$ | $0.60\times10^{-4}$ | $0.85\times10^{-4}$ | $0.29\times10^{-5}$ | $0.48\times10^{-5}$ | $2.32\times10^{-5}$ | $1.78\times10^{-5}$ | $2.42\times10^{-5}$ |
| Co | $0.13\times10^{-4}$ | $0.22\times10^{-4}$ | $0.99\times10^{-4}$ | $0.65\times10^{-4}$ | $0.92\times10^{-4}$ | $0.3\times10^{-5}$ | $0.5\times10^{-5}$ | $2.42\times10^{-5}$ | $1.84\times10^{-5}$ | $2.51\times10^{-5}$ |
| Ni | $0.14\times10^{-4}$ | $0.24\times10^{-4}$ | $1.06\times10^{-4}$ | $0.7\times10^{-4}$ | $0.99\times10^{-4}$ | $0.31\times10^{-5}$ | $0.52\times10^{-5}$ | $2.5\times10^{-5}$ | $1.9\times10^{-5}$ | $2.6\times10^{-5}$ |
| Cu | $0.15\times10^{-4}$ | $0.26\times10^{-4}$ | $1.14\times10^{-4}$ | $0.75\times10^{-4}$ | $1.1\times10^{-4}$ | $0.32\times10^{-5}$ | $0.54\times10^{-5}$ | $2.6\times10^{-5}$ | $1.96\times10^{-5}$ | $2.68\times10^{-5}$ |
| Zn | $0.16\times10^{-4}$ | $0.28\times10^{-4}$ | $1.23\times10^{-4}$ | $0.8\times10^{-4}$ | $1.13\times10^{-4}$ | $0.33\times10^{-5}$ | $0.57\times10^{-5}$ | $2.7\times10^{-5}$ | $2.02\times10^{-5}$ | $2.77\times10^{-5}$ |
| Ga | $0.21\times10^{-4}$ | $0.37\times10^{-4}$ | $1.61\times10^{-4}$ | $1.04\times10^{-4}$ | $1.48\times10^{-4}$ | $0.38\times10^{-5}$ | $0.66\times10^{-5}$ | $3.07\times10^{-5}$ | $2.27\times10^{-5}$ | $3.15\times10^{-5}$ |
| Ge | $0.27\times10^{-4}$ | $0.48\times10^{-4}$ | $2.1\times10^{-4}$ | $1.31\times10^{-4}$ | $1.88\times10^{-4}$ | $0.43\times10^{-5}$ | $0.76\times10^{-5}$ | $3.46\times10^{-5}$ | $2.5\times10^{-5}$ | $3.51\times10^{-5}$ |
| As | $0.34\times10^{-4}$ | $0.61\times10^{-4}$ | $2.55\times10^{-4}$ | $1.61\times10^{-4}$ | $2.32\times10^{-4}$ | $0.48\times10^{-5}$ | $0.86\times10^{-5}$ | $3.84\times10^{-5}$ | $2.73\times10^{-5}$ | $3.86\times10^{-5}$ |
| Se | $0.41\times10^{-4}$ | $0.75\times10^{-4}$ | $3.09\times10^{-4}$ | $1.92\times10^{-4}$ | $2.79\times10^{-4}$ | $0.53\times10^{-5}$ | $0.96\times10^{-5}$ | $4.22\times10^{-5}$ | $2.93\times10^{-5}$ | $4.2\times10^{-5}$ |
| Br | $0.49\times10^{-4}$ | $0.9\times10^{-4}$ | $3.69\times10^{-4}$ | $2.27\times10^{-4}$ | $3.31\times10^{-4}$ | $0.58\times10^{-5}$ | $1.07\times10^{-5}$ | $4.59\times10^{-5}$ | $3.13\times10^{-5}$ | $4.54\times10^{-5}$ |
| Kr | $0.58\times10^{-4}$ | $1.08\times10^{-4}$ | $4.33\times10^{-4}$ | $2.63\times10^{-4}$ | $3.88\times10^{-4}$ | $0.63\times10^{-5}$ | $1.18\times10^{-5}$ | $4.96\times10^{-5}$ | $3.31\times10^{-5}$ | $4.86\times10^{-5}$ |
| Rb | $0.03\times10^{-6}$ | $0.06\times10^{-6}$ | $0.36\times10^{-6}$ | $0.21\times10^{-6}$ | $0.32\times10^{-6}$ | $0.13\times10^{-7}$ | $0.25\times10^{-7}$ | $1.56\times10^{-7}$ | $1.07\times10^{-7}$ | $1.59\times10^{-7}$ |
| Sr | $0.05\times10^{-6}$ | $0.11\times10^{-6}$ | $0.6\times10^{-6}$ | $0.35\times10^{-6}$ | $0.54\times10^{-6}$ | $0.17\times10^{-7}$ | $0.33\times10^{-7}$ | $2.01\times10^{-7}$ | $1.37\times10^{-7}$ | $2.04\times10^{-7}$ |
| Y | $0.06\times10^{-6}$ | $0.12\times10^{-6}$ | $0.66\times10^{-6}$ | $0.38\times10^{-6}$ | $0.59\times10^{-6}$ | $0.18\times10^{-7}$ | $0.34\times10^{-7}$ | $2.12\times10^{-7}$ | $1.44\times10^{-7}$ | $2.15\times10^{-7}$ |

**TABLE 3.3** Continued

| Atom | $\chi^{SC}_{[0]}$ | $\chi^{SC}_{[1]}$ | $\chi^{SC}_{[11]}$ | $\chi^{SC}_{[111]}$ | $\chi^{SC}_{[1111]}$ | $\eta^{SC}_{[0]}$ | $\eta^{SC}_{[1]}$ | $\eta^{SC}_{[11]}$ | $\eta^{SC}_{[111]}$ | $\eta^{SC}_{[1111]}$ |
|---|---|---|---|---|---|---|---|---|---|---|
| Zr | $0.07\times10^{-6}$ | $0.13\times10^{-6}$ | $0.73\times10^{-6}$ | $0.42\times10^{-6}$ | $0.65\times10^{-6}$ | $0.19\times10^{-7}$ | $0.36\times10^{-7}$ | $2.22\times10^{-7}$ | $1.5\times10^{-7}$ | $2.25\times10^{-7}$ |
| Nb | $0.07\times10^{-6}$ | $0.15\times10^{-6}$ | $0.8\times10^{-6}$ | $0.46\times10^{-6}$ | $0.72\times10^{-6}$ | $0.2\times10^{-7}$ | $0.38\times10^{-7}$ | $2.33\times10^{-7}$ | $1.57\times10^{-7}$ | $2.35\times10^{-7}$ |
| Mo | $0.08\times10^{-6}$ | $0.16\times10^{-6}$ | $0.87\times10^{-6}$ | $0.5\times10^{-6}$ | $0.78\times10^{-6}$ | $0.21\times10^{-7}$ | $0.4\times10^{-7}$ | $2.43\times10^{-7}$ | $1.64\times10^{-7}$ | $2.45\times10^{-7}$ |
| Tc | $0.09\times10^{-6}$ | $0.17\times10^{-6}$ | $0.95\times10^{-6}$ | $0.55\times10^{-6}$ | $0.85\times10^{-6}$ | $0.22\times10^{-7}$ | $0.42\times10^{-7}$ | $2.53\times10^{-7}$ | $1.71\times10^{-7}$ | $2.56\times10^{-7}$ |
| Ru | $0.09\times10^{-6}$ | $0.19\times10^{-6}$ | $1.03\times10^{-6}$ | $0.59\times10^{-6}$ | $0.92\times10^{-6}$ | $0.23\times10^{-7}$ | $0.44\times10^{-7}$ | $2.64\times10^{-7}$ | $1.77\times10^{-7}$ | $2.66\times10^{-7}$ |
| Rh | $0.1\times10^{-6}$ | $0.21\times10^{-6}$ | $1.12\times10^{-6}$ | $0.64\times10^{-6}$ | $1\times10^{-6}$ | $0.23\times10^{-7}$ | $0.46\times10^{-7}$ | $2.74\times10^{-7}$ | $1.84\times10^{-7}$ | $2.76\times10^{-7}$ |
| Pd | $0.11\times10^{-6}$ | $0.22\times10^{-6}$ | $1.2\times10^{-6}$ | $0.69\times10^{-6}$ | $1.07\times10^{-6}$ | $0.24\times10^{-7}$ | $0.47\times10^{-7}$ | $2.85\times10^{-7}$ | $1.9\times10^{-7}$ | $2.86\times10^{-7}$ |
| Ag | $0.12\times10^{-6}$ | $0.24\times10^{-6}$ | $1.29\times10^{-6}$ | $0.74\times10^{-6}$ | $1.15\times10^{-6}$ | $0.25\times10^{-7}$ | $0.49\times10^{-7}$ | $2.95\times10^{-7}$ | $1.96\times10^{-7}$ | $2.96\times10^{-7}$ |
| Cd | $0.13\times10^{-6}$ | $0.26\times10^{-6}$ | $1.39\times10^{-6}$ | $0.79\times10^{-6}$ | $1.24\times10^{-6}$ | $0.26\times10^{-7}$ | $0.51\times10^{-7}$ | $3.05\times10^{-7}$ | $2.03\times10^{-7}$ | $3.06\times10^{-7}$ |
| In | $0.17\times10^{-6}$ | $0.34\times10^{-6}$ | $1.83\times10^{-6}$ | $1.03\times10^{-6}$ | $1.63\times10^{-6}$ | $0.3\times10^{-7}$ | $0.6\times10^{-7}$ | $3.5\times10^{-7}$ | $2.3\times10^{-7}$ | $3.5\times10^{-7}$ |
| Sn | $0.21\times10^{-6}$ | $0.44\times10^{-6}$ | $2.33\times10^{-6}$ | $1.3\times10^{-6}$ | $2.07\times10^{-6}$ | $0.34\times10^{-7}$ | $0.68\times10^{-7}$ | $3.95\times10^{-7}$ | $2.56\times10^{-7}$ | $3.92\times10^{-7}$ |
| Sb | $0.27\times10^{-6}$ | $0.55\times10^{-6}$ | $2.9\times10^{-6}$ | $1.61\times10^{-6}$ | $2.56\times10^{-6}$ | $0.38\times10^{-7}$ | $0.77\times10^{-7}$ | $4.39\times10^{-7}$ | $2.81\times10^{-7}$ | $4.34\times10^{-7}$ |
| Te | $0.32\times10^{-6}$ | $0.68\times10^{-6}$ | $3.52\times10^{-6}$ | $1.94\times10^{-6}$ | $3.1\times10^{-6}$ | $0.42\times10^{-7}$ | $0.86\times10^{-7}$ | $4.83\times10^{-7}$ | $3.05\times10^{-7}$ | $4.75\times10^{-7}$ |
| I | $0.39\times10^{-6}$ | $0.81\times10^{-6}$ | $4.2\times10^{-6}$ | $2.29\times10^{-6}$ | $3.68\times10^{-6}$ | $0.46\times10^{-7}$ | $0.95\times10^{-7}$ | $5.27\times10^{-7}$ | $3.29\times10^{-7}$ | $5.16\times10^{-7}$ |
| Xe | $0.46\times10^{-6}$ | $0.97\times10^{-6}$ | $4.94\times10^{-6}$ | $2.68\times10^{-6}$ | $4.32\times10^{-6}$ | $0.5\times10^{-7}$ | $1.04\times10^{-7}$ | $5.71\times10^{-7}$ | $3.51\times10^{-7}$ | $5.56\times10^{-7}$ |

*All values are in eV (electron-volts) (Putz, 2007).

For both semiclassical electronegativity and chemical hardness fourth order scales there is observed a better regularization of their increasing trend along periods, a feature more apparent for the actual chemical hardness scale, see Figure 3.1. More, for chemical hardness a phenomenological rule would demand to have lower values than that of the associated electronegativity. This observation is based on the secondary order effects that chemical hardness controls throughout its basic definition (3.3) as the derivative of the electronegativity. However, this rule not always obeyed for the finite difference $\eta^{FD}$ values (for instance, see the elements He, Ne, Ar, Kr, Xe) is well satisfied with the present semiclassical ones $\eta^{SC}_{|IV}$ compared with their counterpart electronegativities, $\chi^{FD}$ and $\chi^{SC}_{|IV}$, respectively. There is thus hope that the actual chemical hardness scale to furnish a better frame of analysis and testing of bonding and reactivity through application of the chemical hardness principles (Sen & Mingos, 1993; Pearson, 1997).

For electronegativity, the present $\chi^{SC}_{|IV}$ values seem to respect almost all empirical criteria for acceptability (Murphy et al., 2000). For instance, the atoms N, O, F, Ne, and He have the highest electronegativities among the main groups; the electronegativity of N is by far greater than that of Cl—a situation that is not met in the finite-difference approach; the Si rule demanding that most metals to have EN values which are less than or equal to that of Si is as well widely satisfied; the considered metalloid elements (B, Si, Ge, As, Sb, Te) clearly separates the metals by nonmetals; along periods the highest EN values belong to the noble elements—a rule as well not fulfilled by the couples (Cl, Ar), (Br, Kr), and (I, Xe) from the finite difference scheme, see Table 3.1 (Putz, 2007).

The electronegativities of chalcogens (O, S, Se, Te) reveal great distinction between the chemistry of oxygen and the rest elements of VIA group; the transitional metals are grouped in a distinct contracted region thus closely emphasizing on the d-orbitals effects, a criteria almost not fulfilled by the finite-difference values of electronegativities of Table 3.1, see also Figure 3.1. Finally, we have to point that the systematic decrease of orders of magnitude of electronegativity and hardness semiclassical scales of Table 3.3 and Figure 3.1 has a fundamental consequence, namely stands as the computational proof that the electronegativity and hardness are indeed pure quantum indices. As such, they do not manifest with the same intensity among all elements of the Periodic System but having values that tend to considerably diminish as the frontier electrons are farer

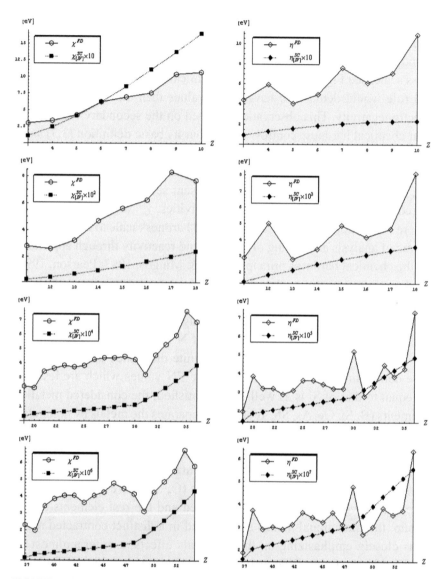

**FIGURE 3.1**  Comparative trend of rescaled electronegativity (left panel) and chemical hardness (right panel) fourth order semiclassical values of Table 3.3 respecting their finite-difference counterparts of Table 3.1, for the second, third, fourth, and fifth periods of elements, from top to bottom, respectively (Putz, 2007).

and feel less and less the quantum influence (potential and force) of the nucleus and of the core electrons, in accordance with the electronic localization principles in an atom, see the Chapter 5 of the present volume.

After all, the present electronegativity and chemical hardness values establish new viable scales, grounded on intrinsic quantum properties of the atoms.

The analysis of data represented in Figure 3.1 leads to three major conclusions for the use of the Green function propagators in evaluating the reactivity indices of electronegativity and chemical hardness (Putz, 2007):

- The striking difference in terms of orders of magnitudes observed between elements down groups is the main characteristic of the actual atomic scales of electronegativity and chemical hardness; however, due to the fact the actual definition of electronegativity and chemical hardness reflects the *holding* power with which the whole atom attracts valence electrons to its center—this is not a surprising behavior;
- As the atom is richer in core electrons down groups the attractive force is lesser on the outer electrons from the center of the atom. In this regard, the actual scales mirror the *atomic stability* of the valence shell;
- The systematic decrease of orders of magnitude of electronegativity and hardness semiclassical scales may stand as the computational proof that the electronegativity and hardness behave like pure quantum/structural indices, although not both with a clearly demonstrated observable character—see Section 3.2. As such, they are not manifesting themselves with the same intensity among all elements of the Periodic Table, while displaying values that tend to considerably diminish as the frontier electrons are farther and feel less and less the quantum influence (the force) of the nucleus and of the core electrons.

## 3.5   CONCLUSION

The main lessons to be kept for the further theoretical and practical investigations of the quantum-propagators in the periodical physical atom that are presented in the present chapter pertain to the following:

- Identifying the observational nature of the electronegativity as associated with the atomic frontier propensity to engage electronic interactions;

- Employing the quantum Hilbert/Fock space of 0-to-1 occupancy to reveal the identity between the orbital/density based electronegativity with the minus of the global chemical potential of a given atomic system;
- Writing the same Fock occupancy formalism for the second order of the atom's frontier interaction, i.e., referring to the chemical hardness observability, with a not-determinate result, so that assuming it as indeed merely belonging to the "chemical" (fuzzy) feature of a chemical system in general and of an atom in special, departing from electronegativity behavior from observability perspective;
- Dealing with path integral formalism for characterizing the physical atom, i.e., as a modern—observationally related—methodology;
- Characterizing the physical atom by the quantum amplitude instead of the customary wave function and fashioned orbitals with the configurational consequences;
- Understanding the electronic movement in physical atomic as being driven by the connected and correlated functions especially by the (temporally) causal Green-function/quantum propagators;
- Describing the physical atom as a semiclassical description of quantum motion, i.e., merely quantum than classical yet with certain orders of Planck constant contributions in electronic orbits in atom;
- Learning the difference between the second and the fourth order of path integral expansion of the quantum amplitude of electronic orbits as quantifies in the associated partition functions;
- Treating the temporal characterization of the electronic orbits in physical atom by Wick rotation towards the temperature characterization of a given quantum state;
- Solving the physical atom by combining the semiclassical path integral expansion for causal electronic motion with the Bohrian prescription for space-time characterizing the closed orbits;
- Formulating the electronegativity and chemical hardness working semiclassical expressions in physical atom by invoking the spectroscopically accepted Slater rules in characterizing the atomic shells in general and of the valence one in special;
- Interpreting the electronegativity as an integral rather than a derivative effect of electronic behavior in the physical atom, in close agreement with the path integral present formulation;
- Connecting the integral interpretation of electronegativity with its actual interpretation as the *power of holding electrons in the*

*valence shell opposing to that exercised upon them from the center of atom;*

- Developing the chemical hardness as derivative of the path integral (quantum amplitude) related electronegativity, in accordance with the local/intra-structure chemical hardness behavior not associated with a cutting-observational measure, as above shown;
- Finding applications of path integral related electronegativity and related chemical hardness by illustration of physical atomic periodicity across the periodic system, yet with fundamental finding in placing chemical hardness in the quantum fluctuation regime, respecting electronegativity, with less and less observational effect (magnitude of its absolute value) down the groups while systematically increasing along the periods.

## KEYWORDS

- **chemical hardness**
- **connected correlation functions**
- **connected Green functions**
- **electronegativity**
- **semiclassical chemical hardness**
- **semiclassical electronegativity**
- **semiclassical propagator**
- **semiclassical quantum analysis**

## REFERENCES

### AUTHOR'S MAIN REFERENCES

Putz, M. V. (2016). *Quantum Nanochemistry. A Fully Integrated Approach: Vol. I. Quantum Theory and Observability*. Apple Academic Press & CRC Press, Toronto-New Jersey, Canada-USA.

Putz, M. V. (2011). Electronegativity and chemical hardness: different patterns in quantum chemistry. *Current Physical Chemistry* 1(2), 111–139 (DOI: 10.2174/1877946811101020111).

Putz, M. V. (2010). Chemical hardness: quantum observable? *Studia Universitatis Babeş-Bolyai—Seria Chemia* 55(Tom I), 47–50.

Putz, M. V. (2009a) Electronegativity: quantum observable. *Int. J. Quantum Chem.* 109, 733–738 (DOI: 10.1002/qua.21957).

Putz, M. V. (2009b) Path integrals for electronic densities, reactivity indices, and localization functions in quantum systems. *International Journal of Molecular Sciences* 10(11), 4816–4940 (DOI: 10.3390/ijms10114816).

Putz, M. V. (2007). Semiclassical electronegativity and chemical hardness. *J. Theor. Comput. Chem.* 6, 33–47 (DOI: 10.1142/S0219633607002861).

## SPECIFIC REFERENCES

Allen, L. C. (1989). Electronegativity is the average one-electron energy of the valence-shell electrons in ground-state free atoms. *J. Am. Chem. Soc.* 111, 9003–9014.

Askamani, R., Manjula, R. (1989). Correlation between electronegativity and superconductivity. *Phys. Rev. B* 39, 4217–4221.

Ballhausen, C. J. (1979). Quantum mechanics and chemical bonding in inorganic complexes. III. The spread of ideas, *J. Chem. Educ.* 56, 357–361.

Bartolotti, L. J., Gadre, S. R., Parr, R. G. (1980). Electronegativities of the elements from the simple Xα theory. *J. Am. Chem. Soc.* 102, 2945–2948.

Bohr, N. (1921). *Abhandlungen über Atombau aus des Jaren 1913–1916*, Vieweg&Son, Braunschweig.

Cyrański, M. K., Krygowski, T. M., Katritzky, A. R., Schleyer, P. V. R. (2002). To what extent can aromaticity be defined uniquely? *J. Org. Chem.* 67, 1333–1338.

Dachen, R., Hasslacher, B., Neveu, A. (1974). Nonperturbative methods and extended-hadron models in field theory. I. Semiclassical functional methods. *Phys. Rev. D* 10, 4114–4129.

Drago, R. S., Kabler, R. A. (1972). Quantitative evaluation of the HSAB (hard-soft acid-base) concept. *Inorg. Chem.* 11, 3144–3145.

Drago, R. S., Wong, N., Bilgrien, C., Vogel, C. (1987). E and C parameters from Hammett substituent constants and use of E and C to understand cobalt-carbon bond energies. *Inorg. Chem.* 26, 9–14.

Feynman, R. P., Hibbs, A. R. (1965). *Quantum Mechanics and Path Integrals*, McGrow Hill, New York.

Garza, J., Robles, J. (1993). Density functional theory softness kernel. *Phys. Rev. A* 47, 2680–2685.

Gázquez, J. L., Ortiz, E. (1984). Electronegativities and hardness of open shell atoms. *J. Chem. Phys.* 81, 2741–2748.

Gordy, W. (1946). A new method of determining electronegativity from other atomic properties. *Phys. Rev.* 69, 604–607.

Grosche, C. (1993). Path integration via summation of perturbation expansions and applications to totally reflecting boundaries, and potential steps. *Phys. Rev. Lett.* 71, 1–4.

Hinze, J., Jaffe, H. H. (1962). Electronegativity. I. Orbital electronegativity of neutral atoms. *J. Am. Chem. Soc.* 84, 540–546.

Ichikawa, S. (1989). High-Tc superconductors and weighted harmonic mean electronegativities. *J. Phys. Chem.* 93, 7302–7304.

Iczkowski, R. P., Margrave, J. L. (1961). Electronegativity. *J. Am. Chem. Soc.* 83, 3547–3551.

Katritzky, A. R., Karelson, M., Sild, S., Krygowski, T. M., Jug, K. (1998). Aromaticity as a quantitative concept. 7. Aromaticity reaffirmed as a multidimensional characteristic. *J. Org. Chem.* 63, 5228–5231.

Kleinert, H. (1986). Path integral for second-derivative Lagrangian $L = (k/2)\bar{x}^2 + (m/2)\dot{x}^2 + (k/2)x^2 - j(\tau)x(\tau)$. *J. Math. Phys.* 27, 3003–3013.

Kleinert, H. (1995). *Path Integrals in Quantum Mechanics, Statistics, and Polymer Physics*, 2nd ed., World Scientific, Singapore.

Kleinert, H. (2004). *Path Integrals in Quantum Mechanics, Statistics, Polymer Physics, and Financial Markets*, 3rd Ed., World Scientific, Singapore.

Klopman, G. (1965). Electronegativity. *J. Chem. Phys.* 43, S124-S129.

Lackner, K. S., Zweig, G. (1983). Introduction to the chemistry of fractionally charged atoms: Electronegativity. *Phys. Rev. D* 28, 1671–1691.

Manning, R. S., Ezra, G. S. (1994). Regularized semiclassical radial propagator for the Coulomb potential. *Phys. Rev. A* 50, 954–966.

Mulliken, R. S. (1934). A new electroaffinity scale: together with data on valence states and an ionization potential and electron affinities. *J. Chem. Phys.* 2, 782–793.

Murphy, L. R., Meek, T. L., Allred, A. L., Allen, L. C. (2000). Evaluation and test of Pauling's electronegativity scale. *J. Phys. Chem. A* 104, 5867–5871.

Parr, R. G. (1985). Density functional theory in chemistry. In: *Density Functional Methods in Physics*, Dreizler, R. M., Providencia, J. D. (Eds.), Plenum Press, New York.

Parr, R. G., Bartolotti, L. J. (1983). Some remarks on the density functional theory of few-electron systems. *J. Phys. Chem.* 87, 2810–2815.

Parr, R. G., Donnelly, R. A., Levy, M., Palke, W. E. (1978). Electronegativity: the density functional viewpoint, *J. Chem. Phys.* 68, 3801–3807.

Parr, R. G., Pearson, R. G. (1983). Absolute hardness: companion parameter to absolute electronegativity. *J. Am. Chem. Soc.* 105, 7512–7516.

Pauling, L. (1932). The nature of the chemical bond IV. The energy of single bonds and the relative electronegativity of atoms. *J. Am. Chem. Soc.* 54, 3570–3582.

Pearson, R. G. (1972). [Quantitative evaluation of the HSAB (hard-soft acid-base) concept]. Reply to the paper by Drago and Kabler (1972). *Inorg. Chem.* 11, 3146.

Pearson, R. G. (1997). *Chemical Hardness*, Wiley-VCH, Weinheim.

Robles, J., Bartolotti, L. J. (1984). Electronegativities, electron affinities, ionization potentials, and hardnesses of the elements within spin polarized density functional theory. *J. Am. Chem. Soc.* 106, 3723–3727.

Sanderson, R. T. (1988). *Chemical Bonds and Bond Energy*, 2nd edition, Academic Press, New York.

Schleyer, P.v.R. (2005). Introduction: delocalization–pi and sigma. *Chem. Rev.* 105, 3433–3435.

Sen, K. D., Jørgensen, C. K., Eds. (1987). *Electronegativity*, Springer-Verlag, Berlin.

Sen, K. D., Mingos, D. M. P., Ed. (1993). *Chemical Hardness*, Springer Verlag, Berlin.

Slater, J. C. (1930). Atomic shielding constants. *Phys. Rev.* 36, 57–64.

Surján, P. (1989). *Second Quantized Approach to Quantum Chemistry*, Springer, Berlin.

White, H. E. (1934). *Introduction to Atomic Spectra*, McGraw-Hill, New York.

# CHAPTER 4

# PERIODICITY BY PERIPHERAL ELECTRONS AND DENSITY IN CHEMICAL ATOM

## CONTENTS

## ABSTRACT

Aiming to affirm specific physical-chemical quantities of electronegativity and hardness as the major electronic indicators of structure and reactivity their systematic definition are presented and discussed, for valence atomic region, by Bohmian quantum mechanics and by the associated density functionals, along introducing their related reactivity index as electrophilicity, within conceptual density functional theory in general and for softness bilocal to global quantum observability in special; this enterprise may serve for further analytical studies of periodicity for atomic properties (atomic radii, diamagnetic susceptibility, or polarizability)—here undertaken, as well as for future understanding and chemical bonding, reactivity, aromaticity, up to the biological activity modeling of atoms in molecules and in nanostructures—in the forthcoming volumes of this five-volume work.

## 4.1   INTRODUCTION

As the classical quantum chemistry had proposed a series of principles and rules to operate in describing the atomic, molecular samples and the reaction mechanisms (Bredow & Jug, 2005) also the modern quantum physical-chemistry likes to unitarily characterize the quantum nature of the chemical and biochemical bonding and transformation on the base of the electronic density (Burresi & Sironi, 2004).

While searching for an adequate expressing of the electronic density al atomic and polyatomic forms of matter through an entire arsenal of quantitative techniques such as are the computational methods of the self-consistent field, of the pseudopotentials, of the matrices and their combinations, and of the graph theory (Vishveshwara et al., 2002; Fujita, 2005), the resulted electronic densities can be then properly integrated or differentiated to provide density functionals, e.g., the total energy, the bond energy, the promoting energy, the solvation energy, reactivity indices, etc. (Tomasi, 2004), as well as the localization functions or the electronic basins of stability (Berski et al., 2003; Kohout et al., 2004).

However, the conceptual chemistry evolves through developing specific objects expressing the reality of chemical reactivity, eventually at the valence levels of by means of the frontier electron movements. In this context, the electronegativity stands as the benchmark as well as the forefront of the modern conceptual quantum chemistry since it may be related and correlated in principle with any many-electronic systems behavior in isolated and reactive environment. Moreover, considered jointly with its companion as hardness, arisen as the second order controlling factor of the total or valence energy expansion, constitute one of the most powerful conceptual binomial in Chemistry with which help either chemical bonding or the reactivity or even biological activity may be modeled in an elegant yet efficient analytical framework (Tarko & Putz, 2010; Putz, 2011a).

However, the striking difference between an atom as a physical entity, with an equal number of electrons and protons (thus in equilibrium), and the same atom as a chemical object, with incomplete occupancy in its periphery quantum shells (thus attaining equilibrium by changing accepting or releasing electrons), is closely related to the electronegativity phenomenology in modeling chemical reactivity. Moreover, this difference triggers perhaps the most important debate in conceptual chemistry: the ground vs. valence state definition of an atom.

As such, the present chapter unfolds the most intriguing aspects of electronegativity and hardness, from their basic definitions and principles to density functional forms, when clarifying their absolute and chemical systematic realizations; the observability character was already approached in the Chapter 3 by means of the second quantization formalism according which electronegativity is indeed revealed as the minus of the chemical potential or even more as the negative of the eigen-energies for fully occupied states thus affirming the plenty of observability, while the hardness still preserves the "quantum hidden" character for such circumstance along the vacuum state, with proved no observability for factionary occupancy; such electronegativity vs. chemical hardness behavior will be also by this chapter explored, and further combined in the new concept of electrophilicity, while also driving the related atomic periodicity characteristics as radii, diamagnetic susceptibility and or polarizability, while showcasing the reasonable quantitative correlation and prediction across the main periods of the Periodic Table (Putz, 2008a).

## 4.2   ELECTRONEGATIVITY AND ITS QUANTUM PRINCIPLES

### 4.2.1   GENERAL DEFINITIONS AND PRINCIPLES

A series of definitions and principles, some based on a purely qualitative logic, others given as a consequence, some refining equivalents revealing of new aspects and reformulating in other terms the fundamental problem of quality—quantitative defining and implementation the electronegativity concept, perhaps the most celebrated concept of Chemistry along the chemcial bonding, will be here enounced and commented (Komorowski, 1983a–c, 1987a,b ).

*DEFINITION EN1*: It will mean by the electronegativity of a system— atom, ion, molecule or radical—that global property which determines how the bonding electrons will be distributed between this system and another, when the two systems become connected by a chemical bond.

Naturally, the basic entity from where the chemical reactivity starts and the chemistry as a science itself, is the atom with its electron cloud. Ultimately, from this entity and from its properties should start the basic study for defining the global properties of the electronic distribution and the tendency of evolution of this distribution defined through the electronegativity.

*PRINCIPLE EN1*: A successful evaluation of the atomic electronegativity as a global property of the electronic distribution should be based on the compaction degree of the electronic cloud around the atomic nucleus.

It can be followed the logical transposition of this principle in the schematic representation in Figure 4.1, where the properties of the atom in question devolve from the occupation degree with electrons of the valence shell.

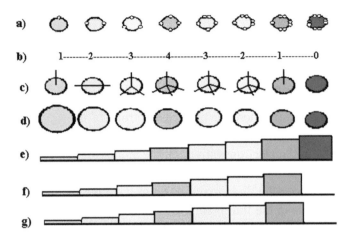

**FIGURE 4.1**    The modification of the atomic structure properties: (a) the occupancy degree of the valence states, (b) with the consequence of modifying the the covalence; (c) the bond angles; (d) the covalent radius; (e) the effective nuclear charge; (f) the electronegativity; and (g) the energy of homonuclear bond; after Sanderson (1988a,b).

It is noted that for a progressive occupation of the valence shell from one electron to eight electrons occur the characteristic changes regarding:

1. the effective nuclear charge, the fraction of the nuclear charge that can be "felt" by the electrons from the valence orbitals, increases with the increasing of the valence electrons number;
2. the covalent radius, correlated with the degree of electron cloud compaction decreases with the increasing of the number of the valence electrons;
3. the electronegativity, correlated with the tendency of the electronic compaction on the valency stratum also increases with the increasing of the number of the electrons in this shell;
4. *homonuclear bond energy, depends on the electronegativity and follow its tendency*;
5. the covalence, which corresponds to the capacity to put together a number of electrons in order to satisfy a more stable compaction of the electronic cloud around the nucleus;
6. the bond angles, formed between the bonding electrons and the pairs of the non-participating electrons.

All these properties naturally devolve from the charges modifications of the atomic structure and are directly correlated from the electronic distribution, i.e., in relation with the electronegativity (Matsunaga, 1969; Matcha, 1983; Haasnoot, 1980).

Of course, they appear also in a few particular cases of coordination, where are necessary to formulate some principles of influence of the specific electronegativity.

*PRINCIPLE EN2*: When the valence electrons of the atoms in the main groups are left unused in the compounds formation, they act in a way that reduces the initial electronegativity of the atom.

Is the case of the "inert pair effect" manifested through the tendency of elements from the three main group to have the oxidation state I and of the elements in the four main group to have the oxidation state II. For example, in case of thallium oxide (I), $Tl_2O$, the "inert pair" reduces the electronegativity from 2.25 for thallium (III) to 0.99 for thallium (I), so that the compound becomes more strongly bonded than is expected.

*PRINCIPLE EN3*: For the transitional elements, the deeper $d$ electrons which are not directly involved in the bonds, also reduce the electronegativity, so that the 3d electrons have a much higher effect in bond than those of the 4d and 5d orbitals.

This is the case, for example, of the atoms of Cr(VI) witch have an electronegativity of 3.37, much higher than that of the W(VI) atoms, which because of the 3d internal orbitals have an electronegativity of 1.67 and therefore render the combination with the oxygen more stable in the $WO_3$ than for $CrO_3$.

*PRINCIPLE EN4* (of transferability): When all the normal valence electrons are involved in a bond, there is not (yet) any restriction for the electronegativity to not be the same, even for different oxidation states.

For example, in the nitrogen case, has the same electronegativity, both in the ammonia compounds and nitrogen oxides.

Perhaps the most important principle regarding the nature of the electronegativity is that one which tells what happens with it when two atoms (electronic systems), initially different, are combining in order to form a compound:

*PRINCIPLE EN5* (of electronegativity equalization): In a compound, all the atoms (the electronic systems) are adjusted to a intermediate electronegativity, global for the respective compound

Here, it will be given a qualitative argument, proposed by Sanderson in 1951, which will be followed by an analytical justification in a further special paragraph, dedicated to this principle. As example, for two atoms, the generalization being immediate, initially found in neutral electric states, but which have different electronegativities, as a result of their coordination, occurs a charge transfer from the less electronegative to the most electronegative, resulting a complex from this combination in which the first atom became partially positively charged, and the second partially negatively charged (Sanderson, 1988a,b).

The form atom the complex which is partially negatively charged, actually presents a growth in the average of the electronic population, which leads to the increasing of the interelectronic repulsion and to the reduction of the effective nuclear charges. From Figure 4.1, this means an increase of the electronic cloud radius, a reduction in the electronic compactness and ultimately leads to a decrease of the electronegativity.

In the complex, the atom partially positively charged, presents a reduction of the interelectronic repulsion, therefore an increase of the effective nuclear charges, an increase of the compactness of the electronic cloud remained around the nucleus and consecutively leads to the increase of the electronegativity (Mullay, 1987a,b).

Therefore, the initially atom more electronegative becomes less electronegative and that one less electronegative becomes more electronegative, when two initial atoms enter in a chemical combination, involving an electronic transfer until the equalization of the atoms electronegativities in the formed compound.

*PRINCIPLE EN6*: The intermediate electronegativity, resulted from the equalization of the initial electronegativities of the subsystems which compose a chemical assembly, is quantitatively equal to the average of the electronegativities of the initial systems.

Parr and Pearson showed in 1983–1989 that (Parr & Pearson, 1983; Pearson, 1986, 1987, 1988a-*b*, 1989):

*PRINCIPLE EN7*: The electronegativity corresponds to the chemical potential, taken with a change sign.

This fact is as simple formulated as efficient. The association is even natural, as long as the electronegativity can be understood as a virtual "seizure" of electrons, and the chemical potential can be interpreted as the potency of "propel" electrons. From the two "opposite directions," but on the same "phenomenological direction" the Principle EN7 is flowing naturally. However, they have very important effects and an analytical efficacy that will be revealed during this study. As a first consequence, the electronegativity equalization principle, the Principle EN5, results immediately in terms of chemical potentials, from the thermodynamic considerations of the states which characterize a new compound, formed by its initial components, which requires the equilibrium potential equalization, and thus of the electronegativities.

## 4.2.2  *CLASSICAL AND QUANTUM PICTURES FOR ELECTRONEGATIVITY*

The studies made for over the 70 years to characterize in a quite realistic way the electronegativity concept can be broadly grouped into two distinct periods.

A period called classic, in which it was predominantly tried to define the atomic electronegativity and the second period, the modern stage, which pursued the association of some electronegativities for a group of atoms in molecules or of some molecular fragments so that their electronegativity should be (almost) identically transferred when the relative group changes its coordination partners, in other complexes. Therefore, in the second period, the concern was focused on the molecular electronegativity calculation of group, orbital, together with the determination of the partial charge distributed between the atoms of the molecular structure.

The moment in which the electronegativity has been approached in terms of the density functionals theory was thanks to the effort of Parr and his collaborations from 1978 (Parr et al., 1978). However, there are some common characteristics of the different approaches that can be compressed into a principle of the molecular formation:

*PRINCIPLE EN8*: (a) The molecules consist of atoms kept together by chemical bonds. (b) The chemical bonds involve an electronic distribution between the molecular atoms. (c) The electrons are not always equally distributed between the atoms.

The main contributions in defining the electronegativity for the various scales and physical images are summarized in the following section.

### 4.2.2.1  Pauling and Mulliken Scales

In 1932 Pauling formulated the first idea referring to the explanation of the chemical bond nature by introducing the electronegativity (Pauling, 1932):

*PRINCIPLE EN9 (of Pauling)*: (a) The normal covalent character between two atoms A and B is transposed by the bond energies additivity of involved atoms:

$$E^c(AB) = \frac{E(AA) + E(BB)}{2} \qquad (4.1)$$

(b) The ionic character of the chemical bond between the atoms *A* and *B* involves an inequality in the electrons distribution between the atoms and brings a contribution of the normal covalent energy proportional to the electronegativities difference of the atoms involved in bond:

$$E(AB) = E^c + E^i = E^c + (\chi_A - \chi_B)^2 \qquad (4.2)$$

Using thermochemical data for the bonds formation and based on the principle just expressed, Pauling was able to calculate the electronegativity for the main elements from the periodic table (see Table 4.1).

In Table 4.1 are presented the electronegativities calculated by Pauling, where for the rare gases the values were completed by the Allen's contribution (Allen & Huheey, 1980). It is noted the increase of the electronegativity as a general tendency once with the increase of the groups, by the correlation with the electronic compactness, especially of the valence layers, concomitant with a decrease in the groups once with the periods increase, due to the decrease of the effective nuclear charge, the nuclear charge being shielded by more and more electronic layers, the tendency to "seize" electrons decreases.

This method deficiency consists in the fact that the association of a single electronegativity value for each atom do not cover all the cases of coordination, in this case, the effects due to the hybridizations and also the charges distributed to atoms in molecule, are not counted.

**TABLE 4.1**    The Electronegativities Values Calculated by Pauling Method (Pauling, 1932)

| | | | | | | | | | | | | | | | | |
|---|---|---|---|---|---|---|---|---|---|---|---|---|---|---|---|---|
| **H** | 2.1 | | | | | | | | | | | | | | **He** | 5.2 |
| **Li** | 1.0 | **Be** | 1.5 | **B** | | 2.5 | **C** | 2.5 | **N** | 3.0 | **O** | 3.5 | **F** | 4.0 | **Ne** | 4.5 |
| **Na** | 0.9 | **Mg** | 1.2 | **Ai** | | 1.5 | **Si** | 1.8 | **P** | 2.1 | **S** | 2.5 | **Cl** | 3.0 | **Ar** | 3.2 |
| **K** | 0.8 | **Ca** | 1.1 | **Sc** | | 1.3 | **Ge** | 1.8 | **As** | 2.0 | **Se** | 2.4 | **Br** | 2.8 | **Kr** | 2.9 |
| **Rb** | 0.8 | **Sr** | 1.0 | **Y** | | 1.2 | **Sn** | 1.8 | **Te** | 1.9 | **Te** | 2.1 | **I** | 2.5 | **Xe** | 2.4 |
| **Cs** | 0.7 | **Ba** | 0.9 | **La-Lu** | 1.1 | **Pb** | 1.8 | **Po** | 2.0 | **Po** | 2.0 | **At** | 2.2 | **Rn** | 2.1 | | |
| **Fr** | 0.7 | **Ra** | 0.9 | **Ac** | | 1.1 | | | | | | | | | | |

Therefore, it was imposed the introduction of a more comprehensive image, better accepted as a reference in particular in the theoretical calculations, due to Mulliken's and formulated for the first time in 1934 (Mulliken, 1934):

*PRINCIPLE EN10 (of Mulliken)*: (a) A chemical bond between two atomic systems (molecular) can be seen as a competition between them for the electrons pairs. (b) The electronegativity of such a system ($S$) involved in the chemical bond is expected to represent the system's ability to compete for the electrons and has the general expression given by:

$$\chi_M \equiv \frac{IP + EA}{2} \tag{4.3}$$

where $IP$, respectively $EA$, represents the ionization potential and the electronic affinity, in agreement with the competition for the electrons: the system tends to keep an electron (the resistance to become positive ion), but also tends to simultaneously acquire a second electron (becoming a negative ion), see Figure 4.2.

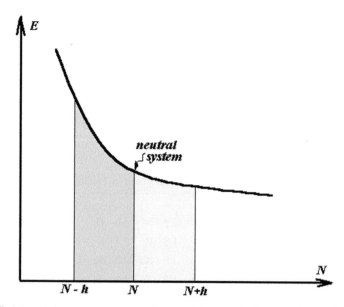

**FIGURE 4.2**  Schematically representation of the parabolic dependence of the total energy of an N-electronic system, connecting its ionic ($\pm$ h) states (Putz, 2003).

For an atom ($A$), the $IP_A$ and $EA_A$ values can be calculated in any state: fundamental state, excited states or valence states. The calculated values may be modeled by linear correlation:

$$\chi_A^P = 0.168\left(\chi_A^M - 1.23\right) \qquad (4.4)$$

where the numerical values of the coefficients actually adjusts the Pauling (P) scale of the electronegativity, defined by the radical of the energy transferred in the chemical bond to the Mulliken (M) scale in which the electronegativity unit is the electron-volt.

Also the Mulliken scale presents an increasing complication of the electronegativity expression for the atoms involved in hybridizations, during their complexity.

### 4.2.2.2    Allred-Rochow and Gordy Scales

The Allred-Rochow scale is, after the one of Pauling, the most commonly used among the experimental studies and is based on the following principle (Allred & Rochow, 1958; Allred, 1961):

*PRINCIPLE EN11* (of Allred-Rochow): The electronegativity of an atom is given by the attraction force between the shielded nucleus and the electron found at a distance equal with the covalent radius and is expressed as follows:

$$\chi_A = 0.36\frac{Z^*}{r^2} + 0.74 \qquad (4.5)$$

This principle states the introduction of the force concept in explaining the chemical bonds and in defining the electronegativity. In the last relation the effective charge of the shielded nucleus, $Z^* = Z_{eff} = Z - s$, is calculated according to the *Slater's rules* (see Section 3.4.1).

A modification of Allred-Rochow relation in the way proposed by Huheey $Z^* \to Z^* - 3\delta$, with $\delta$ the partial atomic charge of the atom in molecule, is insignificant in terms of concept.

Boyd and Markus, in 1981, took the concept of force in defining the atomic electronegativity and by nonempirical calculations reached the expression:

$$\chi_A = 69.4793\frac{Z_A}{r_A^2}\left(1 - \int_0^{r_A}\rho(r)dr\right) \qquad (4.6)$$

where $Z_A$ and $r_A$ are the atomic number and respectively the relative radius of the atom $A$, and $\rho(r)$ is the radial density function.

A similar formalism will be developed also in this study, in order to adjust the electronegativity scales, in one of the next chapter sections, differing only by the physical image through which the electronegativity will associate, not with the force but with the respective potential.

Also in the terms of the potential image Gordy described the electronegativity:

*PRINCIPLE EN12 (of Gordy)*: The electronegativity may be associated with the electrostatic potential which act at the distance of the shielded covalent radius and is given by the expression (Gordy, 1946, 1956):

$$\chi_A = 0.62\frac{Z'}{r} + 0.50 \tag{4.7}$$

where $Z'$ *is* the nuclear charge shielded by Gordy's method: the complete electronic layers shielding totally and the valence electrons ($V$) have a shielding factor of 0.5. Therefore, the Gordy nuclear charge, can be considered to be shielded as

$$Z' = V - 0.5(V - 1) = 0.5(V + 1) \tag{4.8}$$

through which the electronegativity relation becomes

$$\chi_A = 0.31\frac{V+1}{r} + 0.50 \tag{4.9}$$

It should be noted that for a Slater nuclear charge instead of the ones of Gordy, the electronegativity correspondence with those calculated by the Pauling's method is not actually good. The Gordy numerical coefficients were thus calibrated in order to allow an immediate reporting to the Pauling scale.

Also basing on the idea of the electrostatic potential, St. John and Bloch in 1974 were introduced the atomic electronegativity in correlation with the orbital electronegativity for the valence orbitals,

$$X_l = \frac{1}{r_l} \tag{4.10}$$

by the formula:

$$\chi_A = 0.43\sum_{l=0}^{2} X_l + 0.24 \tag{4.11}$$

where $X_0$, $X_1$, $X_2$ correspond to the electronegativities of the orbitals $s$, $p$, and $d$, respectively.

### 4.2.2.3  Sanderson Scale

The Sanderson electronegativity concept is related to the electronic compactness around the nucleus, but through the very intuitive image, specific for the chemical approaches, opens the way for the concept of partial electric charge distributed to the atoms in molecule (Sanderson, 1983a,b, 1986a,b, 1988a,b).

*PRINCIPLE EN13 (of Sanderson)*: (a) The atomic electronegativity is a measure of the compactness of the electrons around the respective atoms, even if the atoms are molecular constituents. (b) The atomic electronegativity is defined as the stability ratio (*RS*) between the electronic density of the isolated atom (or involved in the bond) $\rho$ and the isoelectronic density of the inert atom $\rho^0$, where

$$\rho(r) = \frac{Z}{4.19r^3} \tag{4.12}$$

with $r$ the covalent radius and $Z$ the of the electrons number from atom.

From the way of defining is noted the dimensionless image of the electronegativity and the bond with the Pauling electronegativities scale is rendered by the bond:

$$\chi_A = [0.21(RS) + 0.77]^2 \tag{4.13}$$

Sanderson had used this image of the electronegativity corroborated with the Principle for Electronegativity Atomic Equalization in a chemical compound, Principle EN5, as the average of the electronegativity of the compound from the constituent atoms (Principle EN6). Therefore, he had obtained a very simple scheme for the atomic charge calculation in molecule:

*PRINCIPLE EN14*: The partial charge associated to an atom in a molecule represents the ratio between the modification of the individual atomic electronegativity toward the molecular intermediate electronegativity and the variation of stability ratio of the respective atom which corresponds with the acquisition of the charge unit (+ or −) toward the rest of the molecule:

$$\delta_A = -\frac{(RS)_A - \sqrt{\prod_{A_i}(RS)_{A_i}}}{\Delta(RS)_A}, A_i \in \{A,...A_i,...\} \equiv molecule \qquad (4.14)$$

where $(RS)_A$ is actually the electronegativity of the atom $A$ in Sanderson scale,

$$\Delta(RS)_A = 2.08\sqrt{(RS)_A} \qquad (4.15)$$

The problem of determining the group electronegativity and the partial charges of the atoms in molecule is the stated goal of the modern approaches regarding the transfer electronegativity of the atoms groups in various and complexes molecular combinations. The results obtained by Sanderson are well correlated with the experimental data.

Simons et al. (1976) had also developed a dimensionless image of the electronegativity, based on the consideration of the mobile spherical orbitals of Gauss, with the following of the molecular formation principle (Simons et al., 1976):

*PRINCIPLE EN15*: For an electronic description in the Gauss orbitals, the chemical bond is described by those orbitals which are located between the atoms which form the bond and correspond to a minimum orbital energy.

Based on this principle can be defined an orbitalic factor,

$$f_{AB} = \frac{R_A}{R_A + R_B} \qquad (4.16)$$

where $R_A$ and $R_B$ are the distances from the atoms $A$ and $B$ until the center of the bonding orbital. If $f_{AB} = 0.5$, both atoms equally attract the bonding orbital, and if $f_{AB} < 0.5$ then the atom $A$ attract more than $B$. In these conditions can be introduced the atomic electronegativity by the difference:

$$\chi_A - \chi_B = k(f_{AB} - 0.5) \qquad (4.17)$$

in which the parameter $k$ can be determined from the imposed conditions, for example $\chi_{Li} = 1, \chi_F = 4.0$.

### 4.2.2.4 Iczkowski-Margrave-Huheey Picture

A new approach of the electronegativity, which allows also the partial charges calculation of the atoms in molecule, is opened by considering as a starting point the bond energy associated with the atom in cause (Iczkowski & Margrave, 1961; Huheey, 1978).

*PRINCIPLE EN16*: For an expression of the bond energy of the atom A written as a polynomial of the charge function, $q = N_A^0 - N_A$, with $q = 0$ for the neutral atom,

$$E_A(q) = E_A^0 + aq + bq^2 + cq^3 + dq^4 \qquad (4.18)$$

where $a$, $b$, $c$, and $d$ are constants dependent of the atom and its valence state, are associated to the neutral atom electronegativity, the report:

$$\chi_A^0 = \frac{dE_A}{dq}\Big|_{q=0} = a \qquad (4.19)$$

Under these conditions, the electronegativity variation of the isolated atom becomes (Iczkowski-Margrave relation):

$$\chi_A = \chi_A^0 + 2bq \qquad (4.20)$$

if from the parabolic expression above are retained only the first three terms.

The linear expression in the electric charge associated with the electronegativity equalization principle (Principle EN5), allowed to Huheey to calculate the group electronegativity ($-AB_n$), respectively the partial charges distributed to each atom in the group, by solving the system:

$$\begin{cases} \chi_G = \chi_A(q_A) = a_A + 2b_A q_A = a_{B_1} + 2b_{B_1} q_{B_1} = \dots = \chi_{B_n}(q_{B_n}) \\ q_A + q_{B_1} + \dots + q_{B_n} = q_G = 0 \vee 1 \vee -1 \end{cases} \qquad (4.21)$$

where the second equation of the system, the charge conservation equation, is written with those values of a radical group, anionic or cationic.

The method has, beyond the simplicity of the image of the charge and electronegativity conservation (through equalization) within the group, some disadvantages.

Basically, this method do not allow the separate treatment of the isomers ($-CH_2CFH_2$ toward $-CFHCH_3$, for example), do not allow a simple treatment of the multiple bonds and uses the electronegativity equalization principle in its total form, FEOE (Full Equalization Orbital Electronegativity).

A simplifying change was offered by Huheey in 1984, rewriting the charge relation of the electronegativity under the form:

$$\chi_A = a_A(1 + q_A) \tag{4.22}$$

which integrated into the system of the equalization electronegativity with the condition of radical group, $q_G = 0$, allows the writing of the following expressions for the immediate calculation of the *group electronegativity*, of the equalized electronegativity and partial charge (Bratsch, 1984, 1985):

$$\chi_G = \frac{N_G}{\sum_i \left(\frac{n_i}{\chi_i}\right)} \tag{4.23}$$

$$\chi_{eg} = \frac{N_i}{\sum_G \left(\frac{N_G}{\chi_G}\right)} \tag{4.24}$$

$$q_G = N_G \frac{\chi_{eg} - \chi_G}{\chi_G} \tag{4.25}$$

where $N_G$ is the number of the atoms in the group, $n_i$ the number of identical atoms in the group with the specific electronegativity and respectively $N_i$ the total number of atoms in molecule.

To exemplify this calculation, otherwise very simple and important, we will consider the values of the atomic electronegativity given in the Pauling's Table 4.1, considering the methanol, $CH_3OH$ as working molecule:

$$\chi_{CH_3} = \frac{4}{\dfrac{1}{2.5} + \dfrac{3}{2.1}} = 2.18, \; \chi_{OH} = \frac{2}{\dfrac{1}{3.5} + \dfrac{3}{2.1}} = 2.62, \; \chi_{eg} = \frac{6}{\dfrac{4}{2.18} + \dfrac{2}{2.62}} = 2.31$$

$$q_{CH_3} = 4\frac{2.31 - 2.18}{2.18} = +0.25, \; q_{OH} = 2\frac{2.31 - 2.62}{2.62} = -0.23 \tag{4.26}$$

### 4.2.2.5 Klopman's Picture

In 1965, Klopman formulated a picture of the atomic and group electronegativity, starting from the physical correspondent of the relation energy-electronegativity for the total energy of a free atom in a particular configuration (Klopman, 1965, 1968):

$$E_A = \sum_i E_A^l + \frac{1}{2}\sum_{ij} E_A^+ \delta_{ij} + \frac{1}{2}\sum_{ij} E_A^-(1 - \delta_{ij}) \tag{4.27}$$

where $E_A^l$ is the electron $i$ energy in the spin-orbital with the azimuthally quantum number $l$, $E_A^+$, $E_A^-$ are the energies of interaction between the electrons with the same, respectively with opposite, spins orientation, and $\delta_{ij}$ is the Kronecker symbol.

It is noted that if the electrons are in the same orbital (the spins are necessarily opposites, $E_A^+ = E_A^- \equiv e_A$ then it can be rewritten:

$$E_{A,j} = n_{A,j} E_A^l + \frac{1}{2} n_{A,j} \left( n_{A,j} - 1 \right) e_A \qquad (4.28)$$

where $n_{A,j}$ represents the occupancy number of the orbital $j$ in the atom $A$.

*PRINCIPLE EN17*: For the atomic energy, expressed in terms of orbital occupation numbers, is associated the orbital electronegativity with the variation of this energy at the modification of the occupancy number of the orbital involved in the chemical bond.

$$\chi_{A,j} = -\frac{\partial E_A}{\partial n_{A,j}} \qquad (4.29)$$

Beside, a chemical bond at equilibrium involves the orbital electronegativity equalization toward the molecular energy equalization $E_M$, thus defining the molecular-orbital electronegativity:

$$\frac{\partial E_M}{\partial n_j^i} = \frac{\partial E_M}{\partial n_k^i} \qquad (4.30)$$

where $n_j^i$ is referring to the occupancy degree of the electron $i$ and the orbital $j$.

From the relation energy-electronegativity application under the form of the first order partial derivative, immediately results the expression of the orbital electronegativity such as (Klopman)

$$\chi_{A,j} = \left( -E_A^l + \frac{1}{2} e_A \right) - e_A n_{A,j} \qquad (4.31)$$

which found the linear form in charge, if the properly notations are made.

Klopman had shown that its relation can be adapted in order to include also the hybridization characteristics of the orbitals for which the electronegativity is written through the algorithm:

$$\chi_{A,j} = \chi_{A,j}^0 \left( 1 + 1.5 \sum_i q_{A,i} \right) \qquad (4.32)$$

$$\sum_i q_{A,j} = q_A = 1.5\frac{\chi_{eg}}{\chi_A} - 1 \tag{4.33}$$

$$\chi_{eg} = \frac{N_i + 1.5q_G}{\sum_G \left(\dfrac{N_G}{\chi_G}\right)} \tag{4.34}$$

$$\chi_{A,j}^0 = 1.67G_j \frac{Z_{eff}^2}{n_{eff}^2} + 0.41 \tag{4.35}$$

where $G_j$ is a linear function which shows the $p$ hybridization percentage of the orbitals $j$, $n_{eff}$ and $Z_{eff}$, respectively, are the effective principal quantum number and the effective nuclear charge, calculated by the Slater rules (see Section 3.4.1); note that the constants in last equations are in Pauling units. Note that Slater arbitrarily defined *effective* principal quantum number $n_{eff} = n*$ by the rule that the sequence of the principal quantum numbers

$$n = 1, 2, 3, 4, 5, 6 \tag{4.36a}$$

correspond to

$$n_{eff} = n* = 1, 2, 3, 3.7, 4.0, \text{ and } 4.2 \tag{4.36b}$$

respectively, such that calculated atomic energies fit to experimental data.

The presented method has the disadvantage of lacking the reference orbitals in these equations, which are valid only if each atom uses the same bonding orbitals in each bond.

However, the fact that the electronegativity image is here an orbital one, allows the expansion of the isolated atomic electronegativity relation to a form which includes in a much more realistic way the intra-atomic electronic repulsion for the valence orbitals:

$$\chi_{A,j} = \chi_{A,j}^0 \left( 1 + 0.5 \sum_{i \neq j} q_{A,j} + 1.5q_{A,j} \right) \tag{4.37}$$

relation which allows the treatment of a bond $AB$ in a group of atoms or in a molecule, according to the principle of orbital equalization and charge conservation:

$$\chi_{A,j} = \chi_{B,j} \tag{4.38}$$

$$q_{A,j} + q_{B,j} = 0 \qquad (4.39)$$

### 4.2.2.6  Hinze-Witehead-Jaffe Picture

This image develops the concept of orbital electronegativity (Hinze & Jaffe, 1962, 1963a-b; Hinze et al., 1963; Hinze, 1968):

*PRINCIPLE EN18*: For an atom involved in bonds with the hybrid orbitals which compete on the chemical bond, is associated for the bonding electronegativity the set of orbital electronegativities, defined as follows:

$$\chi_i(q_i) = \frac{\partial E}{\partial q_i}\Big|_{n_0} = -\frac{\partial E}{\partial n_i}\Big|_{n_0} \qquad (4.40)$$

where $E$ is the atom energy, $q_i$ is the negative occupancy number of the orbital $i$, $0 > q_i > 2$, $n_i$ mean the charges number from the considered orbital $i$ and $n_0$ has the sense that all the other occupied orbitals are maintained with a fixed occupancy number.

By using this physical image for an atom valence state, characterized for example by the wave function $\psi$, can be rigorously determined the orbital electronegativity working in the space Fock $N$—electronic (see Section 3.2)

$$F_N = \left(H_{(0)} \oplus H_{(1)}\right) \otimes \left(H_{(0)} \oplus H_{(1)}\right) \otimes \ldots \otimes \left(H_{(0)} \oplus H_{(1)}\right)_{of\ N\ times} \qquad (4.41)$$

factorizing on the Hilbert spaces of uniparticle and vacuum

$$H_{(1)} = \left\{\alpha|\phi\rangle \,\big|\, \alpha \in C, \langle\phi|\phi\rangle = \hat{1}\right\} \qquad (4.42)$$

$$H_{(0)} = \left\{|0\rangle \,\big|\, \langle\phi|\phi\rangle = \hat{1}\right\} \qquad (4.43)$$

There are considered the independent particles and the creation and annihilation operators associated as in Eqs. (3.4) and (3.5), defined on the vacuum states and uniparticle of the electronic space, whose completeness relation is given by Eq. (3.6), and where the operators satisfy the anticommutation relations:

$$\left\{a, a^+\right\} = \hat{1} \qquad (4.44)$$

$$\left\{a, a\right\} = \hat{0} \qquad (4.45)$$

$$\left\{a^+, a^+\right\} = \hat{0} \tag{4.46}$$

Under these conditions, the number of orbital occupancy in the state $\psi$ is given by:

$$n_{0,i} = \left\langle \psi \left| a_i^+ a_i \right| \psi \right\rangle \tag{4.47}$$

and the orbital state energy

$$E = \langle E \rangle = \left\langle \psi \left| \hat{H} \right| \psi \right\rangle \tag{4.48}$$

is given by the extreme value expected

$$\delta \left\langle \psi \left| \hat{H} \right| \psi \right\rangle = 0 \tag{4.49}$$

given that

$$\left\langle \psi \middle| \psi \right\rangle = 1 \tag{4.50}$$

It should be accentuated that the treating of the valence state requires the consideration of some constraints, similar to treating of the excited states, they not being conventional stationary states.

In this sense it will be considered an infinitesimal displacement of the Fock subspace of $N$ particles to the one of $(N-1)$ with the aid of the annihilation operator and through an infinitesimal parameter, $\lambda$.

$$\left| \psi \right\rangle \rightarrow \left(1 + \lambda a\right) \left| \psi \right\rangle \tag{4.51}$$

Worth remarking the actual Hinze difference respecting the Eqs. (3.13) and (3.14), for which it will be evaluated the expression modified by the parameter $\lambda$ in order to characterize the orbital valence state it will be evaluated:

$$\frac{\partial \langle E \rangle}{\partial n} = \frac{\partial \langle E \rangle}{\partial \lambda^2} \frac{\partial \lambda^2}{\partial n} \tag{4.52}$$

where, for the simplicity of the calculation proceeding, one temporarily renounced to the orbital indices writing.

Using these relations, immediately result:

$$n(\lambda) = \frac{\left\langle \psi \left| \left(1 + \lambda a^+\right) a^+ a \left(1 + a\lambda\right) \right| \psi \right\rangle}{\left\langle \psi \left| \left(1 + \lambda a^+\right) \left(1 + a\lambda\right) \right| \psi \right\rangle} = \frac{n_0}{1 + \lambda^2 n_0} \tag{4.53}$$

from where the concerned derivative is obtained,

$$\frac{\partial \lambda^2}{\partial n} = -\frac{1}{n^2(\lambda)} = -\left(\frac{1 + \lambda^2 n_0}{n_0}\right)^2 \qquad (4.54)$$

while for the energy variation by parametrical derivative one gets:

$$\frac{\partial \langle E \rangle}{\partial \lambda^2} = \frac{\partial}{\partial \lambda^2} \frac{\langle \psi \|(1 + \lambda a^+)\hat{H}(1 + \lambda a)\| \psi \rangle}{\langle \psi \|(1 + \lambda a^+)(1 + \lambda a)\| \psi \rangle} = \frac{\langle \psi | a^+ \hat{H} a | \psi \rangle - n_0 \langle \psi | \hat{H} | \psi \rangle}{(1 + \lambda^2 n_0)^2} \qquad (4.55)$$

resulting for Hinze electronegativity the expression:

$$-\chi = \frac{\partial \langle E \rangle}{\partial n}\bigg|_{n_0} = \frac{1}{n_0} \langle \psi | \hat{H} | \psi \rangle - \frac{1}{n_0^2} \langle \psi | a^+ \hat{H} a | \psi \rangle \qquad (4.56)$$

which gives the first form of the Hinze's orbital electronegativity. Moreover, using the commutator of the Hamiltonian with the annihilation operator, there can be written:

$$-\chi = \frac{\partial \langle E \rangle}{\partial n}\bigg|_{n_0} = \frac{1}{n_0^2} \left[ \langle \psi \|(n_0 - a^+ a)\hat{H}\| \psi \rangle + \langle \psi | a^+ (a, \hat{H}) | \psi \rangle \right] \qquad (4.57)$$

In this expression is expected that the first term annuls itself in two circumstances: if $\psi$ is a proper function of the Hamiltonian or of the operatorial product $a+a$. In these conditions it can be put in other form the orbital electronegativity, by specifying the Hamiltonian (3.29) where the coefficients of creation-annihilation operators products are respectively the orbital integrals of uni and bi-particle:

$$h_{pq} = \langle p | h | q \rangle, \ h = -\frac{1}{2}\nabla^2 - \frac{Z}{x} \qquad (4.58)$$

$$g_{pq,rs} = \left\langle p \langle r | \frac{1}{x_{12}} | s \rangle q \right\rangle \qquad (4.59)$$

With these relations helping and also using the properties of the anti-commutator of the electronic creation-annihilation operators, previously exposed, is finally obtained for the orbital electronegativity, the expression:

$$-\chi_i = \frac{\partial \langle E \rangle}{\partial n_i}\bigg|_{n_0} = \frac{1}{n_0^2} \langle \psi | a_i^+ [a_i, \hat{H}] | \psi \rangle = \frac{1}{n_0^2} \sum_q \left[ h_{iq} \gamma_{iq} + \sum_{rs} g_{iq,rs} \Gamma_{iq,rs} \right] \qquad (4.60)$$

where the reduced elements of density matrix are given, respectively by:

$$\gamma_{iq} = \langle \psi | a_i^+ a_q | \psi \rangle \tag{4.61}$$

$$\Gamma_{iq,rs} = \langle \psi | a_i^+ a_r^+ a_s a_q | \psi \rangle \tag{4.62}$$

Of course, there can be noted the fact that the orbital energy $\varepsilon_{ii}$ from the last relation is of $n_0$ times higher than the conventional one, the difference coming from the fact that in this approach the operators Fock have not been normalized to the unoccupied orbital state.

The presented method can be generalized even more, if considering the mediation spin in addition to the spatial one in the orbital occupation between 0 and 2, case in which can be analogous deduced the valence state characterization, using the displacement between the Fock subspaces of particles, given by:

$$|\psi\rangle \to \left[1 + \lambda\left(a_\alpha + a_\beta\right)\right]|\psi\rangle \tag{4.63}$$

and density operators of free spin,

$$\gamma_{pq} = \sum_\sigma a_{p\sigma}^+ a_{q\sigma} \tag{4.64}$$

$$\Gamma_{pq,rs} = \sum_{\sigma\rho} a_{p\sigma}^+ a_{r\rho}^+ a_{s\rho} a_{q\sigma} \tag{4.65}$$

Hinze and co-workers had also deduced the electronegativities corresponding to the three types of orbital occupancies, empty orbital, and single and double occupied, starting from an expression of the atomic energy analogous to its parabolic expression:

$$E(q_i) = a_i + b_i q_i + c_i q_i^2 \tag{4.66}$$

for which the orbital electronegativity becomes:

$$\chi_i(q_i) = b_i + 2c_i q_i \tag{4.67}$$

For the valence state of the considered orbital $i$, the IP and the electronic affinity will be written as:

$$IP_i = E_i(0) - E_i(-1) \tag{4.68}$$

$$EA_i = E_i(-1) - E_i(-2) \tag{4.69}$$

system from which result the coefficients that interfere in writing the orbital electronegativity:

$$b_i = \frac{3IP_i - EA_i}{2} \tag{4.70}$$

$$c_i = \frac{IP_i - EA_i}{2} \tag{4.71}$$

With these expressions, it can be immediately calculated the orbital electronegativity for different occupations (Bergmann & Hinze, 1987):

$$\chi_i(-1) = \frac{IP_i + EA_i}{2} \tag{4.72}$$

$$\chi_i(0) = \frac{3IP_i - EA_i}{2} \tag{4.73}$$

$$\chi_i(-2) = \frac{3EA_i - IP_i}{2} \tag{4.74}$$

from which only the first relation corresponds to the Mulliken electronegativity, from the Principle EN10.

In order to determine the energy transfer on the molecular formation, it is applied the principle of electronegativity equalization, EN5 Principle, such as:

$$\chi_i(q_i^0 + \Delta q) = \chi_j(q_j^0 - \Delta q) \tag{4.75}$$

and which represents the charges transfer $\Delta q$ from the orbital $i$ to the orbital $j$. Given the expression of the charge of the orbital electronegativity, results the transferred charge:

$$\Delta q = \frac{\chi_j(q_j^0) - \chi_i(q_i^0)}{2(c_j - c_i)} = \frac{\chi_j(q_j^0) - \chi_i(q_i^0)}{IP_j - EA_j + IP_i + EA_i} \tag{4.76}$$

Through this equation which the energy decreases due to the charges transfer and is given by:

$$\Delta E = \int_0^{\Delta q} \left[ \left( \frac{\partial E_i}{\partial q_i} \right)_{q_i^0 + q} - \left( \frac{\partial E_j}{\partial q_j} \right)_{q_j^0 - q} \right] dq = \int_0^{\Delta q} \left[ \chi_i(q_i^0 + q) - \chi_j(q_j^0 - q) \right] dq \tag{4.77}$$

resulting the *extra-ionic resonance energy (Pauling)*

$$\Delta E = -\frac{\left[ \chi_j(q_j^0) - \chi_i(q_i^0) \right]^2}{4(c_j + c_i)} \tag{4.78}$$

In these expressions the constant $c_i$ corresponds right to the chemical strength (in this case orbital) here defined by the presented formalism:

$$\eta_i = \frac{1}{2}\left(\frac{\partial^2 E}{\partial q_i^2}\right)_{n_0} = \frac{1}{2}\frac{\partial \chi_i}{\partial q_i} = c_i = \frac{IP_i - EA_i}{2} \tag{4.79}$$

In the situations in which the units used for the energy and potentials are the electron-volts, the measurement unit for the orbital strength is Volt per electron, i.e., the same size as the electric capacity. This analogy allows a better intuition of the physical—chemical significance of the chemical hardness in the charge transfer processes and the bonds formation.

It is natural the extension of the electronegativity concept and the equalization principle to which is subject, in forming the multiple bonds of an atom in a molecule. For this, is considered the total net charge of the original neutral atom with all the orbitals involved in a charge transfer process of $\Delta q_k$ with $k = 1, \ldots, m$:

$$Q = \sum_{k=1}^{m} \Delta q_k \tag{4.80}$$

If is focused the attention on an orbital $i$ contribution to the bonds formation of the atom in molecule, after the direct participation to the molecular bond with the charge $\Delta q_i$, it remains a rest charge that may engage in indirect transfers in bond, by transfers to the others atomic orbitals—and, from there, in the molecular bond; the rest charge will be:

$$r_i = \sum_{k \neq i} \Delta q_k = Q - \Delta q_i \tag{4.81}$$

In these circumstances, by the direct and implicit transfer, the orbital electronegativity will depend also by the rest charge:

$$\chi_i(q_i, r_i) = b_i(r_i) + 2c_i(r_i)q_i \tag{4.82}$$

In this expression the parameters depend on the rest charge on the respective orbital. But for the coefficient $c_i(r_i)$, can be considered the independence of the rest $r_i$, through the nature of constant of the atomic or orbital species of the chemical hardness with which is associated. For the coefficient $c_i(r_i)$, taking into account the specific relations previously found, it can be written:

$$b_i = IP_i + c_r \tag{4.83}$$

being able to propose the linear dependence on the rest of charge, as follows:

$$b_i(r_i) = b_i^0 + b_i^1 r_i \tag{4.84}$$

$$b_i^0 = IP_i + c_i \tag{4.85}$$

$$b_i^1 = \frac{\partial IP_i}{\partial r_i} \tag{4.86}$$

With these, the orbital electronegativity becomes:

$$\chi_i(q_i, r_i) = b_i^0 + b_i^1 r_i + 2c_i q_i \tag{4.87}$$

and the principle of electronegativity equalization, Principle EN5, involves:

$$\chi_i(q_i^0 + \Delta q_i, r_i) = \chi_j(q_j^0 - \Delta q_j, r_j) \tag{4.88}$$

For explaining the equations system, which will have as solutions the transferred orbital charges and the equalized electronegativity of the atomic orbitals in the molecular bond, is noted that:

$$\Delta q_{ij} = sign(i - j)\Delta q_i = sign(i - j)\Delta q_j \tag{4.89}$$

through which it can be written:

$$Q_A = \sum_{k \in A} sign(k - l_k)\Delta q_{[\max\{k, J_k\}, \min\{k, J_k\}] = [k, J_k]} \tag{4.90}$$

$$r_{Ai} = \sum_{\substack{k \in A \\ k \neq i}} sign(k - l_k)\Delta q_{[k, J_k]} \tag{4.91}$$

With these, the orbital electronegativities equalization becomes the proto-type equation:

$$b_i^0 + 2c_i q_i^0 + 2c_i \Delta q_{ij} + \sum_{\substack{k \neq i \\ k \in A}} sign(k - l_k)b_i^1 \Delta q_{[k, J_k]}$$

$$= b_j^0 + 2c_j q_j^0 + 2c_j \Delta q_{ij} + \sum_{\substack{k \neq j \\ k \in B}} sign(k - l_k)b_j^1 \Delta q_{[k, J_k]} \tag{4.92}$$

By recognizing the standard orbital electronegativity:

$$\chi_i^0 = \chi_i(q_i^0, r_i = 0) = b_i^0 + 2c_i q_i^0 \tag{4.93}$$

the type equations from above become:

$$2\left(c_i + c_j\right)\Delta q_{ij} + \sum_{\substack{k \neq i \\ k \in A}} sign(k - l_k) b_i^1 \Delta q_{[k,l_k]} + \sum_{\substack{k \neq j \\ k \in B}} sign(k - l_k) b_j^1 \Delta q_{[k,l_k]} = \chi_j^0 - \chi_i^C \quad (4.94)$$

which is the type of the searched equation. Adding one of such equation for each bond $i > j$ is obtained the equations set for $N$ unknown $\Delta q_{ij}$ corresponding to a given molecule with $N$ double centers located ($2N$ atomic orbitals for the molecular bonds double electronic).

With all the arguments presented, it can be concluded that the image of the electronegativity in the Hinze–Whitehead–Jaffe approach refines Mulliken image, but also brings an accentuation regarding the orbital electronegativities, once with the extension of the Principle EN5 of electronegativity equalization to an orbital level:

*PRINCIPLE EN19:* For a molecular bond in which the atoms contribute with all the orbitals, the individual orbital electronegativities decided by the Principle EN18 are equalized by the multiple chemical bond and correspond to a system of equations of orbital form.

The molecular electronegativity of equilibrium, respectively the individual transferred inter-orbital charges and the energies corresponding to the mutual transfers, immediately result by solving this kind of system.

### 4.2.2.7 Quantum Semiempirical Picture

Based on the idea of orbital electronegativity introduced by Hinze–Whitehead–Jaffe, in 1981 Ponec had introduced the orbital electronegativity based on the semiempirical approximation CNDO ("complete neglecting differential overlapping", the complete neglect of the differential-orbital overlap even for the orbitals belonging to the same atom) (Ponec, 1981).

Thus, in the limits CNDO, which essentially means to consider only the diagonal integrals, the configuration energy of an atom A is written as:

$$E\left(A, 2s^m, 2p^n\right) = mU_{ss} + nU_{pp} + C_{m+n}^2 \gamma_{AA} \quad (4.95)$$

with $U_{jj}$ the monoatomic integral representing the energy of the electron from the orbital $j$ in the field of the atomic body $A$,

$$U_{jj} = \int \psi_j^*(1)\left[-\frac{1}{2}\Delta_1 - V_A(1)\right]\psi_j(1)d\tau_1 \quad (4.96)$$

also the monoatomic repulsion integral written in the condition of local spatial invariance of the orbitals system at the central rotation on the atom A:

$$\left(s_A s_A \middle| s_A s_A\right) = \left(s_A s_A \middle| px_A \, px_A\right) = ... = \left(px_A \, px_A \middle| px_A \, px_A\right) = ...$$

$$= \left(px_A \, px_A \middle| pz_A \, pz_A\right) = \gamma_{AA} = \int\int \psi_s(1)\psi_s(1)\frac{1}{r_{12}}\psi_s(2)\psi_s(2)d\tau_1 d\tau_2 \qquad (4.97)$$

Using the expression of CNDO energy is directly calculated the ionization energy and the affinity for the electrons of the orbitals of atom $A$,

$$IP_p = E\left(A^+, 2s^m, 2p^{n-1}\right) - E\left(A, 2s^m, 2p^n\right) \qquad (4.98)$$

$$-EA_p = E\left(A^-, 2s^m, 2p^{n+1}\right) - E\left(A, 2s^m, 2p^n\right) \qquad (4.99)$$

$$IP_s = E\left(A^+, 2s^{m-1}, 2p^n\right) - E\left(A, 2s^m, 2p^n\right) \qquad (4.100)$$

$$-EA_s = E\left(A^-, 2s^{m+}, 2p^n\right) - E\left(A, 2s^m, 2p^n\right) \qquad (4.101)$$

which can be summarized as:

$$IP_j = -U_{jj} - \left(Z_A - 1\right)\gamma_{AA} \qquad (4.102)$$

$$-EA_j = U_{jj} + Z_A\gamma_{AA} \qquad (4.103)$$

where it had been noted with $ZA = m + n$, the charge of the atomic core $A$. Combining the ionization energy and the electronic affinity in a semi-sum results the expression of orbital electronegativity in CNDO version:

$$\chi_j^A = \left.\frac{IP_j + EA_j}{2}\right|_A = -U_{jj}^A - \left(Z_A - \frac{1}{2}\right)\gamma_{AA} \qquad (4.104)$$

or more general,

$$\chi_j^A = -U_{jj}^A - \left(P_A - \frac{1}{2}\right)\gamma_{AA} \qquad (4.105)$$

where

$$P_A = \sum_{j\in A} P_{jj} = \sum_{j\in A}\sum_v 2c_{jv}c_{jv} \qquad (4.106)$$

represent the total electronic population (density) of the atom $A$ obtained by summing the electronic populations (densities) from all the atom orbitals, without considering the overall populations (CNDO), and the coefficients of $c_{jv}$ are the coefficients of LCAO development ("linear combination atomic orbitals") for the molecular orbitals writing

$$\Psi_v = \sum_{j \in A} c_{jv} \psi_j,$$ (4.107)

Thus, the global electronegativity of the atom $A$ in molecule can be evaluated as:

$$\chi_A = \frac{\sum p_{jj} \chi_j^A}{\sum p_{jj}}$$ (4.108)

where $p_{jj}$ represents the charge density of the orbital $j$ in the atom $A$.

This approach allows numerical evaluations of the orbital electronegativity so by various approximations of the integrals used, as by a combination of the experimental data, for example, with the information which came from the X-ray analysis of the electronic distribution in molecule.

Another method, essentially empirical, was proposed by Jørgensen in 1970, in direct correlation with the electrons transfer from the spectra of transition metal complexes, MX. Thus, he had established the relationship between the electronegativity difference (called optical) associated to the ligand (X) and respectively to the metal (M) and the photon energy transferred for the the electrons transfer from the metal-ligand system in the first Laporte band (Ponec, 1981; Mullay, 1987)

$$h\upsilon = \left[ \chi_{op}(X) - \chi_{op}(M) \right] \cdot 3 \times 10^4 \, cm^{-1}$$ (4.109)

While chaning the the constant, $3 \times 10^4 \, cm^{-1} \sim 3.7 \, eV$, the optical electronegativity of the halogens in Pauling units is obtained. The linearity relationship between the optical electronegativity and the difference of the proper energies of the analyzed systems can be rationalized in a theoretical approach, one of this possible approaches being the density functional theory.

### 4.2.3 QUANTUM BOHMIAN CHALLENGES FOR PARABOLIC CHEMICAL REACTIVITY

#### 4.2.3.1 Subquantum Electronegativity?

The specific measure of chemical reactivity, electronegativity ($\chi$), which lacks a definite quantum operator but retains an observable character through its formal identity with the macroscopic chemical potential $\chi = -\mu$ (Parr et al., 1978; Parr & Yang, 1989), was tasked with carrying quantum information within the entanglement environment of Bohmian mechanics (Bohm, 1952; Bohm & Vigier, 1954; Cushing, 1994) and has thus far been identified with the square root of the so-called quantum potential (Boeyens & Levendis, 2008)

$$\chi = V_Q^{1/2} \qquad (4.110)$$

In next we explicate such possibility with the actual refining. Actually, since of the need to reduce Copenhagen's indeterminacy for quantum phenomena, i.e., by associating it the quantum description of "Newtonian" forms of motion, though by preserving probability densities, quantum averages, etc., the so-called "minimalist" quantum theory may be formulated following the Bohm quantum mechanical program as follows (Putz, 2012a).

One begins with the general eikonal wave-function form (Bohm, 1952)

$$\psi(r,t) = R(r)\exp\left(\frac{i}{\hbar}S(r,t)\right) \qquad (4.111)$$

which represents the mid-way between wave and particle mechanics because it contains both information regarding Hamilton-Jacobi theory and the Wentzel-Kramers-Brillouin (WKB) approximation, see the Volume I of the present five-volume set, through the principal phase function $S(r, t)$ while preserving the amplitude relationship with the systems' quantum density:

$$\rho(r) = \psi^2(r) = R^2(r) \qquad (4.112)$$

In this framework, the Schrödinger equation,

$$i\hbar\frac{\partial}{\partial t}\psi(r,t) = -\frac{\hbar^2}{2m}\nabla_r^2\psi(r,t) + V(r)\psi(r,t) \qquad (4.113)$$

decomposes into real and imaginary parts, The real part can be expressed as follows:

$$\frac{\partial S(r,t)}{\partial t} + \frac{(\nabla_r S(r,t))^2}{2m} - \frac{\hbar^2}{2m}\frac{\nabla_r^2 R(r)}{R(r)} + V(r) = 0 \qquad (4.114)$$

representing a continuous "fluid" of particles driven by the "guidance" momentum:

$$mv = p = \nabla_r S(r,t) \qquad (4.115)$$

moving under a joint external potential $V(r)$ as well as under the so-called quantum potential influence:

$$V_\varrho(r) = -\frac{\hbar^2}{2m}\frac{\nabla_r^2 R(r)}{R(r)}$$

(4.116)

The consequences are nevertheless huge. For example, this methodology allows for the interpretation of the trajectories orthogonal to constant surfaces, by canceling the Laplacian of the wave fronts

$$\nabla_r^2 S(r,t) = 0$$

(4.117a)

which are obtained from Eqs. (4.114) and (4.115) as the quantum equation of motion:

$$\frac{\partial p}{\partial t} = -\nabla_r\left[V_\varrho(r) + V(r)\right]$$

(4.117b)

Equation (4.117) resembles the classical Newtonian acceleration-force relationship only in a formal way; in fact, it generalizes it: it prescribes acceleration motion even in the absence of an external classical potential. This is essential in explaining why the inter-quark forces increase with the increase in inter-quark distances, no matter how great a separation is considered (a specific quantum effect), due to the presence of a quantum potential that does not fall off with distance as $V$ does. It also nicely explains the observed interference patterns in double-slit experiments in the absence of classical forces. Alike, Eq. (4.117) also appears suited for modeling chemical reactivity for the *valence atoms as free particles* in a virtually infinite potential environment to characterize their reactive behavior. In this regard, it is worth considering for such atoms the uniform motion by having $\partial p / \partial t = 0$ through the time-constant associated wavefront condition and action $S(r = cnst., t) = cnst$ (equivalent with Lagrangian constancy), in all given chemical space-points (atomic basins within molecule complex) (Guantes et al., 2004). This picture is also equivalently to have

$$\frac{\partial S(r,t)}{\partial t} = 0$$

(4.118)

applied to Eq. (4.114). By doing so, one obtains

$$\frac{(\nabla_r S(r,t))^2}{2m} = -\left[-\frac{\hbar^2}{2m}\frac{\nabla_r^2 R(r)}{R(r)} + V(r)\right]$$

(4.119)

which can be rearranged as follows:

$$T = -V_Q - V(r) \tag{4.120}$$

such that the total energy of a the valence system is now entirely driven by the quantum potential:

$$E_Q = T + V(r) = -V_Q \tag{4.121}$$

At this point, one can see that when turning to electronegativity and combining Eq. (4.121) with DFT definition (3.1), one obtains *a generalization* of the previous Boeyens formulation (Boeyens & Levendis, 2008):

$$\chi_{Q\text{-}DFT} = -\left(\frac{\partial V_Q}{\partial N}\right)_{V(r)} \tag{4.122}$$

which is the variation in the quantum potential with electron exchange under a constant classical or external potential.

However, for a quantum characterization of the valence state, we are interested in how the energy described by Eq. (4.121) varies under a quantum potential (4.116)

$$E_Q(a.u.) = \frac{\nabla_r^2 R(r)}{2R(r)} = \frac{\nabla_r^2 \rho^{1/2}(r)}{2\rho^{1/2}(r)}$$

$$= \frac{1}{4}\frac{\nabla_r^2 \rho(r)}{\rho(r)} - \frac{1}{8}\frac{[\nabla_r \rho(r)]^2}{\rho^2(r)} \tag{4.123}$$

when the above relations (4.112) and (4.116) are substituted into Eq. (4.121).

It is worth noting that although we obtained the total energy (4.123) in the Bohmian mechanics context, it showcases a clear electronic density dependency, not under a density functional (as DFT would require) but merely as a spatial function, which is a direct reflection of the entanglement behavior of Bohmian theory through the involvement of a quantum potential. However, in most cases, and especially for atomic systems, Eq. (4.123) will yield numerical values under custom density function realizations.

## 4.2.3.2   Physical or Chemical Atom?

The striking difference between an atom as a physical entity, with an equal number of electrons and protons (thus in equilibrium), and the same atom as a chemical object, with incomplete occupancy in its periphery quantum shells (thus attaining equilibrium by changing accepting or releasing electrons), is closely related to the electronegativity phenomenology in modeling chemical reactivity. Moreover, this difference triggers perhaps the most important debate in conceptual chemistry: the ground vs. valence state definition of an atom (Putz, 2012a).

The difficulty may be immediately revealed by considering the variation in the total energy (of the ground and/or valence state—see below for an explanation of their difference) around the physical equilibrium (neutral atom) attained between the release (by ionization, $I$) and receipt (through affinity, $A$) of electrons toward chemical equilibrium (in molecules, chemical bonding). Accordingly, the curve passing through these points apparently only behaves as shown in Figure 4.3(a), while in all systems (with numerical $I$ and $A$), the obtained interpolating curve presents a minimum toward accepting electrons, see Figure 4.3(b), thus confirming the electronegativity concept as a chemical reality, although with a predicted fractional charge (for example, the critical charge $N^*$) on an atom at chemical equilibrium (i.e., not reducible/comprehensible to/by an ordinary physical description of atoms) (Putz, 2012a).

However, the physical-to-chemical paradox continues in an even more exciting fashion as follows. When, in light of the above discussion, electronegativity is recognized with the two-point limits shown in Figure 4.3(b), namely (Parr & Yang, 1989; von Szentpály, 2000)

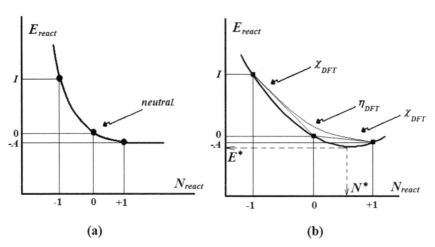

(a)                              (b)

**FIGURE 4.3**    The two energy curves (thick lines) for the quantum atom in: (a) the apparent or *reactive ground state* and (b) the shifted or *critical ground state* (Putz, 2012a).

$$\chi = \begin{cases} IP & ...N_{react} \in [-1,0) \\ EA & ...N_{react} \in (0,+1] \end{cases} \qquad (4.124)$$

the limits represent tangents to a curve that does not describe chemical equilibrium but an excited state driven by the parabolic form, since combining Eqs. (3.1) and (3.3) in a single energy expansion around equilibrium charged system

$$E_{DFT} = -\chi N + \eta N^2 \qquad (4.125)$$

which happens to correspond to the celebrated density functional theory (DFT) working energy expression (Parr & Yang, 1989; Ayers & Parr, 2000; Geerlings et al., 2003; Putz, 2003) written in terms of electronegativity and chemical hardness, respectively (Parr, 1983; Parr & Pearson, 1983; Putz, 2006, 2008a).

The point is that curve (4.125) is not chemically minimized, although it is very often assumed to be in the DFT invoked by the chemical reactivity literature (Chattaraj & Parr, 1993; Chattaraj & Sengupta, 1996; Chattaraj & Maiti, 2003; Chattaraj & Duley, 2010); however, the curve cannot be considered indicative of a sort of ground state (neither reactive nor critical states of Figure 4.3). Additionally, by comparing the curves of Figures 4.3(a) and 4.3(b), the curve of Eq. (4.125) occurs above both the reactive and critical curves of Figure 4.1; it thus should represent the *chemical valence state* with which to operate. Therefore, much caution should be taken when working with Eq. (4.125) in assessing the properties of atoms, molecules, atoms in molecules, etc. Nevertheless, this is another case of chemistry not being reducible to physics and should be treated accordingly. It is worth noting that Parr, the "father" of Eq. (4.125) and a true pioneer of conceptual DFT (Ayres & Parr, 2001; Kohn et al., 1996), had tried to solve this dichotomy by taking the "valence as the ground state of an atom in a perturbed environment." This statement is not entirely valid because perturbation is not variation such that it may be corrected by applying the variational principle to Eq. (4.125), for example. In fact, using such variation should be considered a double variational technique that is necessary to arrive at the celebrated chemical reactivity principles of electronegativity and chemical hardness, as recently shown (Putz, 2011b).

The current line of work takes a step forward by employing the double variation of the parabolic energy curve of type (4.125) to provide the *quantum* (DFT) *valence charge* of an atom (say, $N^{**}$) and to compare it either quantitatively and qualitatively with the chemical critical charge $N^*$. The goal of these efforts is to gain new insight into the valence state and chemical reactivity at the quantum level. To this end, the relation of Bohmian mechanics to the concept of the golden ratio will be essential and will be introduced in the following.

The consequences of the joint consideration of Bohmian mechanics and the golden ratio for the main atomic systems will be explored, and the *quantum chemical valence state* will be accordingly described alongside the so-called universal electronegativity and chemical hardness, refining the work of (Parr & Bartolotti, 1982) as well as generalizing the previous Bohmian-Boeyens approach (Boeyens, 2005, 2011).

However, for practical implementation, density is considered a "goldmine" in current computational and conceptual quantum chemistry due to its link with observable quantities, energy density functionals in particular, as celebrated by DFT (Hohenberg & Kohn, 1964; Putz, 2008b). However, to quantitatively approach the chemical phenomenology presented in Figure 4.3, involving the ionization-to-affinity atomic description, the general Slater (Parr & Bartolotti, 1982) density (involving the orbital parameter $\xi$ dependency) will be here employed for the first trial on modeling the combined Bohmian and gold-ratio features of valence atom; it assumes the general (trough still crude) working form:

$$\rho(r,\xi) = \rho_0 \exp[-2\xi r] \qquad (4.126)$$

For the reactivity at the valence atomic level, or for some outer shell ($n$) considered at the atomic frontier, one may assume almost electronic free motion or at least electronic motion under almost vanishing nuclear potential $V(r)$; this way the density (4.126), while entering the quantum potential (4.116) recovers the negative kinetic energy by the virial identity (4.120). Analytically, since Eqs. (4.112), (4.116) and (4.126), one has

$$\nabla_r^2 \rho^{1/2} = \xi^2 \rho^{1/2} \qquad (4.127)$$

and the actual valence atomic virial realization looks like (Putz, 2012a)

$$V_Q(r) = -\frac{\hbar^2 \xi^2}{2m} \ldots = -T = -\frac{p^2}{2m} \qquad (4.128)$$

In the conditions of circular orbits, Eq. (4.128) leaves with the identity:

$$\hbar\xi = p \qquad (4.129)$$

which may be seen as an orbital effective realization of the Heisenberg observability too (Putz, 2010a); It may be also further rewritten with the help of the atomic Bohr-de Broglie (2.2) relationship (see Section 2.2) to provide the atomic frontier radii shell-dependency

$$r_{frontier} = \frac{n}{\xi} \tag{4.130}$$

Remarkably, the same result is obtained when employing a far more reach atomic shell structure description, namely when starting with the full atomic radial Schrödinger density (see also the similar forms of Sections 4.7.1 and 4.7.4) (Putz, 2006)

$$\rho_n(r,\xi) = 4\pi r^2 \frac{(2\xi)^{2n+1}}{(2n)!} r^{2n-2} \exp[-2\xi r] \tag{4.131}$$

and imposing the null-gradient condition (Ghosh & Biswas, 2002),

$$\nabla_r \rho_n(r,\xi) = 0 \tag{4.132}$$

in accordance with the celebrated Bader condition of electronic flux of atoms-in-molecules (Bader, 1990), to yield:

$$r_{max} = \frac{n}{\xi} \tag{4.133}$$

The identity between Eqs. (4.130) and (4.133) gives sufficient support to the present Slater density approach Eq. (4.126) in modeling the valence atoms or the atoms at their frontiers approaching reactivity (i.e., atoms-in-molecules complexes by chemical reactions).

Once convinced by the usefulness of the Slater density form (4.131) for the present valence atomic analysis, one will next employ it under the so-called Parr-Bartolotti form (Parr & Bartolotti, 1982)

$$\rho(r,\xi) = N \frac{\xi^3}{\pi} \exp[-2\xi r] \tag{4.134}$$

such that to obey the $N$-normalization condition, as required by DFT (Parr & Yang, 1989, Putz, 2011a),

$$\int_0^\infty 4\pi \, r^2 \rho(r,\xi) dr = N \tag{4.135}$$

by applying the Slater integral recipe

$$I_k(m) = \int_0^\infty r^k e^{-mr} dr$$

$$= \int_0^\infty \frac{d^k(e^{-mr})}{d(-m)^k} dr = (-1)^k \frac{d^k}{dm^k} \int_0^\infty e^{-mr} dr = (-1)^k \frac{d^k}{dm^k} \left[ -\frac{1}{m} e^{-mr} \right]_0^\infty$$

$$= (-1)^k \frac{d^k}{dm^k}(m^{-1}) = (-1)^k(-1)^k k! m^{-1-k} = \frac{k!}{m^{k+1}} \qquad (4.136)$$

It nevertheless showcases the parametric—dependency that can be smeared out by considering the variational procedure

$$\frac{\partial E[\xi]}{\partial \xi} = 0 \qquad (4.137)$$

upon applying the total atomic energy

$$E[\xi] = T[\xi] + V_{ee}[\xi] + V_{ne}[\xi] \qquad (4.138)$$

where the components are individually evaluated within a radial atomic framework with the respective results for (Parr, 1972; Putz, 2003):

- kinetic energy

$$T[\xi] = \int_0^\infty 4\pi r^2 \left\{ -\frac{1}{2}\left[ \frac{1}{r^2} \frac{\partial}{\partial r}\left( r^2 \frac{\partial}{\partial r} \right) \right] + \frac{\xi}{2r} \right\} \rho(r,\xi) dr = N\frac{\xi^2}{2} \qquad (4.139)$$

- nucleus-electronic interaction

$$V_{ne}[\xi] = -\int_0^\infty 4\pi \, r^2 \frac{\rho(r,\xi)}{r} dr = -N\xi \qquad (4.140)$$

- inter-electronic interaction

For the inter-electronic interaction, see Figure 4.4; in evaluating $V_{ee}[\xi]$ the two-electronic density is approximated by the Coulombic two mono-electronic density product, thus neglecting the second-order density matrix effects associated with the exchange-correlation density. However, for the analytical evaluation of the electron–electron repulsion energy using the density (4.134), much care must be taken. For instance, one has to use the electrostatic Gauss theorem, which states that the classical electrostatic potential outside a uniform spherical shell of charge is just what it would be if that charge were localized at the center of the shell and that the potential everywhere inside such a shell is that at the surface (Parr, 1972; Putz, 2003), see Figure 4.4. Therefore, the electronic repulsion energy becomes

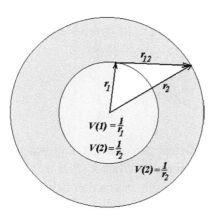

**FIGURE 4.4**   Representation of the space regions of the 1st and 2nd electrons, their potential influences and reciprocal interaction (Parr, 1972; Putz, 2003).

$$V_{ee}[\xi] = \frac{N-1}{2N} \iint \frac{\rho(1)\rho(2)}{r_{12}} dv(1)dv(2)$$

$$= \frac{N-1}{2N} \int_0^\infty 4\pi r_1^2 N \frac{\xi^3}{\pi} \exp(-2\xi r_1)dr_1 \left[ \int \frac{\rho(2)}{r_{12}} dv(2) \right]$$

$$= \frac{N-1}{2N} \int_0^\infty 4 r_1^2 N \xi^3 \exp(-2\xi r_1)dr_1 \left\{ 4\pi N \frac{\xi^3}{\pi} \left[ \begin{array}{c} \int_0^{r_1} \frac{r_2^2 \exp(-2\xi r_2)}{r_1} dr_2 \\ + \int_{r_1}^\infty \frac{r_2^2 \exp(-2\xi r_2)}{r_2} dr_2 \end{array} \right] \right\}$$

$$= \frac{N-1}{2N} N^2 16\xi^6 \int_0^\infty r_1^2 \exp(-2\xi r_1)dr_1 \left\{ \left[ \begin{array}{c} \int_{r_2 \to r_1}^{r_2 \to \infty} \left( \frac{1}{r_1} \equiv \frac{1}{r_2} \right) r_2^2 \exp(-2\xi r_2)dr_2 \\ + \int_{r_1}^\infty r_2 \exp(-2\xi r_2)dr_2 \end{array} \right] \right\}$$

$$= \frac{N-1}{2N} N^2 32\xi^6 \int_0^\infty r_1^2 \exp(-2\xi r_1)dr_1 \left[ \int_{r_1}^\infty r_2 \exp(-2\xi r_2)dr_2 \right]$$

$$= \frac{N-1}{2N} N^2 \frac{32\xi^6}{(2\xi)^5} \int_0^\infty (2\xi r_1)^2 \exp(-2\xi r_1)d(2\xi r_1)$$

$$\times \left[ \int_{2\xi r_1}^\infty 2\xi r_2 \exp(-2\xi r_2)d(2\xi r_2) \right]$$

$$\equiv \frac{N-1}{2N} N^2 \xi \int_0^\infty s^2 \exp(-s) \left[ \int_s^\infty t \exp(-t)dt \right] ds$$

$$= \frac{N-1}{2N} N^2 \xi \int_0^\infty s^2 \exp(-s) \left[(1+s)\exp(-s)\right] ds$$

$$= \frac{N-1}{2N} N^2 \xi \left( \frac{2!}{2^3} + \frac{3!}{2^4} \right)$$

$$= (N^2 - N) \frac{5}{16} \xi \tag{4.141}$$

Note that the electron-electron repulsion term was written by also consid-ering the Fermi-Amaldi $(N-1)/N$ factor (Parr & Yang, 1989; Putz, 2003), which ensures the correct self-interaction behavior: when only one elec-tron is considered, the self-interaction energy must be zero, $V_{ee}(N \to 1) \to 0$.

With these results, the optimum atomic parameter is quantified by the electronic number as follows:

$$\xi = \frac{21 - 5N}{16} \tag{4.142}$$

which immediately releases the working electronic density

$$\rho^0(r, \xi) = \frac{N}{\pi} \left( \frac{21 - 5N}{16} \right)^3 \exp\left[ -\frac{21 - 5N}{16} 2r \right] \tag{4.143}$$

Having the completely analytical density in terms of number of reactive electrons as in Eq. (4.143), worth pointing here on the so-called sign prob-lem relating with its variation, e.g., its gradient, the gradient of its square root, etc.. Although this problem usually arises in density functional the-ory when specific energy functionals are considered in gradient forms, see for instance (Cohen et al., 2012), there is quite instructive discussing the present behavior and its consequences.

For instance, one can adapt either Eqs. (4.130) or (4.133) through con-sidering the present form (4.142) for the orbital exponent to be (Putz, 2012a)

$$\frac{r}{n} = \frac{16}{21 - 5N_{bonding}} \tag{4.144}$$

Here, one combines the frontier and maximum atomic radii with atoms-in-molecules phenomenology, as above indicated, to arrive to the present identification for the number of valence electrons possible to be involved in the same chemical bonding state as being $N_{bonding}$ in Eq. (4.144).

Accordingly, the Figure 4.5 reveals interesting features of the present Slater-Parr-Bartolotti atomic density with quantum potential (Putz, 2012a):

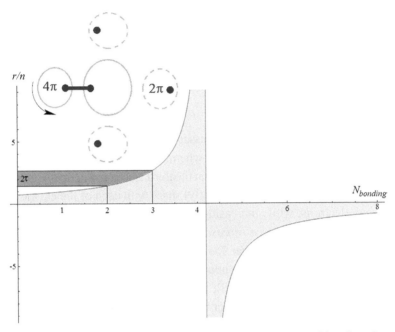

**FIGURE 4.5**    Representation of the bonding length as a function of bonding electrons from valence atoms in molecule(s), based on Eq. (4.144), while marking the double golden ratio $2\tau$ gap between the bonding lengths of the second and third order, as well as the forbidden chemical bonding region for $N_{bonding} \geq 21/5$ for the electrons participating in the same bonding. Further connection of chemical bonding and the 4D space to model it is suggested by the inset picture illustrating the 2-fold ($4\pi$) spinning symmetry of the adduct atom respecting the bonding direction (Putz, 2012a).

- the fact that the (covalent) bond length is proportional to the atomic radii and in inverse correlation with bonding order is well known (Petrucci et al., 2007), and this it is also nicely reflected in Eq. (4.144); however, changing the sign to negative radii as surpassing the threshold 21/5 and fixing in fact the limit $N_{bonding} = 4$, is consistent with maximum bond order met in Nature; it is also not surprising this self-released limit connects with golden ratio by the golden-spiral optimization of bond-order (Boeyens & Levendis, 2012); more subtle, it connects also with the $4\pi$ symmetry of two spherical valence atoms making a chemical bond (Figure 4.5, inset): such "spinning" reminds of the graviton symmetry (Hawking, 2001) (the highest spherical symmetry in Nature, with spin equal 2) and justifies the recent treatments of chemical bonding by means of the quasi-particles known as bondons (Putz, 2010b; Putz & Ori, 2012), as well as the use of the 4D complex projective geometry in modeling the chemical space as a non-Euclidian one, eventually with a time-space metrics including specific "gravitational effects" describing the bonding (Boeyens & Levendis, 2012);

- the "gap" between the atomic systems contributing 2 to 3 electrons to produce chemical bond is about double of the golden ratio

$$(r/n)_{N_{bonding}=3} - (r/n)_{N_{bonding}=2} = 2\tau \tag{4.145}$$

therefore, this gap marks the passage from the space occupied by a pair of electrons and that required when the third electron is added on the same bonding state: it means that the third electron practically needs one golden measure ($\tau$) to (covalently) share with each of the existing pairing electrons, while increasing the bond order to the maximum three; it is therefore a *space representation of the Pauli exclusion principle* itself, an idea also earlier found in relation with dimensionless representation of a diatomic bonding energy ($2\tau$) at its equilibrium bonding distance ($\tau$)(Boeyens, 2010); when the fourth electron is coming into the previous system, in order the maximum fourth order of bonding to be reach the chemical bonding space is inflating about five times more, yet forbidding further forced incoming electrons into the same space of bonding state as the bonding radius becomes negative in sign.

Having revealed the chemical bonding information carried by the density (4.143) when considered for combined valence atoms-in-molecules, it is next employed on energetically describing the atomic reactivity as a propensity for allowing electronic exchanging and bonding. As such, it leaves the total quantum (Bohmian) energy in Eq. (4.123) with the compact form (Putz, 2012a)

$$E_{Q1}(a.u.) = \frac{25}{512} N^2 - \frac{105}{256} N + \frac{441}{512} = \frac{1}{512}(21 - 5N)^2 \tag{4.146}$$

Note that the actual working total energy is not that obtained by replacing the density (4.143) in Eqs. (4.139)–(4.141) and then in total energy (4.138) because here the double-variational procedure was considered; that is, the first optimization condition was considered as in Eq. (4.137), and the resulting (optimum) density (4.143) was then employed in the quantum energy (4.123), which in turn was obtained by applying the variational Eq. (4.118) to the perceived phase transition in the Bohm eikonal wave-function (4.111). To emphasize the accuracy of Eq. (4.123) over that of Eq. (4.138) with density (4.143), when one considers the last case, Eq. (4.138) yields the following non-quadratic form for energy (Putz, 2012a):

$$E_1(a.u.) = -\frac{25}{512}N^3 + \frac{105}{256}N^2 - \frac{441}{512}N$$

$$= -\frac{N}{512}(21 - 5N)^2 \tag{4.147}$$

which is not appropriate for describing the valence state of an atom, as Eq. (4.125) prescribes, despite being similar in form to the Bohmian-based result of Eq. (4.146). Thus, the previous limitation of the Parr-Bartolotti conclusion (Parr & Bartolotti, 1982) and the paradox raised in describing the valence (parabolic) state with the optimized atomic density (4.143) are here solved by the double (or the orthogonal) variational implementation, as recently proved to be customary for chemical spaces (Putz, 2011b). In the light of this remark one may explain also the sign difference between the "physical" energy (4.147) and that obtained for the "chemical" situation (4.146): through simple variational procedure for "physical" energy (4.138) the result (4.147) is inherently negative—modeling systems stability in agreement with the upper branch of Eq. (2.9), whereas the double variational algorithm employing optimized density (4.143) into the Bohmian shaped energy (4.123) it produces the positive output (4.146) associated with activation energy characteristic for chemical reactivity corresponding to the lower branch of Eq. (2.9).

Therefore, to be accurate, one should consider the quantum potential related optimized energy (4.146) instead of simply the orbital optimized one of Eq. (4.147). Therefore, assuming that Eq. (4.146) appropriately describes the atomic valence state in DFT (see the upper/reactive curve in Figure 4.1b), the next task is to search for the quantum valence charge for which the valence energy approaches its optimum value (or the "ground state" of the atomic chemical-reactivity, i.e., the previously golden-ratio quantification of the valence atomic state); to this aim, at this point, one can employ the golden ratio relationship (2.7) and first rewrite Eq. (4.146) as (Putz, 2012a)

$$E_{Q1}(a.u.) = \frac{5^2}{512}(6.80\tau - N)^2 \tag{4.148}$$

which is minimized at the value

$$N = 6.80\tau \tag{4.149}$$

However, one must again apply the double-variational procedure, now in terms of number of electrons, i.e., reconsidering Eq. (4.149) with the golden ratio at the reactive (chemical) electronic level of Eq. (2.9) such that a second equation is formed

$$N = 6.80 \frac{Z}{N} \tag{4.150}$$

with the positive solution (Putz, 2012a)

$$N_{REACT} \equiv N^{**} = 2.60768 \sqrt{Z} \tag{4.151}$$

This expression avails of the significance of the maximum number of electrons, for a given atom, possibly engaged in a reactive environment by either (or both) accepting or (and) ceding electrons to or from its valence state, see Table 4.2.

The result of this process is different from the expected physical result ($N_{STABIL} = Z$) according to the upper branch of Eq. (2.9), which is higher than the physical one until reaching the carbon system ($Z_{INTERCHANGE} = 6.8$), while continuing below it thereafter (see Figure 4.6).

The above interchange (effective) atomic number through which the chemical (reactive) state is associated with lower charge respecting the physical state may be also be found at the energetic level based on quantum equation (4.143), as specialized for the two branches of Figure 4.6 for the $N(Z)$ dependence. Thus, the chemical (reactive) state takes the analytic form (Putz, 2012a)

$$E_{Q1}(N_{REACT} \rightarrow 2.60768\sqrt{Z})$$
$$= 0.861328 - 1.06956\sqrt{Z} + 0.332031 Z \tag{4.152}$$

and interchanges with the ground state $E_{Q1}(N_{STABLE} \rightarrow Z)$ at the points {3.5, 6.8}, as observed also from Figure 4.7; however, the interchanging point beyond which all chemical atomic systems are more stable in the chemical or reactive state than in the physical ground state is consistently recovered.

Nevertheless, the energetic analysis also reveals the atomic systems Be, B and C to be situated over the corresponding physical stable states; this may explain why boron and carbon present special chemical phenomenology (e.g., triple electronic bonds and nanosystems with long C-bindings, respectively), which is not entirely explained by ordinary physical atomic paradigms (March, 1991; Wentorf, 1965; Eremets et al., 2001; van Setten et al., 2007; Widom & Mihalkovic, 2008; Putz, 2011c).

**TABLE 4.2**    Synopsis of the Critical Charges in the Physical Ground State (N*) As Well As for Chemical Reactive (Valence) State (N**) for Atoms of the First Four Periods of the Periodic Table of Elements, As Computed From the Minimum Point of Associated Interpolations of Ionization Potential (IP) and Electronic Affinities (EA) (Parr & Bartolotti, 1982) and of Eq. (4.151), respectively (Putz, 2012a)

| Atom | Z | IP [eV] | EA [eV] | N* | N** |
|------|---|---------|---------|------|------|
| H  | 1  | 13.595 | 0.7542 | 0.558735 | 0.607681 |
| Li | 3  | 5.390  | 0.620  | 0.629979 | 0.516636 |
| B  | 5  | 8.296  | 0.278  | 0.534672 | 0.830952 |
| C  | 6  | 11.256 | 1.268  | 0.626952 | 0.387488 |
| O  | 8  | 13.614 | 1.462  | 0.620309 | 0.375636 |
| F  | 9  | 17.42  | 3.399  | 0.742422 | 0.823043 |
| Na | 11 | 5.138  | 0.546  | 0.618902 | 0.648699 |
| Al | 13 | 5.984  | 0.442  | 0.579755 | 0.402127 |
| Si | 14 | 8.149  | 1.385  | 0.70476  | 0.757049 |
| P  | 15 | 10.484 | 0.7464 | 0.576651 | 0.0995049 |
| S  | 16 | 10.357 | 2.0772 | 0.750876 | 0.430724 |
| Cl | 17 | 13.01  | 3.615  | 0.884779 | 0.751744 |
| K  | 19 | 4.339  | 0.5012 | 0.630596 | 0.366618 |
| V  | 23 | 6.74   | 0.526  | 0.584648 | 0.505999 |
| Cr | 24 | 6.763  | 0.667  | 0.609416 | 0.774976 |
| Fe | 26 | 7.90   | 0.164  | 0.5212   | 0.296616 |
| Co | 27 | 7.86   | 0.662  | 0.59197  | 0.549908 |
| Ni | 28 | 7.633  | 1.157  | 0.67866  | 0.798551 |
| Cu | 29 | 7.724  | 1.226  | 0.688673 | 0.0427917 |
| Se | 34 | 9.75   | 2.0206 | 0.761417 | 0.205262 |
| Br | 35 | 11.84  | 3.364  | 0.896885 | 0.427249 |
| Rb | 37 | 4.176  | 0.4860 | 0.631707 | 0.861904 |
| Zr | 40 | 6.84   | 0.427  | 0.566584 | 0.492423 |
| Nb | 41 | 6.88   | 0.894  | 0.649348 | 0.697305 |
| Mo | 42 | 7.10   | 0.747  | 0.617582 | 0.899704 |
| Rh | 45 | 7.46   | 1.138  | 0.680006 | 0.492856 |
| Pd | 46 | 8.33   | 0.558  | 0.571796 | 0.686153 |
| Ag | 47 | 7.574  | 1.303  | 0.707782 | 0.87736 |
| Sn | 50 | 7.342  | 1.25   | 0.705187 | 0.439089 |
| Sb | 51 | 8.639  | 1.05   | 0.638358 | 0.622567 |
| Te | 52 | 9.01   | 1.9708 | 0.779975 | 0.804255 |
| I  | 53 | 10.454 | 3.061  | 0.91404  | 0.984204 |

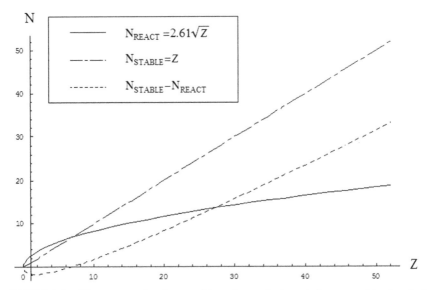

**FIGURE 4.6**    The comparative shapes of the valence electrons to be engaged in chemical reactivity (continuous curve) computed using Eq. (4.151) based on the combined optimal Bohm total energy (4.148) with the golden ratio imbalance of Eq. (2.9), respecting the stable physical case (dot-dashed curve), and of their differences (dashed curve); all originate at the 0th atom (the neutron, $Z=0$) (Putz, 2012a).

The energetic discourse may be complete with the electronegativity and chemical hardness evaluations by applying the DFT definitions (3.1) and (3.3) to physical and chemical energies, respectively. In the first case, expression (4.147) is applied to provide the following so-called "universal" forms of (Parr & Bartolotti, 1982):

$$\chi_{PB1} = -\left(\frac{\partial E_1}{\partial N}\right)_{N=1} = \frac{3}{16}(a.u.) = 5.1(eV) \qquad (4.153)$$

$$\eta_{PB1} = \frac{1}{2}\left(\frac{\partial^2 E_1}{\partial N^2}\right)_{N=1} = \frac{135}{512}(a.u.) = 7.17(eV) \qquad (4.154)$$

The result, nevertheless, appears to be an unusually higher increase in chemical hardness than in electronegativity, which certainly cannot be used to model a reactive-engaged tendency because it is more stable (by chemical hardness) than reactive (by electronegativity); it is, however, consistent with the physical stability of the system, provided by the single variational procedure through which Eq. (4.147) was produced.

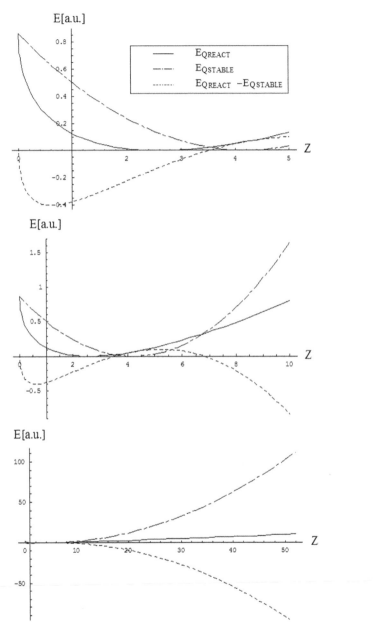

**FIGURE 4.7**    The same comparative shapes shown in Figure 4.6, here at the level of energy (4.146) specialized for the reactive and the stable $N(Z)$ dependencies of Figure 4.6; the various plots successively display increasingly large atomic Z-ranges to better emphasize the chemical vs. physical behavior (see text) (Putz, 2012a).

Instead, to chemically model reactivity, the double variation procedure is applied and Eq. (4.146) is substituted into Eqs. (3.1) and (3.3), though by considering also the double reactive procedure for charge as well, i.e., by considering Eq. (4.151) with the golden ratio information of (2.9) to respectively yield the results (Putz, 2012a),

$$\chi_{PB2} = -\left(\left(\left(\frac{\partial E_{Q1}}{\partial N}\right)_{N=2.60768\sqrt{Z}}\right)_{Z=\pi N}\right)_{N=1} \tag{4.155}$$
$$= 0.209963\,(a.u.) = 5.713\,(eV)$$

$$\eta_{PB2} = \frac{1}{2}\left(\frac{\partial^2 E_{Q1}}{\partial N^2}\right) = \frac{25}{512}(a.u.) = 1.3286\,(eV) \tag{4.156}$$

Remarkably, the actual electronegativity of Eq. (4.155) obtained by the quantum Bohm and golden ratio double procedure yields sensible results similar to those of the single variational approach (4.153); however, the chemical hardness of Eq. (4.156) is approximately 5-fold lower than its "stable" counterpart (4.154), affirming therefore the manifestly reactive framework it produces—one described by a quadratic equation (4.146) instead of a cubic one (4.147).

### 4.2.3.3   Chemical Reactivity by Charge Waves

One considers the chemical reactivity discussion as based on the gauge reaction that equilibrates the chemical bond by symmetrical bond polarities (Putz, 2006)

$$A^- + B^+ = A - B = A^+ + B^- \tag{4.157}$$

such that the reactive electrons are varied on the reunited intervals of Eq. (4.124); such analysis was previously employed to fundament systematic electronegativity and chemical hardness definitions by the averaging (through the integration) factor

$$0.5 = \frac{1}{\int_{-1}^{+1} dN} \tag{4.158}$$

along the reaction path accounting for the acidic (electron accepting, $0 \leq N \leq +1$) and basic (electron donating, $-1 \leq N \leq 0$) chemical behaviors.

In this scaled (gauge) context of reactivity, the foregoing discussion is dedicated to investigating the link between the critical ground state charge ($N^*$) and the valence or reactive state ($N^{**}$). While the first appears as a consequence of naturally fitting the three points in Figure 4.3 (the ionization, neutral and affinity states), with the effect of biasing the minimum of the energetic curve in Figure 4.3b with respect to the apparent Parr-DFT curve in Figure 4.3a, and is thus derived graphically (see Figure 4.8), the valence charge is based on the combined quantum energy and golden ratio information in Eq. (4.151). Both

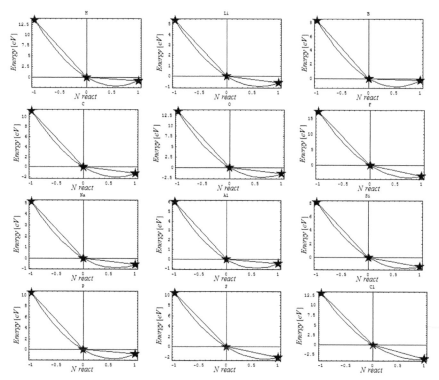

**FIGURE 4.8**    A graphical interpolation for selected elements of Table 4.2 in terms of their ionization, neutral and affinity states, aiming to determine the critical (displaced) charge of the DFT ground state, as prescribed by Figure 4.3b (Putz, 2012a).

are reported for the indicated number of atomic systems of the periodic table of elements in Table 4.2. One notes, for instance, that while the critical ground state charge $N^*$ always lies in the range $[0.5, 1]$, the valence charge $N^{**}$ may span the interval $[0, 1]$; one may interpret such behavior as being associated with the difference between the fraction ½ and integer "1" in driving the principles of chemical reactivity, actually by relaxing the normalization to acid-base behavior interval, being such scaling equivalent with above acidic-basic gauge averaging of Eq. (4.158), see Eq. (4.158) of Section 4.1.3.3 as well as the discussion of Eq. (4.252) in Section 4.4. This way, the valence charge problem may be extended to the interval $[0,2]$, at its turn seen as a gauge transformation of the chemical reactivity charge domain $[-1, +1]$, where one reencounters the challenging problem of whether the "One electron is less than half what an electron pair is" (Ferreira, 1968), the response to which is generally complex but may here be approached through the following steps (Putz, 2012a).

First, by employing the data presented in Table 4.2, one constructs the so-called "continuous" ground and valence charge states by appropriately fittingw over the first four periods of elements, here restrained to 10th-order polynomials. This is performed by interpolating every three points of the 32 elements presented in Table 4.2, although by spanning the atomic number range $Z \in [1, 53]$, thus yielding (see also the allied representations of Figure 4.9) (Putz, 2012a):

$$
\begin{aligned}
NC^* = {} & 0.677771 - 0.193006\,Z + 0.104303\,Z^2 \\
& - 0.0242757\,Z^3 + 0.00302359\,Z^4 \\
& - 0.000220204\,Z^5 + 9.79371 \cdot 10^{-6}\,Z^6 \\
& - 2.6894 \cdot 10^{-7}\,Z^7 + 4.4448 \cdot 10^{-9}\,Z^8 \\
& - 4.05002 \cdot 10^{-11}\,Z^9 + 1.56254 \cdot 10^{-13}\,Z^{10}
\end{aligned}
\tag{4.159}
$$

$$
\begin{aligned}
NC^{**} = {} & 0.768074 - 0.224502\,Z + 0.076654\,Z^2 \\
& - 0.0107337\,Z^3 + 0.000667575\,Z^4 \\
& - 0.000012746\,Z^5 - 6.0246 \cdot 10^{-7}\,Z^6 \\
& + 4.01411 \cdot 10^{-8}\,Z^7 - 9.49568 \cdot 10^{-10}\,Z^8 \\
& + 1.05449 \cdot 10^{-11}\,Z^9 - 4.58533 \cdot 10^{-14}\,Z^{10}
\end{aligned}
\tag{4.160}
$$

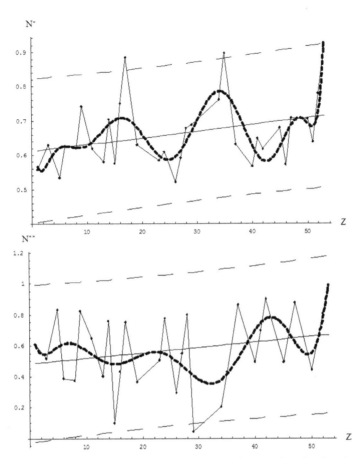

**FIGURE 4.9** The critical ground state and valence charge points for the elements of Table 4.2 and their 10th-order continuous interpolations according to Eqs. (4.159) and (4.160) (Putz, 2012a).

Equations (4.159) and (4.160) are then combined into a sort of special charge wave function based on their difference on the golden ratio scale (see Figure 4.9 for graphical representation)

$$\Psi_Z = \tau\left(NC_Z^* - NC_Z^{**}\right) \tag{4.161}$$

with the peculiar property that its square-integrated form over the Z-range of interpolation gives

$$\int_1^{53} |\Psi_Z|^2 dZ = 0.667233 \approx \tau \tag{4.162}$$

The result of Eq. (4.162) has the following conceptual fundamental quantitative interpretation: the difference between the ground and valence optimum charges is regulated by the golden ratio scale, or in other terms,

$$\int_{1}^{53} \left( NC_Z^* - NC_Z^{**} \right)^2 dZ \approx \frac{1}{\tau} = 1 + \tau \qquad (4.163)$$

such that it provides a sort of normalization corrected by the golden ratio value; it also fulfills the interesting relationship:

$$\sqrt{\int_{1}^{53} \left( NC_Z^* - NC_Z^{**} \right)^2 dZ} \approx 2\tau \qquad (4.164)$$

In any case, the present analysis provides the qualitative result that the difference between the critical ground state and optimal valence charges is more than half of an electronic pair, giving rise to the significant notion that chemical reactivity is not necessarily governed by a pair of electrons but governed by no less than half of a pair and is related to the golden ratio.

However, fractional values in general and those related to the golden ratio particular, may be interpreted as a consistent manifestation of the quantum mechanical (i.e., wave functional) approach of chemical phenomena, here at the reactivity level. Moreover, the quadratic critical charge function (4.161), as shown in Figure 4.10, clearly reveals that a higher contribution to electronic pair chemistry is given by the third period of elements and by the third and fourth transitional elements in particular, a result that nicely agrees with the geometrical interpretation of the chemical bond, particularly the crystal ligand field paradigm of inorganic chemistry (Bader, 1990).

Also a local analysis of the type of charge that is dominant in atomic stability, i.e., the critical physical ground state or the chemical valence reactive state based on Eqs. (4.159) and (4.160), respectively, may be of considerable utility in refined inorganic chemistry structure-reactivity analysis. To the same extent, it depends on the degree of the polynomials used to interpolate the critical and valence charges over the concerned systems; however, through the present endeavor, we may assert that the analysis should be of the type (4.163), which in turn remains a sort of integral version of the *imbalance equation* (2.5), in this case for the ground-valence charge states of a chemical system (Putz, 2012a).

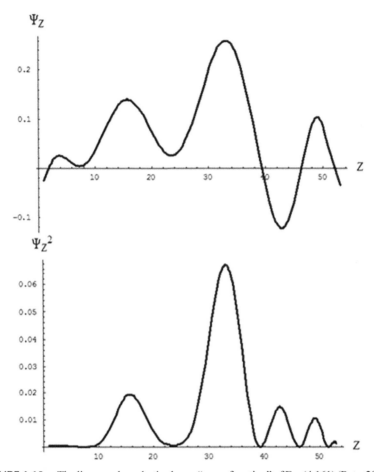

**FIGURE 4.10**    The linear and quadratic charge "wave function" of Eq. (4.161) (Putz, 2012a).


## 4.3   CHEMICAL REACTIVITY BY SOFTNESS DENSITY FUNCTIONAL THEORY

### 4.3.1   *CHEMICAL HARDNESS-SOFTNESS DENSITY RELATED HIERARCHIES*

The essence of the density function theory consist in writing the total energy of an $N$-electronic system as a function of the electronic density function of the fundamental state associated to the system (Kohn et al., 1996):

$$E_V[\rho] = \int \rho(\mathbf{r}) V(\mathbf{r}) d\mathbf{r} + F_{HK}[\rho] \qquad (4.165)$$

where $V$ is the extern potential apply to the electrons system (as example, the nucleus potential). The measurement $F_{HK}[\rho]$ is the universal functional Hohenberg-Kohn (exactly analytically unknown as a density functional) expresses as a sum of the functional density of the kinetic and electronic repulsion energy:

$$F_{HK}[\rho] = T[\rho] + V_{ee}[\rho] \qquad (4.166)$$

and $\rho(\mathbf{r})$ is the unielectronic effective density (spin free) expressed in terms of the system wave function (Parr, 1983):

$$\rho(\mathbf{r}) = N \int |\Psi(1,2,\dots,N)|^2 ds_1 d\tau_2 \dots d\tau_N \qquad (4.167)$$

with $d\tau_i = ds_i dr_i (\mathbf{r}_i \in \Re)$ as being the volume space—spin element, and the $N$ number seen also as a functional density

$$N = N[\rho] = \int \rho(\mathbf{r}) d\mathbf{r} \qquad (4.168)$$

The univocal relationship between the external potential applied to the electronic system and the electronic density is assured by the Hohenberg-Kohn theorems (Hohenberg & Kohn, 1964). Besides, one of the theorem also state the inequality between the energy as a functional density of a random electronic state $E[\rho']$ and the corrected energy of the fundamental electronic state of the system $E[\rho]$:

$$E_V[\rho'] \geq E_V[\rho] \qquad (4.169)$$

inequality equivalent with the existence of the stationary equation:

$$\delta\{E_V[\rho'] - \mu N[\rho']\} = 0 \qquad (4.170)$$

where $\mu$ is the Lagrange multiplier associated to the electrons total number from the system. For the right density of the fundamental state the $\mu$ multiplier represent the chemical potential associate to the system and is expressed by the functional derivate:

$$\mu = \left( \frac{\delta E[\rho]}{\delta \rho(\mathbf{r})} \right)_{\rho=\rho(V;)} \tag{4.171}$$

which can de reduced, from the stationary principle (4.170), to a partial derivative respecting the electrons number from the fundamental state:

$$\mu = \left( \frac{\partial E}{\partial N} \right)_V \tag{4.172}$$

and which, besides, is associating with the electronegativity with changed sign: $\mu = -\chi$ (Parr et al., 1978). By calculating the functional derivative respecting the external potential $V(r)$ in Eq. (4.165), there is obtained:

$$\rho(\mathbf{r}) = \left( \frac{\delta E[\rho]}{\delta V(\mathbf{r})} \right)_{\rho=\rho(V;)} \tag{4.173}$$

from where, along with the relation (4.172), the total difference of the functional energy (4.165), $E = E(N,V)$, it can be expressed under the general form (Ayers & Parr, 2001):

$$dE = \mu dN + \int \rho(\mathbf{r}) dV(\mathbf{r}) d\mathbf{r} \tag{4.174}$$

This is the first equation of the chemical transformation in TFD, because it correlate the total energy variation of an electronic state (atomic or molecular) with the charge exchanged to which is compiled, and with the potential variation which governed the electronic state. By applying the Maxwell relations to the equation results the next row of identities which define the Fukui functions $f(\mathbf{r})$:

$$\left( \frac{\delta \mu}{\delta V(\mathbf{r})} \right)_N = \left( \frac{\partial \rho(\mathbf{r})}{\partial N} \right)_V \equiv f(\mathbf{r}) \tag{4.175}$$

Analogously, it can be expresed the total diferention of the chemical potential it can be expressed as a function of the electrons number from the system and the applied external potential:

$$d\mu = \left( \frac{\partial \mu}{\partial N} \right)_V dN + \int \left( \frac{\delta \mu}{\delta V(\mathbf{r})} \right)_N \delta V(\mathbf{r}) d\mathbf{r} \tag{4.176}$$

In the second term from the right of Eq. (4.176), it can be recognize the identity (4.175), and the prime term from the right of Eq. (4.176) is recognized as the chemical hardness associated to the electronic system, according to the definitions from the Sections 3.2 and 4.2:

$$2\eta = \left(\frac{\partial^2 E}{\partial N^2}\right)_V = \left(\frac{\partial \mu}{\partial N}\right)_V \qquad (4.177)$$

With relations (4.175) and (4.177) the total differential can be rewritten in the form (Ayers & Parr, 2000):

$$d\mu = 2\eta dN + \int f(\mathbf{r})dV(\mathbf{r})d\mathbf{r} \qquad (4.178)$$

This is the second equation of the chemical transformations in DFT, equivalent with the first one, (4.174), but rewritten on another level. The Eq. (4.178) correlate the chemical potential change (the electronegativity with reverse sign) of an electrochemical state (atomic, molecular) with the charge change and with potential modification, through the chemical hardness on this variation, and of the Fukui frontier function, having a promoting role of the frontier orbital charge variation.

With relation (4.171) in (4.165), a new Euler-Lagrange equation of the electronic system is obtained, involving the chemical potential

$$\mu = \left(\frac{\delta E}{\delta \rho(\mathbf{r})}\right)_V = V(\mathbf{r}) + \frac{\delta F_{HK}[\rho]}{\delta \rho(\mathbf{r})} \qquad (4.179)$$

equation which satisfies also the differential form (4.178).

These are the chemical transformations equation, in terms of functional densities which will be employed and transformed in order to be applied in describing the open electronic evolution systems and of their participation to the chemical reactions.

Next the connection between the local and global sensitivity indices on the DFT will be exposed in a manner which should allow an explicit implementation of the electronic densities. It is start from a generalized form of the Euler-Lagrange equation (4.165):

$$\Xi = \int \rho(\mathbf{r})V(\mathbf{r})d\mathbf{r} + F_{HK}[\rho] - \mu \int \rho(\mathbf{r})d\mathbf{r} = \int \rho(\mathbf{r})u(\mathbf{r})d\mathbf{r} + F_{HK}[\rho] \qquad (4.180a)$$

through which minimization,

$$\delta\Xi = 0 \qquad (4.180b)$$

in report with the electronic density for the external fixed potential, it is obtaining a relationship equivalent with the expression (4.179):

$$-u(\mathbf{r}) = \mu - V(\mathbf{r}) = \frac{\delta F_{HK}[\rho]}{\delta \rho(\mathbf{r})} \qquad (4.181)$$

For an open electronic system found under the influence on the potential

$$u(\mathbf{r}) = V(\mathbf{r}) - \mu \qquad (4.182)$$

which satisfy the Eq. (4.181), are defining in analogy with Eq. (4.177), but in functional manner (De Proft et al., 1997; Gázquez et al., 1990; Berkowitz & Parr, 1988; Senet, 1996, 1997), the chemical hardness nucleus associated to the system:

$$2\eta(\mathbf{r}, \mathbf{r}') = -\frac{\delta u(\mathbf{r})}{\delta \rho(\mathbf{r}')} = \frac{\delta^2 F_{HK}}{\delta \rho(\mathbf{r}) \delta \rho(\mathbf{r}')} \qquad (4.183)$$

and the local chemical hardness for the considered system:

$$\eta(\mathbf{r}) = \frac{1}{N} \int \eta(\mathbf{r}, \mathbf{r}') \rho(\mathbf{r}') d\mathbf{r}' \qquad (4.184)$$

By the way of defining the potential $u(\mathbf{r})$ it can be remarked how this include all the system properties: those connected to the external potential $V(\mathbf{r})$ in which evolves, and also those characterized by the chemical potential, the electronegativity with reverse sigh of the system. If there is the functional derivative of the electronic density $\rho(\mathbf{r})$ in report with $u(\mathbf{r})$:

$$\frac{\delta \rho(\mathbf{r})}{\delta u(\mathbf{r})} < \infty, \forall \mathbf{r} \in \Re \qquad (4.185)$$

this existence allow the introduction of a new measurements which characterize an electronic system (open).

It can be defined the chemical softness kernels associated to the electronic system:

$$s(\mathbf{r}, \mathbf{r}') = -\frac{\delta \rho(\mathbf{r})}{\delta u(\mathbf{r}')} \qquad (4.186)$$

and the local chemical softness characterized to the electronic system:

$$s(\mathbf{r}) = \int s(\mathbf{r}, \mathbf{r}') d\mathbf{r}' \qquad (4.187)$$

If we combine the definition relations (4.183), (4.184), (4.186), (4.187), it can be obtained also the bonding relation between the introduced measurements. It is considered the integer-deferential identity:

$$\int \frac{\delta\rho(\mathbf{r})}{\delta u(\mathbf{r}')}\frac{\delta u(\mathbf{r}')}{\delta\rho(\mathbf{r}'')}d\mathbf{r}' = \delta(\mathbf{r}''\text{-}\mathbf{r}) \qquad (4.188)$$

whit which help, if it is considered the (4.183), (4.186), it can be obtain the equivalent expression:

$$2\int s(\mathbf{r},\mathbf{r}')(\mathbf{r},\mathbf{r}')\eta(\mathbf{r},\mathbf{r}'')d\mathbf{r}' = \delta(\mathbf{r}''\text{-}\mathbf{r}) \qquad (4.189)$$

From this relation it is concluded that, the chemical softness and hardness nucleus for a considered electronic system are inverse measurements. The relation (4.189) can be written also on another localization level; when it is multiplied with $\rho(\mathbf{r}'')$, the result it is then integrated upon $\mathbf{r}''$ and the Eq. (4.184) is taking into consideration-one obtains:

$$2\int s(\mathbf{r},\mathbf{r}')\eta(\mathbf{r}')d\mathbf{r}' = \frac{1}{N}\rho(\mathbf{r}) \qquad (4.190)$$

from where, following the further integration upon $x$ while taking into account of Eq. (4.187), the next identity can be written:

$$2\int s(\mathbf{r})\eta(\mathbf{r})d\mathbf{r} = 1 \qquad (4.191)$$

Besides, also other relations can be obtained if the transformation succession is considered:

$$dp(\mathbf{r}) = \int \frac{\delta\rho(\mathbf{r})}{\delta u(\mathbf{r}')}du(\mathbf{r}')d\mathbf{r}' = -\int s(\mathbf{r},\mathbf{r}')du(\mathbf{r}')d\mathbf{r}'$$
$$= -\int s(\mathbf{r},\mathbf{r}')[dV(\mathbf{r}') - d\mu]d\mathbf{r}' = \int s(\mathbf{r},\mathbf{r}')d\mathbf{r}' d\mu - \int s(\mathbf{r},\mathbf{r}')dV(\mathbf{r}')d\mathbf{r}' \qquad (4.192)$$

If the Eq. (4.178) it is taken into consideration under the form:

$$d\mu = 2\eta dN + \int f(\mathbf{r}'')dV(\mathbf{r}'')d\mathbf{r}'' \qquad (4.193)$$

the expression (4.192) will also the form:

$$dp(\mathbf{r}) = 2s(\mathbf{r})\eta dN + \int [-s(\mathbf{r},\mathbf{r}'') + s(\mathbf{r})f(\mathbf{r}'')]dV(\mathbf{r}'')d\mathbf{r}'' \qquad (4.194)$$

Note that the factor "2" appearing in above equations is due to the chemical hardness definition bearing "1/2" in the custom definition (3.3) and can be skipped (as will be done below) when the chemical hardness will be considered as not-normalized to its acid-base behavior, see Eq. (4.158) of Section 4.2.3.3 as well as the discussion of Eq. (4.252) in Section 4.5.

A general expression of the electronic density variation (in a physical-chemical process) can be written also from the functional dependency $\rho = \rho[N,V]$ correlated with the way of defining the Fukui function (4.175):

$$d\rho(\mathbf{r}) = f(\mathbf{r})dN + \int \left( \frac{\delta\rho(\mathbf{r})}{\delta V(\mathbf{r}'')} \right)_N dV(\mathbf{r}'')d\mathbf{r}'' \tag{4.195}$$

By comparing the expressions (4.194) and (4.195) there are obtaining the identities:

$$s(\mathbf{r}) = \frac{f(\mathbf{r})}{2\eta} = f(\mathbf{r})S = \left( \frac{\partial\rho(\mathbf{r})}{\partial N} \right)_V \left( \frac{\partial N}{\partial \mu} \right)_V = \left( \frac{\partial\rho(x)}{\partial \mu} \right)_V \tag{4.196}$$

In the second identity (4.196) it was practically introduced the definition of global chemical softness $S$, and can be further used in writing the relations

$$\kappa(\mathbf{r},\mathbf{r}'') \equiv \left( \frac{\delta\rho(\mathbf{r})}{\delta V(\mathbf{r}'')} \right)_N \tag{4.197}$$

$$= -s(\mathbf{r},\mathbf{r}'') + \frac{s(\mathbf{r})s(\mathbf{r}'')}{S} \tag{4.198}$$

The relations (4.197) and (4.198) are very important because they correlate the sensitivity descriptors on the density functional level with the *linear response function* $\kappa(\mathbf{r},\mathbf{r}')$.

## 4.3.2　LONG-RANGE LOCAL-TO-GLOBAL BEHAVIOR

Given the form of the linear response kernel of Eq. (4.198) (Ayers, 2001; Sablon et al., 2010; Yang et al., 2012) one has the problem to formulate the local form of the softness kernel $s(\mathbf{r},\mathbf{r}')$ fulfilling the Berkowitz-Ghosh-Parr relationship (4.197) (Berkowitz et al., 1985; Berkowitz & Parr, 1988) within conceptual DFT. To this aim one starts with rearranging the Eq. (4.198) as an integral form along the chemical reaction path (Putz & Chattaraj, 2013)

$$\int_{\substack{REACTION \\ PATH}} \delta\rho(\mathbf{r}) = \int_{\substack{REACTION \\ PATH}} \kappa(\mathbf{r},\mathbf{r}')\delta V(\mathbf{r}') \tag{4.199}$$

with the finite general result

$$\Delta\rho(\mathbf{r}) = \int \kappa(\mathbf{r},\mathbf{r}')\Delta V(\mathbf{r}')d\mathbf{r}' \tag{4.200}$$

Observe, for instance, that when integrating both sides of Eq. (4.200) with respect to "$\mathbf{r}$," one may use the fundamental DFT normalization constraint (4.168) that along the integration result of Eq. (4.198) leaves with the useful constrain

$$\int \kappa(\mathbf{r},\mathbf{r}')d\mathbf{r} = 0 \qquad (4.201)$$

which correctly establishes the charge conservation condition, $\Delta N = 0$, in the chemical reaction/transformation course. However, when specialized to valence picture of chemical reactivity, one may use for Eq. (4.200) the "nucleus-to-valence state" settings

$$\Delta\rho(\mathbf{r}) \cong \rho(\mathbf{r})-0 = \rho(\mathbf{r}) \qquad (4.202a)$$

$$\Delta V(\mathbf{r}') \cong 0-V(\mathbf{r}') = -V(\mathbf{r}') \qquad (4.202b)$$

so that the working form of the so-called valence or long-range density solution albeit approximately is obtained

$$\rho(\mathbf{r}) = -\int \kappa(\mathbf{r},\mathbf{r}')V(\mathbf{r}')d\mathbf{r}' \qquad (4.203)$$

Actually, the entire present development stays under the valence or long-range regime of electrons in atoms and molecules in various forms and approximations of conceptual DFT. The minus sign in Eq. (4.203) agrees with the opposite phenomenological behavior in density and potential variation, as provided by Poisson equation—for instance (Putz et al., 2005), and is in accordance with alternative derivation based on chemical action principle and virial theorem (Putz, 2009a).

Next, one uses the integral linking the hardness and softness kernels through the reciprocity relation of Eq. (4.189) without the chemical hardness factor ("2" vs. "1/2"), see Eq. (4.158) of Section 4.2.3.3 as well as the discussion of Eq. (4.252) in Section 4.5

$$\delta(\mathbf{r}-\mathbf{r}'') = \int s(\mathbf{r},\mathbf{r}')\eta(\mathbf{r}',\mathbf{r}'')d\mathbf{r}' \qquad (4.204)$$

to be integrated to the actual chemical hardness free factor version of Eq. (4.190)

$$\rho(\mathbf{r}) = N\int s(\mathbf{r},\mathbf{r}')\eta(\mathbf{r}')d\mathbf{r}' \qquad (4.205)$$

through employing the local vs. kernel chemical hardness hierarchy (4.184). Note that Eq. (4.184) may be seen as a special case of a general integral

$$\eta(\mathbf{r}) = \int \eta(\mathbf{r}, \mathbf{r}') f(\mathbf{r}') d\mathbf{r}' \qquad (4.206)$$

involving the *shape form* of the Fukui function (Funtealba, 1998)

$$f(\mathbf{r}) = \frac{\rho(\mathbf{r})}{N} \qquad (4.207)$$

Worth noting that despite shape function is conceptually different from the genuine Fukui function (4.175) they both fulfill the normalization condition

$$1 = \int f(\mathbf{r}) d\mathbf{r} \qquad (4.208)$$

in the light of above DFT density constraint, Eq. (4.168). Yet, one should be warned the dichotomy of shape function vs. Fukui function may be regarded as another manifestation of the long-range vs. general electronic density behavior, respectively, and corresponds with the (inherently ambiguous) local frontier chemical hardness believed to behave as a global quantity (Parr & Bartolotti, 1993; Ayers, 2000; Ayers & Parr, 2008a); this was ultimately extended through the softness relationship with exchange-correlation density and (global) electronegativity (Matito & Putz, 2012), while the direct chemical hardness long-range behavior and consequences will be analyzed below.

Now, by comparing Eqs. (4.203) and (4.205) another relationship between linear response and softness kernels is obtained, namely

$$\kappa(\mathbf{r}, \mathbf{r}') = -N \frac{s(\mathbf{r}, \mathbf{r}') \eta(\mathbf{r}')}{V(\mathbf{r}')} \qquad (4.209)$$

Together, Eqs. (4.197) and (4.209) it gives the hint for considering the actual symmetrical long-range local form for the softness kernel (Putz & Chattaraj, 2013):

$$s(\mathbf{r}, \mathbf{r}') = \frac{s(\mathbf{r}) s(\mathbf{r}')}{S} \left[ \frac{V(\mathbf{r})}{V(\mathbf{r}) - N\eta(\mathbf{r})} + \frac{V(\mathbf{r}')}{V(\mathbf{r}') - N\eta(\mathbf{r}')} \right] \qquad (4.210)$$

It fulfills also the kernel softness condition by reducing to $s(\mathbf{r})s(\mathbf{r}')/S$ in the long-range (electrostatic) interaction when one has approximately

$$\eta(\mathbf{r}) = -\frac{V(\mathbf{r})}{N} \qquad (4.211)$$

The long-range consistency of local chemical hardness (4.211) with softness kernel (4.210) may be proved also by checking the key softness hierarchy condition in passing from kernel to local dependency:

$$\int s(\mathbf{r},\mathbf{r}')d\mathbf{r}' = \int \frac{s(\mathbf{r})s(\mathbf{r}')}{S} \frac{V(\mathbf{r})}{V(\mathbf{r})-N\eta(\mathbf{r})}d\mathbf{r}'$$
$$+ \int \frac{s(\mathbf{r})s(\mathbf{r}')}{S} \frac{V(\mathbf{r}')}{V(\mathbf{r}')-N\eta(\mathbf{r}')}d\mathbf{r}'$$
$$= \frac{s(\mathbf{r})V(\mathbf{r})}{V(\mathbf{r})-N\eta(\mathbf{r})} \frac{1}{S}\underbrace{\int s(\mathbf{r}')d\mathbf{r}'}_{=1} + \frac{s(\mathbf{r})}{S}\int \frac{s(\mathbf{r}')V(\mathbf{r}')}{V(\mathbf{r}')-N\eta(\mathbf{r}')}d\mathbf{r}'$$
$$\underset{\substack{\eta(\mathbf{r})=-V(\mathbf{r})/N \\ \eta(\mathbf{r}')=-V(\mathbf{r}')/N}}{\longrightarrow} \frac{1}{2}s(\mathbf{r}) + \frac{1}{2}s(\mathbf{r})\frac{1}{S}\underbrace{\int s(\mathbf{r}')d\mathbf{r}'}_{1}$$
$$= s(\mathbf{r}) \qquad (4.212)$$

through considering both the long-range chemical hardness limit (4.211) and the consecrated local to global softness integral constraint as appeared from (4.196) with (4.208):

$$S = \int s(\mathbf{r})d\mathbf{r} \qquad (4.213)$$

All in all, the long-range related softness kernel (4.210) fulfills the main new softness conditions as follows (Pérez et al., 2008)

   I.  Is symmetric with respect to exchange of its coordinates $\mathbf{r} \leftrightarrow \mathbf{r}'$, by construction;

  II.  It converges to $s(\mathbf{r})s(\mathbf{r}')/S$ when for each terms of (4.210) the electrostatic (long-range) regime is considered through the condition (4.211);

 III.  It attains the hierarchy from non-local to local quantities in the long-range regime (4.211).

### 4.3.3  ANALYTICAL CHEMICAL SOFTNESS BY DENSITY AND APPLIED POTENTIAL

#### 4.3.3.1  Translational Chemical Softness

By the way in which the local sensitivity descriptors were introduced, it can de seen the importance of the universal Honenberg-Kohn functional $F[\rho]$. It still can be developed a method which avoid this dependency (Garza & Robles, 1993), having in mind that an exact analytic expression of this functional is not (yet) known. This inconvenient can be transpose in the unrecognizing the exact functional relation between the electronic density $\rho(\mathbf{r})$ and the global potential $u(\mathbf{r})$, and this new inconvenience can be eluded by considering the translational invariance of the external potential applied to the electrons system, respectively of the density of them in atomic or molecular system. The explicit bound between those types of variations (of the global potential and of the electronic density), for an open system can took the forms:

$$\nabla \rho(\mathbf{r}) = \int d\mathbf{r}' \frac{\delta \rho(\mathbf{r})}{\delta u(\mathbf{r}')} \nabla' u(\mathbf{r}') \tag{4.214}$$

$$\nabla u(\mathbf{r}) = \int d\mathbf{r}' \frac{\delta u(\mathbf{r})}{\delta \rho(\mathbf{r}')} \nabla' \rho(\mathbf{r}') \tag{4.215}$$

These relations can be re-transcribed in the virtue of expressions (4.183) and (4.186) so that:

$$\nabla \rho(\mathbf{r}) = -\int d\mathbf{r}' s(\mathbf{r}, \mathbf{r}') \nabla' u(\mathbf{r}') \tag{4.216}$$

$$\nabla u(\mathbf{r}) = -2 \int d\mathbf{r}' \eta(\mathbf{r}, \mathbf{r}') \nabla' \rho(\mathbf{r}') \tag{4.217}$$

The expression (4.216) can be easily developed if is considered the chemical softness nucleus as being de(compose) (in) by a local contribution $L(\mathbf{r})$ and a nonlocal contribution of the type:

$$s(\mathbf{r}, \mathbf{r}') = L(\mathbf{r}')\delta(\mathbf{r} - \mathbf{r}') + t(\mathbf{r})\rho(\mathbf{r}') \tag{4.218}$$

whit which help the relation (4.216) becomes:

$$\nabla \rho(\mathbf{r}) = -L(\mathbf{r})\nabla u(\mathbf{r}) - t(\mathbf{r})\int d\mathbf{r}' \rho(\mathbf{r}') \nabla' u(\mathbf{r}') \tag{4.219}$$

The last term from Eq. (4.219) identically vanishes by the virtue of

$$\nabla V(\mathbf{r}) = \nabla u(\mathbf{r}) \tag{4.220}$$

and of the Hellmann-Feynman theorem. So that the local contribution $L(\mathbf{r})$ will be given by the form:

$$L(\mathbf{r}) = -\frac{\nabla \rho(\mathbf{r}) \cdot \nabla u(\mathbf{r})}{|\nabla u(\mathbf{r})|^2} = -\frac{\nabla \rho(\mathbf{r}) \cdot \nabla V(\mathbf{r})}{|\nabla V(\mathbf{r})|^2} \tag{4.221}$$

in order to determine the nonlocal contribution it will be firs evaluate the chemical softness (4.218) on different localization levels, by successively integrations, meaning:

$$s(\mathbf{r}) = L(\mathbf{r}) + Nt(\mathbf{r}) \tag{4.222}$$

$$S = \int s(\mathbf{r}) d\mathbf{r}$$
$$= \int [L(\mathbf{r}) + Nt(\mathbf{r})] d\mathbf{r} = \int L(\mathbf{r}) d\mathbf{r} + N \int t(\mathbf{r}) d\mathbf{r} \equiv a + N \int t(\mathbf{r}) d\mathbf{r} \tag{4.223}$$

Now the expression (4.198) can be as well evaluated with the result:

$$\kappa(\mathbf{r}, \mathbf{r}') = -L(\mathbf{r}')\delta(\mathbf{r} - \mathbf{r}') - t(\mathbf{r})\rho(\mathbf{r}') + \frac{[L(\mathbf{r}) + Nt(\mathbf{r})][L(\mathbf{r}') + Nt(\mathbf{r}')]}{\int [L(\mathbf{r}) + Nt(\mathbf{r})] d\mathbf{r}} \tag{4.224}$$

If the number of the system electrons $N$ is fixed, from the normalization condition of the linear response function in Eq. (4.201) applied to relation (4.224), one will obtain the integral equation:

$$\int t(\mathbf{r}) d\mathbf{r} = \frac{N}{\rho(\mathbf{r})} t(\mathbf{r}) \tag{4.225}$$

with the simple solution $t(\mathbf{r}) = \rho(\mathbf{r})$. With this result, the expression of the chemical softness nucleus will have the functional form:

$$s(\mathbf{r}, \mathbf{r}') = -\frac{\nabla \rho(\mathbf{r}) \cdot \nabla V(\mathbf{r})}{|\nabla V(\mathbf{r})|^2} \delta(\mathbf{r} - \mathbf{r}') + \rho(\mathbf{r}) \rho(\mathbf{r}') \tag{4.226}$$

From now on, the analytically implementation is immediate, for an expression of an atomic or molecular electronic explicit density.

### 4.3.3.2   Long-Range Chemical Softness

For practical implementation we provide further softness and chemical hardness dependence on the electronic density and potential, alternatively to Eq. (4.226). To this aim we use a local-nonlocal ansatz for softness kernel within the present long-range/shape function approximation of Eq. (4.207), namely as (Garza & Robles, 1993; Putz, 2009a)

$$s(\mathbf{r},\mathbf{r}') = -\frac{\rho(\mathbf{r}')}{V(\mathbf{r}')}\delta(\mathbf{r}-\mathbf{r}') + \rho(\mathbf{r}')\rho(\mathbf{r}) \qquad (4.227)$$

so that the local softness immediately turns out to be by Eq. (4.187)

$$s(\mathbf{r}) = \int s(\mathbf{r},\mathbf{r}')d\mathbf{r}' = \rho(\mathbf{r})\left[N - \frac{1}{V(\mathbf{r})}\right] \qquad (4.228)$$

while the global softness (4.213) becomes

$$S = \int s(\mathbf{r})d\mathbf{r} = \int \rho(\mathbf{r})\left[N - \frac{1}{V(\mathbf{r})}\right]d\mathbf{r} \qquad (4.229)$$

For the local hardness determination one may use the identity, somehow similar with that of Eq. (4.196), yet here at the complete local level, namely

$$s(\mathbf{r})\eta(\mathbf{r}) = f(\mathbf{r}) \qquad (4.230)$$

Equation (4.230) can be proved within long-range density interaction by starting from Eq. (4.205): its integral over "$\mathbf{r}$" coordinate is taken first, then, by using the basic DFT constraint (4.168) and kernel-to-local softness integral relationship (4.213) one arrives at the "1/2-chemcial hardness" modified form of Eq. (4.191), as

$$\int s(\mathbf{r})\eta(\mathbf{r})d\mathbf{r} = 1 = \int f(\mathbf{r})d\mathbf{r} \qquad (4.231)$$

with which occasion the frontier normalization (4.208) was also considered on its right hand side; this way the relationship (4.230) is roughly recovered by equating integrands on the left and right sides.

This way, Eq. (4.230) allows the local chemical hardness in the density-potential formulation:

$$\eta(\mathbf{r}) = \frac{f(\mathbf{r})}{s(\mathbf{r})} = \frac{1}{N\left[N - \frac{1}{V(\mathbf{r})}\right]} \qquad (4.232)$$

Note that Eq. (4.232) can be rearranged so that it can be compared with Eq. (4.211) that is (Putz & Chattaraj, 2013)

$$\eta(\mathbf{r}) = -\frac{V(\mathbf{r})}{N} \frac{1}{1 - NV(\mathbf{r})} \cong -\frac{V(\mathbf{r})}{N} \left[1 + NV(\mathbf{r}) + ...\right] \qquad (4.233)$$

so that it really tends to Eq. (4.211) under the long-range regime of vanishing potential and/or the number of electrons, so proving the reliability of the presented working form and softness density functional based formalism at large.

## 4.4 ELECTROPHILICITY: LONG RANGE CHEMICAL SOFTNESS FORMULATION

Since most of the chemical reactions may be characterized in terms of electrophilic/nucleophilic action of charge transfer through accepting/ donating electrons, the concept of electrophilicity (Chattaraj & Sarkar, 2006) becomes essential and its quantification (Parr et al., 1999), especially within the DFT, see (Parr & Yang, 1989; Geerlings et al., 2003), has been considered to be important mainly due to its reliability in modeling a variety of physico-chemical phenomena such as site selectivity (Pérez et al., 2002; Chamorro et al., 2003), molecular vibrations and rotation (Parthasarathi et al., 2005), intramolecular and intermolecular reactivity patterns (Domingo et al., 2002a,b), solvent and external field effects (Meneses et al., 2006; Torrent–Sucarrat et al., 2008, 2010; Gál et al., 2011) as well as biological activity and toxicity (Miller & Miller, 1977; Ashby & Tennant, 1991; Parthasarathi et al., 2004, 2005; Padmanabhan et al., 2006a,b; Rong et al., 2007; Roy et al., 2007). The definition of electrophilicity relies on two other basic chemical concepts, viz., as electronegativity ($\chi$) and hardness ($\eta$) as follows (Putz & Chattaraj, 2013)

$$\omega = \frac{\chi^2}{2\eta} = \frac{1}{2}\chi^2 S \qquad (4.234)$$

where the second form employs the reciprocal relationship between global chemical hardness and softness ($S$) (Parr & Yang, 1989). However, the first form of Eq. (4.234), comes through an electrostatic correspondence with potential and resistance (Parr et al., 1999), respectively, thereby providing electrophilicity the meaning of the "power" of attracting electrons by an electrophile. However, due to the discontinuity in the energy ($E$) vs. number of electrons ($N$) curve (Perdew et al., 1982; Ayers, 2008), separate electrophilicity indices for charge donation ($\omega^+$) and charge acceptance ($\omega^-$) are introduced (Gazquez et al., 2007). This picture is extended to the local electrophilicity by connection with Fukui function (Parr & Yang, 1984) $f(\mathbf{r})$, for the nucleophilic (+), electrophilic (–), and radical (0) attacks to model respective local electrophilicities $\omega^{(+,-,0)} = \omega f^{(+,-,0)}$ which may in turn be condensed to a specific site in a molecule (Chattaraj et al., 2003; Funtealba & Parr, 1991; Geerlings & De Proft, 2008; Cárdenas et al., 2009; Senet, 1996). Any local function $\omega(\mathbf{r})$ which integrates to $\omega$ may be considered to be a candidate for local electrophilicity. Projection via $f(\mathbf{r})$ is in the spirit of the definition of local softness. Therefore, there remains an open field to assess a continuous analytical form for the local electrophilicity, $\omega(\mathbf{r})$, as well as for its kernel version, $\omega(\mathbf{r},\mathbf{r}')$, and to explore the consequences from atoms to molecules in the light of global elemental periodicity and of chemical reactivity principles of electronegativity and chemical hardness (Parr & Zhou, 1993; Chattaraj & Maiti, 2001), respectively.

In this context, we can make a step forward in advancing general recipe for developing local and kernel electrophilicity forms in terms of local softness $s(\mathbf{r})$ and kernel softness $s(\mathbf{r}, \mathbf{r}')$ measures of a chemical species as follows

$$\omega(\mathbf{r}) = \frac{1}{2}\chi^2 s(\mathbf{r}) \qquad (4.235)$$

$$\omega(\mathbf{r},\mathbf{r}') = \frac{1}{2}\chi^2 s(\mathbf{r},\mathbf{r}') \qquad (4.236)$$

being consistent with global electrophilicity index through the integral relationships

$$\omega = \int \omega(\mathbf{r})d\mathbf{r} = \iint \omega(\mathbf{r},\mathbf{r}')d\mathbf{r}d\mathbf{r}' \qquad (4.237)$$

according to the associated softness kernel hierarchy (Berkowitz et al., 1985). The local electrophilicity $\omega(\mathbf{r})$ is formally same as that defined in Chattaraj et al. (2003) while there are presumably even higher order local electrophilicities, e.g., the hierarchy of hypersoftness and hyperhardness kernels (Ayers & Parr, 2008b), within higher-order conceptual DFT developments (Chattaraj et al., 2003; Funtealba & Parr, 1991;

Geerlings & De Proft, 2008; Cardenas et al., 2009; Senet, 1996). One may obtain a condensed-to-atom variant and also for the electrophilic, nucleophilic and radical attacks in the usual way. Moreover, the inverse of $\omega(\mathbf{r},\mathbf{r}')$ may generate a hierarchy of nucleophilicity kernel. Unlike the previous formulations, the overall treatment here is general and analytic with hardly any bearing on the explicit form of E(N). The traditional operational definition of local softness and hardness contain the same "potential information" and they should be interpreted as the "local abundance" or "concentration" of their corresponding global properties.

A similar situation happens with the local electrophilicity. Then, it is necessary to derive new expressions, and conditions that the correct local counterparts of the global softness, hardness, and electrophilicity must fulfill. This work can also be considered as a possible alternative to the traditional expressions of local softness and electrophilicity, see Torrent–Sucarrat et al. (2008, 2010); Gál et al. (2011).

Within the actual framework of *long-range softness DFT*, the concepts of local and kernel electrophilicity were developed toards the local to global hierarchical criteria such as bilocal symmetry, asymptotic behavior, and integral local to global relationships. Accordingly, with (4.210) in Eq. (4.236), the associated electrophilicity kernel looks like (Putz & Chattaraj, 2013)

$$\omega(\mathbf{r},\mathbf{r}') = \chi^2 \frac{s(\mathbf{r})s(\mathbf{r}')}{2S}\left[\frac{V(\mathbf{r})}{V(\mathbf{r})-N\eta(\mathbf{r})} + \frac{V(\mathbf{r}')}{V(\mathbf{r}')-N\eta(\mathbf{r}')}\right] \qquad (4.238)$$

At this point, aiming to formulate a local version for electrophilicity a working local softness expression is needed; to this end, by inspecting the Eq. (4.212) on its intermediate form, while remembering the local-to-global ratio definition for Fukui function according with Eq. (4.196) here under the form

$$\frac{s(\mathbf{r})}{S} = f(\mathbf{r}) \qquad (4.239)$$

one may easily recognize the form (Putz & Chattaraj, 2013)

$$s(\mathbf{r}) = 2Sf(\mathbf{r})\frac{V(\mathbf{r})}{V(\mathbf{r})-N\eta(\mathbf{r})} \qquad (4.240)$$

First of all, it is apparent that Eq. (4.240) is a rearrangement of Eq. (4.211) when Eq. (4.239) applies thus revealing its long-range nature. Then, within the present long-range framework, one may further check the fulfillment of hierarchy (4.213):

$$\int s(\mathbf{r})d\mathbf{r} = 2S \int f(\mathbf{r}) \frac{V(\mathbf{r})d\mathbf{r}}{V(\mathbf{r}) - N\eta(\mathbf{r})} \xrightarrow{\eta(\mathbf{r}) = -V(\mathbf{r})/N} S \underbrace{\int f(\mathbf{r})d\mathbf{r}}_{=1} = S \quad (4.241)$$

when recalling Eq. (4.211) and normalization of the Fukui/shape function (4.208). To finally prove the full consistency of the actual long-range approach of electronic density, one may reciprocally consider the hierarchy (4.213) as granted by the local softness (4.240) within the long-range form of the local hardness (4.211), i.e., through integrating both sides of (4.240) with respect to "**r**" one arrives at identity

$$\frac{1}{2} = \int \frac{f(\mathbf{r})V(\mathbf{r})}{V(\mathbf{r}) - N\eta(\mathbf{r})} d\mathbf{r} \quad (4.242)$$

which equivalently rewrites with basic DFT Eq. (4.168) as:

$$\frac{1}{2N} \int \rho(\mathbf{r})d\mathbf{r} = \int \frac{f(\mathbf{r})V(\mathbf{r})}{V(\mathbf{r}) - N\eta(\mathbf{r})} d\mathbf{r} \quad (4.243)$$

leaving with the approximate local relationship

$$2Nf(\mathbf{r})V(\mathbf{r}) = \rho(\mathbf{r})V(\mathbf{r}) - N\rho(\mathbf{r})\eta(\mathbf{r}) \quad (4.244)$$

From Eq. (4.244), it is clear that in the condition of Eq. (4.211) the shape function (4.207) results, and vice-versa, this way proving the present long-range consistency.

With these, the local electrophilicity uses the local softness (4.240) to yield (Putz & Chattaraj, 2013)

$$\omega(\mathbf{r}) = \chi^2 \frac{s(\mathbf{r})V(\mathbf{r})}{V(\mathbf{r}) - N\eta(\mathbf{r})} \quad (4.245)$$

It is now clear that Eq. (4.238) with Eqs. (4.245) and (4.234) are related by consistent hierarchical integration (4.237), within the long-range regime of electronic interactions, i.e., through involving local hardness (4.211) and the related local softness (4.240).

Regarding the chemical reactivity implications of the actual working expressions, the softness and electrophilicity kernels (4.210) and (4.238), and the local electrophilicity (4.245) one may have interesting insight by considering the series expansion of the potential dependent terms (Putz & Chattaraj, 2013)

$$\frac{V(\mathbf{r})}{V(\mathbf{r})-N\eta(\mathbf{r})} \xrightarrow{\eta(\mathbf{r})\to 0} 1+N\frac{\eta(\mathbf{r})}{V(\mathbf{r})}+N^2\left[\frac{\eta(\mathbf{r})}{V(\mathbf{r})}\right]^2+... \xrightarrow{\eta(\mathbf{r})=-V(\mathbf{r})/N} 1-1+1-... \quad (4.246)$$

for small local chemical hardness contributions, i.e., higher chemical reactivity local behavior. Within the long-range regime of interaction, Eq. (4.211) however, is consistent with actual local hardness limit $\eta(\mathbf{r}) \to 0$, one obtains that the terms of type (4.246) vanishes in the odd orders and reduces to unity for even orders of electronic charge; in usual conceptual DFT language this means that when chemical reactivity is driven by first order of electronic charge (transfer) no softness and electrophilicity either at bilocal or local long-range is manifested while they are strengthened by a factor of two when parabolic electronic charge models the chemical reactivity; this way the present approach has as the limiting case the parabolic view of the chemical reactivity.

## 4.5 ABSOLUTE ELECTRONEGATIVITY AND CHEMICAL HARDNESS

According to the Coulson's famous metaphor, to solve the electronic frontier (valence or bonding) problems on the basis of the many-electronic ground state stands for quantum-chemists almost as "to determine the weight of the captain of a large ship by weighing the ship when he is and when he is not on board" (Coulson, 1960). That is the case of electronegativity (EN or $\chi$), a concept that, being introduced from the Berzelius classification of atoms as electronegative or electropositive, had achieved over the years many definitions and interpretations concerning atomic reactivity and bonds formation (Huheey, 1978). Among many important contributions to this concept (Sen & Jørgensen, 1987), we are remembering the landmark contribution of Pauling (1932), which established an empirical scale of $\chi$ based on bond energies. In his view, the electronegativity means "the power of an atom in a molecule to attract electrons to itself" (Pauling, 1960). Just as Pauling, being interested on a scale to estimate the polarity of the chemical bonds, Mulliken, through the seminal works of 1934–1935 (Mulliken, 1934), averaged the $IP$ and the $EA$ for the valence states of an atom or radical to define his electronegativity as in Section 4.2.2.1, Eq. (4.3).

Many years after, a new emerging form of quantum mechanics, the DFT, appears as the modern quantum frame in which a chemical system (an atom, an ion, a radical, a molecule or several molecules) can be treated in a state of interaction (Parr & Yang, 1989). In this modern context, the cornerstone EN definition of Parr as the minus of the chemical potential ($\mu$) of a system in a grand canonical ensemble at zero temperature ($T$) was formulated (in atomic units), see (Parr et al., 1978), $\chi_P \equiv -\mu$ as in Eq. (3.1), when the ground state energy $E_N$ is assumed to be a smooth function of the total number of electrons $N$.

Due to the Hohenberg and Kohn theorems (1964), all properties of the ground state are functions only on $N$ and $V(\mathbf{r})$. Therefore, much chemistry is comprised in above EN, as $\mu = \mu[N, V(\mathbf{r}')]$ measures the escaping tendency of an electronic cloud from the equilibrium system.

However, if the finite-difference ($FD$) approximation of Parr EN (3.1) is employed, around the referential integer total number of electrons $N_0$ the original Mulliken formula (4.3) for EN is formally recovered

$$\chi_P \equiv -\left(\frac{\partial E_N}{\partial N}\right)_{V(r)} \cong \frac{IP + EA}{2} \equiv \chi_{FD} \qquad (4.247)$$

Nevertheless, at this point an opportunity for confusion and misunderstanding can arise. This because there is a conceptual difference between the DFT values of *IP* and *EA*, that are for the *ground state* of a system, in definition (4.247), and their averaged values on the supposed *valence or excited states*, as displayed in Eq. (3.1). Such dichotomy can be transposed at the potential level: in the Parr picture, since $V(\mathbf{r})$ is a non-zero constant the almost *vertical* values are involved, whereas in the Mulliken approach the almost *adiabatic* case is fixed by setting $V(\mathbf{r}) = 0$, as no further electrons are attached to the system.

The reconciliation of these two extremes in defining electronegativity between the valence and bounded electronic states, i.e., free and constant ground state potential limits, respectively, is the main purpose of this work. The consequences of the new emerging electronegativity ansatz are exposed in relation with associated reactivity indices and applied on the atomic scales.

Searching for the relation between the Mulliken valence- and Parr ground state-electronegativities, there was already proved that the former can be seen as the average of the last one (Komorowski, 1987; Putz, 2006):

$$\chi_M \equiv \frac{IP + EA}{2} = \frac{\left(E_{N_0-1} - E_{N_0}\right) + \left(E_{N_0} - E_{N_0+1}\right)}{2}$$

$$= \frac{E_{N_0-1} - E_{N_0+1}}{2} = -\frac{1}{2}\int_{N_0-1}^{N_0+1} dE_N \cong -\frac{1}{2}\int_{N_0-1}^{N_0+1}\left(\frac{\partial E_N}{\partial N}\right)_{V(r)} dN$$

$$= \frac{1}{2}\int_{N_0-1}^{N_0+1} \chi_P dN \qquad (4.248)$$

The series of relations (4.248) establishes, however, that the Mulliken and Parr electronegativities can be appropriately called the *absolute* and *chemical* ones, respectively.

Next, just on thermodynamic basis, when a bond $A - B$ is considered, the electronegativity concept comes into play to prove the bond polarity through the gauge reaction (4.157) providing the difference in energy of the right and left products, respectively, as (Putz, 2006):

$$\Delta E = E_{N_0-1}^A + E_{N_0+1}^B - E_{N_0+1}^A - E_{N_0-1}^B$$

$$= \left(E_{N_0-1}^A - E_{N_0}^A\right) + \left(E_{N_0}^A - E_{N_0+1}^A\right) - \left(E_{N_0-1}^B - E_{N_0}^B\right) - \left(E_{N_0}^B - E_{N_0+1}^B\right)$$

$$= IP^A + EA^A - \left(IP^B + EA^B\right)$$

$$= 2\left(\chi^A - \chi^B\right) \qquad (4.249)$$

It is thus clear that in order to proper describe the reactive propensity of the chemical systems the change in electronegativity has to be as well considered. That is to consider also the variation $d\chi_P$ and then to average it against the interval $(N_0 - 1, N_0 + 1)$ to extract the counterpart information respecting electronegativity, that should measure the *resistance* to the reactivity, named hardness ($\eta$) (Sen & Mingos, 1993; Pearson, 1997). In such, formally, the *absolute* hardness is obtained as (Putz, 2006, 2008a):

$$\eta_A = -\frac{1}{2}\int_{N_0-1}^{N_0+1} d\chi_P \cong -\frac{1}{2}\int_{N_0-1}^{N_0+1}\left(\frac{\partial \chi_P}{\partial N}\right)_{V(r)} dN \qquad (4.250)$$

that unfortunately cannot give particular information as far $\chi_P(N)$ remains unknown.

Still, if the Parr-Pearson or *chemical* hardness is abstracted from Eq. (4.250) as the derivative of the Parr's electronegativity (3.1) (Parr & Pearson, 1983), as in Eq. (3.3) presented, an operational definition of it from the method of finite differences can be achieved:

$$\eta_{PP} = \frac{1}{2}\left(\frac{\partial^2 E_N}{\partial N^2}\right)_{V(r)} \cong \frac{E_{N_0+1} - 2E_{N_0} + E_{N_0-1}}{2} = \frac{IP - EA}{2} \equiv \eta_{FD} \qquad (4.251)$$

recovering the Hinze formulation (4.79), while assisting the well-known principles of chemistry: the hard and soft acids and bases (HSAB) and the maximum hardness (MH) principles, see Volume III of the present five-volume work, and Pearson (1973, 1990); Chattaraj & Schleyer (1994); Chattaraj & Maiti (2003); Chattaraj et al. (1991, 1995).

Worth noting that when considering the factor (4.158), now rewritten as

$$\frac{1}{2} = \frac{1}{\int\limits_{N_0-1}^{N_0+1} dN} \qquad (4.252)$$

as well in Eq. (4.250) and as in Eq. (3.3), it accounts for the average of the acidic (electron accepting, $N_0 \le N \le N_0 +1$) and basic (electron donating, $N_0 - 1 \le N \le N_0$) behaviors, being therefore an inherent part of the hardness definition at large. Nevertheless, the Parr-Pearson electronegativity-hardness formulations, (4.3) and (4.251), respectively, provide a drastic unification of the $\chi$-theories and the HSAB rules in terms of IPs and electronic affinities (Putz, 2006, 2008a). Actually, beside the fact that the theoretical approaches and scales for Parr electronegativity (3.1) and Parr-Pearson hardness (3.3) under their chemical counterpart forms Eqs. (4.3) and (4.251), respectively, are available (Robles & Bartolotti, 1984), it is essential to consider the experimental values of $IP_E$ and $EA_E$, as displayed in the Table 4.3 (Lackner & Zweig, 1983), from where the finite-differences chemical $\chi_{FD}$ and $\eta_{FD}$ are abstracted at the atomic level.

The inspection of the Table 4.3 reveals, apart of the periodic trends for the chemical electronegativity, hardness, and IP that, across elements, the electron affinities poses as well positive as negative signs. This situation suggests that, when searching for a theoretical rationalization of these numbers, a proper combination of the adiabatic (valence) and vertical (ground state) approaches should apply. For the shake of generalization, the Parr electronegativity and the Parr-Pearson hardness definitions as derivatives, Eqs. (3.1) and (3.3), are labeled from now as $\chi_D$ and $\eta_D$, respectively. In this respect the confusion with their chemical counterparts, Eqs. (4.3) or (4.247) and (4.251), as the finite difference definitions, is avoided. However, from above discussion about absolute reactivity indices, the following present working definitions will be considered (Putz, 2006):

**TABLE 4.3** Experimental First Ionization Potential $IP_E$, First Electron Affinity $EA_E$, and Their Associated Chemical Electronegativity $\chi_{FD}$ and Chemical Hardness $\eta_{FD}$, Calculated Upon the Finite Difference Equations (4.3) or (4.247) and (4.251), Respectively, For the Ordinary Elements (Lackner & Zweig, 1983)

**Legend:** *Symbol of Element*
Finite Difference Electronegativity: $\chi_{FD}$
Finite Difference Hardness: $\eta_{FD}$
Experimental Ionization Potential: $IP_E$
Experimental Electron Affinity: $EA_E$

| Element | $\chi_{FD}$ | $\eta_{FD}$ | $IP_E$ | $EA_E$ |
| --- | --- | --- | --- | --- |
| H | 7.18 | 6.45 | 13.62 | 0.73 |
| He | 12.27 | 12.48 | 24.65 | -0.21 |
| Li | 3.02 | 4.39 | 5.41 | 0.62 |
| Be | 3.43 | 5.93 | 9.36 | -2.5 |
| B | 4.26 | 4.06 | 8.32 | 0.21 |
| C | 6.24 | 4.99 | 11.34 | 1.25 |
| N | 6.97 | 7.59 | 14.56 | -0.62 |
| O | 7.59 | 6.14 | 13.62 | 1.46 |
| F | 10.4 | 7.07 | 17.47 | 3.33 |
| Ne | 10.71 | 10.92 | 21.63 | -0.31 |
| Na | 2.80 | 2.89 | 5.02 | 0.52 |
| Mg | 2.6 | 4.99 | 7.7 | -2.39 |
| Al | 3.22 | 2.81 | 6.03 | 0.42 |
| Si | 4.68 | 3.43 | 8.22 | 1.25 |
| P | 5.62 | 4.89 | 10.5 | 0.73 |
| S | 6.24 | 4.16 | 10.4 | 2.08 |
| Cl | 8.32 | 4.68 | 13 | 3.64 |
| Ar | 7.7 | 8.11 | 15.81 | -0.42 |
| K | 2.39 | 1.98 | 4.37 | 0.52 |
| Ca | 2.29 | 3.85 | 6.14 | -1.66 |
| Sc | 3.43 | 3.22 | 6.55 | 0.21 |
| Ti | 3.64 | 3.22 | 6.86 | 0.42 |
| V | 3.85 | 2.91 | 6.76 | 0.94 |
| Cr | 3.74 | 3.12 | 6.76 | 0.62 |
| Mn | 3.85 | 3.64 | 7.49 | 0.31 |
| Fe | 4.26 | 3.64 | 7.90 | 0.62 |
| Co | 4.37 | 3.43 | 7.90 | 0.94 |
| Ni | 4.37 | 3.22 | 7.7 | 1.14 |
| Cu | 4.47 | 3.22 | 7.8 | 1.25 |
| Zn | 4.26 | 5.2 | 9.46 | -0.94 |
| Ga | 3.22 | 2.81 | 6.03 | 0.42 |
| Ge | 4.58 | 3.33 | 7.90 | 1.25 |
| As | 5.3 | 4.47 | 9.78 | 0.83 |
| Se | 5.93 | 3.85 | 9.78 | 2.08 |
| Br | 7.59 | 4.26 | 11.86 | 3.43 |
| Kr | 6.86 | 7.28 | 14.04 | -0.42 |

**TABLE 4.3** Continued

| Rb | Sr | Y | Zr | Nb | Mo | Tc | Ru | Rh | Pd | Ag | Cd | In | Sn | Sb | Te | I | Xe |
|---|---|---|---|---|---|---|---|---|---|---|---|---|---|---|---|---|---|
| 2.29 | 1.98 | 3.43 | 3.85 | 4.06 | 4.06 | 3.64 | 4.06 | 4.26 | 4.78 | 4.47 | 4.16 | 3.12 | 4.26 | 4.89 | 5.51 | 6.76 | 5.82 |
| 1.87 | 3.74 | 2.91 | 3.02 | 2.91 | 3.12 | 3.64 | 3.43 | 3.22 | 3.64 | 3.12 | 4.78 | 2.70 | 3.02 | 3.85 | 3.54 | 3.74 | 6.34 |
| 4.16 | 5.72 | 6.45 | 6.86 | 6.86 | 7.18 | 7.28 | 7.38 | 7.49 | 8.42 | 7.59 | 9.05 | 5.82 | 7.38 | 8.63 | 9.05 | 10.5 | 12.17 |
| 0.52 | -1.77 | 0.52 | 0.83 | 1.14 | 1.04 | 0.00 | 0.62 | 1.04 | 1.25 | 1.35 | -0.62 | 0.31 | 1.25 | 1.04 | 1.98 | 3.12 | -0.42 |

*All units are in eV (electron-volts) (Putz, 2006, 2008a).

$$\chi_A = -\frac{1}{2}\int_{N_0-1}^{N_0+1} dE_N \qquad (4.253)$$

$$\eta_A = -\frac{1}{2}\int_{N_0-1}^{N_0+1} d\chi_D \qquad (4.254)$$

$$IP_A = \int_{N_0}^{N_0-1} dE_N \qquad (4.255)$$

$$EA_A = \int_{N_0+1}^{N_0} dE_N \qquad (4.256)$$

as the *absolute* electronegativity, hardness, IP, and EA, respectively.

The definitions (4.253)–(4.256) are displayed in their most general form since the changes in the ground state energy ($dE_N$) as well in the chemical potential ($d\chi_D = -d\mu_D$) provide the fundamental DFT equations for treating the chemical reactivity (Putz, 2006, 2008a). Remarkably, if the semi-sum of the absolute *IP* and *EA* of Eqs. (4.255) and (4.256), respectively, is performed the absolute EN (4.253) is recovered. This way, the unification of the Mulliken and Parr approaches of electronegativity, through the relations (4.3) and (3.1), respectively, is consecrated whatever form $E_N$ of the total energy is assumed, due to actual integral absolute definitions. Instead, since the electronegativity-hardness (4.254) correlation involves the differential electronegativity rather that the total energy dependence there is expected that the absolute hardness to do not be reduced as the semi-difference of the absolute *IP* and *EA* of Eqs. (4.255) and (4.256), respectively. In this respect the actual integral absolute hardness (4.254) provides a generalization picture of the previous chemical hardness counterpart (4.251).

The unification of the Mulliken and the Parr electronegativity approaches stands, at the valence level, once the number of electrons $N_0$ is further identified with the number of the valence electrons $N_v$, while the ground state information are comprised through the involvement of the total energy of the system $E_N$, before the limits through the Eqs. (4.253), (4.255), and (4.256) definitions to be performed.

Nevertheless, from the relations (4.253)–(4.256) there is remarked that the absolute hardness (4.254) do not enter in such combined scheme as far as its valence nature, through the limits over the differential EN rather than on the total energy $E_N$, is emphasized. This conceptual behavior for the absolute hardness (4.254), respecting its absolute electronegativity counterpart (4.253), predicts special features for its computations' effects at whatever atomic or molecular levels.

However, the absolute $\chi$, $\eta$, $IP$, and $EA$ analytical density functionals based on (4.253)–(4.256) formulations, respectively, can be achieved within conceptual DFT chemistry, as will be further exposed. More, because of the Hohenberg-Kohn theorem prescription, the relations (4.253)–(4.256) open the possibility of the systematic treatment of the absolute $\chi$, $\eta$, $IP$, and $EA$ indices when either or both functional dependences on $N$ and $V(r)$ of the $E_N$ and $\chi_D$ are assumed; such systematics will be next exposed.


## 4.6   DENSITY FUNCTIONAL ELECTRONEGATIVITY AND CHEMICAL HARDNESS

### 4.6.1   FIRST ORDER VARIATION IN CHARGE

In considering the analytical realization of the present model, the case of the first order in charge expansion of the differential electronegativity $\chi_D$ and total energy $E_N$ is firstly treated (Putz, 2006).

Actually, for the differential EN we have the identities:

$$d\chi_D^{[1]} = \left( \frac{\partial \chi_D^{[1]}}{\partial N} \right)_{V(r)} dN = -\frac{1}{S} dN = -\frac{1}{a+N^2} dN \qquad (4.257)$$

that can immediately be integrated furnishing the simple first case-density functional for $\chi_D$:

$$\chi_D^{[1]} = \int_0^{N_v} d\chi_D^{[1]} = -\frac{1}{\sqrt{a}} \arctan\left( \frac{N_v}{\sqrt{a}} \right) \qquad (4.258)$$

with the given parameter of Eq. (4.223), $a = \int L(\mathbf{r}) d\mathbf{r}$. Worth remarking that the above integration procedure corresponds with the valence picture in that the valence shell is filled from 0 up to the pertinent number of valence electrons $N_v$. In this respect, $N_v$ cannot exceed a reasonable low number of electrons so assuring the stability of the concerned chemical system.

Within the same approach the absolute hardness of (4.254) can be as well evaluated to give its density functional, yielding the form:

$$\eta_A^{[1]} = -\frac{1}{2}\int_{N_v-1}^{N_v+1} d\chi_D^{[1]} = \frac{1}{2\sqrt{a}}\left[\arctan\left(\frac{N_v+1}{\sqrt{a}}\right) - \arctan\left(\frac{N_v-1}{\sqrt{a}}\right)\right] \quad (4.259)$$

Nevertheless, having computed $\chi_D^{[1]}$ as (4.258) it is useful in writing the total energy expansion accordingly:

$$dE_N^{[1]} = \left(\frac{\partial E_N^{[1]}}{\partial N}\right)_{V(r)} dN = -\chi_D^{[1]}(N)dN \quad (4.260)$$

Note that although the *physical* functional dependence of $\chi_D^{[1]}$ is of $N_v$ as resulting from (4.258), the $N$-dependence of $\chi_D^{[1]}$ in Eq. (4.260) was emphasized in order to be consistent with the *mathematical* expression of $dE_N^{[1]}$ in the view of further integration between the $N_v$ related limits. Nevertheless, such physical-mathematically switching between $N_v$ and $N$ dependences is circumscribed within the present valence-ground state unified approach of electronegativity and related indices.

There is now easy to recognize that the absolute EN (4.253) will be particularized by the appropriate integration of Eq. (4.260) when Eq. (4.258) is considered, giving the density functional (Putz, 2006):

$$\chi_A^{[1]} = -\frac{1}{2}\int_{N_v-1}^{N_v+1} dE_N^{[1]} = \frac{1}{2}\left\{\frac{1}{\sqrt{a}}\left[\begin{array}{c}(N_v-1)\arctan\left(\frac{N_v-1}{\sqrt{a}}\right)\\ -(N_v+1)\arctan\left(\frac{N_v+1}{\sqrt{a}}\right)\end{array}\right] + \frac{1}{2}\ln\left[\frac{a+(N_v+1)^2}{a+(N_v-1)^2}\right]\right\}$$

$$(4.261)$$

whereas, for the absolute *IP* of Eq. (4.255), by applying its specific limits of integration around the reference valence number of electrons $N_v$, the density functional results as:

$$IP_A^{[1]} = \int_{N_v}^{N_v-1} dE_N^{[1]} = \frac{1}{\sqrt{a}}\left[\begin{array}{c}(N_v-1)\arctan\left(\frac{N_v-1}{\sqrt{a}}\right)\\ -N_v\arctan\left(\frac{N_v}{\sqrt{a}}\right)\end{array}\right] + \frac{1}{2}\ln\left[\frac{a+N_v^2}{a+(N_v-1)^2}\right] \quad (4.262)$$

Finally, without being necessary to perform a separate integration, the density functional for the absolute *EA* of (4.256) will follow from the present absolute EN and *IP* ones, (4.261) and (4.262), respectively, through the relation:

$$EA_A^{[1]} = \int_{N_v+1}^{N_v} dE_N^{[1]} = 2\chi_A^{[1]} - IP_A^{[1]} \tag{4.263}$$

as already anticipated.

## 4.6.2  SECOND ORDER VARIATION IN CHARGE

In the same manner as previous, the expansion up to second order in charge will lead with the same steps, apart of arising of an additional term, in analyzing the consequences of the differential $\chi_D$ and $N_v$ variations.

Practically, the present $d\chi_D^{[2]}$, within the exposed approximate DFT softness hierarchy, expands under the form (Putz, 2006):

$$d\chi_D^{[2]} = \left(\frac{\partial \chi_D^{[2]}}{\partial N}\right)_{V(r)} dN + \frac{1}{2}\left[\frac{\partial}{\partial N}\left(\frac{\partial \chi_D^{[2]}}{\partial N}\right)_{V(r)}\right]_{V(r)} dNdN$$

$$= -\frac{1}{S}dN - \left[\frac{\partial}{\partial N}\left(\frac{1}{S}\right)\right]dNdN = -\frac{1}{a+N^2}dN + \frac{N}{\left(a+N^2\right)^2}dNdN \tag{4.264}$$

Then, from Eq. (4.264) the definite density functional $\chi_D^{[2]}$ is evaluated to be:

$$\chi_D^{[2]} = \int_0^{N_v} d\chi_D^{[2]} = -\frac{1}{\sqrt{a}}\arctan\left(\frac{N_v}{\sqrt{a}}\right) + \frac{N_v - \sqrt{a}\arctan\left(N_v/\sqrt{a}\right)}{2a} \tag{4.265}$$

while the associate absolute hardness takes from Eq. (4.254) the actual density functional expression:

$$\eta_A^{[2]} = -\frac{1}{2}\int_{N_v-1}^{N_v+1} d\chi_D^{[2]} = \frac{1}{4\sqrt{a}}\left\{\begin{array}{c}\arctan\left(\frac{N_v-2}{\sqrt{a}}\right) + \arctan\left(\frac{N_v+2}{\sqrt{a}}\right) \\ \left[-2\left[\arctan\left(\frac{N_v-1}{\sqrt{a}}\right) + \arctan\left(\frac{N_v}{\sqrt{a}}\right) - \arctan\left(\frac{N_v+1}{\sqrt{a}}\right)\right]\right]\end{array}\right\}$$

$$\tag{4.266}$$

Next, calling $\chi_D^{[2]}$ when going to the second order in charge expansion for the total energy $E_N$,

$$dE_N^{[2]} = \left(\frac{\partial E_N^{[2]}}{\partial N}\right)_{V(\mathbf{r})} dN + \frac{1}{2}\left(\frac{\partial^2 E_N^{[2]}}{\partial N^2}\right)_{V(\mathbf{r})} dNdN$$

$$= -\chi_D^{[2]}(N)dN - \frac{1}{2}\left(\frac{\partial \chi_D^{[2]}(N)}{\partial N}\right)_{V(\mathbf{r})} dNdN \qquad (4.267)$$

as before, the proper integrations of Eqs. (4.253), (4.255) and (4.256) will provide the density functionals for the actual absolute electronegativity (Putz, 2006):

$$\chi_A^{[2]} = -\frac{1}{2}\int_{N_v-1}^{N_v+1} dE_N^{[2]} = \frac{3}{16a}\left\{4N_v + 3a\ln\left[\frac{a+(N_v+1)^2}{a+(N_v-1)^2}\right] + 6\sqrt{a}\left[(N_v-1)\arctan\left(\frac{N_v-1}{\sqrt{a}}\right) - (N_v+1)\arctan\left(\frac{N_v+1}{\sqrt{a}}\right)\right]\right\} \qquad (4.268)$$

for the absolute $IP$:

$$IP_A^{[2]} = \int_{N_v}^{N_v-1} dE_N^{[2]} = \frac{3}{8a}\left\{2N_v - 1 + 3a\ln\left[\frac{a+N_v^2}{a+(N_v-1)^2}\right] + 6\sqrt{a}\left[(N_v-1)\arctan\left(\frac{N_v-1}{\sqrt{a}}\right) - N_v\arctan\left(\frac{N_v}{\sqrt{a}}\right)\right]\right\} \qquad (4.269)$$

and for the absolute $EA$:

$$EA_A^{[2]} = \int_{N_v+1}^{N_v} dE_N^{[2]} = 2\chi_A^{[2]} - IP_A^{[2]} \qquad (4.270)$$

respectively.

### 4.6.3  FIRST ORDER VARIATION IN CHARGE AND POTENTIAL

Further improvement in our systematic applies when also the potential variation is taken into account. Such picture is particularly meaningful for assigning reactivity indices based on external potential direct influence (Ayers & Parr, 2001).

Under the first order in charge and potential variation, the differential EN expansion primarily looks like (Putz, 2006):

$$d\chi_D^{[3]} = \left( \frac{\partial \chi_D^{[3]}}{\partial N} \right)_{V(r)} dN + \int \left( \frac{\delta \chi_D^{[3]}}{\delta V(r)} \right)_N \delta V(\mathbf{r}) d\mathbf{r} \qquad (4.271)$$

from where there is observed, that in order to make a closed contact with softness DFT picture, the functional derivative $\left( \delta \chi_D^{[3]} / \delta V(r) \right)_N$ have to be rearranged firstly. In doing so, worth introducing also the first order expansion of the total energy respecting the charge and potential:

$$dE_N^{[3]} = \left( \frac{\partial E_N^{[3]}}{\partial N} \right)_{V(r)} dN + \int \left( \frac{\delta E_N^{[3]}}{\delta V(r)} \right)_N \delta V(\mathbf{r}) d\mathbf{r} \qquad (4.272)$$

Since the functional derivative of the $E_N$ with respect to the potential $V(\mathbf{r})$ may be determined from the fundamental DFT relation (4.173) the Eq. (4.272) rewrites as:

$$dE_N^{[3]} = -\chi_D^{[3]}(N)dN + \int \rho(\mathbf{r})\delta V(\mathbf{r}) d\mathbf{r} \qquad (4.273)$$

from where the Maxwell reciprocal relations can be abstracted as in Eq. (4.175), with the appropriate chain rule (4.196) specific to the actual softness approach. However, with these information, the above (4.271) equation of $d\chi_D^{[3]}$ takes the integrable form:

$$d\chi_D^{[3]} = -\frac{1}{S}dN - \int \frac{s(\mathbf{r})}{S} \delta V(\mathbf{r}) d\mathbf{r} \qquad (4.274)$$

Nevertheless, when performing the integration to get, for instance, the absolute electronegativity of (4.253) the path integral over $\delta V(\mathbf{r})$ is involved. This can be solved between the adiabatic $(V(\mathbf{r}) = 0)$ and vertical $V(\mathbf{r}) = ct \neq 0$ limits. Such treatment corresponds with the physical picture in which an electron can be added to the chemical system from infinity due to its electronegativity (Putz et al., 2003). This approach is consistent also with $IP$ and

*EA* as far it was already established that they are in close relation with EN, even at the absolute (or integral) level. This way, the adiabatic and vertical approaches of defining EN respecting the valence and ground state, respectively, are as well combined through the described potential path integral recipe, reflecting another level of unifying Mulliken and Parr electronegativity views. Under these circumstances the finite EN is now successively calculated from (4.274) until its density functional form (Putz, 2006):

$$
\chi_D^{[3]} = \int_0^{N_v} d\chi_D^{[3]} = -\int_0^{N_v} \frac{1}{a+N^2} dN - \int_0^{V(\mathbf{r})} \int_0^{} \frac{1}{a+N^2} [L(\mathbf{r}) + N\rho(\mathbf{r})] \delta V(\mathbf{r}) d\mathbf{r}
$$

$$
= -\frac{1}{\sqrt{a}} \arctan\left(\frac{N_v}{\sqrt{a}}\right) - \frac{b+NC_A}{a+N^2}\Bigg|_{N \to N_v}
\tag{4.275}
$$

where the additional notation:

$$
b \equiv \int L(\mathbf{r}) V(\mathbf{r}) d\mathbf{r}
\tag{4.276}
$$

beside the introduced chemical action

$$
C_A = \int \rho(\mathbf{r}) V(\mathbf{r}) d\mathbf{r}
\tag{4.277}
$$

However, in performing the potential path integral of (4.275) another assumption was made, namely to consider both $L(\mathbf{r})$ and $\rho(\mathbf{r})$ as "independently" of $V(\mathbf{r})$ as far as they do not pose an explicit dependence on it. Instead, the new introduced $V(r)$-dependent quantities, $b$ and $C_A$ of Eqs. (4.276) and (4.277), respectively, were considered "independent" of $N$, as before was the case in Eq. (4.223).

   In similar manner, the absolute hardness is obtained from the integration of Eq. (4.274) between the limits prescribed in Eq. (4.254) with the density functional result:

$$
\eta_A^{[3]} = -\frac{1}{2} \int_{N_v-1}^{N_v+1} d\chi_D^{[3]}
$$

$$
= \frac{1}{2}\left\{ \frac{2[C_A(1+a-N_v^2)-2bN_v]}{(1+a)^2 + 2(a-1)N_v^2 + N_v^4} + \frac{1}{\sqrt{a}}\left[ \arctan\left(\frac{N_v+1}{\sqrt{a}}\right) - \arctan\left(\frac{N_v-1}{\sqrt{a}}\right) \right] \right\}
\tag{4.278}
$$

Then, back to the total energy differential expansion (4.273), by considering the previous deduced $\chi_D^{[3]}$ with Eq. (4.275) analytic form and making use of the described path integration methodology, one gets from the Eqs. (4.253) and (4.255) definitions the present density functionals (Putz, 2006, 2011a):

$$\chi_A^{[3]} = -\frac{1}{2}\int_{N_v-1}^{N_v+1} dE_N^{[3]} = \frac{1}{2}\left\{\frac{1}{\sqrt{a}}\left[\begin{array}{l}(b+N_v-1)\arctan\left(\dfrac{N_v-1}{\sqrt{a}}\right)\\[2mm]-(b+N_v+1)\arctan\left(\dfrac{N_v+1}{\sqrt{a}}\right)\end{array}\right]\right.$$
$$\left.+\frac{C_A-1}{2}\ln\left[\frac{a+(N_v-1)^2}{a+(N_v+1)^2}\right]\right\}$$

(4.279)

$$IP_A^{[3]} = \int_{N_v}^{N_v-1} dE_N^{[3]} = \frac{1}{\sqrt{a}}\left[\begin{array}{l}(b+N_v-1)\arctan\left(\dfrac{N_v-1}{\sqrt{a}}\right)\\[2mm]-(b+N_v)\arctan\left(\dfrac{N_v}{\sqrt{a}}\right)\end{array}\right] + \frac{C_A-1}{2}\ln\left[\frac{a+(N_v-1)^2}{a+N_v^2}\right]$$

(4.280)

for the absolute EN and IP, respectively, while for the absolute EA of Eq. (4.256) there is maintained the functional relation:

$$EA_A^{[3]} = \int_{N_v+1}^{N_v} dE_N^{[3]} = 2\chi_A^{[3]} - IP_A^{[3]}$$

(4.281)

as already consecrated. As a note, the absolute electronegativity $\chi_A^{[3]}$ under the Eq. (4.279) density functional form have been as well deduced based on the so-called *chemical action* (4.277) functional (Putz, 2008a), within a conceptual variational DFT framework, its reliability being then tested in deriving the atomic radii and the associate size-dependent atomic properties (Putz et al., 2003).

### 4.6.4 SECOND ORDER VARIATION IN CHARGE—FIRST ORDER IN POTENTIAL

The present venture further supports a more rich systematic structure in differential electronegativity and total energy expansions. For instance, when the expansions up to the second order in charge and of the first order in potential are brought together will produce the $d\chi_D^{[4]}$ expression, successively as (Putz, 2006):

$$d\chi_D^{[4]} = \left(\frac{\partial \chi_D^{[4]}}{\partial N}\right)_{V(\mathbf{r})} dN + \frac{1}{2}\left[\frac{\partial}{\partial N}\left(\frac{\partial \chi_D^{[4]}}{\partial N}\right)_{V(\mathbf{r})}\right]_{V(\mathbf{r})} dNdN + \int\left(\frac{\delta \chi_D^{[4]}}{\delta V(\mathbf{r})}\right)_N \delta V(\mathbf{r})d\mathbf{r}$$

$$= -\frac{1}{S}dN - \left[\frac{\partial}{\partial N}\left(\frac{1}{S}\right)\right]dNdN - \int\frac{s(\mathbf{r})}{S}\delta V(\mathbf{r})d\mathbf{r}$$

$$= -\frac{1}{a+N^2}dN + \frac{N}{\left(a+N^2\right)^2}dNdN - \int\frac{L(\mathbf{r})+N\rho(\mathbf{r})}{a+N^2}\delta V(\mathbf{r})d\mathbf{r}$$

$$(4.282)$$

One should note at this point that in above electronegativity expansion, Eq. (4.258), the first order in potential variation brings the frontier contribution to the system though the basic definition of Fukui function of Eq. (4.196) (Yang & Parr, 1985) while the presence of the second order derivation respecting potential in the energy expansion of Eq. (4.257)

$$\int\int\left(\frac{\delta^2 E_N}{\delta V(\mathbf{r})\delta V(\mathbf{r'})}\right)_N \delta V(\mathbf{r})\delta V(\mathbf{r'})d\mathbf{r}d\mathbf{r'} = \int\int \kappa(\mathbf{r},\mathbf{r'})\delta V(\mathbf{r})\delta V(\mathbf{r'})d\mathbf{r}d\mathbf{r'} \quad (4.283)$$

is accounted through the basic DFT relationship (4.173) by the contribution of the so-called *linear response function* of Eq. (4.197) (Chermette, 1999) that sums the negative the kernel softness $s(\mathbf{r}, \mathbf{r'})$ (Berkowitz & Parr, 1988) with local softness coupling averaged by the total softness index. Yet, it apparently does not influence the chemical information contents because it vanishes under the following successive integral transformations (Putz, 2011d)

$$\int\int \kappa(\mathbf{r},\mathbf{r'})\delta V(\mathbf{r})\delta V(\mathbf{r'})d\mathbf{r}d\mathbf{r'} = \int\int\left\{\left(\frac{\delta\rho(\mathbf{r})}{\delta V(\mathbf{r'})}\right)_N \delta V(\mathbf{r})\right\}\delta V(\mathbf{r'})d\mathbf{r}d\mathbf{r'}$$

$$= \int\int\left\{\left(\frac{\delta[\rho(\mathbf{r})\delta V(\mathbf{r})]}{\delta V(\mathbf{r'})}\right)_N - \rho(\mathbf{r})\underbrace{\left(\frac{\delta[\delta V(\mathbf{r})]}{\delta V(\mathbf{r'})}\right)_N}_{=0}\right\}d\mathbf{r}\delta V(\mathbf{r'})d\mathbf{r'}$$

$$= \int\frac{\delta}{\delta V(\mathbf{r'})}\underbrace{\left\{\int\rho(\mathbf{r})\delta V(\mathbf{r})d\mathbf{r}\right\}}_{=0}\delta V(\mathbf{r'})d\mathbf{r'} = 0$$

$$(4.284)$$

since assuming a non-correlated external potential—i.e., in the absence of any quantum entanglement whatsoever (Bruss, 2002)—motivating the first null in Eq. (4.284), and due to the application of the Hellmann-Feynman theorem (Hellmann, 1937; Feynman, 1939) prescribing zero average of the exercised force $F(\mathbf{r})$ of applied external potential around the ground/equilibrium state (Deb, 1974), here employed under the functional derivative form

$$0 = \langle \mathbf{F}(\mathbf{r}) \rangle = \int \rho(\mathbf{r}) \mathbf{F}(\mathbf{r}) d\mathbf{r} = -\int \rho(\mathbf{r}) \delta V(\mathbf{r}) d\mathbf{r} \tag{4.285}$$

for motivating the second null in Eq. (4.284). However, it is worth noting that the term (4.284) also connects with the *chemical action* (4.277) and of its principle (Putz, 2009a)

$$\delta \mathcal{C}_A = 0 \tag{4.286}$$

see Volume III of the present five-volume work. Within the approximations the integration of Eq. (4.282) produces the finite differential EN first (Putz, 2006):

$$\chi_D^{[4]} = \int_0^{N_v} d\chi_D^{[4]}$$

$$= -\frac{1}{\sqrt{a}} \arctan\left(\frac{N_v}{\sqrt{a}}\right) + \frac{N_v - \sqrt{a}\arctan(N_v / \sqrt{a})}{2a} - \frac{b + N_v C_A}{a + N_v^2} \tag{4.287}$$

and then the absolute hardness (4.254) with the actual density functional form:

$$\eta_A^{[4]} = -\frac{1}{2} \int_{N_v-1}^{N_v+1} d\chi_D^{[4]}$$

$$= \frac{1}{4} \left\{ \frac{4\left[C_A\left(1 + a - N_v^2\right) - 2bN_v\right]}{\left(1 + a\right)^2 + 2(a-1)N_v^2 + N_v^4} + \frac{1}{\sqrt{a}}\left[\arctan\left(\frac{N_v-2}{\sqrt{a}}\right) + \arctan\left(\frac{N_v+2}{\sqrt{a}}\right)\right] - \frac{2}{\sqrt{a}}\left[\arctan\left(\frac{N_v-1}{\sqrt{a}}\right) + \arctan\left(\frac{N_v}{\sqrt{a}}\right) - \arctan\left(\frac{N_v+1}{\sqrt{a}}\right)\right] \right\} \tag{4.288}$$

Next, the EN of Eq. (4.287) stands as an important ingredient in writing the total energy expansion (Putz, 2006):

$$dE_N^{[4]} = \left(\frac{\partial E_N^{[4]}}{\partial N}\right)_{V(\mathbf{r})} dN + \frac{1}{2}\left(\frac{\partial^2 E_N^{[4]}}{\partial N^2}\right)_{V(\mathbf{r})} dNdN + \int\left(\frac{\delta E_N^{[4]}}{\delta V(\mathbf{r})}\right)_N \delta V(\mathbf{r})d\mathbf{r}$$

$$= -\chi_D^{[4]}(N)dN - \frac{1}{2}\left(\frac{\partial \chi_D^{[4]}(N)}{\partial N}\right)_{V(\mathbf{r})} dNdN + \int \rho(\mathbf{r})\delta V(\mathbf{r})d\mathbf{r} \qquad (4.289)$$

with the help of which also the absolute density functionals of electronegativity:

$$\chi_A^{[4]} = -\frac{1}{2}\int_{N_v-1}^{N_v+1} dE_N^{[4]}$$

$$= \frac{3}{16a}\left\{ 4N_v + 2\sqrt{a}\left[ \begin{array}{l} (2b+3N_v-3)\arctan\left(\frac{N_v-1}{\sqrt{a}}\right) \\ -(2b+3N_v+3)\arctan\left(\frac{N_v+1}{\sqrt{a}}\right) \end{array} \right] + a(2C_A-3)\ln\left[\frac{a+(N_v-1)^2}{a+(N_v+1)^2}\right] \right\} \qquad (4.290)$$

of $IP$:

$$IP_A^{[4]} = \int_{N_v}^{N_v-1} dE_N^{[4]}$$

$$= \frac{3}{8a}\left\{ 2N_v - 1 + 2\sqrt{a}\left[ \begin{array}{l} (2b+3N_v-3)\arctan\left(\frac{N_v-1}{\sqrt{a}}\right) \\ -(2b+3N_v)\arctan\left(\frac{N_v}{\sqrt{a}}\right) \end{array} \right] + a(2C_A-3)\ln\left[\frac{a+(N_v-1)^2}{a+N_v^2}\right] \right\} \qquad (4.291)$$

and of $EA$:

$$EA_A^{[4]} = \int_{N_v+1}^{N_v} dE_N^{[4]} = 2\chi_A^{[4]} - IP_A^{[4]} \qquad (4.292)$$

are specifically deduced from the definitions (4.253), (4.255), and (4.256), respectively, as usual in this systematic.

Finally, worth specifying that the presented line of systematic formulations of the density functional reactivity indices can be in principle continued when also the expansions containing higher order terms in potential are considered through the nonlinear electronic responses (Senet, 1996, 1997). The recent effects as the spin-philicity and spin-donicity in spin-catalysis phenomena can be rationalized on such generalized analysis (Pérez et al., 2002). Therefore, this way, also a closely diagrammatical theory of the absolute $\chi$, $\eta$, $IP$, and $EA$ can be built up with increasing accuracy in the non-local effects that the softness kernel approximation may induce.

### 4.6.5  NOTE ON DENSITY FUNCTIONAL MULLIKEN ELECTRONEGATIVITY

While the integral approximation in Eq. (4.248) may be rewriting in a more elaborate way (Putz et al., 2005)

$$\chi_M = \frac{E(N-1)-E(N+1)}{2} = -\frac{1}{2}\int_{(N-1)}^{(N+1)} dE_{(N)}[N,V(\mathbf{r})]$$

$$= -\frac{1}{2}\int_{(N-1)}^{(N+1)}\left[\left(\frac{\partial E_{(N)}}{\partial N}\right)_V dN + \int \rho_{(N)}(\mathbf{r})dV_{(N)}(\mathbf{r})d\mathbf{r}\right]$$

$$= \frac{1}{2}\int_{N-1}^{N+1}\chi(N)dN - \frac{1}{2}\left[\int \rho_{(N)}(\mathbf{r})V_{(N+1)}(\mathbf{r})d\mathbf{r} - \int \rho_{(N)}(\mathbf{r})V_{(N-1)}(\mathbf{r})d\mathbf{r}\right] \qquad (4.293)$$

now, according with the Hohenberg–Kohn first theorem (Hohenberg & Kohn, 1964) and of the chemical action principle, see Eq. (4.286) and (Putz, 2009a, 2011b) the two terms in the right hand side bracket of Eq. (4.293) identically vanish since does not optimize the associations of the electronic density of one state with the external potential applied on that state, thus leaving with the identity (4.279) [Putz et al., 2003; Putz, 2006) of the *absolute* electronegativity as the true *chemical* Mulliken density functional (Putz, 2008c).

### 4.6.6  PATH INTEGRAL CONNECTION WITH DENSITY FUNCTIONAL THEORY

As the DFT prescribes, all the required main information about an electronic state is found in its external potential. From the physical point of view this approach is natural based on the fact that the imposed external influence will determine the actual electronic density. A classical link between these two quantities is made by the associated Schrödinger equation for the specific potential and giving the eigenfunction ($\Psi$) as the solution the electronic density will be obtain by making the squared $|\Psi|^2$. However, this quantum classical way introduces the eigenfunction as another intermediate step and we don't follow this route (Putz, 2009b).

Alternatively there is another formalism that gives the electronic density intermediated by the electronic partition function ($Z$) and in which instead to solve a differential equation, as Schrödinger equation is, it is proposed for solving a parametrical integral. This method was initiated by Feynman and now it is called the path integral formalism (Feynman, 1948). From the various formulation and equivalences of the path integral formalism we prefer to consider the quantum statistical picture in which the path integral of partition function has the form (2.21) (Kleinert, 2004), see also see Section 2.3: in expression (2.21) clearly appears the parametrical dependence of the spatial coordinates by the so-called "quantum statistical time" ($\tau \equiv \hbar\beta$) where $\hbar$ states as the Planck's universal constant with $\beta = 1/(k_B T)$ as the inverse of the temperature in terms of the Boltzmann's universal constant $k_B$, $m$ being the electronic mass. Because this $\tau$-parametric dependency the above integral is called path-integral.

However, for the practical evaluation, the general path integral for the electronic partition function it is further transformed and approximated. For such, Feynman and Kleinert have proposed the classical effective potential version of the partition function (2.21) reduces it to the simple form of Eq. (2.64) where the path influence it was comprised within the introduced Feynman centroid (2.20). Thus, the Feynman-Kleinert (FK) path integral approximation scheme of Section 2.5.2 provides a very elegant recipe for partition function calculation using only the external potential dependence. The FK relations furnish the simplicity in the path integral analyticity but still remains at the formal level because of the unspecific analytical form of the classical effective potential as a function of the Feynman centroid, here $\mathbf{r}_0 \equiv x_0$. At this moment it will be introduced another level of approximation within Feynman-Kleinert formalism, controlled by the variational principle.

As it was derived in the previous Section 4.6.5 the Mulliken density functional electronegativity requires the knowledge of the electronic density under the external potential influence. Being exposed all the ingredients for the analytical expression for the partition function with only the external potential dependence, the electronic density computed through out of Feynman-Kleinert path integral algorithm takes the form, see also Eq. (2.11):

$$\rho_1(\mathbf{r}_0) = Z_1^{-1} \frac{1}{\sqrt{2\pi\hbar^2 \beta / m_0}} \exp\left[-\beta W_1(\mathbf{r}_0)\right] \qquad (4.294)$$

Here it is observed that the normalization condition looks like:

$$\int_{-\infty}^{+\infty} \rho_1(\mathbf{r}_0) d\mathbf{r}_0 = 1 \qquad (4.295)$$

instead of standard DFT formulation (4.168). However, this fact is in accordance with the picture of the effective valence electron which will be ionized or will be added under the core influence. This way of computing the electronegativity will serve us as the second scale called *path integral (PI) electronegativity* scale.

For the concrete calculations it is easily observed that the self-consistency of the involved parameters requires extra-input trial information. This supplementary knowledge can be avoided if we consider an additional limit. It is introduced the so-called *Markovian approximation* that regards the Eq. (2.122) limit (see discussion of Section 2.5.5) $\hbar\beta \to 0$. We should mention here that this limit corresponds with the ultra-short correlation of the involved electrons with the applied external potential. This can be motivated by remembering the temporal nature of the quantum statistical quantity $\hbar\beta \propto \Delta t$, see for instance Eq. (3.36). This means that assuming initially $(\Delta t = 0 \Leftrightarrow \beta = 0)$ the electronic system in the free motion such that no potential influence is felt, as the external potential applies it impose an immediately $\Delta t \to 0 \Leftrightarrow \beta \to 0$ for orbit stabilization of the electronic system, see Eq. (2.122). Moreover, this limit introduces also *correlation effects* with medium (Putz, 2009b).

We start performing the above Markovian limit that gives:

$$a^2(\mathbf{r}_0) \overset{\beta \to 0}{=} \frac{\beta}{12} \qquad (4.296)$$

Then, in the smeared-out potential relation there is considered the change of variable in the way that

$$z(\mathbf{r}_0') = (\mathbf{r}_0' - \mathbf{r}_0) / \sqrt{2a^2(\mathbf{r}_0)} \tag{4.297a}$$

and

$$d\mathbf{z} = d\mathbf{r}_0' / \sqrt{2a^2(\mathbf{r}_0)} \tag{4.297b}$$

that permits the writing of the smeared out potential expression successively in the so-called *Wigner expansion* (Putz, 2009b):

$$
\begin{aligned}
V_{a^2}(\mathbf{r}_0) &= \int_{-\infty}^{+\infty} \frac{d\mathbf{r}_0'}{\sqrt{2\pi a^2(\mathbf{r}_0)}} V(\mathbf{r}_0') \exp\left[ -\frac{(\mathbf{r}_0' - \mathbf{r}_0)^2}{2a^2(\mathbf{r}_0)} \right] \\
&= \frac{1}{\sqrt{2\pi a^2(\mathbf{r}_0)}} \sqrt{2a^2(\mathbf{r}_0)} \int_{-\infty}^{+\infty} d\mathbf{z}\, V\left( \mathbf{r}_0 + \sqrt{2a^2(\mathbf{r}_0)}\mathbf{z} \right) \exp\left[ -\mathbf{z}^2 \right] \\
&\overset{\beta \to 0}{=} \frac{1}{\sqrt{\pi}} \int_{-\infty}^{+\infty} d\mathbf{z} \left\{ V(\mathbf{r}_0) + \sqrt{2a^2(\mathbf{r}_0)}\mathbf{z} V'(\mathbf{r}_0) + \frac{1}{2} 2a^2(\mathbf{r}_0)\mathbf{z}^2 V''(\mathbf{r}_0) + ... \right\} \exp\left[ -\mathbf{z}^2 \right] \\
&= \frac{1}{\sqrt{\pi}} \int_{-\infty}^{+\infty} d\mathbf{z} \left\{ V(\mathbf{r}_0) + \frac{1}{2} 2a^2(\mathbf{r}_0)\mathbf{z}^2 V''(\mathbf{r}_0) \right\} \exp[-\mathbf{z}^2] \\
&= V(\mathbf{r}_0) + \frac{1}{2} a^2(\mathbf{r}_0) V''(\mathbf{r}_0)
\end{aligned}
\tag{4.298}
$$

Finally, the optimized frequency (2.84) results in the Markovian limit as:

$$\Omega^2(\mathbf{r}_0) = \frac{1}{m_0} V''(\mathbf{r}_0) \tag{4.299}$$

The smeared out potential can be summed up as an explicit dependency of the bar external potential $V(\mathbf{r})$ and its second derivative. Finally, with partition function and corresponding electronic density the Mulliken electronegativity density functional is computed within path integral picture.

As a note, it should be mention that Parr and Yang (1989) have shown, how the integral formulation of the Kohn-Sham DFT arrives to the electronic density expression performing Wigner semiclassical expansion combined with the short time approximation regarding to the $\beta$ parameter. However, the common tool between their and the actual electronic density excursion states for the introduced effective potentials, that here it was further approximated by the smeared out (effective) potential. Then, all the potential components around the original one $V(\mathbf{r})$ can be formally interpreted as the *exchange-correlation path integral potential* $V_{XC}^{PI}(\mathbf{r})$ with medium. This potential can be derived in any desired potential expansion order as based on Eq. (2.80), yet we will limit ourselves here to the second order with the result (Putz, 2009b)

$$W_1(\mathbf{r}_0)_{\beta \to 0} \cong V(\mathbf{r}_0) + \hbar^2 \frac{\beta}{24m_0} V''(\mathbf{r}_0) \equiv V(\mathbf{r}_0) + V_{XC}^{PI}(\mathbf{r}_0) \qquad (4.300)$$

From the identification applied within the two above forms of effective potentials there is clear that the exchange-correlation path integral with medium

$$V_{XC}^{PI}(\mathbf{r}_0) = \hbar^2 \frac{\beta}{24m_0} V''(\mathbf{r}_0) \qquad (4.301)$$

corrects in a clearly quantum manner the original classical external potential $V(\mathbf{r})$.

It is obvious that for applying the present Mulliken electronegativity formula to the atomic systems it should be know at least the core potential in which the valence electrons are evolved. Fortunately the pseudopotential theory provides such information for each atomic system starting from the Li one. It is natural to choose this way for our purpose because in this case the pseudopotential is seen as the external potential that applies to the valence electrons in agreement with the density functional picture. Then, the different electronic density formulation can be considered.

More, because we are interested in the Mulliken electronegativity (in order to can compare our formula with the available related values) it is significant to consider the unity normalization condition, see also Eq. (4.295), in the valence state (Putz, 2009b):

$$\int [\rho(\mathbf{r})]_{valence} d\mathbf{r} = 1 \qquad (4.302)$$

Such *path integral-PI* normalization leads with the well chemical interpretation that both ionization that act to extract one valence electron and the electronic affinity that tend to attract another electron to form a stabilized valence structure are motivated in terms of the one *effective valence electron*.

Within the pseudopotential methods we arrive at two possibilities for the electronic density and thus the electronegativity evaluations. First one consider only the pseudopotentials into the path integral formalism that gives the electronic density in the quantum statistical manner as it was described in the previous section. This way has a strong physical meaning because all the information about the electronic density and electronegativity are comprised (and dictated) only by the pseudopotential. The problem that arises in this approach is that the electronic density depends on the $\beta$ parameter. This parameter will be fixed so that the electronic density to fulfill the *path integral* normalization condition. Additionally, the search of the $\beta$ parameter must be done in the Markovian limit ($\beta \to 0$) for which the path integral formalism was performed.

The second approach takes beyond to the pseudopotential data also the valence basis and the electronic densities are then computed in the classical quantum manner. At this point we need to consider the working orbital type for the atomic systems and we will chose the s-basis set because its spherical symmetry. This method resembles the previous discussed pseudopotential approach however here combined with PI-effective smeared out potential.

As it was presented, both electronic density approaches have their own parametric dependency. Then these densities with fixed parameters are implemented for computation of the electronegativity. This implies that also the final electronegativity will have fixed scaling parameters. In both cases the appearance of the fixed parameters has the scaling effect to the electronic density in order to reduce the valence electronic system at one effective valence electron for which its density normalize to one. Making this assumption at beginning of computation we should recover in the final electronegativity the real (many) electronic valence state by an adequate re-scaling in terms of the specific values for the fixed ($\beta$ and $q$) parameters.

Remarkably, the Feynman-Kleinert variational algorithm in path integrals may be viewed as providing the calculation of the effective electronic density by constructing the constraint-searched partition function picture as the Levy constraint-search formalism (Levy, 1982) does in seeking the electronic density from the trial wave function. More clearly, Levy's recipe prescribes that the ground state energy minimization scheme—within the second (Hohenberg & Kohn, 1964) theorem—involves, in fact, two steps: one over all wave functions ($\Psi$) that give the same density (the inner minimization step) followed by the minimization throughout all density classes (the outer minimization step) (Putz, 2009b)

$$
\begin{aligned}
E_0 &= \min_{\rho}\left[\min_{\Psi \to \rho}\langle\Psi|(T + V_{ee} + V)|\Psi\rangle\right] \\
&= \min_{\rho}\left[\min_{\Psi \to \rho}\langle\Psi|(T + V_{ee})|\Psi\rangle + \int\rho(x)V(x)dx\right] \\
&= \min_{\rho}\left(F_{HK}[\rho] + C_A[\rho]\right) \\
&= \min_{\rho}\left(E[\rho]\right)
\end{aligned}
\tag{4.303}
$$

through recalling the density functional basic functionals of Eqs. (4.165), (4.166), and (4.277).

Such equivalence between the path integral Feynman-Kleinert formalism and the density functional Levy's one recommends the use of the Feynman-Kleinert density/densities for being implemented in density functionals with chemical relevance, the electronegativity for instance.

In this regard, the density functional for Mulliken electronegativity is in next computed and exemplified for atomic scales.

## 4.7 ATOMIC SCALES AND PERIODICITY BY REACTIVITY DFT INDICES

### 4.7.1 QUANTUM ATOM ON VALENCE STATE

A natural basic test for the just obtained systematic absolute $\chi$, $\eta$, *IP*, and *EA* density functionals applies for the atomic systems.

Nevertheless, when many electronic systems are involved inherent approximations have to be considered. According with a pre-quantum mechanical thinking (Bohr, 1921), each electron in many-electronic atoms should have its own one-electron function or orbital, a picture suggested as well by the currently self-consistent methods (Hartree, 1957), being fairly rationalized by many spectroscopic facts since the energy levels are identified with those of a single electron moving in a central field (White, 1934). Therefore, the one-electron picture stands as the general frame in which also the present analysis is situated (Putz, 2006).

The starting point will be a hydrogenic atom containing an atomic nucleus of charge Z and an electron interacting by means of the Coulomb potential (in atomic units):

$$V(r) = -\frac{Z}{r} \tag{4.304}$$

Since the interaction (4.304) is central, the associate wave equation may be separated in spherical polar coordinates to produce the normalized radial function. For the bound states hydrogenic atoms in the case of an infinitely heavy nucleus it looks like (Bransden & Joachain, 1983):

$$R_{n,l}(r) = -\left[ (2\xi)^3 \frac{(n-l-1)!}{2n[(n+l)!]^3} \right]^{1/2} (2\xi r)^l \exp(-\xi r) L_{n+l}^{2l+1}(2\xi r) \tag{4.305}$$

which depends on the principal quantum number $n$, on the orbital angular momentum quantum number $l$, on the parameter:

$$\xi = \frac{Z}{n} \tag{4.306}$$

and on the associated Laguerre polynomials (Margenau & Murphy, 1964):

$$L_{n+l}^{2l+1}(2\xi r) = \sum_{k=0}^{n-l-1} (-1)^{k+1} \frac{[(n+l)!]^2}{(n-l-1-k)!(2l+1+k)!} \frac{(2\xi r)^k}{k!} \tag{4.307}$$

of degree

$$n_r = n - l - 1 \tag{4.308}$$

that sets out also the number of radial nodes (or zeros). Therefore, the radial distribution functions:

$$\rho_{n,l}(r) = r^2 |R_{n,l}(r)|^2 \tag{4.309}$$

exhibit $(n-1)$ maxima.

At this point worth noting that, for a given $n$, there is only one maximum when the quantum orbital number takes its largest value:

$$l = n - 1 \tag{4.310}$$

leading from (4.308) with

$$n_r = 0 \tag{4.311}$$

condition.

It follows that, when studying the valence properties of an atom, at large distance from nucleus, the nodes in the radial part of the wave function are found to be unimportant (Ghosh & Biswas, 2002). Such case corresponds to the Slater's asymptotic large distance picture for a hydrogenic-like wave function of an *effective principal quantum number n** in the field of an *effective nuclear charge Z**,

$$V^*(r) = V_{eff}(r) = -\frac{Z^*}{r} \tag{4.312}$$

due to the screening effects of the inner electrons. As a direct consequence, the orbital parameter (4.306) becomes now the so-called *orbital exponent*:

$$\xi^* = \frac{Z^*}{n^*} \tag{4.313}$$

Note that the values of $n^*$ and $Z^*$ are computed upon specific rules set up so that the associate energy levels to check fairly with experiment (Slater, 1930). Going to extract from the Eqs. (4.305) and (4.307) the actual working radial distribution function, within the Eqs. (4.310) and (4.311) conditions, we first get out from the Eq. (4.307), the expression:

$$L_{2n-1}^{2n-1}(2\xi^* r) = -(2n-1)! \tag{4.314}$$

that substituted in Eq. (4.305) leads with the result:

$$R_{n,n-1}(r) = \frac{(2\xi^*)^{n+1/2}}{[(2n)!]^{1/2}} r^{n-1} \exp(-\xi^* r) \tag{4.315}$$

However, because the radial function (4.315) is normalized as:

$$\int_0^\infty 4\pi r^2 |R_{n,n-1}(r)|^2 dr = 1 \tag{4.316}$$

the associate radial density abstracted upon definition (4.309) takes the working form recalling (4.131):

$$\rho_{n,n-1}^*(r) = N_v \frac{(2\xi^*)^{2n+1}}{(2n)!} r^{2n} \exp(-2\xi^* r) \tag{4.317}$$

has to be conciliated with the DFT many-electronic constrain (4.168) so that the desired effective radial density at the valence shell level, when $N_v$ electrons are involved fulfilles the constraint

$$\int_0^\infty 4\pi \, \rho_{n,n-1}(r)dr = N_v \tag{4.318}$$

by applying the Slater integral recipe of Eq. (4.136).

Remarkably, the radial density (4.317) corresponds with the case in which the most probable radial distance from nucleus is achieved through its stationary equation,

$$d/dr\left[\rho_{n,n-1}^*(r)\right] = 0 \tag{4.319}$$

with the value:

$$r^* = \frac{2n}{2\xi^*} \tag{4.320}$$

leaved under the form (4.320) for reasons to be revealed in a moment below.

From now we have at our side a simple analytical radial density frame in which the valence properties of the many-electron atomic systems can be fairly treated. In such, the Table 4.4 presents the particular constants that are relevant for the atomic systems under actual study.

To apply the presented radial density approach on the actual absolute $\chi$, $\eta$, IP and EA density functionals the evaluation of the quantities $C_A$, $a$, and $b$ of relations (4.277), (4.223), and (4.276), respectively, it is firstly demanded.

Doing so, for the so-called chemical action (4.277), one gets by straight replacement (Putz, 2006):

$$C_A^*\left(N_v, n, \xi^*, Z^*\right) = 4\pi \int_0^\infty \rho_{n,n-1}^*(r)V^*(r)r^2 dr$$

$$= -4\pi Z^* N_v \frac{\left(2\xi^*\right)^{2n+1}}{(2n)!} \int_0^\infty r^{2n+1} \exp\left(-2\xi^* r\right)dr \tag{4.321}$$

**TABLE 4.4**  The Periodic Input Parameters used in the Actual Study: the Total Number of $s+p$ Electrons $N_v$, the Principal Quantum Number $n$, the Orbital Exponent $\zeta^*$ and the Effective Charge $Z^*$, Calculated Upon the Slater Method (Slater, 1930), for the Valence Shells of the Ordinary Elements (Putz, 2006)

**Legend:** *Symbol of Element*

*Number of s +p valence electrons: $N_v$*

*Valence principal quantum number: $n$*

*Orbital exponent: $\zeta^*$*

*Effective charge: $Z^*$*

| Symbol | $N_v$ | $n$ | $\zeta^*$ | $Z^*$ |
|---|---|---|---|---|
| H | 1 | 1 | 1 | 1 |
| He | 2 | 1 | 1.7 | 1.7 |

| Symbol | $N_v$ | $n$ | $\zeta^*$ | $Z^*$ |
|---|---|---|---|---|
| Li | 1 | 2 | 0.65 | 1.30 |
| Be | 2 | 2 | 0.98 | 1.95 |
| B | 3 | 2 | 1.3 | 2.60 |
| C | 4 | 2 | 1.63 | 3.25 |
| N | 5 | 2 | 1.95 | 3.90 |
| O | 6 | 2 | 2.28 | 4.55 |
| F | 7 | 2 | 2.6 | 5.2 |
| Ne | 8 | 2 | 2.93 | 5.85 |

| Symbol | $N_v$ | $n$ | $\zeta^*$ | $Z^*$ |
|---|---|---|---|---|
| Na | 1 | 3 | 0.73 | 2.20 |
| Mg | 2 | 3 | 0.95 | 2.85 |
| Al | 3 | 3 | 1.17 | 3.50 |
| Si | 4 | 3 | 1.39 | 4.15 |
| P | 5 | 3 | 1.6 | 4.80 |
| S | 6 | 3 | 1.8 | 5.45 |
| Cl | 7 | 3 | 2.03 | 6.10 |
| Ar | 8 | 3 | 2.25 | 6.75 |

| Symbol | $N_v$ | $n$ | $\zeta^*$ | $Z^*$ |
|---|---|---|---|---|
| K | 1 | 4 | 0.59 | 2.20 |
| Ca | 2 | 4 | 0.77 | 2.85 |
| Sc | 2 | 4 | 0.81 | 3.00 |
| Ti | 2 | 4 | 0.85 | 3.15 |
| V | 2 | 4 | 0.89 | 3.30 |
| Cr | 2 | 4 | 0.93 | 3.45 |
| Mn | 2 | 4 | 0.97 | 3.60 |
| Fe | 2 | 4 | 1.01 | 3.75 |
| Co | 2 | 4 | 1.05 | 3.90 |
| Ni | 2 | 4 | 1.09 | 4.05 |
| Cu | 2 | 4 | 1.14 | 4.20 |
| Zn | 2 | 4 | 1.18 | 4.35 |
| Ga | 3 | 4 | 1.35 | 5.00 |
| Ge | 4 | 4 | 1.53 | 5.65 |
| As | 5 | 4 | 1.70 | 6.30 |
| Se | 6 | 4 | 1.88 | 6.95 |
| Br | 7 | 4 | 2.05 | 7.60 |
| Kr | 8 | 4 | 2.23 | 8.25 |

**TABLE 4.4** Continued

| Rb | Sr | Y | Zr | Nb | Mo | Tc | Ru | Rh | Pd | Ag | Cd | In | Sn | Sb | Te | I | Xe |
|---|---|---|---|---|---|---|---|---|---|---|---|---|---|---|---|---|---|
| 1 | 2 | 2 | 2 | 2 | 2 | 2 | 2 | 2 | 2 | 2 | 2 | 3 | 4 | 5 | 6 | 7 | 8 |
| 5 | 5 | 5 | 5 | 5 | 5 | 5 | 5 | 5 | 5 | 5 | 5 | 5 | 5 | 5 | 5 | 5 | 5 |
| 0.55 | 0.71 | 0.75 | 0.79 | 0.83 | 0.86 | 0.9 | 0.94 | 0.98 | 1.01 | 1.05 | 1.09 | 1.25 | 1.41 | 1.58 | 1.74 | 1.9 | 2.06 |
| 2.20 | 2.85 | 3.00 | 3.15 | 3.30 | 3.45 | 3.60 | 3.75 | 3.90 | 4.05 | 4.20 | 4.35 | 5.00 | 5.65 | 6.30 | 6.95 | 7.60 | 8.25 |

Instead, for computing the sensitivity indices (4.223) and (4.276) the local response function (4.221) is primarily written:

$$L^*\left(r, N_v, n, \xi^*, Z^*\right) = \frac{\left[\nabla \rho^*_{n,n-1}(r)\right]\left[-\nabla V^*(r)\right]}{\left[-\nabla V^*(r)\right]^2}$$

$$= 2\frac{N_v}{Z^*}\frac{\left(2\xi^*\right)^{2n+1}}{(2n)!}\left(\xi^* r^{2n+2} - nr^{2n+1}\right)\exp\left(-2\xi^* r\right) \qquad (4.322)$$

Then, it is used to lay down their respective expressions:

$$a^*\left(N_v, n, \xi^*, Z^*\right) = 4\pi \int_0^\infty L^*\left(r, N_v, n, \xi^*, Z^*\right) r^2 dr$$

$$= 8\pi \frac{N_v}{Z^*}\frac{\left(2\xi^*\right)^{2n+1}}{(2n)!}\left[\xi^* \int_0^\infty r^{2n+4}\exp\left(-2\xi^* r\right)dr - n\int_0^\infty r^{2n+3}\exp\left(-2\xi^* r\right)dr\right]$$

$$(4.323)$$

$$b^*\left(N_v, n, \xi^*\right) = 4\pi \int_0^\infty L^*\left(r, N_v, n, \xi^*, Z^*\right) V^*(r) r^2 dr$$

$$= 8\pi N_v \frac{\left(2\xi^*\right)^{2n+1}}{(2n)!}\left[\xi^* \int_0^\infty r^{2n+3}\exp\left(-2\xi^* r\right)dr - n\int_0^\infty r^{2n+2}\exp\left(-2\xi^* r\right)dr\right]$$

$$(4.324)$$

In order to evaluate the above integrals by employing the valence shell properties to which they are associated, let's remark that all of them are covered by the integral type:

$$I(\alpha, q) = \int_0^\infty r^\alpha \exp\left(-qr\right)dr, \ \alpha > 0 \qquad (4.325)$$

At this point is useful to further consider the equivalent forms of (4.325) as:

$$I(\alpha, q) = \int_0^\infty \exp\left[\alpha \ln r - qr\right]dr = \int_0^\infty \exp\left[\alpha\left(\ln r - q\frac{r}{\alpha}\right)\right]dr \equiv \int_0^\infty \exp\left[\alpha f(r)\right]dr$$

$$(4.326)$$

from where, it follows that the phase function

$$f(r) = \ln r - qr / \alpha \qquad (4.327)$$

fulfills the stationary condition, $f'(r) = 0$, at the saddle point:

$$r_0 = \frac{\alpha}{q} \tag{4.328}$$

Now, if the quantity (4.328) is particularized to the integrals appearing through the relations (4.321), (4.323), and (4.324) there is easily concluded that they belong to the same family of values (Putz, 2006):

$$r_0 = \frac{2n+4}{2\xi^*} \sim \frac{2n+3}{2\xi^*} \sim \frac{2n+2}{2\xi^*} \sim \frac{2n+1}{2\xi^*} \sim \frac{2n}{2\xi^*} = r^* \tag{4.329}$$

like the most probable radius (4.320) of a given (valence) shell.

This observation is most helpful for our asymptotical treatment of the atomic systems suggesting that the saddle point approximation (Mathews & Walker, 1970), is suitable to fairly analytical perform the involved integrals. According with the saddle point method, to evaluate an integral of type (4.325) the intermediate form (4.326) is approximated by the *saddle-point recipe* (3.154) specialized here as (see also the Appendix of the present volume):

$$I(\alpha, q) \cong \exp[\alpha f(r_0)] \sqrt{-\frac{2\pi}{\alpha f''(r_0)}}$$

$$= \exp\left[\alpha\left(\ln\frac{\alpha}{q} - 1\right)\right] \sqrt{\frac{2\pi\alpha}{q^2}} = \sqrt{2\pi} \exp(-\alpha)\frac{\alpha^{\alpha+\frac{1}{2}}}{q^{\alpha+1}} \tag{4.330}$$

This way, the actual working formulas are (Putz, 2006):

$$C_A^*\left(N_v, n, \xi^*, Z^*\right) = -2\sqrt{2\pi}^{3/2}\frac{N_v Z^*}{\xi^*}\frac{(2n+1)^{2n+3/2}}{(2n)!}\exp\left[-(2n+1)\right] \tag{4.331}$$

$$a^*\left(N_v, n, \xi^*, Z^*\right) = \pi^{3/2}\frac{N_v}{Z^*\xi^{*3}(2n)!}\left(\frac{2}{e}\right)^{2(n+2)}\left[(n+2)^{2n+9/2} - en\left(n+\frac{3}{2}\right)^{2n+7/2}\right] \tag{4.332}$$

$$b^{*}\left(N_{v},n,\xi^{*}\right)=\pi^{3/2}\frac{N_{v}}{\xi^{*2}\left(2n\right)!}\frac{4^{n+2}}{\exp\left(2n+3\right)}\left[en(n+1)^{2n+5/2}-\left(n+\frac{3}{2}\right)^{2n+7/2}\right]$$

$$(4.333)$$

With expressions (4.331)–(4.333) the systematic absolute $\chi$, $\eta$, $IP$, and $EA$ can be computed once their particular density functional formulations are employed upon a certain atomic system with the relevant parameters from the Table 4.4.

Regarding the analytical model further comments may apply. The general theoretical framework stands as the DFT; however, this venture was developed on its conceptual rather than on its computational virtues. This way, the approximate energetic functional approaches (Nalewajski, 1996; Putz, 2008b) were systematically avoided by considering the independent-particle picture of the softness kernel formulation, see Sections 4.6.1–4.6.4.

There was merged out that, actually, the local response function (4.221) plays a crucial role in our systematic. We like to emphasis that since the $L(r)$ dependence on the gradient of density $\nabla\rho(r)$ and on the gradient of the minus external potential $-\nabla V(r)$, instead of the purely density $\rho(r)$ and potential $V(r)$. More, $L(r)$ through its (4.221) definition comprises two quantities, meaningfully as well on atomic and molecular levels. In such, the gradient of the minus potential $-\nabla V(r)$ provides, physically, the force acting on the electronic system, emphasizing therefore on the force concept in chemistry (Deb, 1973), whereas the gradient of density $\nabla\rho(r)$ correlates with the modern fruitful concept of the flux partition of the electron density in mirroring of the chemical bond (Bader, 1990).

### 4.7.2  DISCUSSION ON CHEMICAL REACTIVITY RELATED ATOMIC SCALES

Being the actual numerical model mainly a rationalization of the experimental facts, let's involve other estimable quantities. The $IP$, closely related also with the actual absolute electronegativity (4.253) definition, is a quite accessible spectral atomic index, either theoretical from the limit frequency from which the continuum spectrum is achieved, or experimentally by a Franck-Hertz experiment. Nevertheless, as far as *all* the valence electrons are involved two cases can arise. If after each ionization the resulting system relaxes into a different potential that support further ionization processes the so-called spectroscopic $IP_S$ can be measured or computed for most atoms of the elements, see Table 4.5 (Weast et al., 1989).

On the other side, if the "frozen orbitals" are considered at each valence ionization process, a picture supported by (Koopmans' 1934) theorem, see also the Volume I of the present five-volume work, one can write the so-called generalized Pauling $IP$ (Putz, 2006):

$$IP_P = 13.6 N_v \left(\frac{Z^*}{n}\right)^2 [\text{eV}] \tag{4.334}$$

giving therefore the minimum energy consumed by an isolated atom to release at once all its $N_v$ electrons from the valence shell indexed by the principal quantum number $n$, see Table 4.6. However, it seems useful to add such $N_v$ relating $V(r)$ approaches in our study since, in deriving of the present absolute $\chi$, $\eta$, $IP$, and $EA$ density functionals.

Finally, as the $IP$s are directly related to the atomic size the comparison between their trends and the atomic radii predicted from (4.320), see Table 4.5, gives also a direct measure of the goodness of the actual $IP$ scales. The Periodic Table organization through the change in Z across rows and shell number down columns strongly suggests that the so-called "third dimension" related with the average one-electron energy (energy per electron) is required in order to can explain (or predict) the molecular and solid state pattern organization (Allen, 1989).

Such a third dimension can be assigned to the average energy of atoms being therefore also the frame in which the absolute $\chi$, $\eta$, $IP$, and $EA$ from general definitions (4.253)–(4.256), respectively, are taken into discussion.

**TABLE 4.5** Calibrating the $\chi_A$, $\eta_A$, and $IP_A$ Functionals So That Their Respective Values for the Atomic H, as Given in the Table 4.3, to be Recovered; the Energetic [eV] Pre-Factors Have Been Found, Respectively (Putz, 2006)

| | $\chi_A$ | $\eta_A$ | $IP_A$ |
|---|---|---|---|
| [1] | (27.21)×(−98.363) | (27.21)×(88.542) | (27.21)×(−372.997) |
| [2] | (27.21)×(−65.634) | (27.21)×(89.014) | (27.21)×(−248.72) |
| [3] | (27.2)×(0.765) | (27.2)×(−5.331) | (27.2)×(1.552) |
| [4] | (27.21)×(0.51) | (27.21)×(−5.329) | (27.21)×(1.035) |

**TABLE 4.6**  The Values of the Theoretical Atomic Radii r* ($10^{-11}$ m) Upon the Formula (4.320), of the Calculated Generalized Pauling Ionization Potentials $IP_P$ [eV] based on Eq. (4.334), and of the Spectral Ionization Potentials $IP_S$ [eV] (Weast et al., 1989), for the Valence Shells of the Atoms of the Ordinary Elements (Putz, 2006)

*Legend:* *Symbol of Element*
Atomic Radii: $r^*$
Pauling Ionization Potentials: $IP_P$
Spectral Ionization Potentials: $IP_S$

| Element | $r^*$ | $IP_P$ | $IP_S$ |
| --- | --- | --- | --- |
| H | 5.29 | 13.6 | 13.6 |
| He | 3.11 | 78.61 | 54.42 |
| Li | 16.28 | 5.75 | 5.39 |
| Be | 10.86 | 24.55 | 18.21 |
| B | 8.14 | 68.95 | 37.93 |
| C | 6.51 | 143.7 | 64.49 |
| N | 5.43 | 258.6 | 97.9 |
| O | 4.65 | 422.3 | 138.2 |
| F | 4.07 | 643.6 | 185.2 |
| Ne | 3.62 | 930.9 | 239.1 |
| Na | 21.65 | 7.31 | 5.14 |
| Mg | 16.71 | 24.55 | 15.04 |
| Al | 13.61 | 55.53 | 28.45 |
| Si | 11.5 | 104.1 | 45.14 |
| P | 9.92 | 174.1 | 65.02 |
| S | 8.74 | 269.3 | 88.05 |
| Cl | 7.81 | 393.6 | 114.2 |
| Ar | 7.1 | 550.8 | 143.5 |
| K | 35.6 | 4.11 | 4.34 |
| Ca | 27.5 | 13.81 | 11.87 |
| Sc | 26.11 | 15.3 | 12.80 |
| Ti | 24.86 | 16.87 | 13.58 |
| V | 23.73 | 18.51 | 14.65 |
| Cr | 22.70 | 20.23 | 16.50 |
| Mn | 21.75 | 22.03 | 15.64 |
| Fe | 20.89 | 23.91 | 16.18 |
| Co | 20.08 | 25.86 | 17.06 |
| Ni | 19.34 | 27.88 | 18.17 |
| Cu | 18.65 | 29.00 | 20.29 |
| Zn | 18.00 | 32.17 | 17.96 |
| Ga | 15.66 | 63.75 | 30.71 |
| Ge | 13.86 | 108.5 | 45.71 |
| As | 12.43 | 168.7 | 62.63 |
| Se | 11.27 | 246.3 | 81.70 |
| Br | 10.31 | 343.7 | 103.0 |
| Kr | 9.49 | 462.8 | 126 |
| Rb | 48.11 | 2.63 | 4.18 |
| Sr | 37.14 | 8.84 | 11.03 |
| Y | 35.28 | 9.79 | 12.24 |
| Zr | 33.6 | 10.8 | 13.13 |
| Nb | 32.07 | 11.85 | 14.32 |
| Mo | 30.7 | 12.95 | 16.15 |
| Tc | 29.4 | 14.1 | 15.26 |
| Ru | 28.22 | 15.3 | 16.76 |
| Rh | 27.14 | 16.55 | 18.08 |
| Pd | 26.13 | 17.85 | 19.43 |
| Ag | 25.2 | 19.19 | 21.49 |
| Cd | 24.33 | 20.59 | 16.91 |
| In | 21.2 | 40.8 | 28.03 |
| Sn | 18.73 | 69.46 | 40.73 |
| Sb | 16.8 | 108.0 | 56 |
| Te | 15.23 | 157.7 | 70.7 |
| I | 13.93 | 220.0 | - |
| Xe | 12.83 | 296.2 | - |

Before going into particular analysis let's recall that the properties of the chemical elements reflect a non-periodic character if the atomic core is concerned whereas they show a systematic periodicity as far as the outer electronic shells of atoms are involved. In this respect, the elements belonging to the same group pose *analogy* relationships, based on the same type of configuration of the distinctive electron, while the *homology* is the rule across the elements of a period, having different valence quantum states but monotonic similar energies. This way, the properties of the atoms completing their the last quantum shells with electrons are significantly different respecting the atoms of the transitional elements having the last but one shells in the course of completing. Such general fashions are widely respected by all presented atomic scales through Tables 4.5–5.8. Nevertheless, a close attention to the individual studied chemical index *per* scale is also compulsory. In this respect let's first observe that the results can be grouped like the "absolute 1 ($A^{[1]}$) and 2 ($A^{[2]}$)" as well as the "absolute 3 ($A^{[3]}$) and 4($A^{[4]}$)" scales corresponding with the "d$N$," "d$N$+d$N$d$N$," "d$N$+$\delta V(r)$," and "d$N$+d$N$d$N$+$\delta V(r)$" expansions, respectively. This grouping would also have been anticipated through the complementary signs and values in which the numerical calibrations factors of the working functionals appear in Table 4.5 (Putz, 2006).

In the context of actual assumptions and approximations the presented schemes of computations provide the atomic results grouped on the tables of elements as displayed through the Tables 4.5–4.8. Thus, from now only the absolute scales $A^{[2]}$ and $A^{[4]}$, the most complex ones in their groups, will be considered for discussions.

Regarding the *IP*, in the Figure 4.11 there are collected all the experimental ($IP_E$ from Table 4.3), the spectral ($IP_S$ from Table 4.6), the generalized Pauling ($IP_P$ from Table 4.6), the absolute $A^{[2]}$ ($IP_A^{[2]}$ from Table 4.8), and the absolute $A^{[4]}$ ($IP_A^{[2]}$ from Table 4.10) considered scales, together with that of the atomic radii ($r^*$ from Table 4.6).

From the Figure 4.11 it is evident that all *IP*'s and $r^*$ profiles, beside some deviations, are homomorphic and to a decreasing in radius corresponds an increase of *IP*. This is a natural relationship between these two quantities since a more compact atomic volume requires a higher energy for the ionization. Nevertheless, from the quantitative point of view the present $IP_A^{[2]}$ scale is the most appropriate to adopt as the DFT counterpart,

**TABLE 4.7** The Absolute Electronegativity $\chi_A^{[1]}$, Hardness $\eta_A^{[1]}$, Ionization Potential $IP_A^{[1]}$, and Electron Affinity $EA_A^{[1]}$, Computed Upon the Eqs. (4.259), (4.261), (4.262) and (4.263), Respectively, for the Ordinary Elements

**Legend:** *Symbol of Element*

Absolute Electronegativity: $\chi_A^{[1]}$

Absolute Hardness: $\eta_A^{[1]}$

Absolute Ionization Potential: $IP_A^{[1]}$

Absolute Electron Affinity: $EA_A^{[1]}$

| Element | $\chi_A^{[1]}$ | $\eta_A^{[1]}$ | $IP_A^{[1]}$ | $EA_A^{[1]}$ |
|---|---|---|---|---|
| H | 7.18 | 6.45 | 13.62 | 0.73 |
| He | 58.96 | 25.79 | 169.2 | -51.27 |
| Li | 0.73 | 0.65 | 1.39 | 0.08 |
| Be | 3.75 | 1.69 | 10.68 | -3.17 |
| B | 11.63 | 3.46 | 36.83 | -13.56 |
| C | 28.38 | 6.21 | 94.53 | -37.77 |
| N | 57.07 | 9.59 | 196.2 | -82.06 |
| O | 102.3 | 13.27 | 359.8 | -155.3 |
| F | 162.9 | 16.16 | 583.9 | -258 |
| Ne | 240.6 | 17.87 | 874.6 | -393.5 |
| Na | 0.73 | 0.66 | 1.38 | 0.08 |
| Mg | 2.08 | 0.93 | 5.91 | -1.76 |
| Al | 4.76 | 1.42 | 15.06 | -5.53 |
| Si | 9.44 | 2.10 | 31.36 | -12.48 |
| P | 16.56 | 2.92 | 56.63 | -23.52 |
| S | 26.51 | 3.82 | 92.47 | -39.45 |
| Cl | 41.86 | 5.01 | 148.2 | -64.46 |
| Ar | 61.56 | 6.16 | 220.5 | -97.35 |
| K | 0.2 | 0.18 | 0.37 | 0.02 |
| Ca | 0.56 | 0.25 | 1.60 | -0.48 |
| Sc | 0.69 | 0.31 | 1.96 | -0.58 |
| Ti | 0.84 | 0.38 | 2.38 | -0.71 |
| V | 1.01 | 0.45 | 2.86 | -0.85 |
| Cr | 1.2 | 0.54 | 3.41 | -1.01 |
| Mn | 1.42 | 0.64 | 4.04 | -1.2 |
| Fe | 1.67 | 0.75 | 4.75 | -1.41 |
| Co | 1.96 | 0.88 | 5.56 | -1.65 |
| Ni | 2.27 | 1.02 | 6.45 | -1.92 |
| Cu | 2.69 | 1.21 | 7.66 | -2.27 |
| Zn | 3.09 | 1.39 | 8.79 | -2.61 |
| Ga | 5.31 | 1.59 | 16.8 | -6.17 |
| Ge | 8.72 | 1.94 | 28.96 | -11.52 |
| As | 13.28 | 2.35 | 45.40 | -18.84 |
| Se | 19.68 | 2.87 | 68.60 | -29.23 |
| Br | 27.65 | 3.39 | 97.71 | -42.41 |
| Kr | 38.11 | 3.98 | 136.1 | -59.91 |

**TABLE 4.7** Continued

| Rb | Sr | Y | Zr | Nb | Mo | Tc | Ru | Rh | Pd | Ag | Cd | In | Sn | Sb | Te | I | Xe |
|---|---|---|---|---|---|---|---|---|---|---|---|---|---|---|---|---|---|
| 0.09 | 0.25 | 0.32 | 0.39 | 0.47 | 0.55 | 0.66 | 0.78 | 0.92 | 1.04 | 1.21 | 1.4 | 2.43 | 3.94 | 6.17 | 9.07 | 12.86 | 17.67 |
| 0.08 | 0.11 | 0.14 | 0.17 | 0.21 | 0.25 | 0.29 | 0.35 | 0.41 | 0.47 | 0.55 | 0.63 | 0.73 | 0.88 | 1.10 | 1.34 | 1.62 | 1.92 |
| 0.17 | 0.72 | 0.9 | 1.10 | 1.34 | 1.56 | 1.86 | 2.21 | 2.60 | 2.96 | 3.45 | 4.0 | 7.69 | 13.09 | 21.09 | 31.56 | 45.34 | 62.97 |
| 0.01 | -0.21 | -0.27 | -0.33 | -0.4 | -0.46 | -0.55 | -0.66 | -0.77 | -0.88 | -1.02 | -1.19 | -2.83 | -5.20 | -8.74 | -13.42 | -19.63 | -27.62 |

*All units are in eV (electron-volts) per atom (Putz, 2006).

**TABLE 4.8** The Absolute Electronegativity $\chi_A^{[2]}$, Hardness $\eta_A^{[2]}$, Ionization Potential $IP_A^{[2]}$, and Electron Affinity $EA_A^{[2]}$, Computed Upon the Eqs. (4.268), (4.266), (4.269) and (4.270), Respectively, For the Ordinary Elements

**Legend:** *Symbol of Element*

Absolute Electronegativity: $\chi_A^{[2]}$

Absolute Hardness: $\eta_A^{[2]}$

Absolute Ionization Potential: $IP_A^{[2]}$

Absolute Electron Affinity: $EA_A^{[2]}$

| | **H** | **He** |
|---|---|---|
| $\chi$ | 7.18 | 58.48 |
| $\eta$ | 6.45 | 24.85 |
| $IP$ | 13.62 | 168.4 |
| $EA$ | 0.73 | -51.5 |

| | **Li** | **Be** | **B** | **C** | **N** | **O** | **F** | **Ne** |
|---|---|---|---|---|---|---|---|---|
| $\chi$ | 0.73 | 3.75 | 11.62 | 28.19 | 57.06 | 98.29 | 151 | 210.5 |
| $\eta$ | 0.66 | 1.69 | 3.45 | 6.12 | 9.26 | 12.47 | 14.74 | 15.85 |
| $IP$ | 1.39 | 10.68 | 36.78 | 94.03 | 193.3 | 348 | 546.6 | 776.8 |
| $EA$ | 0.08 | -3.17 | -13.54 | -37.64 | -81.23 | -151 | -244.5 | -355.9 |

| | **Na** | **Mg** | **Al** | **Si** | **P** | **S** | **Cl** | **Ar** |
|---|---|---|---|---|---|---|---|---|
| $\chi$ | 0.73 | 2.08 | 4.76 | 9.42 | 16.48 | 26.26 | 41.11 | 59.68 |
| $\eta$ | 0.66 | 0.94 | 1.43 | 2.10 | 2.90 | 3.77 | 4.89 | 5.94 |
| $IP$ | 1.38 | 5.91 | 15.05 | 31.31 | 56.4 | 91.71 | 145.8 | 214.5 |
| $EA$ | 0.08 | -1.75 | -5.53 | -12.46 | -23.44 | -39.18 | -63.61 | -95.09 |

| | **K** | **Ca** | **Sc** | **Ti** | **V** | **Cr** | **Mn** | **Fe** | **Co** | **Ni** | **Cu** | **Zn** |
|---|---|---|---|---|---|---|---|---|---|---|---|---|
| $\chi$ | 0.2 | 0.56 | 0.69 | 0.84 | 1.01 | 1.2 | 1.42 | 1.67 | 1.96 | 2.27 | 2.69 | 3.09 |
| $\eta$ | 0.18 | 0.25 | 0.31 | 0.38 | 0.45 | 0.54 | 0.64 | 0.75 | 0.88 | 1.02 | 1.21 | 1.39 |
| $IP$ | 0.37 | 1.60 | 1.96 | 2.38 | 2.86 | 3.42 | 4.04 | 4.75 | 5.56 | 6.45 | 7.66 | 8.79 |
| $EA$ | 0.02 | -0.47 | -0.58 | -0.71 | -0.85 | -1.01 | -1.2 | -1.41 | -1.65 | -1.91 | -2.27 | -2.61 |

| | **Ga** | **Ge** | **As** | **Se** | **Br** | **Kr** |
|---|---|---|---|---|---|---|
| $\chi$ | 5.31 | 8.70 | 13.23 | 19.55 | 27.33 | 37.41 |
| $\eta$ | 1.59 | 1.94 | 2.34 | 2.84 | 3.34 | 3.9 |
| $IP$ | 16.79 | 28.91 | 45.26 | 68.19 | 96.7 | 133.8 |
| $EA$ | -6.17 | -11.51 | -18.79 | -29.08 | -42.03 | -59.04 |

**TABLE 4.8** Continued

| Rb | Sr | Y | Zr | Nb | Mo | Tc | Ru | Rh | Pd | Ag | Cd | In | Sn | Sb | Te | I | Xe |
|---|---|---|---|---|---|---|---|---|---|---|---|---|---|---|---|---|---|
| 0.09 | 0.25 | 0.32 | 0.39 | 0.47 | 0.55 | 0.66 | 0.78 | 0.92 | 1.04 | 1.21 | 1.41 | 2.43 | 3.94 | 6.17 | 9.05 | 12.79 | 17.53 |
| 0.08 | 0.12 | 0.14 | 0.18 | 0.21 | 0.25 | 0.3 | 0.35 | 0.41 | 0.47 | 0.55 | 0.63 | 0.73 | 0.89 | 1.10 | 1.34 | 1.61 | 1.91 |
| 0.17 | 0.72 | 0.9 | 1.10 | 1.34 | 1.56 | 1.86 | 2.21 | 2.60 | 2.96 | 3.45 | 4.0 | 7.69 | 13.08 | 21.06 | 31.48 | 45.13 | 62.49 |
| 0.01 | -0.21 | -0.27 | -0.33 | -0.4 | -0.46 | -0.55 | -0.66 | -0.77 | -0.88 | -1.02 | -1.18 | -2.82 | -5.20 | -8.72 | -13.39 | -19.54 | -27.43 |

*All units are in eV (electron-volts) (Putz, 2006).

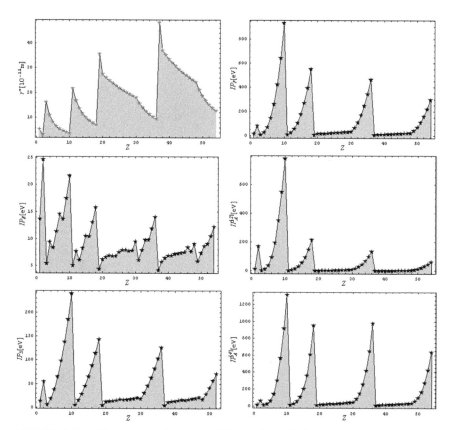

**FIGURE 4.11**    The comparative trend of the atomic radii $r^*$ from Table 4.6 (upper left) with respect to the atomic experimental first ionization potential $IP_E$ from Table 4.6 (middle left), the atomic experimental spectral ionization potential $IP_S$ from Table 4.6 (lower left), the generalized Pauling ionization potential $IP_P$ from Table 4.6 (upper right), the atomic absolute ionization potential $IP_A^{[2]}$ from Table 4.8 (middle right) and the atomic absolute ionization potential $IP_A^{[4]}$ from Table 4.10 (lower right) scales, respectively (Putz, 2006).

within the present approximations, to the generalized Pauling one $IP_P$ even, however, a close relative relationship of both $IP_A^{[2]}$ and $IP_A^{[4]}$ scales with the experimental spectral $IP_S$ one is also remarked (Putz, 2006).

These observations suggest that the $IP_A^{[2]}$ and $IP_A^{[4]}$ atomic scales are merely reflecting the all valence electron ionization process rather than a single-electron releasing, as is commonly reported relating the electro-negativity concept, see Table 4.3.

**TABLE 4.9** The Absolute Electronegativity $\chi_A^{[3]}$, Hardness $\eta_A^{[3]}$, Ionization Potential $IP_A^{[3]}$, and Electron Affinity $EA_A^{[3]}$, Computed Upon the Eqs. (4.279), (4.278), (4.280) and (4.281), Respectively, For the Ordinary Elements

**Legend:** *Symbol of Element*

Absolute Electronegativity: $\chi_A^{[3]}$

Absolute Hardness: $\eta_A^{[3]}$

Absolute Ionization Potential: $IP_A^{[3]}$

Absolute Electron Affinity: $EA_A^{[3]}$

| Element | $\chi_A^{[3]}$ | $\eta_A^{[3]}$ | $IP_A^{[3]}$ | $EA_A^{[3]}$ |
|---|---|---|---|---|
| H | 7.18 | 6.45 | 13.62 | 0.73 |
| He | 32.96 | 45.17 | 60.27 | 5.62 |
| Li | 4.11 | 2.39 | 7.99 | 0.23 |
| Be | 12.04 | 12.13 | 22.67 | 1.41 |
| B | 31.39 | 36.49 | 58.38 | 4.39 |
| C | 75.19 | 80.81 | 140.8 | 9.55 |
| N | 157.6 | 133.7 | 300.5 | 14.82 |
| O | 289.8 | 161 | 564.7 | 14.84 |
| F | 460.1 | 119.1 | 916.6 | 3.58 |
| Ne | 645 | -2.51 | 1309.6 | -19.71 |
| Na | 6.30 | 5.10 | 12.05 | 0.56 |
| Mg | 13.57 | 14.46 | 25.43 | 1.70 |
| Al | 28.47 | 32.63 | 53.03 | 3.91 |
| Si | 57.36 | 62.8 | 107.3 | 7.45 |
| P | 107.8 | 105.2 | 203.5 | 12.14 |
| S | 188.2 | 156.4 | 359.2 | 17.23 |
| Cl | 315 | 211.5 | 608.4 | 21.49 |
| Ar | 486.6 | 247.4 | 951.5 | 21.63 |
| K | 4 | 2.2 | 7.8 | 0.20 |
| Ca | 8.04 | 6.28 | 15.39 | 0.68 |
| Sc | 9.11 | 7.70 | 17.37 | 0.85 |
| Ti | 10.29 | 9.34 | 19.52 | 1.06 |
| V | 11.57 | 11.23 | 21.85 | 1.29 |
| Cr | 12.96 | 13.4 | 24.36 | 1.57 |
| Mn | 14.48 | 15.85 | 27.08 | 1.88 |
| Fe | 16.12 | 18.63 | 30.01 | 2.24 |
| Co | 17.9 | 21.76 | 33.16 | 2.64 |
| Ni | 19.82 | 25.26 | 36.55 | 3.09 |
| Cu | 22.16 | 29.66 | 40.66 | 3.66 |
| Zn | 24.41 | 34.04 | 44.59 | 4.24 |
| Ga | 44.6 | 58.15 | 82.04 | 7.15 |
| Ge | 79.72 | 93.14 | 148.2 | 11.21 |
| As | 134.9 | 137.6 | 253.8 | 16.05 |
| Se | 219.9 | 192.4 | 418.4 | 21.53 |
| Br | 338.8 | 249.9 | 651.2 | 26.39 |
| Kr | 502.2 | 304.5 | 974.8 | 29.51 |

**TABLE 4.9** Continued

| Rb | Sr | Y | Zr | Nb | Mo | Tc | Ru | Rh | Pd | Ag | Cd | In | Sn | Sb | Te | I | Xe |
|---|---|---|---|---|---|---|---|---|---|---|---|---|---|---|---|---|---|
| 3.1 | 5.95 | 6.75 | 7.62 | 8.57 | 9.46 | 10.56 | 11.76 | 13.05 | 14.25 | 15.74 | 17.34 | 30.49 | 52.55 | 88.70 | 142.7 | 221.1 | 330.1 |
| 1.35 | 3.78 | 4.67 | 5.71 | 6.92 | 8.12 | 9.67 | 11.44 | 13.45 | 15.39 | 17.88 | 20.65 | 35.68 | 57.35 | 87.88 | 126.1 | 172.3 | 224.5 |
| 6.09 | 11.52 | 13.01 | 14.63 | 16.38 | 18.02 | 20.03 | 22.2 | 24.52 | 26.69 | 29.34 | 32.18 | 56.69 | 98.3 | 167.2 | 271.3 | 423.7 | 637.2 |
| 0.10 | 0.37 | 0.48 | 0.61 | 0.75 | 0.90 | 1.1 | 1.32 | 1.57 | 1.82 | 2.14 | 2.49 | 4.29 | 6.8 | 10.18 | 14.14 | 18.56 | 22.95 |

*All units are in eV (electron-volts) (Putz, 2006).

**TABLE 4.10**  The Absolute Electronegativity $\chi_A^{[4]}$, Hardness $\eta_A^{[4]}$, Ionization Potential $IP_A^{[4]}$, and Electron Affinity $EA_A^{[4]}$, Computed Upon the Eqs. (4.290), (4.288), (4.291) and (4.292), Respectively, For the Ordinary Elements

**Legend:** *Symbol of Element*

Absolute Electronegativity: $\chi_A^{[4]}$

Absolute Hardness: $\eta_A^{[4]}$

Absolute Ionization Potential: $IP_A^{[4]}$

Absolute Electron Affinity: $EA_A^{[4]}$

| Element | $\chi_A$ | $\eta_A$ | $IP_A$ | $EA_A$ |
|---|---|---|---|---|
| H | 7.18 | 6.45 | 13.62 | 0.73 |
| He | 32.96 | 45.22 | 60.29 | 5.62 |
| Li | 4.11 | 2.39 | 7.99 | 0.23 |
| Be | 12.04 | 12.13 | 22.67 | 1.41 |
| B | 31.39 | 36.49 | 58.38 | 4.39 |
| C | 75.19 | 80.8 | 140.8 | 9.55 |
| N | 157.7 | 133.7 | 300.5 | 14.82 |
| O | 289.8 | 161 | 564.7 | 14.84 |
| F | 460.2 | 119.1 | 916.8 | 3.58 |
| Ne | 645.2 | -2.38 | 1310 | -19.71 |
| Na | 6.30 | 5.10 | 12.05 | 0.56 |
| Mg | 13.57 | 14.46 | 25.43 | 1.70 |
| Al | 28.47 | 32.62 | 53.03 | 3.91 |
| Si | 57.36 | 62.78 | 107.3 | 7.45 |
| P | 107.8 | 105.2 | 203.6 | 12.14 |
| S | 188.2 | 156.4 | 359.2 | 17.23 |
| Cl | 315 | 211.4 | 608.4 | 21.49 |
| Ar | 486.6 | 247.3 | 951.6 | 21.63 |
| K | 4 | 2.2 | 7.8 | 0.20 |
| Ca | 8.04 | 6.28 | 15.39 | 0.68 |
| Sc | 9.11 | 7.70 | 17.37 | 0.85 |
| Ti | 10.29 | 9.34 | 19.52 | 1.06 |
| V | 11.57 | 11.23 | 21.85 | 1.29 |
| Cr | 12.96 | 13.4 | 24.36 | 1.57 |
| Mn | 14.48 | 15.85 | 27.08 | 1.88 |
| Fe | 16.12 | 18.63 | 30.01 | 2.24 |
| Co | 17.9 | 21.75 | 33.16 | 2.64 |
| Ni | 19.82 | 25.25 | 36.55 | 3.09 |
| Cu | 22.16 | 29.65 | 40.66 | 3.66 |
| Zn | 24.41 | 34.03 | 44.59 | 4.24 |
| Ga | 44.6 | 58.13 | 82.04 | 7.15 |
| Ge | 79.72 | 93.11 | 148.2 | 11.21 |
| As | 134.9 | 137.5 | 253.8 | 16.05 |
| Se | 219.9 | 192.3 | 418.4 | 21.53 |
| Br | 338.8 | 249.8 | 651.2 | 26.39 |
| Kr | 502.2 | 304.4 | 974.8 | 29.51 |

**TABLE 4.10** Continued

| Rb | Sr | Y | Zr | Nb | Mo | Tc | Ru | Rh | Pd | Ag | Cd | In | Sn | Sb | Te | I | Xe |
|---|---|---|---|---|---|---|---|---|---|---|---|---|---|---|---|---|---|
| 3.1 | 5.95 | 6.75 | 7.62 | 8.57 | 9.46 | 10.56 | 11.76 | 13.05 | 14.25 | 15.74 | 17.34 | 30.49 | 52.55 | 88.70 | 142.7 | 221.1 | 330.1 |
| 1.35 | 3.78 | 4.67 | 5.71 | 6.92 | 8.11 | 9.67 | 11.44 | 13.44 | 15.39 | 17.87 | 20.65 | 35.67 | 57.33 | 87.85 | 126 | 172.2 | 224.5 |
| 6.09 | 11.52 | 13.01 | 14.63 | 16.38 | 18.02 | 20.03 | 22.2 | 24.52 | 26.69 | 29.34 | 32.18 | 56.69 | 98.3 | 167.2 | 271.3 | 423.7 | 637.2 |
| 0.10 | 0.37 | 0.48 | 0.61 | 0.75 | 0.90 | 1.1 | 1.32 | 1.57 | 1.82 | 2.14 | 2.49 | 4.29 | 6.8 | 10.18 | 14.14 | 18.56 | 22.95 |

*All units are in eV (electron-volts) (Putz, 2006).

As a consequence, worth stressing that further studies employing electronic density of the valence shell containing also the many-electronic correlation and exchange effects, improving therefore the working actual density (4.317) (Cook, 1974), are envisaged to systematically increase also the quantitative accuracy of the computed atomic $IP$'s.

Passing to the $EA$ analysis, in the Figure 4.12 there are comparatively plotted the experimental ($EA_E$ from Table 4.3), with the absolute $A^{[2]}$ ($EA_A^{[2]}$ from Table 4.8), and the absolute $A^{[4]}$ ($EA_A^{[4]}$ from Table 4.10) scales.

First, let's note that experimentally the $EA$ stands as a wild quantity being mostly indirect measured, for instance trough the Haber-Born cycle and only recently improved since the advent of the laser photo-detachment experiments with negative ions (Hotop & Lineberger, 1985) and by the electron transmission spectroscopy (Jordan & Burrow, 1987).

As previously remarked, see Table 4.3 and Figure 4.12, the so-called experimental electron affinities from the $EA_E$ scale admit positive as well negative values. This situation is hard to interpret as far the adiabatic and variational methods provide $EA=0$, whereas the vertical approach, under $V(\mathbf{r}) = ct \neq 0$ condition, suggests $EA<0$ values.

The present $EA_A^{[2]}$ and $EA_A^{[4]}$ scales, through their derivation, i.e., by performing the path integration on $\delta V(\mathbf{r})$ from 0 (adiabatic) up to $V(\mathbf{r}) = ct \neq 0$ (vertical) potential limits, like to solve somehow such dichotomy (Putz, 2006).

From Table 4.8 and in Table 4.10, and clearly evidenced in Figure 4.12, there appears that the $EA_A^{[2]}$ and $EA_A^{[4]}$ values display merely negative or positive scales, respectively. So, both the actual approaches are $V^*(r)$ dependent, although providing different values and scales depending on the complexity of the presence in the effective acting force $-\nabla V^*(\mathbf{r})$ and in its potential. In this respect, the $EA_A^{[4]}$ scale, which is richer in $-\nabla V^*(\mathbf{r})$ and $V^*(r)$ terms through the additional $b$ and $C_A$ functionals, Eqs. (4.276) and (4.277), respectively, appears with a positive rationalized trend in Figure 4.12 and on an acceptable quantitative relative relation with the experimental one from Table 4.3.

Unlike the case of $IP$, the actual $EA$ results suggest that, even considering all valence electrons in present computations, as the $EA$ (4.256) definition is employed the "computational effect" corresponds with that of adding only one electron to the concerned system. Such combined sensitivity further encourages the use of the actual $EA_A^{[2]}$ and $EA_A^{[4]}$ approaches

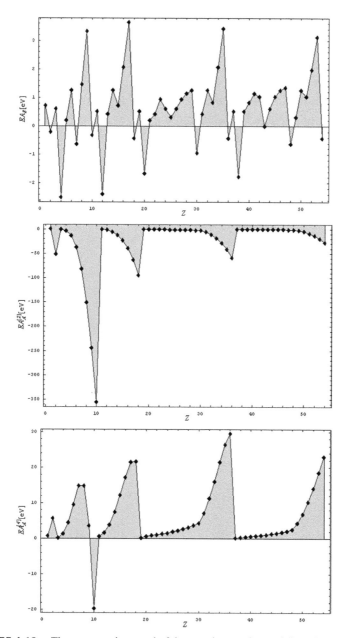

**FIGURE 4.12**    The comparative trend of the atomic experimental first electronic affinity $EA_E$ from Table 4.3 (upper) with respect to the atomic absolute electron affinity $EA_A^{[2]}$ from Table 4.8 (middle), and the atomic electron affinity $EA_A^{[4]}$ from Table 4.10 (lower) scales, respectively (Putz, 2006).

coupled with their correspondent $IP_A^{[2]}$ and $IP_A^{[4]}$ functionals in correlation with the molecular orbital theory (Putz, 2006).

Turning to the non-experimental but matter intrinsic chemical indices, in the Figure 4.13 there are represented side-by-side the atomic scales of electronegativity and of its hardness companion from the finite difference ($\chi_{FD}$ and $\eta_{FD}$ of Table 4.3), the absolute A[2] ($\chi_A^{[2]}$ and $\eta_A^{[2]}$ of Table VI), and the absolute A[4] ( $\chi_A^{[4]}$ and $\eta_A^{[4]}$ of Table 4.10) approaches.

Analyzing the electronegativity scales, Leland Allen has written much on criteria which good models has to fulfill see (Murphy et al., 2000) and the references therein.

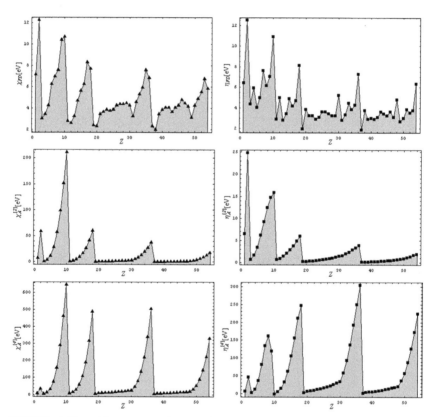

**FIGURE 4.13**    The comparative trend of the atomic finite difference chemical electronegativity $\chi_{FD}$ and chemical hardness $\eta_{FD}$ from Table 4.3 (upper) with respect to the atomic absolute electronegativity $\chi_A^{[2]}$ and and hardness $\eta_A^{[2]}$ from Table 4.8 (middle), and the atomic absolute electronegativity $\chi_A^{[4]}$ and hardness $\eta_A^{[4]}$ from Table 4.10 (lower) scales, respectively (Putz, 2006).

Concerning the acceptability guidelines for the actual $\chi_A^{[2]}$ and $\chi_A^{[4]}$ scales, given in the Tables 4.8 and 4.10, and plotted in the Figure 4.13, their features can be summarized as follow (Putz, 2006).

1. *The free atom definition*: both scales were built up for isolated atoms.
2. *The electronegativity as energy per electron*: the given values are expressed in electron-volts.
3. *The inclusion of all valence electrons*: for all main group atoms and for the most transition metals all valence electrons $N_v$ are included in the working $\chi_A$ definitions, see relations (4.268) and (4.290), and in their computation, see relations (4.331)–(4.333).
4. *The separation and contraction of the main-transition groups*: the close lying values are presented for about 30% of the main groups and for 75% of the transition elements. Among the presented $\chi$ scales in Figure 4.13, all satisfying the last condition, the present ones display absolute values that better separates the main from transitional groups. As a consequence also the contraction of the transitional elements is better emphasized through actual models. Nevertheless, three significant figures are able to distinguish the electronegativities of all the considered elements.
5. *The noble elements highest value*: along the periods the noble elements have the highest electronegativity value. However, the present $\chi_A^{[2]}$ and $\chi_A^{[4]}$ scales highly satisfy this criterion.
6. *The Si rule*: the requirement that all metals must have $\chi$ values which are less than or equal to that of Si is completely satisfied through the Tables 4.8 and 4.10; note that the finite-difference electronegativity picture produces the Pd exception as in Table 4.3 revealed.
7. *The C rule*: the requirement that the $\chi$ value of C have to be greater (or at least equal) with that of the H ($\chi \cong 7.2$ eV) is again completely fulfilled by the actual approaches but not from the finite-difference scheme as it is evident by consulting the Figure 4.13 or the Tables 4.3, 4.8, and 4.10.
8. *The metalloids band*: the six considered metalloid elements (B, Si, Ge, As, Sb, Te) that separate the metals from the non-metals have electronegativity values, which do not allow overlaps between metals and non-metals. However, due actual high values of the actual absolute EN scales this criterion is well followed respecting

the finite-difference scale, compare $\chi_A^{[2]}$ and $\chi_A^{[4]}$ of Tables 4.8 and 4.10, respectively, with $\chi_{FD}$ of Table 4.3.

9. *The covalent-ionic-metallic triangle*: for binary compounds, $AB$, the difference $\chi(A)-\chi(B)$ must quantify the definitions of the ionic, covalent, and metallic bonds. The easiest way to visualize this trend is to construct the Ketelaar's triangle (Ketelaar, 1958), that quantifies the sides of the binary bond by a triangle whose vertices correspond to C (covalent), I (ionic) and M (metallic) limits, see the Figure 4.14. It is immediately from Figure 4.14 that on the diagonal (between M and C bonds) there are recorded the same electronegativity values in agreement with the fact that the covalent and metallic bonds are assigned to "the same basic quantum mechanical maximum overlap-exchange forces" (Slater, 1939). Therefore, along the side C-M we are moving in fact from the right to left in the Periodic Table. This fashion is fully covered by the actual $\chi_A^{[2]}$ and $\chi_A^{[4]}$ scales that as going from right to left in Periodic Table an impressive decrease of electronegativity is recorded, see the Figure 4.13, that strongly emphasize on the passage from the electronegative to the electropositive elements.

10. *The quantum mechanical viable definition*: being an intrinsic property of an atom, that is a quantum object, the electronegativity has to include the quantum nature of the electronic systems to whom is associated. The present scales are accommodated within conceptual DFT by employing the softness realization (4.226), which contains three quantum constraints such the translational invariance condition, the Hellmann-Feynman theorem, and the normalization of the linear response function are.

11. *The systematic behavior in the Periodic Table*: the decreasing of $\chi_A^{[2]}$ and $\chi_A^{[4]}$ values along the group is respected as well as their differences in going from light to heavy atoms of the same period increases left to right across rows. Correctly, the halogen atoms have the highest electronegativity values with respect to their left row neighbors, in all present cases.

After all these criteria it seems that the actual absolute EN density functionals, Eqs. (4.268) and (4.290), and their atomic scales can be taken as reliable when further used in predicting bonding and reactivity.

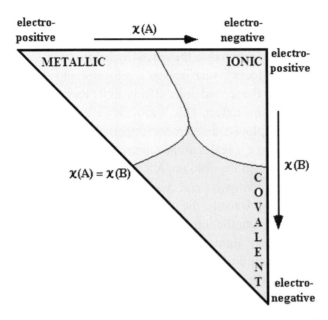

**FIGURE 4.14**    The schematic representation of the Ketelaar triangle for the binary AB bonding tendencies according with their electronegativity differences, between the representative bonds identified as I (ionic), C (covalent), and M (metallic) vertices; after Ketelaar (1958); Putz (2006).

Nevertheless, a feed-back with the DFT Parr EN (3.1) definition worth also: as far as the $\chi_A^{[2]}$ scales gives reliable results, and on the base of its very close relation with the $\chi_A^{[1]}$, compare the $\chi$ values of Tables 4.7 and 4.8, one can equally choose to use the Eq. (4.260) instead of Eqs. (4.267) expansion of $dE_N$, on which the actual absolute EN (4.253) definition is based. But with Eq. (4.260) in Eq. (4.253) the series of transformations given in Eq. (4.248) are recovered, enhancing so far the Mulliken-Parr integral relationship. However, the relation (4.248) is further generalized when the absolute A[4] picture is adopted (Putz, 2006).

Finally, we focus on the absolute $\eta_A^{[2]}$ and $\eta_A^{[4]}$ scales. From the Figure 4.13 there is clear that, respecting the finite-difference energetic values, $\eta_{FD}$ of Table 4.3, the actual $\eta_A^{[2]}$ and $\eta_A^{[4]}$ scales from the Tables 4.8 and 4.10, respectively, show a more rationalized periodic trend. However, also between $\eta_A^{[2]}$ and $\eta_A^{[4]}$ scales a completely different tendency is remarked: at the main group level, in the first case, the decreasing trend of $\eta_A^{[2]}$ values down groups is followed, while the situation is somehow reversed through the $\eta_A^{[4]}$ records. Such situation may add new chemistry when using the late absolute hardness to rationalize the reactivity behavior.

Usually, the hardness is seen only as a companion to electronegativity, i.e., associated with the second order effects, as its Parr-Pearson basic (3.3) and (4.251) definitions reflect. This is also the case when the absolute $A^{[2]}$ actual picture is employed, a statement supported also by the close quantitative atomic $\eta_A^{[2]}$ and $\eta_{FD}$ scales as the representations from the Figure 4.13 indicate. Instead, when the absolute $A^{[4]}$ approach is performed the situation regarding both qualitative and quantitative absolute hardness $\eta_A^{[4]}$ scale completely changes. Actually, the $\eta_A^{[4]}$ values predict a smooth increase of the hardness effects paralleling those of $EA_A^{[4]}$, compare the Figures 4.13 and 4.12, respectively. Remarkably, the linked behavior of $\eta_A^{[4]}$ with $EA_A^{[4]}$ is achieved even their basic definitions, Eqs. (4.254) and (4.256), respectively, are not directly related, as there are, for instance, the definitions of $\chi_A$ and $EA_A$, compare Eqs. (4.253) and (4.256) (Putz, 2006).

It is therefore worth to adopt the Eqs. (4.266) and (4.288) absolute hardness density functionals, the $\eta_A^{[2]}$ and especially the $\eta_A^{[4]}$ one, respectively, as the trial schemes to further test the two major HSAB and MH principles, for the reactions to which even the use of the improved compact finite difference schemes failed to produce the expected order of the reactivity preference, see Volume III of the present five-volume set (Putz, 2016a).

Overall, the computed and represented atomic scales for the reactivity indices $\chi$, $\eta$, $IP$, and $EA$ of this venture display a systematic qualitative periodic trend across the ordinary elements being in a relatively quantitative acceptable ratio with the experimental values, where these are available.

Nevertheless, especially when is about reactivity, having useful tools to predict good qualitative and relative ratios of atomic combinations stands as the "natural tendency of the mind to give to the shape of a graph some intrinsic value...to its ultimate consequences...because qualitative and empirical deduction already gives them sufficient framework for experiment and prediction" (Thom, 1975). This is at the end what the Chemistry is dealing with (Putz, 2006) and this line of analsys will be further employed while treating chemical-biological interaction in the Volume V of the present five-volume work (Putz, 2016b).

### 4.7.3  MULLIKEN ELECTRONEGATIVITY: THE RELATIONSHIP WITH CHEMICAL ACTION

The electrons of the atomic system are distinguished as the core- and the valence-ones within pseudopotential theory of atoms and molecules (Putz et al., 2005; Szasz, 1985; Preuss, 1969); this, because it aims to provide a "valence-only" theory for these systems, while assuring the simplification of the computations. Certainly, the all-electrons picture is also possible through facing with the serious technical problem to assure the *orthogonality* constrains among all wave functions of all electrons of an atom. Moreover, having the valence shell treated separately is relevant for computing electronegativity, because of its definition regarding the added electron to the valence shell under the core influence. Therefore, a wise step is provided by the transformation of a many-valence electronic problem into a one-valence electronic system, so that the canonical density formulations can be at once considered. For achieving this, the link between the exact and density dependent pseudopotential is enforced by the latter's satisfying the *virial theorem* releasing with the radial scaling of the pseudo-orbital (Preuss, 1969)

$$\rho_{PO}^{1/2}(q,r)=\psi(q,r)=q^{3/2}\psi(qr) \tag{4.335}$$

with the scaling factor $q$. Therefore, the scaling factor $q$ is searched in relation with the number of valence electrons, but such to fulfill the normalization condition

$$\int \left|\psi(q,r)\right|^2 dr = 1 \tag{4.336}$$

Next, the effective potential of the core is represented as a pseudopotential employing the Stuttgart/Bonn wave function expansion (Stuttgart Pseudopotentials, 2011)

$$\psi(r)=\sum_i A_i \exp\left(-\alpha_i r^2\right) \tag{4.337}$$

while the Mulliken electronegativity is computed starting from lithium, to assure the existence of the core electrons. For H and He systems the corresponding electronegativity values can be added from other methods of computation (Mulliken, 1934; Hinze & Jaffé, 1962).

Within the pseudopotential methods we arrive at two possibilities for the electronic density and, consequently, for the electronegativity evaluations (Putz et al., 2005, 2009b):

- The first one considers only the pseudo-potentials into the path integral formalism that gives the electronic density in the quantum statistical manner as it was described in the Section 4.6.6. This way, a strong physical meaning is assured because all the information about the electronic density and electronegativity are comprised (and dictated) only by the pseudopotential. Yet, the problem that arises in this approach is that the electronic density depends on the thermal $\beta$ parameter. This parameter will be fixed so that the electronic density to fulfill the *path integral* normalization condition. Additionally, the search of the $\beta$ parameter must be done in the semiclassical (high temperature) limit ($\beta \to 0$) for which the path integral formalism corresponds to the excited (valence) states of atoms.
- The second approach takes beyond to the pseudopotential data also the valence basis and the electronic densities are then computed in the accustomed quantum manner. At this point we need to consider the working orbital type for the atomic systems and we will chose the s-basis set because its spherical symmetry.

Accordingly, it follows that both electronic density approaches have their own parametric dependency. This implies that also the computed electronegativity will feature the scaling effect on the electronic density raised due to the one effective valence electronic approach. With this assumption at the background of density computation we should recover in the provided electronegativity the real (many) electronic valence state by an adequate nomination of the specific values for the $\beta$ and $q$ parameters.

At this point we need a criterion in order to properly control the re-scaling procedure. In order to unveil this criterion we are looking back on differential electronegativity formula (4.275) that should be seen as the kernel function for the Mulliken electronegativity functional (4.279). If we observe the analytical places the introduced chemical response indices $a$, $b$ and the chemical action index $C_A$ appear, respectively, it can be easily seen that only the chemical action is coupled with the total number of electrons in the concerned state.

In this respect, while noting the above scaling condition (4.335) as being quite restrictive for the scaling factor $q$, an additional constrain that takes into consideration the number of valence electrons is to be regarded. This aim is to be accomplished observing that the valence formulation atomic electronegativity fits with the definition of chemical action of Eq. (4.277) for the Coulombic potential (March, 1993)

$$\chi(N,Z)=\left\langle\frac{1}{r}\right\rangle=\int\left\{\rho(N,Z,r)\frac{1}{r}\right\}dr=-\int\{\rho(N,Z,r)V_{Clb}(r)\}dr\equiv-C_A^{1/r}$$

(4.338)

with $Z$ being the nuclear charge.

The results for the chemical action (4.338) computed by the two above-mentioned quantum computational schemes are collected in Table 4.11 with those for the electronegativity (4.279) in Table 4.12, among other significant scales. All data are comparatively represented in Figures 4.15 and 4.16, respectively.

The proposed electronegativity scale follows the general rules for its acceptability [see Section 4.7.2 and Murphy et al. (2000)]. The decreasing of $\chi$ along the group is respected (see, for instance, Ga<Al and Ge<Si) as well as its difference in going from light to heavy atoms of the same group. $\chi$ increases left to right across rows taking into account that for some heavy elements the relativistic effects, which are not considered in the computations, can affect this trend. Correctly, the halogen atoms have the highest electronegativity with respect to their left row neighbors.

The adopted *PI* procedure supports different model potentials. With the aim to test the reliability of the present algorithm for the Coulomb potentials the Bachelet-Hamann-Schülter pseudopotentials (Bachelet et al., 1982) for C, N and O atoms were adopted and their electronegativity values recalculated. The results, both for electronegativity and chemical action, are reported in Table 4.13. The numerical values are only slightly higher than those obtained using Bonn/Stuttgart pseudopotentials and the close relation between electronegativity and chemical action values is present, in all cases.

Analyzing these results, it is clear that for the path integral approach better correlation between the electronegativity and the chemical action trends is obtained as comparing with those arising from the s-basis set implementation. The highest discrepancy between the chemical action and

**TABLE 4.11** The Absolute Atomic Chemical Actions Given by Eq. (4.277)—in Electron-Volts [eV]—Computed by Path Integral (4.294) When Only the Pseudo-Potential is Need (The Upper Value for Each Element) and By Pseudo-Potentials + Basis Set Method (The Lower Value for Each Element) for Valence Electronic Density Computations (Putz et al., 2005, 2009b)

| | | | | | | | | | | | | | | | | | |
|---|---|---|---|---|---|---|---|---|---|---|---|---|---|---|---|---|---|
| **Li** 4.77 / 3.50 | **Be** 6.05 / 3.93 | | | | | | | | | | | **B** 6.77 / 6.07 | **C** 8.69 / 8.44 | **N** 9.73 / 8.95 | **O** 10.93 / 10.72 | **F** 11.84 / 17.80 | **Ne** 10.90 / 17.60 |
| **Na** 4.09 / 3.02 | **Mg** 5.18 / 3.08 | | | | | | | | | | | **Al** 8.73 / 6.20 | **Si** 5.95 / 6.71 | **P** 8.38 / 7.72 | **S** 9.48 / 10.32 | **Cl** 9.94 / 12.07 | **Ar** 9.25 / 13.36 |
| **K** 3.28 / 2.91 | **Ca** 4.41 / 1.76 | **Sc** 2.66 / 1.76 | **Ti** 3.19 / 2.46 | **V** 3.78 / 3.11 | **Cr** 4.71 / 4.58 | **Mn** 5.41 / 5.46 | **Fe** 5.35 / 6.01 | **Co** 5.39 / 6.49 | **Ni** 5.49 / 8.62 | **Cu** 5.83 / 7.01 | **Zn** 4.54 / 9.10 | **Ga** 3.24 / 3.24 | **Ge** 5.12 / 3.58 | **As** 4.53 / 3.89 | **Se** 9.09 / 3.65 | **Br** 9.11 / 5.22 | **Kr** 7.93 / 5.97 |
| **Rb** 1.63 / 1.18 | **Sr** 2.92 / 1.79 | **Y** 2.47 / 1.38 | **Zr** 3.57 / 1.17 | **Nb** 4.34 / 1.12 | **Mo** 5.08 / 1.37 | **Tc** 5.06 / 1.30 | **Ru** 5.36 / 1.23 | **Rh** 5.65 / 1.10 | **Pd** 5.86 / 1.27 | **Ag** 5.86 / 1.39 | **Cd** 4.76 / 1.52 | **In** 5.10 / 2.29 | **Sn** 5.37 / 2.24 | **Sb** 5.05 / 4.55 | **Te** 7.53 / 3.60 | **I** 8.42 / 4.56 | **Xe** 7.37 / 5.40 |

**TABLE 4.12**    The Comparative Atomic Mulliken Electronegativities, in Electron-Volts [eV] (Putz, 2009b)

| Z | Element | Mulliken-Jaffe[a] | Experiment[b] | Xα[c] | Density Functional Equation (4.279) | |
|---|---------|-------------------|---------------|-------|----------------|-----------|
| | | | | | Path Integral | s-Basis Set |
| 3 | Li | 1.8 | 3.01 | 2.58 | 4.11 | 3.02 |
| 4 | Be | 4.8 | 4.9 | 3.80 | 5.64 | 3.40 |
| 5 | B | 5.99 | 4.29 | 3.40 | 5.72 | 5.66 |
| 6 | C | 8.59 | 6.27 | 5.13 | 8.56 | 8.58 |
| 7 | N | 11.21 | 7.27 | 6.97 | 10.13 | 9.77 |
| 8 | O | 14.39 | 7.53 | 8.92 | 11.87 | 12.41 |
| 9 | F | 12.18 | 10.41 | 11.0 | 13.13 | 15.60 |
| 10 | Ne | 13.29 | - | 10.31 | 13.39 | 13.37 |
| 11 | Na | 1.6 | 2.85 | 2.32 | 3.16 | 2.64 |
| 12 | Mg | 4.09 | 3.75 | 3.04 | 4.52 | 3.93 |
| 13 | Al | 5.47 | 3.21 | 2.25 | 5.80 | 5.89 |
| 14 | Si | 7.30 | 4.76 | 3.60 | 6.56 | 6.80 |
| 15 | P | 8.90 | 5.62 | 5.01 | 9.04 | 8.33 |
| 16 | S | 10.14 | 6.22 | 6.52 | 10.09 | 11.88 |
| 17 | Cl | 9.38 | 8.30 | 8.11 | 10.64 | 14.59 |
| 18 | Ar | 9.87 | - | 7.11 | 10.12 | 12.55 |
| 19 | K | 2.90 | 2.42 | 1.92 | 3.15 | 2.48 |
| 20 | Ca | 3.30 | 2.2 | 1.86 | 4.21 | 2.19 |
| 21 | Sc | 4.66 | 3.34 | 2.52 | 2.93 | 1.83 |
| 22 | Ti | 5.2 | 3.45 | 3.05 | 3.52 | 2.28 |
| 23 | V | 5.47 | 3.6 | 3.33 | 4.19 | 2.42 |
| 24 | Cr | 5.56 | 3.72 | 3.45 | 5.23 | 2.72 |
| 25 | Mn | 5.23 | 3.72 | 4.33 | 6.02 | 2.01 |
| 26 | Fe | 6.06 | 4.06 | 4.71 | 5.96 | 3.90 |
| 27 | Co | 6.21 | 4.3 | 3.76 | 6.01 | 3.03 |
| 28 | Ni | 6.30 | 4.40 | 3.86 | 6.12 | 3.48 |
| 29 | Cu | 6.27 | 4.48 | 3.95 | 6.35 | 2.91 |
| 30 | Zn | 5.53 | 4.45 | 3.66 | 5.07 | 3.13 |
| 31 | Ga | 6.02 | 3.2 | 2.11 | 3.49 | 3.30 |
| 32 | Ge | 6.4 | 4.6 | 3.37 | 5.45 | 4.24 |
| 33 | As | 6.63 | 5.3 | 4.63 | 4.87 | 4.94 |

**TABLE 4.12** Continued

| Z | Element | Mulliken-Jaffe[a] | Experiment[b] | Xα[c] | Density Functional Equation (4.279) | |
|---|---------|-------------------|---------------|-------|---------------|-----------|
| | | | | | Path Integral | s-Basis Set |
| 34 | Se | 7.39 | 5.89 | 5.91 | 7.71 | 4.82 |
| 35 | Br | 8.40 | 7.59 | 7.24 | 7.75 | 7.35 |
| 36 | Kr | 8.86 | - | 6.18 | 8.65 | 9.59 |
| 37 | Rb | 2.09 | 2.34 | 1.79 | 1.56 | 1.05 |
| 38 | Sr | 3.14 | 2.0 | 1.75 | 2.87 | 1.63 |
| 39 | Y | 4.25 | 3.19 | 2.25 | 3.33 | 1.76 |
| 40 | Zr | 4.57 | 3.64 | 3.01 | 3.92 | 1.73 |
| 41 | Nb | 5.38 | 4.0 | 3.26 | 4.77 | 1.68 |
| 42 | Mo | 7.04 | 3.9 | 3.34 | 5.59 | 2.07 |
| 43 | Tc | 6.27 | - | 4.58 | 5.57 | 1.96 |
| 44 | Ru | 7.16 | 4.5 | 3.45 | 5.91 | 1.93 |
| 45 | Rh | 7.4 | 4.3 | 3.49 | 6.23 | 1.72 |
| 46 | Pd | 7.16 | 4.45 | 3.52 | 6.46 | 1.98 |
| 47 | Ag | 6.36 | 4.44 | 3.55 | 6.47 | 2.18 |
| 48 | Cd | 5.64 | 4.43 | 3.35 | 5.26 | 2.36 |
| 49 | In | 5.22 | 3.1 | 2.09 | 5.38 | 2.48 |
| 50 | Sn | 6.96 | 4.30 | 3.20 | 5.75 | 2.74 |
| 51 | Sb | 7.36 | 4.85 | - | 5.44 | 6.29 |
| 52 | Te | 7.67 | 5.49 | 5.35 | 6.35 | 4.98 |
| 53 | I | 8.10 | 6.76 | 6.45 | 7.12 | 6.70 |
| 54 | Xe | 7.76 | - | 5.36 | 7.80 | 6.27 |

[a]From Huheey (1978); WebElements (2011).
[b]From Parr and Pearson (1983); Robles and Bartolotti (1984).
[c]From Bartolotti et al. (1980).

the associated electronegativity within the s-basis set computation appears mostly for the first transitional row, see Figure 4.15.

However, we cannot exclude the s-basis set electronegativity scale just through comparison between scales since other similar discrepancies appear (even in the main groups) when the Mulliken-Jaffe and the Xα methods are compared, for instance (see the Table 4.12). In any case, the present s-basis set results may help in judging also the various criteria of validity for an electronegativity scale, see Section 4.7.2. Looking to the transition metal

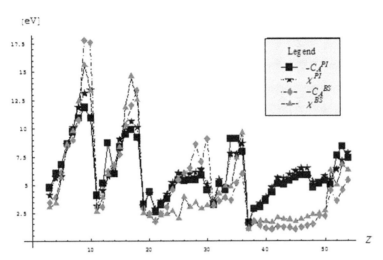

**FIGURE 4.15**   The comparative representation of the atomic electronegativities' values computed upon Eq. (4.279) and the corresponding absolute chemical actions—given by Eq. (4.277)—using the path integral (PI) and basis set (BS) methods (Putz, 2009b).

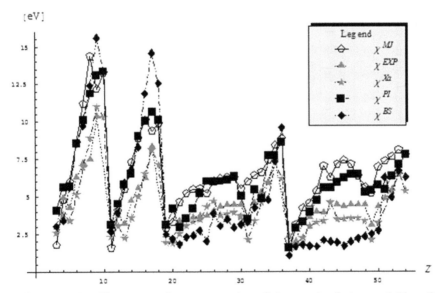

**FIGURE 4.16**   The comparative representation of the atomic electronegativities of Mulliken-Jaffe (MJ), experimental (EXP), and transition state (Xα) respecting the chemical Mulliken one (4.279) with the recorded values of Table 4.12 by path integral (PI) and basis set (BS) methods (Putz, 2009b).

**TABLE 4.13** The Comparative Orbital Electronegativities and the Absolute Chemical Actions for C, N and O Atoms (Putz et al., 2005, 2009b)

| Element | Chemical Information | s Basis Set | s Path Integral | p Basis Set | p Path Integral | sp Basis Set | sp Path Integral | sp² Basis Set | sp² Path Integral | sp³ Basis Set | sp³ Path Integral |
|---|---|---|---|---|---|---|---|---|---|---|---|
| C | *Mulliken-Jaffe's Electronegativity* | *8.59* | | *5.80* | | *10.39* | | *8.79* | | *7.98* | |
| | Electronegativity | 8.58 | 8.56 | 3.11 | 4.04 | 10.73 | 9.89 | 7.53 | 6.99 | 5.77 | 5.71 |
| | Chemical Action | 8.44 | 8.69 | 3.11 | 4.1 | 11.43 | 10.04 | 7.74 | 7.1 | 5.83 | 5.71 |
| N | *Mulliken-Jaffe's Electronegativity* | *11.21* | | *7.39* | | *15.68* | | *12.87* | | *11.54* | |
| | Electronegativity | 9.77 | 10.13 | 5.09 | 6.14 | 16.97 | 17.54 | 11.88 | 12.40 | 9.21 | 10.13 |
| | Chemical Action | 8.95 | 9.73 | 4.80 | 5.9 | 17.99 | 16.86 | 12.34 | 11.92 | 9.35 | 9.73 |
| O | *Mulliken-Jaffe's Electronegativity* | *14.39* | | *9.65* | | *27.25* | | *17.07* | | *15.25* | |
| | Electronegativity | 12.41 | 11.87 | 8.06 | 8.39 | 27.06 | 27.40 | 18.54 | 19.38 | 14.48 | 15.82 |
| | Chemical Action | 10.72 | 10.93 | 7.35 | 7.73 | 28.07 | 25.23 | 19.48 | 17.84 | 14.84 | 14.57 |

*All values are in electron-Volts [eV].

atoms, we underline that the obtained electronegativities fall in a narrow range of values compared with those of the main group atoms. The six considered metalloid elements (B, Si, Ge, As, Sb, Te), that separate the metals from the non-metals, have electronegativity values, which do not allow overlaps between metals and non-metals. Furthermore, looking at the $\chi$ metal values the requirement that they must have electronegativities lower than silicon is satisfied (see for example Ga<Si, Al<Si, Ge<Si) following the so-called silicon rule—see the Section 4.7.2. Finally, we briefly discuss the values obtained for the N, O, F, Ne, Cl, Ar, Br, and Kr elements that present oxidation states lower than their valence electrons. The rule in this case states that $\chi$ parallels the decreasing in valence electrons. The results follow this rule with the exception of the chlorine atom that has a $\chi$ value higher than the nearest noble gas atom Ar. The electronegativity trend for these atoms results in the order Ne>F>O>Cl>N>Ar>Kr>Br.

It is worth to note that the *PI* treatment does not need the orbital type function but only the pseudo-potential representing the field in which the electrons move. In order to verify the influence of the different orbital type we have redone the electronegativity computation for C, O and N atoms by using p-type orbitals and the sp, $sp^2$ and $sp^3$ hybridization states. Results, reported in Table 4.13 and Figure 4.17, show how the actual electronegativity formulation preserves also the orbital hierarchy and is sensitive to the hybrid orbitals as well. Finally, analyzing Table 4.13 and Figure 4.17 we underline that the electronegativity trend from a type of hybridization to another is similar (Putz et al., 2005, 2009b).

However, it remains to show that the employed chemical action criteria (4.338) do not enter in conflict with the type of the orbital choice, when this is properly done. For instance, we consider the atomic systems of C, N, O with the s- and p-orbital type basis set and also the sp, $sp^2$ and $sp^3$ hybridization states.

Then, by applying the re-scaling procedure according with the chemical action—electronegativity rule (4.338) we get the respective electronegativities and chemical actions for both the path integral and basis set implementations using the pseudopotential data (Stuttgart Pseudopotentials, 2011), as shown in Table 4.13 and drawn in Figure 4.17.

They, nevertheless reveal how close the values of the chemical actions and corresponding orbital electronegativities are, in general. Such feature

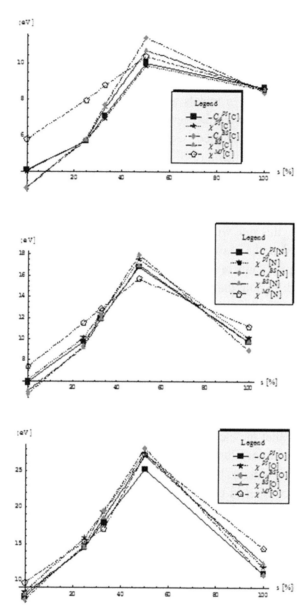

**FIGURE 4.17**    From top to bottom, the representations of the orbital electronegativities and of the absolute chemical actions for C, N and O atoms versus the different percent contribution of s orbital (p: 0%, sp³: 25%, sp²: 33%, sp: 50% and s: 100%) in pseudo-potentials and basis set frameworks of electronic densities computation with path integral (PI), basis set (BS), and Mulliken-Jaffe (MJ) results of Table 4.13 (Putz et al., 2005, 2009b).

is susceptible for extension in treating the chemical bonds as will be in Volume III of the present five-volume work exposed (Putz, 2009a).

## 4.7.4 ELECTROPHILICITY SCALE FOR ATOMS IN LONG-RANGE (VALENCE) STATES

As an application to atomic systems, the effective Slater atomic model will be considered (Slater, 1930; Clementi & Raimondi, 1963; Clementini et al., 1967) that is employing the Slater effective charge and the *valence* electrons through the potential (4.312) and the *radial* electronic density for a given quantum (shell) number $n$ and the orbital exponent $\xi$, related with Eqs. (4.317) and (4.131), here as

$$\rho_n(r,\xi) = N_v \frac{(2\xi)^{2n+1}}{4\pi(2n)!} r^{2n-2} \exp[-2\xi r] \qquad (4.339)$$

nevertheless fulfilling the DFT integration condition to the valence electrons

$$\int_0^\infty 4\pi\, r^2 \rho_n(r,\xi)dr = N_v \qquad (4.340)$$

by applying the Slater integral recipe (4.136).

In these conditions the radial atomic local softness, global softness and local chemical hardness become, from Eqs. (4.228), (4.229) and (4.232), respectively, as (Putz & Chattaraj, 2013):

$$s(r) = \frac{2^{2n-1}\xi^{2n+1}}{\pi(2n)!} N_v \left( N_v + \frac{r}{Z_{eff}} \right) r^{2(n-1)} \exp(-2r\xi) \qquad (4.341)$$

$$S = \int_0^\infty 4\pi r^2 s(r)dr = N_v^2 + \frac{(2n+1)}{2\xi Z_{eff}} N_v \qquad (4.342)$$

$$\eta(r) = \frac{Z_{eff}}{rN_v + N_v^2 Z_{eff}} \qquad (4.343)$$

Within this analytical context the atomic kernel, local and global electrophilicities of Eqs. (4.238), (4.245) and (4.234) respectively become (Putz & Chattaraj, 2013)

$$\omega(r,r') = \chi^2 \frac{2^{4n-2} \xi^{4n+3} N_v}{\pi^2 [(2n)!]^2 Z_{eff} (1 + 2n + 2\xi N_v Z_{eff})}$$
$$\times \left[2N_v^2 Z_{eff}^2 + 4rr' + 3N_v Z_{eff}(r + r')\right]$$
$$\times \frac{(N_v Z_{eff} + r)(N_v Z_{eff} + r')}{(N_v Z_{eff} + 2r)(N_v Z_{eff} + 2r')} (rr')^{2(n-1)} \exp\left[-2\xi(r + r')\right] \quad (4.344)$$

$$\omega(r) = \chi^2 \frac{2^{2n-1} \xi^{2n+1} N_v}{\pi(2n)! Z_{eff}} \frac{(r + N_v Z_{eff})^2}{2r + N_v Z_{eff}} r^{2(n-1)} \exp(-2r\xi) \quad (4.345)$$

$$\omega = \chi^2 N_v \frac{1 + 2n + 2\xi N_v Z_{eff}}{4\xi Z_{eff}} \quad (4.346)$$

For practical implementation one should finally assure the numerical integral hierarchy for each atomic system (Putz & Chattaraj, 2013)

$$\omega = C_\omega \int_0^\infty 4\pi r^2 \omega(r) dr = C_\omega \int_0^\infty \int_0^\infty 16\pi^2 r^2 r'^2 \omega(r,r') dr dr' \quad (4.347)$$

namely through the normalization constant $C_\omega$ appearing to modulate to the correct inter-relationships within the working expressions using the Slater orbital approximations.

It is also worth noting that the two constants in Eq. (4.347) are equal for each chemical species, and this is also numerically confirmed by the present atomic applications as it is implicit for integral hierarchy from kernel to local electrophilicity. However, the same normalization constant is still necessary respecting the global one since the local to global integral identity holds with local chemical hardness of Eq. (4.211), see the comment around Eq. (4.233), while we are currently implementing its generalized version as given by Eq. (4.232) above. Yet, no calibration between reference and actually computed electrophilicities of Eqs. (4.234) and (4.346) are considered (Putz & Chattaraj, 2013).

We present the main results in Table 4.14 where the basic electronegativity and chemical hardness calculated by finite difference approximations in terms of $IP$ and $EA$ definitions are considered (Lackner & Zweig, 1983). This is based on two positive arguments: they are based on the Parr's DFT ground state parabola method that is consistent with definition (4.234) of global electrophilicity; electronegativity under Mulliken

**TABLE 4.14**   Atomic Properties of the First Five Periods of the Periodic System in the Valence Slater States (Slater, 1930; Clementi & Raimondi, 1963; Clementini et al., 1967): Principal Quantum Number (n); Orbital Exponent ($\xi$); Effective Atomic Charge ($Z_{eff}$); Number of Employed [s+p] Valence Electrons ($N_v$) (Ghosh & Biswas, 2002); Electronegativity (4.3) and Chemical Hardness (4.251) based on Experimental Ionization Potential (IP) and Electronic Affinity (EA) (Ghosh & Biswas, 2002), Maximum Radii $R_{max} = n/\xi$ for Vanishing Radial Distribution of Electronic Density (4.339); Reference Electrophilicity $\omega_0$ Calculated Using Eq. (4.234); Electrophilicity $\omega$ Calculated Using Eq. (4.346); Radial Distribution of the Local Electrophilicity of Eq. (4.345) in the Point $R_{max}$; Radial Distribution of the Kernel Electrophilicity of Eq. (4.344) in the Spatial Point ($R_{max}$, $R_{max}$), see Putz & Chattaraj (2013).

| Atom | n | $\xi$ | $Z_{eff}$ | $N_v$ | $\chi$ [eV] | $\eta$ [eV] | $R_{max}$ [a.u.] | $\omega_0$ [eV] | $\omega$ [eV] | $\omega(Y.R_{max})$ [eV] | $\omega(Y.R_{max},R_{max})$ [eV] |
|---|---|---|---|---|---|---|---|---|---|---|---|
| H  | 1 | 1    | 1    | 1 | 7.18  | 6.45  | 1     | 3.996 | 2.368   | 1.086   | 0.470   |
| He | 1 | 1.7  | 1.7  | 2 | 12.27 | 12.48 | 0.588 | 6.032 | 13.938  | 12.486  | 10.702  |
| Li | 2 | 0.65 | 1.30 | 1 | 3.02  | 4.39  | 3.077 | 1.039 | 0.663   | 0.147   | 0.032   |
| Be | 2 | 0.98 | 1.95 | 2 | 3.43  | 5.93  | 2.041 | 0.992 | 1.43    | 0.521   | 0.184   |
| B  | 2 | 1.3  | 2.60 | 3 | 4.26  | 4.06  | 1.538 | 2.235 | 3.741   | 1.871   | 0.913   |
| C  | 2 | 1.63 | 3.25 | 4 | 6.24  | 4.99  | 1.227 | 3.902 | 12.799  | 8.110   | 5.056   |
| N  | 2 | 1.95 | 3.90 | 5 | 6.97  | 7.59  | 1.026 | 3.200 | 23.784  | 18.088  | 13.612  |
| O  | 2 | 2.28 | 4.55 | 6 | 7.59  | 6.14  | 0.877 | 4.691 | 39.634  | 35.286  | 31.193  |
| F  | 2 | 2.6  | 5.2  | 7 | 10.4  | 7.07  | 0.769 | 7.649 | 99.96   | 101.515 | 102.599 |
| Ne | 2 | 2.93 | 5.85 | 8 | 10.71 | 10.92 | 0.683 | 5.252 | 137.356 | 157.225 | 179.354 |
| Na | 3 | 0.73 | 2.20 | 1 | 2.80  | 2.89  | 4.110 | 1.356 | 0.458   | 0.099   | 0.021   |
| Mg | 3 | 0.95 | 2.85 | 2 | 2.6   | 4.99  | 3.158 | 0.677 | 0.818   | 0.241   | 0.069   |
| Al | 3 | 1.17 | 3.50 | 3 | 3.22  | 2.81  | 2.564 | 1.845 | 2.203   | 0.817   | 0.297   |
| Si | 3 | 1.39 | 4.15 | 4 | 4.68  | 3.43  | 2.158 | 3.193 | 7.416   | 3.293   | 1.443   |
| P  | 3 | 1.6  | 4.80 | 5 | 5.62  | 4.89  | 1.875 | 3.229 | 15.832  | 8.118   | 4.123   |
| S  | 3 | 1.8  | 5.45 | 6 | 6.24  | 4.16  | 1.667 | 4.68  | 27.290  | 15.762  | 9.041   |

**TABLE 4.14** Continued

| Atom | n | ξ | $Z_{eff}$ | $N_v$ | χ [eV] | η [eV] | $R_{max}$ [a.u.] | $\omega_0$ [eV] | ω [eV] | ω (Y.$R_{max}$) [eV] | ω (Y.$R_{max}$,$R_{max}$) [eV] |
|---|---|---|---|---|---|---|---|---|---|---|---|
| Cl | 3 | 2.03 | 6.10 | 7 | 8.32 | 4.68 | 1.478 | 7.396 | 64.845 | 42.263 | 27.408 |
| Ar | 3 | 2.25 | 6.75 | 8 | 7.7 | 8.11 | 1.333 | 3.655 | 71.736 | 51.835 | 37.317 |
| K | 4 | 0.59 | 2.20 | 1 | 2.39 | 1.98 | 6.780 | 1.442 | 0.469 | 0.071 | 0.011 |
| Ca | 4 | 0.77 | 2.85 | 2 | 2.29 | 3.85 | 5.195 | 0.681 | 0.781 | 0.161 | 0.033 |
| Sc | 4 | 0.81 | 3.00 | 2 | 3.43 | 3.22 | 4.938 | 1.827 | 1.665 | 0.363 | 0.078 |
| Ti | 4 | 0.85 | 3.15 | 2 | 3.64 | 3.22 | 4.706 | 2.057 | 1.792 | 0.411 | 0.092 |
| V | 4 | 0.89 | 3.30 | 2 | 3.85 | 2.91 | 4.494 | 2.547 | 1.924 | 0.463 | 0.109 |
| Cr | 4 | 0.93 | 3.45 | 2 | 3.74 | 3.12 | 4.301 | 2.242 | 1.749 | 0.441 | 0.109 |
| Mn | 4 | 0.97 | 3.60 | 2 | 3.85 | 3.64 | 4.124 | 2.036 | 1.791 | 0.472 | 0.122 |
| Fe | 4 | 1.01 | 3.75 | 2 | 4.26 | 3.64 | 3.960 | 2.493 | 2.126 | 0.584 | 0.158 |
| Co | 4 | 1.05 | 3.90 | 2 | 4.37 | 3.43 | 3.810 | 2.784 | 2.175 | 0.623 | 0.175 |
| Ni | 4 | 1.09 | 4.05 | 2 | 4.37 | 3.22 | 3.670 | 2.965 | 2.119 | 0.631 | 0.185 |
| Cu | 4 | 1.14 | 4.20 | 2 | 4.47 | 3.22 | 3.509 | 3.103 | 2.159 | 0.674 | 0.207 |
| Zn | 4 | 1.18 | 4.35 | 2 | 4.26 | 5.2 | 3.390 | 1.745 | 1.919 | 0.620 | 0.197 |
| Ga | 4 | 1.35 | 5.00 | 3 | 3.22 | 2.81 | 2.963 | 1.845 | 2.096 | 0.784 | 0.289 |
| Ge | 4 | 1.53 | 5.65 | 4 | 4.58 | 3.33 | 2.614 | 3.150 | 6.970 | 2.967 | 1.251 |
| As | 4 | 1.70 | 6.30 | 5 | 5.3 | 4.47 | 2.353 | 3.142 | 13.989 | 6.628 | 3.118 |
| Se | 4 | 1.88 | 6.95 | 6 | 5.93 | 3.85 | 2.128 | 4.567 | 24.598 | 12.899 | 6.729 |
| Br | 4 | 2.05 | 7.60 | 7 | 7.59 | 4.26 | 1.951 | 6.762 | 54.011 | 30.896 | 17.604 |
| Kr | 4 | 2.23 | 8.25 | 8 | 6.86 | 7.28 | 1.794 | 3.232 | 57.036 | 35.499 | 22.027 |
| Rb | 5 | 0.55 | 2.20 | 1 | 2.29 | 1.87 | 9.091 | 1.402 | 0.534 | 0.0686 | 0.0087 |

**TABLE 4.14** Continued

| Atom | n | ξ | $Z_{eff}$ | $N_v$ | χ [eV] | η [eV] | $R_{max}$ [a.u.] | $\omega_0$ [eV] | ω [eV] | ω (Y.$R_{max}$) [eV] | ω (Y.$R_{max}$,$R_{max}$) [eV] |
|---|---|---|---|---|---|---|---|---|---|---|---|
| Sr | 5 | 0.71 | 2.85 | 2 | 1.98 | 3.74 | 7.042 | 0.524 | 0.680 | 0.1159 | 0.0195 |
| Y | 5 | 0.75 | 3.00 | 2 | 3.43 | 2.91 | 6.667 | 2.021 | 1.922 | 0.347 | 0.0619 |
| Zr | 5 | 0.79 | 3.15 | 2 | 3.85 | 3.02 | 6.329 | 2.454 | 2.293 | 0.436 | 0.0824 |
| Nb | 5 | 0.83 | 3.30 | 2 | 4.06 | 2.91 | 6.024 | 2.832 | 2.428 | 0.488 | 0.0967 |
| Mo | 5 | 0.86 | 3.45 | 2 | 4.06 | 3.12 | 5.814 | 2.642 | 2.334 | 0.487 | 0.1002 |
| Tc | 5 | 0.9 | 3.60 | 2 | 3.64 | 3.64 | 5.556 | 1.82 | 1.800 | 0.394 | 0.0850 |
| Ru | 5 | 0.94 | 3.75 | 2 | 4.06 | 3.43 | 5.319 | 2.403 | 2.157 | 0.494 | 0.112 |
| Rh | 5 | 0.98 | 3.90 | 2 | 4.26 | 3.22 | 5.102 | 2.818 | 2.294 | 0.549 | 0.129 |
| Pd | 5 | 1.01 | 4.05 | 2 | 4.78 | 3.64 | 4.951 | 3.139 | 2.808 | 0.693 | 0.169 |
| Ag | 5 | 1.05 | 4.20 | 2 | 4.47 | 3.12 | 4.762 | 3.202 | 2.384 | 0.613 | 0.155 |
| Cd | 5 | 1.09 | 4.35 | 2 | 4.16 | 4.78 | 4.587 | 1.810 | 2.010 | 0.537 | 0.142 |
| In | 5 | 1.25 | 5.00 | 3 | 3.12 | 2.70 | 4.000 | 1.803 | 2.082 | 0.645 | 0.198 |
| Sn | 5 | 1.41 | 5.65 | 4 | 4.26 | 3.02 | 3.546 | 3.005 | 6.256 | 2.198 | 0.765 |
| Sb | 5 | 1.58 | 6.30 | 5 | 4.89 | 3.85 | 3.165 | 3.105 | 12.199 | 4.812 | 1.885 |
| Te | 5 | 1.74 | 6.95 | 6 | 5.51 | 3.54 | 2.874 | 4.288 | 21.606 | 9.396 | 4.065 |
| I | 5 | 1.9 | 7.60 | 7 | 6.76 | 3.74 | 2.632 | 6.109 | 43.385 | 20.613 | 9.754 |
| Xe | 5 | 2.06 | 8.25 | 8 | 5.82 | 6.34 | 2.427 | 2.671 | 41.447 | 21.356 | 10.969 |
| | | | | | | | $R^2$ | ∟ | 0.4975 | 0.3665 | 0.2633 |
| | | | | | | | RMS | ∟ | 28.977 | 26.943 | 28.03 |

*At the bottom the $R^2$ values and the root mean square (RMS) of errors (residues) between the Parr-Pearson ($\omega_0$) and the actual models (ω, ω(Y.$R_{max}$), ω(Y.$R_{max}$,$R_{max}$)) are respectively provided (Putz & Chattaraj, 2013).

spectroscopic definitions fit with conceptual DFT as earlier established by Parr but also with the softness kernel DFT framework (Putz et al., 2005). So, the so-called experimental electronegativity based on average of IP and EA may be appropriately considered as modeling the actual electrophilicity expressions too.

Worth noting that electronegativity and chemical hardness in Table 4.1 are slightly different values from those considered by Pearson (1988a) and then by Parr and Yang (1989), especially due electronic affinities considered, as provided by Lackner and Zweig (1983) through employing the same IPs (Moore, 1970) yet combined with electronic affinity through a specific interpolation-extrapolation formulation to achieve the iso-electronic series through periods, as relaying on a classical residual charge method (Ginsberg et al., 1958). Better hardness values (Cárdenas et al., 2011) compare favorably (Putz & Chattaraj, 2013).

Table 4.14 presents the global results of atomic reference "0" electrophilicity of Eq. (4.234) along the present computed electrophilicity of Eq. (4.346), and the particular values of the local and kernel electrophilicities of Eqs. (4.344) and (4.345) to the maximum radius ($R_{max}$) of each atomic system obtained through canceling its radial distribution of the valence shell density (Ghosh & Biswas, 2002). Graphical comparison among these quantities is made in Figure 4.18 with remarkable features namely (Putz & Chattaraj, 2013):

- Reference electrophilicity is usually below both electronegativity and chemical hardness on which it relies, according to Eq. (4.234);
- Actual global electrophilicity resembles the reference one especially for the transition elements' contractions with a clear periodical behavior. In general in a period the alkali metal atom is the least electrophilic and the halogen atom is the most electrophilic. Since $\chi$ is not calculated through the Slater approach, two different ways of calculating $\chi$ and $\eta$ may be the possible reason behind the observed unusual $\omega$ trends in certain cases.
- Overall, reference and actual global electrophilicities do not statistically correlate better than the Pearson correlation factor $R^2=0.5$, and systematically goes downwards with the local and kernel forms as of $R^2=0.37$ and $R^2=0.26$, while with root mean squared of residues in the restrained range of 27÷29 [eV] the present models are validated as stable ones, see the bottom of Table 4.14, respectively.

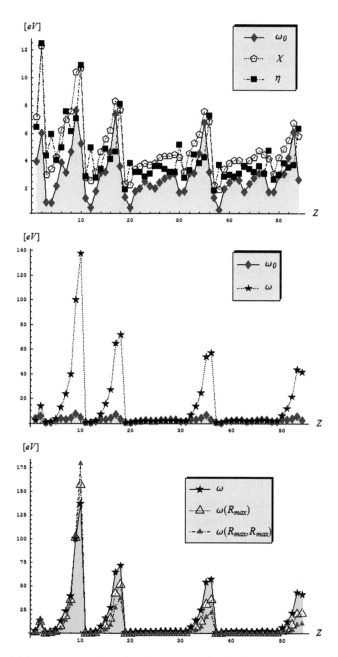

**FIGURE 4.18**     Atomic scales of reference, actual global, local at $R_{max}$ and kernel at $(R_{max}, R_{max})$ electrophilicities, as computed by Eqs. (4.234), (4.346), (4.345), and (4.344), respectively, with the values of Table 4.14 (Putz & Chattaraj, 2013).

However, hierarchical upwards correlation is recorded between actual computed global electrophilicity and the local and kernel ones taken in the $R_{max}$ points of each atomic valence shell

$$R^2[\omega(R_{max},R_{max}), \omega(Y.R_{max})]=0.967 > R^2[\omega(R_{max}), \omega]=0.926 \quad (4.348)$$

according to the successive integration hierarchy of Eq. (4.347), and superior to the double integral hierarchy when the correlation factor is recorded as:

$$R^2[\omega(R_{max},R_{max}), \omega]=0.806 \qquad (4.349)$$

- The kernel and local electrophilicities at $R_{max}$ are always below the global one, as expected from their hierarchical formulae, except for the first period noble element (Ne) and halogen (F) where the values are in reverse order namely

$$\omega_{F/Ne}(R_{max},R_{max}) > \omega_{F/Ne}(R_{max}) > \omega_{F/Ne} \qquad (4.350)$$

due the very strong electrostatic confinement (shrinking the valence electronic cloud) causing the increasing of the electrophilicity kernel as the bilocal holding power of electrons.

The last behavior is checked also for general local and kernel electrophilicity shapes and in Figure 4.19 it is represented for the first period of elements: it follows that in terms of their own global electrophilicities both F and Ne kernel and local electrophilicity surpass unity however producing sharper peaks in radial space representation. Nevertheless the increasing tendency along periods is always fulfilled as well as the decreasing down the groups, e.g., Figure 4.20 for halogens, in radial space representations of local and kernel electrophilicities (Putz & Chattaraj, 2013).

The present approach may be complemented with other works in which also input electronegativity in Eqs. (4.344)–(4.346) is expressed in the same context of DFT softness kernel, with various systematic forms in terms of the atomic valence shell Slater quantities as effective charge and orbital exponent (Putz, 2006). Equally, since the present approach strongly relies on associated chemical hardness, local-to-global hierarchies may be

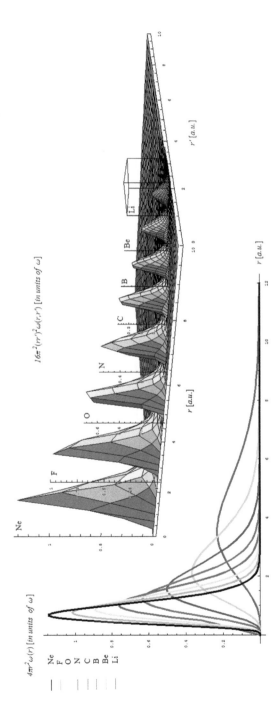

**FIGURE 4.19** Local and kernel radial distributions of electrophilicities of Eqs. (4.345) and (4.344), respectively, for the valence atoms of the first period of elements (Putz & Chattaraj, 2013).

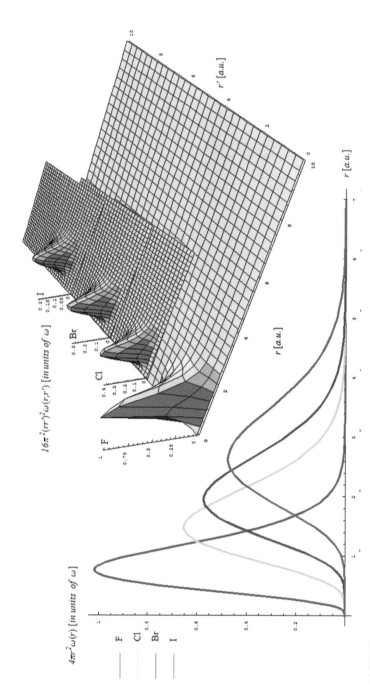

**FIGURE 4.20** Local and kernel radial distributions of electrophilicities of Eqs. (4.345) and (4.344), respectively, for the valence states of halogen atoms (Putz & Chattaraj, 2013).

extracted towards application of specific principles of hard- and soft-acids and bases (Parr & Pearson, 1983; Pearson, 1973, 1990, 1997) and maximum hardness principles (Pearson, 1987, 1993; Parr & Chattaraj, 1991; Chattaraj et al., 1995; Ayres & Par, 2000; Chattaraj et al., 2000) to various levels of site selectivity (local or kernel), so that providing more fundamental working analytical tools for chemical hardness than the global finite difference based one.

## 4.8   ATOMIC SCALES AND PERIODICITY BY ATOMIC RADII

### 4.8.1   RADII-ELECTRONEGATIVITY RELATIONSHIPS

Introducing the atomic orbital radii should be seen more than a definition. That is because this quantity is an effect of the atomic structure being strongly related with the atomic configuration, so to speak with the atomic stability. Rather than to define it is worthwhile to derive the atomic orbital radii from another chemical descriptor that closely takes account of the number of electrons and the stability (the potential equilibrium) of the atomic electronic system. Here it is chosen to take the Mulliken electronegativity $(\chi_M)$ as the basic chemical descriptor for the atomic electronic system founded in an equilibrium structure so that the stability is assumed, from which will be extracted the information regarding the atomic orbital radii $(R)$. Even there are many scales of electronegativity there can be classified into the classical and modern ones. For our purpose here we convene to speak about the class of the electronegativity scales that makes directly the atomic radius dependence and the modern electronegativity scales that mostly are focused on the dependency of the electronic density. Beside their age the classical analytical electronegativity scales provides the first useful tool to compute the atomic orbital radii because their explicit dependence of the spatial coordinate at equilibrium (stabile) electronic structure (Putz, 2012b,c).

The first such scale that is appealed here assumes that electronegativity should be regarded as the electrostatic force between the outermost electron that define the atomic radius and the core of the atomic structure with the effective nuclear charge $(Z_{eff})$, namely *Allred-Rochow* scale with the form, see Eq. (4.5) (Allred & Rochow, 1958)

$$\chi_M^{AR} = \frac{Z_{eff}}{R^2} \tag{4.351}$$

where $Z_{eff}$ are calculated for each individual atomic core structure according to the Slater's rules (Slater, 1930).

A further step of complication is made in the Boyd and Markus (1981) scale, in which the electronegativity remains identified with the electrostatic force but presenting a correction concerning the radial electronic density, $\rho(x)$, with the working form, see also Eq. (4.6)

$$\chi_M^{BM} = \frac{Z_{eff}}{R^2}\left(1 - \int_0^R \rho(r)dr\right) \tag{4.352}$$

where the original atomic number $Z$ is here replaced with the effective nuclear charge in order to emphasis the corrected density term that appears respecting with the simpler Allred-Rochow previous formulation. It should be noted that in the Boyd-Markus scale appears the explicit dependency on the electronic density as a precursor of density functional form, but preserving the force character of the electronegativity.

We turn now to the modern ideas in electronegativity up to the celebrated density functional theory, that assumes that all (necessary) information about an electronic system is comprised in its density, and where the electronegativity is considerate as a potential with the ultimate identification as the negative chemical potential of the system ($\chi = -\mu$) according with the Parr's works (Parr et al., 1978). In this modern framework, first we can make more valuable the study of Ghanty & Ghosh (1996) that has suggested a simple analytical electronegativity dependence of both electronic density and its gradient. Because the above stipulated potential nature of electronegativity we consider the present working formulation of Mulliken electronegativity to be

$$\chi_M^{GG} = -\frac{1}{3}\frac{\nabla_x \rho(r)}{\rho(r)}\Big|_{r \to R} \tag{4.353}$$

It is clear from this formulation that if we read electronegativity equation starting with the right hand of it we get the consistency of the present picture: when the competition between the density and its gradient approaches the atomic radius limit (the HOMO radius) then the equivalent potential of the evolved electrons identifies the Mulliken electronegativity ones. The factor (1/3) was added for the purpose that will be revealed in the next section.

This way the derivation of atomic radius reclaims the gradient counting of the valence electronic density that approach the equilibrium limit around the HOMO radii. We now will consider another Mulliken electronegativity density functional for an $N$-electronic system that moves under the external potential action $V(r)$ that could be identify with the nuclear or atomic core action over the actual electronic system employing the softness density functional form (S-DFT) form (4.279) to the explicit casting (Putz et al., 2003; Putz, 2006):

$$\chi_M^{S-DFT} = \frac{\int_\infty^R L(r)V_{eff}(r)dr + N - 1}{2\sqrt{\int_\infty^R L(r)dr}} \arctan\left(\frac{N-1}{\sqrt{\int_\infty^R L(r)dr}}\right) - \frac{\int_\infty^R L(r)V_{eff}(r)dr + N + 1}{2\sqrt{\int_\infty^R L(r)dr}} \arctan\left(\frac{N+1}{\sqrt{\int_\infty^R L(r)dr}}\right)$$

$$+ \frac{\int_\infty^R \rho(r)V_{eff}(r)dr - 1}{4} \ln\left[\frac{\int_\infty^R L(r)dr + (N-1)^2}{\int_\infty^R L(r)dr + (N+1)^2}\right]$$

(4.354)

so that having also another integral equation for atomic radii once an electronegativity scale is given.

### 4.8.2 ATOMIC RADII FOR UNIFORM VALENCE DENSITY

It is clear from previous section that in order to derive the atomic radii it is necessary to solve the equations regarding the Mulliken electronegativity in the different scales presented. However, from the Allred-Rochow electronegativity scale is quite simple to get the atomic radii with the form (Putz, 2012b,c)

$$R^{AR} = \sqrt{\frac{Z_{eff}}{\chi_M}}$$

(4.355)

The problem appears when we like to solve above electronegativity equations respecting the atomic radii $R$, because the appearance of the unspecified electronic density. To have an analytical general expression for the electronic density in a given external potential states as the main challenge in the quantum chemistry. As much as our main purpose here is not a numerical method for computing electronic density, but having need of an analytical expression of it, we are constrained to restrict ourselves to an approximation. As the density functional theory prescribes, if for the derivation of electronic density we will choose the most simple, classical, way of writing the density appealing the mass-volume relation in an isotropic atomic structure, then the density and the effective potential are seen as uncorrelated quantities that preclude bonding. Thus, our atomic density model assumes the form (March, 1992)

$$\rho(r) = \frac{3A}{4\pi r^3} \qquad (4.356)$$

sytisfying the minimal requires $\rho(0) \to \infty, \rho(r \to \infty) = 0$ with $A$ being the relative atomic weight.

Of course we cannot give rigorous (quantum) arguments for the above proposed density formula, but we can justify it, for instance by replacing it into Ghanty-Ghosh electronegativity that gives us the atomic radii correlated with Mulliken electronegativity under the form

$$R^{GG} = \frac{1}{\chi_M} \qquad (4.357)$$

that recovers the (St. John and Bloch, 1974) electronegativity scale. The obtained the result justifies the anticipated factor (1/3) in the Ghanty-Ghosh equation. This atomic radius derivation claims that the above proposed density working function, even in a simple manner, has its own efficacity. More, it provides a quickly estimation of atomic properties in terms of atomic relative weight.

Returning to the classical electronegativity estimations, Boyd-Markus above formulation can be further developed with the radial electronic density having the above homogeneous form. Arranging the terms as an equation respecting to atomic radii we arrive at the equality:

$$\chi_M R^4 - Z_{eff} R^2 - \frac{3AZ_{eff}}{8\pi} = 0 \qquad (4.358)$$

At this point we like to simplify the fourth order equation, because the small numerical values of atomic radii, but in the present case we are not allowed. This, because canceling the fourth order radius term we loose the connection with the information comprised in the Mulliken electronegativity, and more, the remaining equation will give an imaginary result that is false. Thus we must to solve the entirely fourth-order Boyd-Markus radii equation that will provide the acceptable result as being

$$R^{BM} = \sqrt{\frac{Z_{eff}}{2\chi_M} + \frac{1}{2\chi_M} \sqrt{Z_{eff}\left(Z_{eff} + \frac{3A\chi_M}{2\pi}\right)}} \qquad (4.359)$$

The final atomic radii equation that has to be solved in terms of Mulliken electronegativity is authors' formulation above. However, before to arrive at the effective atomic radius equation, it should be first transformed according to the forms of atomic potential and associate homogeneous density. Performing this substitutions for the main ingredients of S-DFT electronegativity expression (4.354) we get successively (Putz, 2012b,c):

$$L(r) = -\frac{9A}{4\pi Z_{eff}} \frac{1}{r^2} \qquad (4.360)$$

$$\int_{\infty}^{R} L(r)dr = \frac{9A}{4\pi Z_{eff}} \frac{1}{R} \qquad (4.361)$$

$$\int_{\infty}^{R} L(r)V_{eff}dr = -\frac{9A}{8\pi} \frac{1}{R^2} \qquad (4.362)$$

$$\int_{\infty}^{R} \rho(r)V_{eff}(r)dr = \frac{AZ_{eff}}{4\pi} \frac{1}{R^3} \qquad (4.363)$$

With these expressions it is clear that in S-DFT (4.354) equation appears factors of the form:

$$\arctan\left(c_1\sqrt{R}\right), \arctan\left(c_2\sqrt{R}\right), \text{ and } \log\left[\left(1+c_1^2 R\right)/\left(1+c_2^2 R\right)\right] \qquad (4.364a)$$

with

$$c_1 = (N-1)\sqrt{4\pi Z_{eff} R/(9A)} \text{ and } c_2 = (N+1)\sqrt{4\pi Z_{eff} R/(9A)} \qquad (4.364b)$$

From the dimension perspective of atomic radii, we can further apply the first order series expansion respecting to $R \to 0$, that means the truncations

$$\arctan\left(c_{1,2}\sqrt{R}\right) \cong c_{12}\sqrt{R}, \ \log\left[\left(1+c_1^2 R\right)/\left(1+c_2^2 R\right)\right] \cong \left(c_1^2 - c_2^2\right)R \quad (4.364c)$$

With all this considerations, S-DFT (4.354) expression takes the intermediate form

$$\chi_M = -\frac{Z_{eff}}{4}\left[\begin{array}{c}(N-1)\\-(N+1)\end{array}\right]\frac{1}{R} + \frac{Z_{eff}^2}{36}\left[\begin{array}{c}((N-1)^2\\-(N+1)^2\end{array}\right]\frac{1}{R^2} + \frac{Z_{eff}\pi}{9A}\left[\begin{array}{c}(N-1)^2\\-(N+1)^2\end{array}\right]R$$

$$(4.365)$$

and respectively the final one

$$\chi_M = \frac{Z_{eff}}{2}\frac{1}{R} - N\frac{Z_{eff}^2}{9}\frac{1}{R^2} - 4N\frac{Z_{eff}\pi}{9A}R \quad (4.366)$$

In order to analyze properly this formula, first we should remark that in the above used limit $R \to 0$ the third term in last expression does not contribute significantly to the Mulliken electronegativity and will be omitted. The remaining formula (Putz, 2012b,c)

$$\chi_M^{Putz} = \frac{1}{2}\frac{Z_{eff}}{R} - \frac{NZ_{eff}^2}{9}\frac{1}{R^2} \quad (4.367)$$

can be seen as a new Mulliken electronegativity formula, in terms of atomic radii, nuclear efective charge and the total number of electrons in the valence shell. It is obvious that the present result look like a combination between the classical and modern approaches of electronegativity as far as both force and potential terms have a contribution. Also an interesting aspect is that the classical force term is corrected by the modern $N$-dependence in order to can be properly combined with the admited potential modern formulation. This meaningful result can also justify the present semi-classical approach when the modern chemical softness based density functional electronegativity expression is mixed with the classical atomic density model.

However, from S-DFT (4.354) radii equation is now immediately to derive the atomic radii solving an ordinary squared equation with the solutions (Putz, 2012b,c)

$$R_{1,2} = \frac{3Z_{eff} \pm Z_{eff} \sqrt{9 - 16N\chi_M}}{12\chi_M}$$  (4.368)

There are two problems with this solution. The first arise because there are two allowed radii for each set ($N, Z_{eff}, \chi_M$).

The second one regards both radii because in the different value scales of the Mulliken electronegativity it can be possible to obtain an imaginary component when the square root is performed for a given set ($N, \chi_M$). In order to avoid both problems in one step we propose to take as the present atomic radii the combination between the two radii solutions in the form

$$R = \frac{R_1 + R_2}{4}$$  (4.369)

that gives the obvious final S-DFT (4.354) result as

$$R^{Putz} = \frac{Z_{eff}}{8\chi_M}$$  (4.370)

The derived Allred-Rochow, Ghanty-Ghosh, Boyd-Markus and S-DFT relations define the atomic orbital radii scales under attention that will be computed, compared and interpreted for the main atoms from the periodic system in the next discussion.

Our main goal though these derivations were to obtain a list of atomic orbital radii analytically correlated with the electronegativity. Once that this list was laid out, i.e., by the present Allred-Rochow, Ghanty-Ghosh, Boyd-Markus and S-DFT relations, we have the freedom to compute numerical these scales in combination with any available electronegativity atomic scale.

We will choose here two extreme electronegativity scales respectively the sources of their determinations. One is simply the experimental (EXP) atomic electronegativity scale (Pearson, 1988a; Hati & Datta, 1995) that should be always presented as the standard practical comparison. The second one will be choosen in a theoretical fashion to establish the balance between the theoretical and experimental results. At this point we appeal to the analytical S-DFT (4.354) electronegativity where the effective potential is replaced with the corresponding pseudopotential together with the respective basis set, which imposes the associated electronic density for each atomic system for its electronic valence shell (Parr & Pearson, 1983; Robles & Bartolotti, 1984; Pearson, 1997). Performing all needed computations the final numerical Mulliken electronegativities values outlined in this manner will be generically called the basis set (BS) electronegativity scale (Putz et al., 2005, 2009b). The two Mulliken electronegativity numerical scales are displayed in the Table 4.12 for the first four periods of the periodic system of elements.

Further, the atomic radii scales within the basic set and experimental electronegativity scales will be calculated with the results presented correspondingly in the columns <1> to <8> in the Table 4.15. Additionally, there are presented in the columns <9> and <10> of the Table 4.15 the direct experimental evaluation (Web Elements, 2011), and the *ab initio* approaches (Ghanty & Ghosh, 1996), for atomic radii. In this way the respectively atomic radii scales computed indirectly using primary experimental and theoretical (in a pseudopotential manner) electronegativity information are finally compared with the direct experimental and theoretical (in an *ab inition* fashion) atomic radii determinations making this way a complete view of the comparison perspective. For this reasons we will analyze each two atomic radii outlined scales with those direct experimental and theoretical values.

For instance, if we like to compare the Allred-Rochow atomic radii scale within the experimental electronegativity implementation, column <1>, with the experimental direct atomic radii evaluations, column <9>, we see a clear discrepancy in the numbers as well as in their comparative trend, atom by atom. The same conclusions appears from the comparison between the columns <1> and <10>. These comparisons are relevant for our next step in interpretation, because in the Allred-Rochow scale, it was not used the present simple isotropic electronic density, so that no approximation is involved here. Of course here it should be said that the Allred-Rochow scale assumes electronegativity identifies electrostatic force and then the error is explicable. Then, we proceed to the same comparison with the proposed Ghanty-Gosh electronegativity scale that considers electronegativity as the potential, which finally equivalents with the St. John-Bloch electronegativity scale as well. In this case the comparison between the columns <5> and <9> produce the same puzzling one to one uncorrelations and more than that, also the magnitude of the St. John-Bloch atomic radii scale does not fit with the experimental values (column <9>). Similar conclusions are

**TABLE 4.15**  Values of Atomic Orbital Radii in Different Levels of Estimations (Putz, 2012b)

| Sources<br><br>Elements | Allred-<br>Rochow | | Boyd-<br>Markus | | Ghanty-<br>Ghosh | | S-DFT | | literature | |
|---|---|---|---|---|---|---|---|---|---|---|
| | <1> | <2> | <3> | <4> | <5> | <6> | <7> | <8> | <9> | <10> |
| | EXP | BS | EXP | BS | EXP | BS | EXP | BS | EXP | AB INITIO |
| Li | 1.81 | 1.81 | 1.87 | 1.87 | 4.78 | 4.76 | 0.78 | 0.77 | 1.67 | 2.69 |
| Be | 1.74 | 2.09 | 1.82 | 2.16 | 2.94 | 4.23 | 0.71 | 1.03 | 1.12 | 1.79 |
| B | 2.15 | 1.86 | 2.22 | 1.96 | 3.35 | 2.54 | 1.09 | 0.82 | 0.87 | 1.47 |
| C | 1.99 | 1.70 | 2.07 | 1.80 | 2.29 | 1.68 | 0.93 | 0.68 | 0.67 | 1.12 |
| N | 2.02 | 1.74 | 2.12 | 1.86 | 1.98 | 1.47 | 0.97 | 0.72 | 0.56 | 0.91 |
| O | 2.14 | 1.67 | 2.25 | 1.8 | 1.91 | 1.16 | 1.09 | 0.66 | 0.48 | 0.80 |
| F | 1.95 | 1.59 | 2.09 | 1.75 | 1.38 | 0.92 | 0.9 | 0.6 | 0.42 | 0.68 |
| Ne | 2.03 | 1.82 | 2.17 | 1.98 | 1.33 | 1.08 | 0.97 | 0.79 | 0.38 | 0.59 |
| Na | 2.42 | 2.52 | 2.56 | 2.65 | 5.05 | 5.45 | 1.39 | 1.5 | 1.90 | 2.80 |
| Mg | 2.41 | 2.35 | 2.55 | 2.50 | 3.84 | 3.66 | 1.37 | 1.30 | 1.45 | 2.13 |
| Al | 2.88 | 2.13 | 3.02 | 2.30 | 4.48 | 2.44 | 1.96 | 1.06 | 1.18 | 2.29 |
| Si | 2.58 | 2.15 | 2.73 | 2.33 | 3.02 | 2.12 | 1.57 | 1.10 | 1.11 | 1.83 |
| P | 2.55 | 2.09 | 2.72 | 2.29 | 2.56 | 1.73 | 1.54 | 1.03 | 0.98 | 1.54 |
| S | 2.58 | 1.87 | 2.76 | 2.09 | 2.31 | 1.21 | 1.57 | 0.82 | 0.88 | 1.35 |
| Cl | 2.37 | 1.78 | 2.57 | 2.03 | 1.73 | 0.99 | 1.32 | 0.75 | 0.79 | 1.19 |
| Ar | 2.55 | 2.23 | 2.77 | 2.46 | 1.82 | 1.39 | 1.54 | 1.17 | 0.71 | 1.06 |
| K | 2.63 | 2.60 | 2.84 | 2.81 | 5.95 | 5.80 | 1.63 | 1.60 | 2.43 | 3.36 |
| Ca | 3.14 | 3.15 | 3.32 | 3.33 | 6.54 | 6.57 | 2.33 | 2.34 | 1.94 | 2.70 |
| Sc | 2.61 | 3.52 | 2.85 | 3.72 | 4.31 | 7.86 | 1.61 | 2.95 | 1.84 | 2.55 |
| Ti | 2.64 | 3.24 | 2.88 | 3.45 | 4.17 | 6.31 | 1.64 | 2.48 | 1.76 | 2.44 |
| V | 2.64 | 3.22 | 2.90 | 3.44 | 4 | 5.95 | 1.65 | 2.45 | 1.71 | 2.35 |
| Cr | 2.66 | 3.11 | 2.92 | 3.34 | 3.87 | 5.29 | 1.66 | 2.28 | 1.66 | 2.28 |
| Mn | 2.71 | 3.69 | 2.98 | 3.91 | 2.87 | 7.16 | 1.74 | 3.22 | 1.61 | 2.21 |
| Fe | 2.65 | 2.71 | 2.93 | 2.98 | 3.54 | 3.67 | 1.65 | 1.73 | 1.56 | 2.09 |
| Co | 2.63 | 3.13 | 2.92 | 3.39 | 3.34 | 4.75 | 1.63 | 2.32 | 1.52 | 2 |
| Ni | 2.65 | 2.98 | 2.93 | 3.24 | 3.27 | 4.13 | 1.66 | 2.09 | 1.49 | 1.92 |
| Cu | 2.67 | 3.31 | 2.97 | 3.58 | 3.21 | 4.94 | 1.69 | 2.59 | 1.45 | 1.85 |
| Zn | 2.73 | 3.25 | 3.03 | 3.53 | 3.23 | 4.6 | 1.75 | 2.5 | 1.42 | 1.79 |
| Ga | 3.45 | 3.40 | 3.73 | 3.68 | 4.50 | 4.36 | 2.81 | 2.72 | 1.36 | 2.23 |

**TABLE 4.15**    Continued

| Sources | Allred-Rochow | | Boyd-Markus | | Ghanty-Ghosh | | S-DFT | | literature | |
|---|---|---|---|---|---|---|---|---|---|---|
| Elements | <1> | <2> | <3> | <4> | <5> | <6> | <7> | <8> | <9> | <10> |
| | EXP | BS | EXP | BS | EXP | BS | EXP | BS | EXP | AB INITIO |
| Ge | 3.06 | 3.18 | 3.37 | 3.49 | 3.13 | 3.39 | 2.21 | 2.4 | 1.25 | 1.88 |
| As | 3.01 | 3.12 | 3.33 | 3.43 | 2.71 | 2.91 | 2.14 | 2.29 | 1.14 | 1.65 |
| Se | 3 | 3.31 | 3.33 | 3.63 | 2.44 | 2.99 | 2.12 | 2.59 | 1.03 | 1.49 |
| Br | 2.76 | 2.81 | 3.12 | 3.16 | 1.90 | 1.96 | 1.8 | 1.86 | 0.94 | 1.35 |
| Kr | 3 | 2.96 | 3.35 | 3.31 | 2.06 | 2 | 2.12 | 2.06 | 0.88 | 1.24 |
| Rb | 2.68 | 3.99 | 3.06 | 4.29 | 6.15 | 13.7 | 1.69 | 3.76 | 2.65 | 3.54 |
| Sr | 3.29 | 3.65 | 3.64 | 3.97 | 7.19 | 8.83 | 2.56 | 3.14 | 2.19 | 2.93 |
| Y | 2.68 | 3.6 | 3.07 | 3.94 | 4.51 | 8.18 | 1.69 | 3.07 | 2.12 | 2.72 |
| Zr | 2.57 | 3.72 | 2.98 | 4.05 | 3.95 | 8.31 | 1.56 | 3.28 | 2.06 | 2.60 |
| Nb | 2.51 | 3.87 | 2.93 | 4.19 | 3.60 | 8.56 | 1.48 | 3.53 | 1.98 | 2.51 |
| Mo | 2.59 | 3.56 | 3.02 | 3.92 | 3.69 | 6.95 | 1.59 | 3 | 1.90 | 2.43 |
| Tc | - | 3.74 | - | 4.09 | - | 7.34 | - | 3.3 | 1.83 | 2.37 |
| Ru | 2.52 | 3.85 | 2.96 | 4.2 | 3.2 | 7.46 | 1.5 | 3.49 | 1.78 | 2.25 |
| Rh | 2.63 | 4.15 | 3.07 | 4.49 | 3.35 | 8.37 | 1.63 | 4.08 | 1.73 | 2.16 |
| Pd | 2.63 | 3.95 | 3.08 | 4.31 | 3.23 | 7.27 | 1.60 | 3.68 | 1.69 | 2.08 |
| Ag | 2.68 | 3.83 | 3.14 | 4.20 | 3.24 | 6.60 | 1.70 | 3.46 | 1.65 | 2.02 |
| Cd | 2.77 | 3.75 | 3.23 | 4.14 | 3.32 | 6.1 | 1.81 | 3.31 | 1.61 | 1.96 |
| In | 3.50 | 3.92 | 3.92 | 4.30 | 4.64 | 5.80 | 2.9 | 3.62 | 1.56 | 2.41 |
| Sn | 3.16 | 3.96 | 3.61 | 4.35 | 3.35 | 5.25 | 2.36 | 3.71 | 1.45 | 2.10 |
| Sb | 3.14 | 2.76 | 3.60 | 3.25 | 2.97 | 2.29 | 2.34 | 1.8 | 1.33 | 1.89 |
| Te | 3.10 | 3.76 | 3.58 | 4.19 | 2.62 | 3.85 | 2.28 | 3.34 | 1.23 | 1.73 |
| I | 2.93 | 2.94 | 3.42 | 3.43 | 2.13 | 2.15 | 2.02 | 2.04 | 1.15 | 1.60 |
| Xe | 3.22 | 3.16 | 3.70 | 3.65 | 2.37 | 2.29 | 2.44 | 2.37 | 1.08 | 1.49 |

outlined when whatever from the columns <1> to <8> are compared with the columns <9> and <10> (Putz, 2012b,c).

At this stage we can't conclude that the electronegativity scales are not correctly chosen, even that their analytical forms display an approximation *per se*, but the method one-to-one of comparison between the different scales can be improved. Then is better to talk about the global trend of this scales from where we can easily see which of the presented analytical discourse approaches closer reality.

In order to have a suitable tool for validation of a given atomic radii scale, we can appeal the regression methods for representing the polynomial fit of the individual atomic radii scale respecting with the atomic number. In order not to skip the consistency of the comparison, even when we deal with the polynomial regression method, we like to reproduce both the global and local trend of each atomic radii scale. To work this, so-called, auto-correction method a high degree in polynomial regression is required up to that the best cover for all the local one-after-one trend in atomic radii scales is obtained (Putz, 2012b,c).

For the atomic radii given in the Table 4.15, for each scale in each method, was found that the 20 degree polynomials get the best fit with the local atomic radii trend. For having reasonably comparison is compulsory to deal with the same degree in all polynomials that fit the scales of atomic radii data (Putz, 2012b,c).

Regarding the global atomic radii trend this should be simply modeled by the first degree in polynomial regression, namely the celebrated linear regression method. Both 20 and one degree polynomials are presented in the Figure 4.21 to Figure 4.25 performing such regressions for each atomic radii scale in both (direct and indirect) experimental and theoretical approaches. The dotted lines around the linear regression fittings comprise mainly atomic radii data and associated local polynomial with the 95% confidence. The linear regression fits are summarized for all atomic radii scales on their respective linear equations displayed on the top of each pictures. In each case these linear equations can be seen as the regression definition of the atomic radii scale, because the first order dependency in atomic numbers, the same proportionality that is obtained when the atomic radii are derived from the electronegativity as the potential quantity (Putz, 2012b,c).

Looking now to Figures 4.21–4.25 only at the local polynomial fits we better see the acceptable trend of each atomic radii scales that smoothly increase between periods and decrease inside them within the system of elements. Thus, from the polynomial regression methods' point of view the presented atomic radii throughout the analytical electronegativity scales have all necessary qualities (global and local) to constitute the working atomic scale whenever the analytical reasons demands (Putz, 2012b,c).

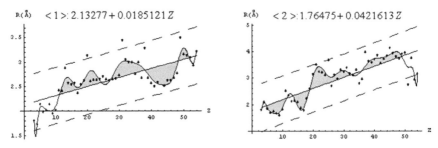

**FIGURE 4.21**   Global (linear) and local (high order polynomial) regressions plots for atomic radii using Allred-Rochow electronegativity's scale with experimental (left) and basis set pseudopotential (right) Mulliken values; after Putz (2012c).

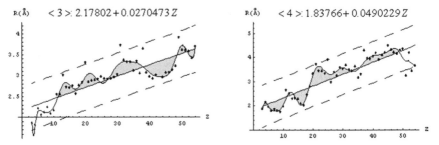

**FIGURE 4.22**   Global (linear) and local (high order polynomial) regressions plots for atomic radii using Boyd-Markus electronegativity's scale with experimental (left) and basis set pseudopotential (right) Mulliken values; after Putz (2012c).

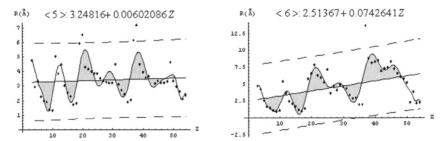

**FIGURE 4.23**   Global (linear) and local (high order polynomial) regressions plots for atomic radii using St. John-Bloch electronegativity's scale with experimental (left) and basis set pseudopotential (right) Mulliken values; after Putz (2012c).

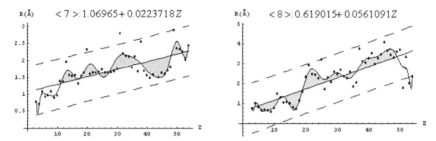

**FIGURE 4.24**    Global (linear) and local (high order polynomial) regressions plots for atomic radii using the S-DFT (4.354) derived electronegativity's scale with experimental (left) and basis set pseudopotential (right) Mulliken values; after Putz (2012c).

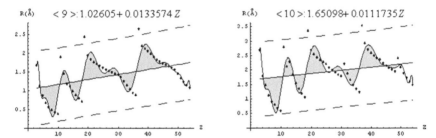

**FIGURE 4.25**    Global (linear) and local (high order polynomial) regressions plots for atomic radii abstracted from experimental (left) and ab initio (right) direct determinations; after Putz (2012c).

Finally we like to conclude about the comparison between present analytical scales with experimental and ab initio ones using the regression plots.

In order to do this we present the Figures 4.26–4.29, where both linear and 20 degree polynomials are compared two-by-two for each atomic radii (intermediates by electronegativities) with those directly evaluated. Starting with the analyze of local polynomials there is turning out that with little discrepancies the local regression trends, all atomic radii scales are quite similar despite their range diversity. This can be best visualized in the Figure 4.26 where Ghanty-Ghosh (St. John-Bloch) atomic radii (using experimental and basis set electronegativity) scales are compared with the direct (experimental and *ab initio*) ones. Passing to the linear (global) regression plots another class of common interesting features are revealed. First, between all global plots for the numerical values computed with the help of analytical atomic radii scales, <1> and <2>, <3> and <4>, <5> and

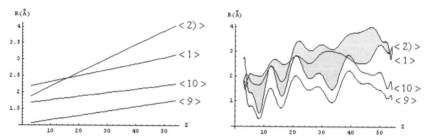

**FIGURE 4.26**    Comparison between atomic radii scales derived from Allred-Rochow (<1> and <2>) electronegativity scale and those direct derived by experimental (<9>) and *ab initio* (<10>) methods concerning their global (left) and local (right) regressions' trend ; after Putz (2012c).

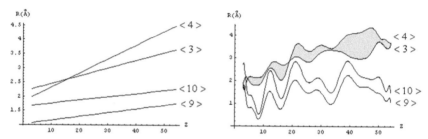

**FIGURE 4.27**    Comparison between atomic radii scales derived from Boyd-Markus (<3> and <4>) electronegativity scale and those direct derived by experimental (<9>) and *ab initio* (<10>) methods concerning their global (left) and local (right) regressions' trend ; after Putz (2012c).

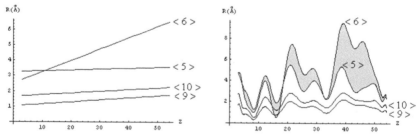

**FIGURE 4.28**    Comparison between atomic radii scales derived from Ghanty-Ghosh (or St. John-Bloch) (<5> and <6>) electronegativity scale and those direct derived by experimental (<9>) and *ab initio* (<10>) methods concerning their global (left) and local (right) regressions' trend; after Putz (2012c).

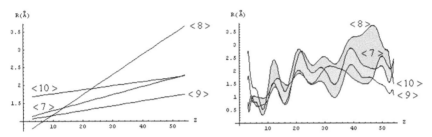

**FIGURE 4.29** Comparison between atomic radii derived from the S-DFT (4.354) scales (<3> and <4>) and those direct derived by experimental (<9>) and *ab initio* (<10>) methods concerning their global (left) and local (right) regressions' trend; after Putz (2012c).

<6>, <7> and <6>, appears a crossing in the atomic number range (Ghosh & Biswas, 2002). In all this cases, the basis set atomic radii scales display global plots with a higher slope compared with ones based on the experimental electronegativity values (Putz, 2012b,c).

A special interesting feature shows the actual S-DFT (4.354) atomic radii scale represented in the Figure 4.27 as a regression in a comparative way with those of direct experimental <9> and *ab initio* <10> founded atomic radii scales. More precisely, the global fit <7> of the atomic radii scale, based on the experimental electronegativity values, displays an intermediate trend comprised between the corresponding linear fits <9> and <10> of the reference direct experimental and *ab initio* atomic radii scales, respectively. This means that the actual atomic S-DFT (4.354) radii formula is situated at least from a global regression point of view between the experimental values and the most powerful *ab initio* output values when also the experimental electronegativity data are used as the primary information. Moreover, the advantage of the present S-DFT (4.354) atomic radii formula is that it correlates directly with the electronegativity in the base on an analytical deduction within the DFT (Putz, 2012b,c).

### 4.8.3 RADII SCALES WITH SLATER DENSITY FOR VALENCE ATOMIC STATES

The iso-electronic density picture has revealed in the previous section quite simple and analytical relations for computing atomic radii, but with the price of a lot of approximations and considerations. In this section we propose a more rigorous treatment of both electronic density and atomic radii, within the Slater orbital description, but instead with the price that in most cases one analytical expression cannot be given, being necessary a numerical solutions. Starting with evaluations, it is certain that Allred-Rochow scale cannot support such Slater particularization. Boyd-Markus formulation permits instead the evaluation of density integral with the implementation of electronic Slater expression.

As previously mentioned, in the approach that correlates atomic radii with the electronegativity it is of fundamental importance to know the atomic electron density of a given system. In this respect, a suitable treatment is based on the Slater orbital electronic picture that produces the normalized distribution functions under the radial form (Slater, 1964):

$$4\pi r^2 \rho_{STO}(r) = \frac{4\pi x^{2n}(2\xi)^{2n+1}}{(2n)!} \exp\left[-2\xi r\right] \tag{4.371}$$

with $n$ being the principal quantum number and $\xi$ the orbital exponent. Performing the integral between the same limits as in the previously, we get (Putz, 2012b,c):

$$\int_{\infty}^{R} 4\pi r^2 \rho_{STO}(r)dr = \frac{4\pi\left(2n\Gamma[1+2n] - 4n^2\Gamma[2n] - (1+2n)\Gamma[1+2n,2R\xi]\right)}{(1+2n)(2n)!} \tag{4.372}$$

where $\Gamma[z]$ stands for Euler gamma function and $\Gamma[z, y]$ for the incomplete gamma function. However, we further proceed with the allowed limit $R\to0$ that also transforms expressions into more suitable ones for algebraic manipulation. Thus, we get the equivalent expressions:

$$\chi_M = \frac{Z_{eff}}{R^2}\left[1 - \int_{\infty}^{R} 4\pi r^2 \rho_{STO}(r)dr\Big|_{R\to0}\right] \tag{4.373}$$

$$\chi_M^{BM-STO} = \frac{Z_{eff}}{R_{<3>}^2}\left\{1 - \frac{4\pi\left[(2R_{<3>}\xi)^2 - 4n^2\Gamma[2n] - \Gamma[1+2n]\right]}{(1+2n)(2n)!}\right\} \tag{4.374}$$

where the last identity can be seen as the definition equation for the STO version of atomic radii from Boyd-Markus electronegativity formulation.

A more simplified expression is getting out if we combine the STO densities with the electronegativity formulation suggested by Ghanty-Ghosh.

In this case it is found another analytical expression for the atomic radii scale, namely (Putz, 2012b,c):

$$R_{<5>}^{GG-STO} = \frac{n}{\xi - 0.5\chi_M} \tag{4.375}$$

At last, it is exploited the S-DFT (4.354) relation regarded as the electronegativity scale derived from the density functional first principles. Again, because of the complexity of expression, it seems that no analytical formulation is available for the atomic radii definition. However, an equation can be formulated following the same procedure as for the atomic radii based on the Boyd-Markus electronegativity picture. For doing this it is enough to identify the main ingredients that appear in the chemcical softnes based electronegativity density functional.

With the help of STO expressions we can integrate and take the first order expansion in the limit $R \to 0$ out these terms, classified as chemical response indices (Putz, 2012c),

$$a_{(STO)}(R_{<7>}) \equiv \int_{\infty}^{R} L(r)dr\Big|_{R\to 0} = \frac{2\pi(2n\Gamma[2+2n,2R\xi]-\Gamma[3+2n,2R\xi])}{Z_{eff}\,\xi(2n)!}\Big|_{R\to 0}$$

$$\cong \frac{2\pi}{Z_{eff}\,\xi(2n)!}\left\{4(R_{<7>}\xi)^{2(1+n)}\left(\frac{2^{1+2n}R_{<7>}\xi}{3+2n}-\frac{2^{2n}n}{1+n}\right)+2n\Gamma[2+2n]-\Gamma[3+2n]\right\} \tag{4.376}$$

$$b_{(STO)}(R_{<7>}) \equiv \int_{\infty}^{R} L(r)V(r)dr\Big|_{R\to 0} = \frac{4\pi(-2n\Gamma[1+2n,2R\xi]+\Gamma[2+2n,2R\xi])}{(2n)!}\Big|_{R\to 0}$$

$$\cong \frac{4\pi}{(2n)!}\left\{\frac{(2R_{<7>}\xi)^{1+2n}}{(1+n)(1+2n)}[2n^2-R_{<7>}\xi+2n(1-R_{<7>}\xi)]-2n\Gamma[1+2n]+\Gamma[2+2n]\right\} \tag{4.377}$$

and the chemical action index:

$$C_{A(STO)}(R_{<7>}) \equiv \int_{\infty}^{R} 4\pi x^2 \rho_{STO}(x)V(x)dx\Big|_{R\to 0} = \frac{8\pi Z_{eff}\,\xi\Gamma[2n,2R\xi]}{(2n)!}\Big|_{R\to 0}$$

$$\cong \frac{4\pi Z_{eff}\,\xi((2n\Gamma[2n]-(2R_{<7>}\xi)^{2n})}{n(2n)!} \tag{4.378}$$

Finally, the desired equation for the atomic radii relating with the electro-negativity has the actual form:

$$\chi_M^{S-DFT-STO} = \frac{b_{(STO)}(R_{<7>}) + N - 1}{2\sqrt{a_{(STO)}(R_{<7>})}} \arctan\left(\frac{N-1}{\sqrt{a_{(STO)}(R_{<7>})}}\right)$$

$$-\frac{b_{(STO)}(R_{<7>}) + N + 1}{2\sqrt{a_{(STO)}(R_{<7>})}} \arctan\left(\frac{N+1}{\sqrt{a_{(STO)}(R_{<7>})}}\right)$$

$$+\frac{C_{A(STO)}(R_{<7>}) - 1}{4} \ln\left[\frac{a_{(STO)}(R_{<7>}) + (N-1)^2}{a_{(STO)}(R_{<7>}) + (N+1)^2}\right] \tag{4.379}$$

The last equation will give the atomic radii scale in the framework of density functional approach with the STO electronic density implementation.

Summarizing, the working atomic radii scales within the STO electronic density picture in atoms are generated equating the expressions Boyd-Markus and S-DFT and respectively by analytical Ghanthy-Ghosh expression.

Let's start with the observation that the STO electronic density was considerate to be normalized to but within the STO atomic radii scales (related electronegativity) group, <3> for Boyd-Markus, <5> for Ghanty-Ghosh and <7> for S-DFT scales, the $N$-dependence was explicitly and particularly nominated for each atomic system, according with the Table 4.16 above. This procedure is correct because the considerate STO density was associated with the outermost electron (taken from infinity and becoming the last electron in the $N$-electronic atom) under the influence of the core atomic system ($Z_{eff}$). This approach describes reasonable the one-electron behaviors in the $N$-electronic bath (last electron included) (Putz, 2012c).

Turning to the individual analyze of STO atomic radii scales, it is obvious from the columns <3>, <5> and <7> in the Table 4.16, that both local and global periodicity in all STO scales is fulfilled.

However, arriving at the moment of comparison with another STO density non-related electronegativity method, we mention the recent work of Ghosh & Biswas (2002). In that work using the STO electronic representation it was derived the atomic radii scales trough the equation:

$$\nabla_r\left[4\pi r^2 \rho_{STO}(r)\right]\big|_{r=} = 0 \tag{4.380}$$

**TABLE 4.16**    Atomic Radii Scales as Computed From Related Electronegativity Methods, within the Slater (STO) Electronic Density Picture (Putz, 2012c)

| Sources Elements | <3> Boyd-Markus STO | <5> Ghanty-Ghosh STO | <7> S-DFT STO | <8> Ghosh-Biswas STO |
|---|---|---|---|---|
| Li | 1.07 | 1.78 | 1.24 | 1.63 |
| Be | 0.71 | 1.19 | 0.82 | 1.09 |
| B | 0.54 | 0.87 | 0.62 | 0.81 |
| C | 0.43 | 0.70 | 0.49 | 0.65 |
| N | 0.36 | 0.58 | 0.41 | 0.54 |
| O | 0.31 | 0.49 | 0.35 | 0.46 |
| F | 0.27 | 0.44 | 0.31 | 0.41 |
| Ne | 0.24 | 0.39 | 0.27 | 0.36 |
| Na | 1.23 | 2.34 | 1.36 | 2.16 |
| Mg | 0.95 | 1.80 | 1.05 | 1.67 |
| Al | 0.77 | 1.43 | 0.85 | 1.36 |
| Si | 0.65 | 1.22 | 0.72 | 1.15 |
| P | 0.56 | 1.06 | 0.62 | 0.99 |
| S | 0.50 | 0.94 | 0.55 | 0.87 |
| Cl | 0.44 | 0.84 | 0.49 | 0.78 |
| Ar | 0.40 | 0.75 | 0.44 | 0.71 |
| K | 1.87 | 3.88 | 2.03 | 3.56 |
| Ca | 1.43 | 2.9 | 1.56 | 2.75 |
| Sc | 1.36 | 2.83 | 1.48 | 2.61 |
| Ti | 1.30 | 2.69 | 1.40 | 2.49 |
| V | 1.24 | 2.57 | 1.35 | 2.37 |
| Cr | 1.19 | 2.46 | 1.29 | 2.27 |
| Mn | 1.14 | 2.35 | 1.23 | 2.18 |
| Fe | 1.09 | 2.26 | 1.19 | 2.09 |
| Co | 1.05 | 2.18 | 1.14 | 2.01 |
| Ni | 1.01 | 2.10 | 1.10 | 1.93 |
| Cu | 0.97 | 2 | 1.05 | 1.86 |
| Zn | 0.94 | 1.93 | 1.02 | 1.8 |
| Ga | 0.82 | 1.64 | 0.89 | 1.57 |
| Ge | 0.72 | 1.46 | 0.78 | 1.39 |
| As | 0.65 | 1.32 | 0.70 | 1.24 |

**TABLE 4.16** Continued

| Sources Elements | <3> Boyd-Markus STO | <5> Ghanty-Ghosh STO | <7> S-DFT STO | <8> Ghosh-Biswas STO |
|---|---|---|---|---|
| Se | 0.59 | 1.19 | 0.64 | 1.13 |
| Br | 0.54 | 1.11 | 0.58 | 1.03 |
| Kr | 0.50 | 1.01 | 0.54 | 0.95 |
| Rb | 2.36 | 5.22 | 2.54 | 4.81 |
| Sr | 1.84 | 3.93 | 1.97 | 3.71 |
| Y | 1.74 | 3.83 | 1.87 | 3.53 |
| Zr | 1.65 | 3.66 | 1.77 | 3.36 |
| Nb | 1.57 | 3.50 | 1.69 | 3.21 |
| Mo | 1.52 | 3.36 | 1.63 | 3.07 |
| Tc | 1.45 | 3.24 | 1.55 | 2.94 |
| Ru | 1.39 | 3.09 | 1.49 | 2.82 |
| Rh | 1.33 | 2.94 | 1.43 | 2.71 |
| Pd | 1.29 | 2.85 | 1.39 | 2.61 |
| Ag | 1.24 | 2.73 | 1.33 | 2.52 |
| Cd | 1.20 | 2.62 | 1.28 | 2.43 |
| In | 1.05 | 2.22 | 1.12 | 2.12 |
| Sn | 0.93 | 1.99 | 0.99 | 1.87 |
| Sb | 0.82 | 1.77 | 0.89 | 1.68 |
| Te | 0.75 | 1.61 | 0.80 | 1.52 |
| I | 0.69 | 1.49 | 0.74 | 1.39 |
| Xe | 0.63 | 1.36 | 0.68 | 1.28 |

that leads with the simple atomic radii scale under the form:

$$R_{<8>}^{GB} = \frac{n}{\xi} \qquad (4.381)$$

However, despite of the simple above form, as well as of the correct atomic radii trend obtained (because the periodicity of the atomic parameters involved in Ghosh-Biswas formulation), the above equations is not full meaning for atomic radii determination. This because its originating equation is in fact an extremum equation for electronic density and not for the atomic radii. Then, only such condition is not enough to furnish the correct derivation of atomic radii.

Suppose that the considered optimum condition it gives one minimum density, it is not unique in atoms because the nodes presence; instead, if the density that follows from extreme condition is a maximum, is hard to accept that the most electronic population are concentrated far from nucleus (because the quantum occupation). Thus, the Ghosh-Biswas extreme equation does not contain the full information about the atomic radii.

This is clearly out for the transitional metals where the choice of principal quantum number (between $s$- or $d$-orbitals) should be accompanied by atomic radii related information. This should be electronegativity, as a unique value (because the equalization of orbital electronegativities in atoms), to be involved as the missing information.

With the help of the Table 4.3, the atomic Ghosh-Biswas radii scale is displayed in the column <8> in Table 4.16 and reproduces the early results (Ghosh & Biswas, 2002).

To arrive at the electronegativity is simply to consider the competition between Ghosh-Biswas condition and the supplemented one:

$$\rho_{STO}(r)\big|_{r=R} = 0 \tag{4.382}$$

at the atomic radii frontier. In the case that we perform the ratio between the two above conditions to get a more comprehensible atomic radii determination, as a parameter at which both gradient of density as well as the density itself vanish (immediately near, for increasing sense, of the atomic radii).

The indeterminate ratio,

$$\lim_{\substack{\nabla_r\rho(r)\to 0, \\ \rho(r)\to 0}} \frac{\nabla_r\rho(r)}{\rho(r)}\bigg|_{r=R} \tag{4.383}$$

is solved by the electronegativity, under the electronegativity Ghanthy-Ghosh scale, that for STO representation gives the Ghanthy-Ghosh-STO atomic radii scale, with the values given in column <5> in Table 4.16 (Putz, 2012c).

Is now clear that the Ghosh-Biswas corrected atomic radii scale has the electronegativity as an additional ingredient, with the effect in completing the atomic information at the level of the outermost atomic shell.

Having completed the group of the actual STO computed atomic radii with the results derived from Boyd-Markus, Ghanthy-Ghosh, S-DFT, and Ghosh-Biswas equations, the comparative plot is draw in the upper part of Figure 4.30.

To have a symmetric approach with the previous ISO analyze, we proceed also here with the linear regressions of STO atomic radii in terms of electronic atomic charge. We get the following relations (Putz et al., 2003, 2012b,c):

$$R_{<3>}^{Boyd-Markus-STO} = 0.653473 + 0.0116798\,Z \qquad (4.384)$$

$$R_{<5>}^{Ghanty-Ghosh-STO} = 1.11907 + 0.0323337\,Z \qquad (4.385)$$

$$R_{<7>}^{Mulliken-Putz-STO} = 0.739578 + 0.0116477\,Z \qquad (4.386)$$

$$R_{<8>}^{Ghosh-Biswas} = 1.03772 + 0.0300192\,Z \qquad (4.387)$$

which are represented together in the lower part of Figure 4.30. It is clear also from the linear representation that equations Ghanty-Ghosh, with numerical values in column <5>, and Ghosh-Biswas with column <8> as particularization, belongs to the same class of atomic radii scales, having close range and orientations, in the top of the linear representations. Instead, the Boyd-Markus and the S-DFT (4.354) atomic radii STO scales, produce numerical values, columns <3> and <7>, that permit to classify them together in bottom of the linear representations, but separately from Ghanty-Ghosh-Biswas linear fits, as well in range as in orientation.

However, is good to mention here that from all working electronegativity and atomic radii scales presented only the one based on DFT first principles, namely S-DFT (4.354) formulation and its corresponding STO version follows a coherent and less approximate picture. It comprises the biggest amount of information in terms of potential, electronegativity, and number of electrons, electronic density and its gradient.

In an approach that correlates atomic radius with electronegativity is of fundamental importance to know the electronic density distribution of a given system and in this respect a suitable treatment is based on the Slater orbital electronic picture.

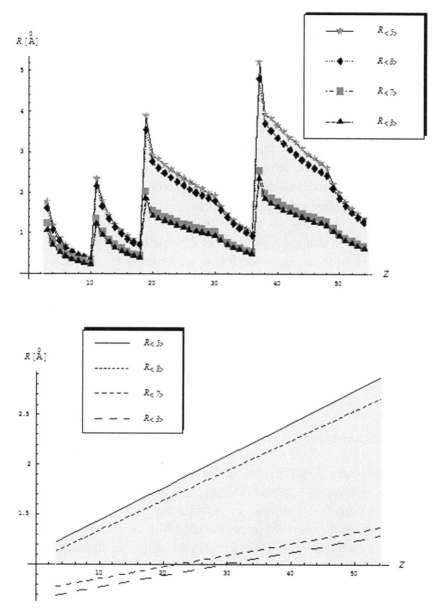

**FIGURE 4.30**   The plots as local representation (upper draws) and linear regressions like global representation (lower draws) for the atomic radii scales abstracted from Slater electronic density picture, both in related and non-related electronegativity methods, as indicated in the brackets < > referred to the Table 4.16; after Putz et al. (2003, 2012b,c).

## 4.9   SCALES AND PERIODICITY BY ATOMIC SIZE RELATED DESCRIPTORS

An electronic system in presence of an external field (nuclear, magnetic, and electric) presents different responses according with the nature of the action.

Correspondingly, the specific sensitivity parameters can be introduced: diamagnetic susceptibility as the response under applied magnetic field, static dipole polarizability that accounts for the electronic cloud deformation under applied electric perturbation and the chemical hardness associated with the compactness of the electronic cloud by the nuclear influence, and possible applied electric perturbation.

Therefore, these are the descriptors that are sensitive to the external applied fields so that can give through the suitable experiments the useful structure information. In the follow we quantify these structure indices in terms of atomic radii performing the correspondence between the full atomic quantities (total number of electrons, nuclear charge and the spatial averages) and those related with atomic radii (the outer electrons, the effective charge of the outer shell and the atomic radii itself) (Putz et al., 2003, 2012b,c):

$$f\big(N, Z, \langle g(r) \rangle\big) \rightarrow f\big(N_{outer}, Z_{eff}, g(R)\big) \qquad (4.388)$$

### 4.9.1   DIAMAGNETIC SUSCEPTIBILITY

When a magnetic field is applied to the atomic system the electronic cloud responds creating an opposite magnetic moment in order to decrease the external perturbation. This behavior is called the diamagnetic effect and can be detected via the observed (measured) diminishment of the applied magnetic field (Selwood, 1956). The diamagnetic effect was introduced at the turn of twentieth century by Langevin as a change in the magnetic moment of a material caused by the deceleration of the orbiting electrons on the basis of the electric field induced by an applied external magnetic field. Here we expose the quantitative formulation of this effect (Putz et al., 2003, 2012b,c).

Let's consider an electron with elementary charge $e$ and mass $m$ orbiting with the velocity $v$ at the distance $r$ of nucleus. Its movement produces a microscopic electrical current:

$$I = \frac{e}{2\pi\left(\dfrac{r}{v}\right)} \tag{4.389}$$

Being the orbital area $A = \pi r^2$ that associates with the orbital current, the corresponding magnetic moment can be written as their product:

$$\mu_m \equiv IA = \frac{er}{2}v \tag{4.390}$$

When an external magnetic field ($\mu_0 H$, with $\mu_0$ the permeability of vacuum) is applied it produced a flux through the orbital area ($\phi = \mu_0 HA$) that induces a potential according with the Faraday's law:

$$V_e \equiv -\frac{d\phi}{dt} = -\frac{d}{dt}\left(\mu_0 HA\right) = -\pi r^2 \mu_0 \frac{dH}{dt} \tag{4.391}$$

The induced electric field can be determined as the ratio of induced potential per orbit length:

$$E = \frac{V_e}{2\pi r} = -\frac{\mu_0}{2} r \frac{dH}{dt} \tag{4.392}$$

Furthermore, the induced electric field will act as the force ($eE$) to gives raise the orbiting deceleration:

$$a = \frac{eE}{m} = -\frac{e\mu_0}{2m} r \frac{dH}{dt} \tag{4.393}$$

This equation can be integrated in the interval in which the magnetic field strength increases from 0 to $H$, with the resulting change in electron velocity:

$$\Delta v = \int_0^t a \, dt = -\frac{e\mu_0}{2m} r \int_0^H dH = -\frac{e\mu_0}{2m} rH \tag{4.394}$$

Next, the change in velocity produces the *diamagnetic* change of the magnetic moment to be:

$$\Delta\mu_m = \frac{er}{2}\Delta v = -\frac{e^2\mu_0}{4m} r^2 H \tag{4.395}$$

However, the obtained expression assumes so far that the applied magnetic field is perpendicular to the orbit plane. In reality the orbit plane varies constantly so that the spatial average has to be taken into account:

$$\left\langle r^2 \right\rangle = \frac{2}{3} \left\langle R^2 \right\rangle \qquad (4.396)$$

being $R$ the spherical orbital radii.

Considering the average quantities it follows that the average diamagnetic change in magnetic moment for *an* electron in a single atom has the expression:

$$\left\langle \Delta \mu_m \right\rangle = -\frac{e^2 \mu_0}{6m} \left\langle R^2 \right\rangle H \qquad (4.397)$$

For an atom containing $N$ electrons and for a material containing $z$ atoms per unit volume the magnetization due to diamagnetic effect is then given by:

$$M_{dia} = zN \left\langle \Delta \mu_m \right\rangle \equiv \chi_{dia} H \qquad (4.398)$$

from where the diamagnetic susceptibility ($\chi_{dia}$) follows to be:

$$\chi_{dia} = -\frac{z e^2 \mu_0}{6m} N \left\langle R^2 \right\rangle \qquad (4.399)$$

There is clearly that diamagnetic susceptibility is negative, the produced magnetization is such that is against of the applied magnetic field. In general the magnitude of $\chi_{dia}$ is $10^{-28}$ cm$^3$ per mole.

The practical link of the diamagnetic susceptibility with atomic radii can be reached performing the stipulated correspondence: when from the total number of electrons, $N$, only the outer (valence) electrons in the atomic systems are considered, $N_{outer}$, while the average $\left\langle R^2 \right\rangle$ becomes the absolute atomic radii itself, $R^2$. Therefore, the actual atomic working diamagnetic susceptibility formula in terms of atomic radii becomes (Putz et al., 2003, 2012b,c):

$$\chi_{dia} = -N_{outer} R^2 [0.56 \times 10^{-28} cm^3 /mol] \qquad (4.400)$$

## 4.9.2 ATOMIC POLARIZABILITY

### 4.9.2.1 The Functional Formulation of Polarizability

The electric dipole polarizability, $\alpha$, describes *the linear response* of the electron cloud of a chemical system to an external electric field that is less than that one needed to ionize the system. Therefore, the polarizability states as the linear coefficient between the induced dipole moment $d$ and the applied electric field $\mathbf{E}$:

$$\mathbf{d} = \alpha \mathbf{E} \tag{4.401}$$

The interaction energy between the induced dipole and the applied field is given by the term:

$$\varepsilon = -\mathbf{d} \cdot \mathbf{E} = -\alpha E^2 \tag{4.402}$$

In general, the electric field $E$ is an oscillating ones, but when is constant it correlates with the induced dipole through the static polarizability. Moreover, for atomic systems, assuming the spherical coordinates with the radius $r$, a given direction of the field $E$ will produces the perturbation potential along the azimuth direction:

$$U(r) = E r \cos \theta \tag{4.403}$$

being $\theta$ the azimuth angle. For an $N$-electronic system with density $\rho(r)$ and the nuclear potential $V(r)$, in terms of perturbative potential and the *linear response function* (4.197) (Hohenberg & Kohn, 1964), the interaction energy assumes also the form:

$$\varepsilon = \iint U(r)\kappa(r,r')U(r')drdr' = \left[\iint \kappa(r,r')rr'\cos^2\theta\, drdr'\right]E^2 \tag{4.404}$$

By comparison between relations of $\varepsilon$ it follows the atomic static dipole polarizability to be (Garza & Robles, 1993):

$$\alpha = -\iint \kappa(r,r')rr'\cos^2\theta\, drdr' \tag{4.405}$$

Any evaluation of the polarizability demands the knowledge of the linear response function. The general expression for this function in DFT can be formulated as in Eq. (4.197) with in terms of softness kernel, the local and the global softness, respectively, all related with the introduced local response function $L(r)$. However, for atomic systems a very sensitive approximation consists in neglecting all non-local contributions in the linear response function (Putz et al., 2003, 2012b,c):

$$\kappa(r,r') \cong -L(r')\delta(r-r') \tag{4.406}$$

For the central nuclear potential $V(r) = -Z/r$, the static dipole polarizability becomes successively:

$$\alpha = \iint L(r')\delta(r-r')rr'\cos^2\theta dr dr'$$

$$= 2\pi\int_0^\infty\int_0^\pi L(r)r^4\cos^2\theta\sin\theta\, dr d\theta = 2\pi\int_0^\infty L(r)r^4 dr\int_1^{-1}\cos^2\theta\, d(-\cos\theta)$$

$$= \frac{4\pi}{3}\int_0^\infty -\frac{r^2}{Z}\frac{\partial\rho(r)}{\partial r}r^4 dr = -\frac{4\pi}{3Z}\int_0^\infty r^6\frac{\partial\rho(r)}{\partial r}dr \tag{4.407}$$

Furthermore, using the partial derivation decomposition:

$$r^q\frac{\partial\rho(r)}{\partial r} = \frac{\partial[r^q\rho(r)]}{\partial r} - qr^{q-1}\rho(r) \ , \ q\in\Re \tag{4.408}$$

when $q = 6$, and the fact that the density $\rho(r)$ assumes the exponential decaying form, the atomic static dipole polarizability can be simplified to yield (Putz et al., 2003, 2012b,c):

$$\alpha = \frac{4\pi}{3Z}\int_0^\infty 6r^5\rho(r)dr = \frac{2}{Z}\langle r^3\rangle \tag{4.409}$$

That's clear that the polarizability direct correlates with the atomic volume, assuming so the usual dimension, cm$^3$. As in the diamagnetic case, the transcription of above relation in the atomic radii language assumes the correspondences that identify the average $\langle r^3\rangle$ with the cube atomic radii, $R^3$, and the nuclear charge $Z$ with the effective one $Z_{eff}$ (Putz et al., 2003, 2012b,c):

$$\alpha = \frac{2}{Z_{eff}}R^3[10^{-24}\,\text{cm}^3] \tag{4.410}$$

From the dependence of atomic polarizability by the cubic atomic radii rises immediately the correspondence with the atomic volume also. Therefore, atomic polarizability states as a fully atomic related property and have to closely follow the periodicity of atomic volumes, the Lothar Meyer's periodic curve.

### 4.9.2.2    The Quantum Formulation of Polarizability by Bethe Rules

Starting from the consecrated second order perturbation energy (see Volume I/Section 3.7 of the present five-volume set)

$$E^{(2)} = \sum_{k \neq n} \frac{\left| \langle n | \hat{H}_1 | k \rangle \right|^2}{E_k - E_n} \qquad (4.411)$$

is specialized for the Stark potential produced by the applied external electric field with the amplitude $\varepsilon$ in the $0x$ direction

$$\hat{H}_1 = V(\hat{x}) = -\hat{x} Z e_0 \varepsilon \qquad (4.412)$$

under the form

$$E^{(2)} = -\frac{1}{2} \alpha \varepsilon^2 \qquad (4.413)$$

that allows for $\alpha$–polarizability in Eq. (4.413) the general hydrogenic ($Z$-dependent) formulation

$$\alpha = 2 e_0^2 Z^2 \sum_{k \neq n} \frac{\left| \langle n | \hat{x} | k \rangle \right|^2}{E_n - E_k} \qquad (4.414a)$$

where

$$e_0^2 = \frac{e^2}{4 \pi \varepsilon_0} \qquad (4.414b)$$

is the reduced squared elementary charge.

Now, going to evaluate the atomic polarizability in terms of the quantum basic information contained within the atomic quantum numbers (e.g., $n$, $k$), one starts recognizing the general operatorial identity over the complete set of quantum (eigen) states (Putz, 2010c).

$$\sum_k \left| \langle n | \hat{O} | k \rangle \right|^2 = \sum_k \langle n | \hat{O} | k \rangle \langle k | \hat{O} | n \rangle = \langle n | \hat{O} \underbrace{\left\{ \sum_k | k \rangle \langle k | \right\}}_{1} \hat{O} | n \rangle = \langle n | \hat{O}^2 | n \rangle \qquad (4.415)$$

Equation (4.415) is eventually known as the sum rule of Bethe and Jackiw (Bethe & Jackiw, 1968; Jackiw, 1967), while its simplest dipole matrix element sum rule casts as

$$\sum_k \left| \langle n|\hat{x}|k \rangle \right|^2 = \langle n|\hat{x}^2|n \rangle \qquad (4.416)$$

On the other hand, recalling the basic quantum commutation rule of momentum with space coordinate

$$[\hat{p},\hat{x}] = \frac{\hbar}{i} \qquad (4.417)$$

along the companion energy-coordinate commutator

$$[\hat{H},\hat{x}] = \left[ \frac{\hat{p}^2}{2m} + V(\hat{x}), \hat{x} \right] = \frac{1}{2m}[\hat{p}^2,\hat{x}] = \frac{\hbar}{mi}\hat{p} \qquad (4.418)$$

there can be inferred the quantum relationship

$$\frac{\hbar}{i} = \langle n|(\hat{p}\hat{x} - \hat{x}\hat{p})|n \rangle = \sum_k \{ \langle n|\hat{p}|k \rangle \langle k|\hat{x}|n \rangle - \langle n|\hat{x}|k \rangle \langle k|\hat{p}|n \rangle \} \qquad (4.419)$$

upon inserting of the above quantum closure relation over the complete set of eigen-states. The first term in the right-hand side of the last expression may be reformulated as

$$\langle n|\hat{p}|k \rangle = \langle n|\frac{mi}{\hbar}[\hat{H},\hat{x}]|k \rangle = \frac{mi}{\hbar}\langle n|(\hat{H}\hat{x} - \hat{x}\hat{H})|k \rangle = \frac{mi}{\hbar}(E_n - E_k)\langle n|\hat{x}|k \rangle \qquad (4.420)$$

and along the similar relation that springs out from the second term in Eq. (4.419) one gets the equation (Putz, 2010c)

$$\frac{\hbar}{i} = \frac{mi}{\hbar}\sum_k [(E_n - E_k) - (E_k - E_n)]\left| \langle n|\hat{x}|k \rangle \right|^2 \qquad (4.421)$$

that can be rearranged under the so-called Thomas-Reiche-Kuhn (TRK) energy-weighted sum rule (Thomas, 1925; Reiche, 1925; Kuhn, 1925)

$$\frac{\hbar^2}{2m} = \sum_k (E_k - E_n)\left| \langle n|\hat{x}|k \rangle \right|^2 \qquad (4.422)$$

Remarkably, the expansion (4.422) may be also obtained by requiring that the Kramers-Heisenberg dispersion relation reduce to the Thomas scattering formula at high energies; indeed, through re-writing Eq. (4.422) in the form

$$\sum_k \frac{2m(E_k - E_n)}{\hbar^2}|\langle n|\hat{x}|k\rangle|^2 = \sum_k f_{n,k} = 1 \qquad (4.423)$$

it provides an important theoretical support for the experimental checks of the oscillator strengths ($f_{n,k}$) as a confirmation of early quantum results (Mehra & Rechenberg, 1982; Bethe, 1997).

Now, returning to the evaluation of polarizability given by (4.414) one can use the recipe (4.422) to facilitate the skipping out of the energy-singularity towards the all-eigen-state summation (4.416) with the successive results (Putz, 2010c)

$$\alpha = 2e_0^2 Z^2 \sum_{k \neq n} \frac{|\langle n|\hat{x}|k\rangle|^2}{E_n - E_k} = 2\frac{2me_0^2}{\hbar^2}Z^2 \frac{\hbar^2}{2m}\sum_{k \neq n} \frac{|\langle n|\hat{x}|k\rangle|^2}{E_n - E_k}$$

$$= \frac{4me_0^2}{\hbar^2}Z^2 \left\{\sum_k (E_k - E_n)|\langle n|\hat{x}|k\rangle|^2\right\}\sum_{k \neq n} \frac{|\langle n|\hat{x}|k\rangle|^2}{E_n - E_k}$$

$$\xrightarrow{all\ k} 2\frac{4me_0^2}{\hbar^2}Z^2 \left(\sum_k |\langle n|\hat{x}|k\rangle|^2\right)^2 = 8\frac{me_0^2}{\hbar^2}Z^2|\langle n|\hat{x}^2|n\rangle|^2 = 8\frac{Z^2}{a_0}|\langle n|\hat{x}^2|n\rangle|^2$$

$$(4.424)$$

where we recognized the first Bohr radius expression (2.119).

Finally, the obtained expression (4.424) is unfolded through replacing the coordinate observation with the atomic radius quantum average displacement respecting its instantaneous value (Putz, 2010c)

$$x \to r - \langle r\rangle_{nl} \qquad (4.425)$$

It allows the immediate formation of the squared coordinate expression

$$x^2 = r^2 - 2r\langle r\rangle_{nl} + \langle r\rangle_{nl}^2 \qquad (4.426)$$

of which the observed quantum average looks like

$$\langle n|\hat{x}^2|n\rangle \to \langle x^2\rangle_{nl} = \langle r^2\rangle_{nl} - \langle r\rangle_{nl}^2 \qquad (4.427)$$

The replacement of Eq. (4.427) in the polarizability (4.424) produces its radial averages' dependency

$$\alpha = 8\frac{Z^2}{a_0}\left[\left\langle r^2\right\rangle_{nl} - \left\langle r\right\rangle_{nl}^2\right]^2 \tag{4.428}$$

Knowing the first and second order quantum averages for the atomic radius of a Hydrogenic system written in terms of the principal and azimuth quantum numbers $n$ and $l$, respectively (Morse & Feshbach, 1953)

$$\left\langle r\right\rangle_{nl} = \frac{1}{2}\left(\frac{a_0}{Z}\right)\left[3n^2 - l(l+1)\right] \tag{4.429}$$

$$\left\langle r^2\right\rangle_{nl} = \frac{1}{2}\left(\frac{a_0}{Z}\right)^2 n^2\left[5n^2 - 3l(l+1)+1\right] \tag{4.430}$$

the static atomic polarizability (4.428) takes the analytical form (Putz, 2010c)

$$\alpha_{nl}(Z) = \frac{a_0^3}{2Z^2}\left[n^2\left(2+n^2\right) - l^2\left(1+l\right)^2\right]^2 \tag{4.431}$$

recovering the exact result for the Hydrogen limiting case

$$\alpha_{n=1,l=0}(Z=1) = \frac{9}{2}a_0^3 \tag{4.432}$$

Worth noting that the present derivation relays on the second order perturbation energy (4.411) while the final expression (4.431) is assumed to be exact through the Hydrogen checking case (4.432), although different by the other reported also as valid formulations, see McDowell (1976); Delone & Krainov (1994); Krylovetsky (1997). Nevertheless, the present atomic polarizability, either under expressions (4.424) or (4.431) is to be further tested for reliability in modeling of atomic (or ionic) and molecular systems.

### 4.9.3 THE FUNCTIONAL ATOMIC CHEMICAL HARDNESS

Among the various ways of chemical hardness ($\eta$) formulations, here will be considered that one that connects the global hardness of an electronic system to its global softness ($S$) by the inverse relation (4.196):

$$\eta^S = \frac{1}{2S} \tag{4.433}$$

Since the chemical hardness describes the electronic cloud's inertia to deform under the nuclear (or applied weak electrical field) the global softness accounts for such deformations. The global softness is then expressed for atomic systems through assuming the forms (Putz et al., 2003, 2012b,c):

$$S = 4\pi \int_0^\infty L(r) r^2 dr + N^2 = -\frac{4\pi}{Z} \int_0^\infty \frac{\partial \rho(r)}{\partial r} r^4 dr + N^2 = \frac{4}{Z}\langle r \rangle + N^2 \quad (4.434)$$

The last expression was obtained making use of the above rule of decomposition adapted for the present situation ($q=4$). Therefore, the global softness provides the global hardness with expression:

$$\eta^S = \frac{Z}{2(4\langle r \rangle + ZN^2)} \quad (4.435)$$

However, to be situated in the same type of approximation as for the atomic polarizability was made, the non-local contribution to the global softness, the $N^2$ term, has to be neglected. In practice this approximation is extended to neglect the term $ZN^2$ instead of $N^2$ only. This way, the approximation was performed in the hardness formula not in the softness one. This extension will be useful in applying this kind of approximation to another model for chemical hardness derivation. Finally, the hardness derived from softness displays the simple formula:

$$\eta^S = \frac{Z}{8\langle r \rangle} \quad (4.436)$$

The usually units for hardness are the electron-Volts.

Next, performing the same type of correspondences that was considered for the diamagnetic and polarizability formulas, the atomic hardness can be rewritten in terms of atomic radii and yields:

$$\eta^S = \frac{Z_{eff}}{8R}[eV] \quad (4.437)$$

and the previous approximation is turning now in neglecting the term $Z_{eff} N_{outer}^2$. Nevertheless the relation for chemical hardness displays the same format as the previous formulations do (Ghosh & Biswas, 2002).

Now can be determined also the relation between global hardness and atomic polarizability. Eliminating the atomic radii from polarizability we arrive at the relation (in atomic units) (Putz et al., 2003, 2012b,c):

$$\eta^S = \frac{Z_{eff}^{2/3}}{2^{8/3}\alpha^{1/3}} \tag{4.438}$$

From this equation clearly appears the inverse correspondence between the chemical hardness and polarizability and also the fact that this relation does not reflect the direct inversion of these quantities as the hardness and softness s behaves, for instance.

Another way of quantifying the chemical hardness in DFT is to use its relation with electronegativity of Eq. (3.3). Noting the fact that the present formula for electronegativity contains all the non-local terms in the linear response function, it appears that the present DFT electronegativity formulation can provide an important improvement in chemical hardness analytic expression. Performing the partial derivation in both sides of *chemical density-functional electronegativity* respecting with the total number of electrons, the chemical hardness takes the form (4.278) where, for atomic systems we have that (Putz et al., 2003, 2012b,c):

$$a = \int L(r)dr = \frac{4}{Z}\langle r \rangle \rightarrow \frac{4}{Z_{eff}}R \tag{4.439}$$

$$C_A = \int \rho(r)V(r)dr = -Z\left\langle \frac{1}{r} \right\rangle \rightarrow -Z_{eff}\frac{1}{R} \tag{4.440}$$

while adapted for atomic radii when the effective one replaces nuclear charge. Since last relations are plugged into the chemical hardness (4.278) also the total number of electrons have to be reduced to the $N_{outer}$, as it was always considerate when the passage from spatial averages to the atomic radii was made. With all these considerations the working chemical hardness formulation in terms of atomic radii becomes (Putz et al., 2003, 2012b,c):

$$\eta_\chi = \frac{Z_{eff}}{8R}\left\{ \frac{4Z_{eff}\left[2R(N_{outer}-2)+Z_{eff}\left(N_{outer}^2-1\right)\right]}{16R^2+8RZ_{eff}\left(1+N_{outer}^2\right)+Z_{eff}^2\left(N_{outer}^2-1\right)^2} \right.$$

$$\left. +\sqrt{\frac{R}{Z_{eff}}}\left[\arctan\left(\frac{N_{outer}+1}{2\sqrt{R}}\sqrt{Z_{eff}}\right)-\arctan\left(\frac{N_{outer}-1}{2\sqrt{R}}\sqrt{Z_{eff}}\right)\right] \right\} \tag{4.441}$$

This chemical hardness formula involves definitively more complex combination of atomic information than its softness-related counterpart.

Nevertheless, in order to can proper compare these hardness formulations the same kind of approximation has to be considered. In arriving to
obtained result the term $Z_{eff} N_{outer}^2$ was neglected. These means that also the
terms $N_{outer} \sqrt{Z_{eff}}$, which falls in the same class, have to be neglected. With
these considerations, the actual chemical hardness assumes the actual
form (Putz et al., 2003, 2012b,c):

$$\eta_C^\chi = \frac{Z_{eff}}{8R} \left\{ \frac{4Z_{eff}\left[2R\left(N_{outer}-2\right)-Z_{eff}\right]}{16R^2 + 8RZ_{eff} + Z_{eff}^2\left(N_{outer}^2 - 1\right)^2} \right.$$

$$\left. + \sqrt{\frac{R}{Z_{eff}}} \left[ \arctan\left(\frac{1}{2}\sqrt{\frac{Z_{eff}}{R}}\right) - \arctan\left(-\frac{1}{2}\sqrt{\frac{Z_{eff}}{R}}\right) \right] \right\} \qquad (4.442)$$

We have to remark that in the chemical hardness based on electronegativity, not all the non-local contributions was canceled but only that ones
based on global softness. We claim that the actual chemical hardness
reveals a more complex relation among the structural atomic information.

Nevertheless, eliminating atomic radii from polarizability it follows
also that the expression of the present chemical hardness in terms of
atomic polarizability yielding (Putz et al., 2003, 2012b,c):

$$\eta_\chi = \frac{Z_{eff}^{2/3}}{2^{8/3}\alpha^{1/3}} \left\{ \frac{4Z_{eff}\left[2^{2/3}Z_{eff}^{1/3}\alpha^{1/3}\left(N_{outer}-2\right)-Z_{eff}\right]}{2^{10/3}Z_{eff}^{2/3}\alpha^{2/3} + 2^{8/3}Z_{eff}^{4/3}\alpha^{1/3} + Z_{eff}^2\left(N_{outer}^2-1\right)^2} \right.$$

$$\left. + \frac{\alpha^{1/6}}{2^{1/6}Z_{eff}^{1/3}} \left[ \arctan\left(\frac{Z_{eff}^{1/3}}{2^{5/6}\alpha^{1/6}}\right) - \arctan\left(-\frac{Z_{eff}^{1/3}}{2^{5/6}\alpha^{1/6}}\right) \right] \right\} \qquad (4.443)$$

It is obvious that even the inverse relations between chemical hardness and
atomic polarizability was mainly preserved, it displays now richer analytical
structural information than that one directed derived from global softness.

### 4.9.4 DISCUSSION OF ATOMIC RADII BASED PERIODICITIES

The above discussed response indices of an atomic electronic system to
the environment influence can be directly particularized once an atomic
radii scale is available. However, here we will focus on the present S-DFT

(4.354) deduced atomic radii, $R_{(Putz)-DFT}$. Using these radii values (see the second column in the Table 4.17) the diamagnetic susceptibility and the static dipole polarizability are evaluated while their behaviors once change

**TABLE 4.17** For the First Four Periods Of Elements, Using the DFT Atomic Radii (Second Column), There Are Displayed the Calculated Diamagnetic Susceptibility, Static Dipole Polarizability, the Softness and Electronegativity Related Chemical Hardness and the Experimental First Ionization Potentials, From the Left to the Right Columns, Respectively (Putz et al., 2003, 2012b,c)

| Element | $R_{(Putz)-DFT}$ [Å] | $-\chi_{dia}$ $(0.56\times10^{-28}$ cm$^3$/mol) | $\alpha$ $(10^{-24}$ cm$^3$) | $\eta_s$ [eV] | $\eta_\chi$ [eV] | $I_1$ (Expt.) [eV] |
|---------|------|------|------|------|------|------|
| Li | 1.24 | 1.54 | 2.93 | 1.89 | 1.28 | 5.4 |
| Be | 0.82 | 1.34 | 0.57 | 4.28 | 3.23 | 9.4 |
| B | 0.62 | 1.15 | 0.18 | 7.54 | 6.44 | 8.29 |
| C | 0.49 | 0.96 | 0.07 | 11.93 | 9.62 | 11.27 |
| N | 0.41 | 0.84 | 0.03 | 17.11 | 12.88 | 14.5 |
| O | 0.35 | 0.735 | 0.02 | 23.38 | 16.4 | 13.61 |
| F | 0.31 | 0.673 | 0.01 | 30.17 | 19.85 | 17.42 |
| Ne | 0.27 | 0.58 | 0.007 | 38.97 | 23.88 | (4.257) |
| Na | 1.36 | 1.85 | 2.29 | 2.91 | 1.48 | 5.15 |
| Mg | 1.05 | 2.2 | 0.81 | 4.88 | 3.52 | 7.65 |
| Al | 0.85 | 2.17 | 0.35 | 7.41 | 6.35 | 5.89 |
| Si | 0.72 | 2.07 | 0.18 | 10.37 | 8.58 | 8.14 |
| P | 0.62 | 1.92 | 0.1 | 13.92 | 10.93 | 10.5 |
| S | 0.55 | 1.815 | 0.06 | 17.82 | 13.31 | 10.36 |
| Cl | 0.49 | 1.68 | 0.04 | 22.39 | 15.88 | 12.98 |
| Ar | 0.44 | 1.54 | 0.02 | 27.59 | 18.58 | 15.8 |
| K | 2.03 | 4.12 | 7.6 | 1.95 | 1.3 | 4.34 |
| Ca | 1.56 | 4.87 | 2.66 | 3.29 | 2.67 | 6.2 |
| Sc | 1.48 | 4.38 | 2.16 | 3.65 | 2.88 | 6.54 |
| Ti | 1.40 | 3.92 | 1.74 | 4.05 | 3.11 | 6.82 |
| V | 1.35 | 3.645 | 1.49 | 4.4 | 3.29 | 6.7 |
| Cr | 1.29 | 3.33 | 1.24 | 4.81 | 3.49 | 6.78 |
| Mn | 1.23 | 3.03 | 1.03 | 5.26 | 3.7 | 7.44 |
| Fe | 1.19 | 2.83 | 0.9 | 5.67 | 3.87 | 7.87 |
| Co | 1.14 | 2.6 | 0.76 | 6.15 | 4.07 | 7.9 |

**TABLE 4.17**    Continued

| Element | $R_{(Putz)\text{-}DFT}$ [Å] | $-\chi_{dia}$ $(0.56\times10^{-28}$ cm$^3$/mol) | $\alpha$ $(10^{-24}$ cm$^3)$ | $\eta_s$ [eV] | $\eta_\chi$ [eV] | $I_1$ (Expt.) [eV] |
|---|---|---|---|---|---|---|
| Ni | 1.10 | 2.42 | 0.66 | 6.62 | 4.25 | 7.65 |
| Cu | 1.05 | 2.2 | 0.55 | 7.19 | 4.46 | 7.73 |
| Zn | 1.02 | 2.08 | 0.49 | 7.67 | 4.62 | 9.39 |
| Ga | 0.89 | 2.38 | 0.28 | 10.1 | 8.15 | 6.1 |
| Ge | 0.78 | 2.43 | 0.17 | 13.03 | 10.32 | 8 |
| As | 0.70 | 2.45 | 0.11 | 16.19 | 12.33 | 9.8 |
| Se | 0.64 | 2.46 | 0.07 | 19.53 | 14.29 | 9.76 |
| Br | 0.58 | 2.35 | 0.05 | 23.57 | 16.51 | 11.81 |
| Kr | 0.54 | 2.33 | 0.04 | 27.48 | 18.53 | 14 |
| Rb | 2.54 | 6.45 | 14.9 | 1.56 | 1.14 | 4.19 |
| Sr | 1.97 | 7.76 | 5.36 | 2.6 | 2.23 | 5.7 |
| Y | 1.87 | 6.99 | 4.36 | 2.89 | 2.42 | 6.38 |
| Zr | 1.77 | 6.27 | 3.52 | 3.2 | 2.62 | 6.85 |
| Nb | 1.69 | 5.71 | 2.92 | 3.51 | 2.81 | 7 |
| Mo | 1.63 | 5.31 | 2.51 | 3.81 | 2.97 | 7 |
| Tc | 1.55 | 4.8 | 2.07 | 4.18 | 3.17 | 7.28 |
| Ru | 1.49 | 4.44 | 1.76 | 4.53 | 3.35 | 7.5 |
| Rh | 1.43 | 4.09 | 1.5 | 4.91 | 3.53 | 7.46 |
| Pd | 1.39 | 3.86 | 1.33 | 5.24 | 3.69 | 8.34 |
| Ag | 1.33 | 3.54 | 1.12 | 5.68 | 3.88 | 7.58 |
| Cd | 1.28 | 3.28 | 0.96 | 6.11 | 4.06 | 8.99 |
| In | 1.12 | 3.76 | 0.56 | 8.03 | 6.78 | 5.9 |
| Sn | 0.99 | 3.92 | 0.34 | 10.26 | 8.51 | 7.35 |
| Sb | 0.89 | 3.96 | 0.22 | 12.73 | 10.17 | 8.65 |
| Te | 0.80 | 3.84 | 0.15 | 15.62 | 11.99 | 9.01 |
| I | 0.74 | 3.83 | 0.11 | 18.47 | 13.68 | 10.45 |
| Xe | 0.68 | 3.7 | 0.08 | 21.82 | 15.57 | 12.13 |

from one atom to another in the periodic table are indicated in upper of Figure 4.31. As the experimental periodic reference it was chosen the first ionization potential. This quantity has the phenomenological advantage that direct accounts for the outer shell electronic structures, thus being

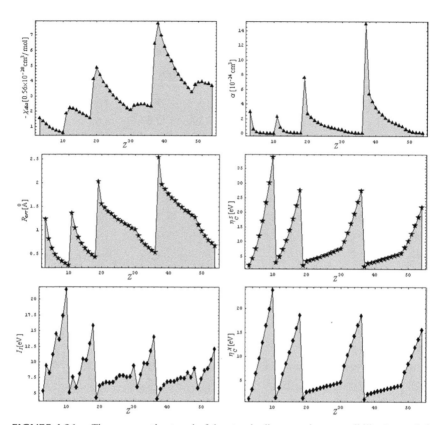

**FIGURE 4.31**  The comparative trend of the atomic diamagnetic susceptibility (upper left representation), static dipole polarizability (upper right representation) respecting to the atomic radii (middle left representation) and the experimental first ionization potential (middle right representation) along the softness and electronegativity related chemical hardness (in lower left and right) elemental periodical trend, respectively (Putz et al., 2003, 2012b,c).

a suitable indicator for the periodicity of the atomic response to the external influences in terms of outer shell properties as diamagnetic and polarizability state (Putz et al., 2003, 2012b,c).

The values in the Table 4.17 are relatively to the used formulas in terms of the atomic radii scale that is considered. There are also relatively to the assumed approach to replace all average dependencies with the atomic radii dependencies so that all the inner structure properties for an atomic system was concentrated its outer shell properties. This is naturally justified is looking to the global atomic response to the external influences where the outermost electrons are the mainly involved.

However, this theoretical picture is absolute, in sense that depends on the atomic radii that in its turn was deduced on the basis only about the DFT first principles and the considerate active electronic shell. As a general remark, all the computed quantities are sensitive to the periodic behavior. In the Figure 4.31 the comparative trend for the first four periods was displayed among the (absolute) diamagnetic susceptibility, atomic radii and the first ionization potential (Putz et al., 2003, 2012b,c).

Atomic radii states as the theoretical reference while the first ionization as the experimental ones. The diamagnetic global shape follows the atomic radii correspondence and is in good agreement to the first ionization value.

As the atomic radii decrease in periods the diamagnetic response diminishes as well due to the increase in compactness (decrease in orbiting area) of the electronic system requiring a higher first ionization potential. Among the groups is remarked the constantly increase in atomic radii, diamagnetic susceptibility with the consequence in the reduction of the first ionization potential. The analysis of atomic polarizability is enlightened by the maximum hardness principle and minimum polarizability principle (Putz, 2008a), respectively. There follows the chemical intuition that as an atom presents a closer-shell structure as the more stable, harder and much polarizable is.

The values in Table 4.17 and the shapes represented in Figure 4.31 turns out that the polarizability decrease along the periods, while the hardness has to increase along the periods. Such trend can sustain once more (here, at atomic level) the maximum hardness and minimum polarizability principles.

However, for elements that belong to the same group but different periods these principles are not relevant because it's difficult to apply these principles to systems with almost the same degree of occupancy in the outer shell. Moreover, the general trend of polarizability is to increase and of hardness is to decrease as the number of shells is increasing due to the screening effects.

However, a closer look to the values of Table 4.17 and Figure 4.31 indicates that previous pointed out periodic trend for polarizability and hardness is better reflected when the periods are analyzed two by two: the first two only with the principal groups and the last two containing the transitional elements. There follows that the presence of the transitional groups, especially in the third period, shifts the values to the lower polarizability and (more evident) to the higher hardness values of the principal elements of the third period respecting with that ones belonging to the second period (Putz et al., 2003, 2012b,c).

This behavior is attenuated in the fourth period, both respecting to the third and previous periods, on the basis of the accentuated global effects as the increasing in screening and in atomic radii. A closer look to the values of Table 4.17 and Figure 4.31 indicates that previous pointed out periodic trend for polarizability and hardness is better reflected when the periods are analyzed two by two: the first two only with the principal groups and the last two containing the transitional elements. It follows that the presence of the transitional groups, especially in the third period, shifts the values to the lower polarizability and (more evident) to the higher hardness values of the principal elements of the third period respecting with that ones belonging to the second period. This behavior is attenuated in the fourth period, both respecting to the third and previous periods, on the basis of the accentuated global effects as the increasing in screening and in atomic radii (Putz et al., 2003, 2012b,c).

## 4.10  QUANTITATIVE PROPERTY-PERIODICITY CORRELATIONS

There raises the idea that the atomic number has to be related, in principle, with all physical and chemical properties an atomic structure carries, or in a more phenomenological order, it appears as an effect or as a consequence of a certain existential elemental property. From this remark until the endeavor of viewing "Z" as the "atomic activity/property" that may be cast in terms of a plethora of structural indices is just a step and this is to be unveil in this communication, while testing one particular quantitative structure-property relationship (QSPR) for certain element leads, in fact, with testing the elemental periodicity of the Periodic System along a given period (Putz et al., 2011).

The streamline of QSPR analysis resides in evaluating the coefficients of the expansion

$$Y = a_0 + \sum_{i=1}^{M} a_i X_i \qquad (4.444)$$

between the dependent variable (Y) and the independent coordinates (X), either by traditional statistical models (Manly, 1986) or by recent advanced S-SAR algebraically ones, see Volume V of the present five-volumes set. Still, when about atomic QSPR one has to note that the above relation has to be restrained, at least in the first instance, to the level of periods, since the general multi-linear relations are not able to model the relatively complicated inter-periods (or down groups) behavior along the Periodic System. As such the present approach will restrict only to the mono-variate case

$$Y = a_0 + a_1 X_1 \qquad (4.445)$$

that is applied to the second, third and fourth periods for which a series of physicochemical properties, i.e., the atomic mass, atomic radius (R), melting point (MP), boiling point (BP), density ($\rho$) (Horovitz et al., 2000), finite difference electronegativity ($\chi$-FD) and chemical hardness ($\eta$-FD), experimental ionization potential (EXP-IP) and electronic affinity (EXP-EA) (Putz, 2008a), and their DFT third order (DFT[3]) counterpart (Putz, 2006), see Table 4.18.

The results are presented in Tables 4.19–4.21, where the obtained QSPR equations are supplemented with the associate correlation factors (r) and tested for the remaining element of each considered period: Ne (Z=10), Ar (Z=18), Sc (Z=21) for the second, third and transitional metals from the forth period, respectively. Individually, the predicted Z in Tables 2.19–4.21 reveals the important feature of the atomic structure that the electronegativity and chemical hardness related indices are the most appropriate for modeling the periodicity, beside the expected Z=Z(A) relationship, when combined both the closest computational result respecting the observed one with the highest correlation factors (Putz et al., 2011).

Conversely, there was shown that the so-called "macroscopic" physicochemical structural features such as melting or boiling point or density are not the best indicators for rational ordering of the elemental periodicity; this perhaps in all these cases all atomic electrons and levels are perturbed or regarded as equivalent when averaging for density, in contrast with the electronic frontier behavior quantified by electronegativity and related reactivity indices.

The fundamental issue of elemental periodicity is here addresses through quantitative-structure-property-relationship (QSPR) by assuming the atomic number as the atomic activity/property to be correlated with structural indicators, among which those relating with outermost orbitals, electronegativity, chemical hardness, ionization potential and electronic

**TABLE 4.18** Synopsis for the Structural Parameters Employed in this Work for the Second, Third and Fourth Period Transitional Metals: Atomic Number (Z), Atomic Mass (A), Atomic Radius (R), Melting Point (MP), Boiling Point (BP), Density (ρ) (Horovitz et al., 2000), Finite Difference Electronegativity (χ-FD) and Chemical Hardness (η-FD), Experimental Ionization Potential (EXP-IP) and Electronic Affinity (EXP-EA) (Putz, 2008a), and Their Density Functional Theory Third Order (DFT[3]) Counterpart (Putz, 2006)

| Element | Z | A | R (pm) | MP (K) | BP (K) | ρ (g/cm³) | χ-FD [eV] | η-FD [eV] | EXP-IP [eV] | EXP-EA [eV] | χ-DFT [3] [eV] | η-DFT [3] [eV] | DFT [3]-IP [eV] | DFT [3]-EA [eV] |
|---|---|---|---|---|---|---|---|---|---|---|---|---|---|---|
| Li | 3 | 6.941 | 156 | 454 | 1615 | 0.53 | 3.02 | 4.39 | 5.41 | 0.62 | 4.11 | 2.39 | 8 | 0.23 |
| Be | 4 | 9.012 | 113 | 1551 | 2745 | 1.848 | 3.43 | 5.93 | 9.36 | -2.5 | 12 | 12.1 | 22.6 | 1.41 |
| B | 5 | 10.81 | 97 | 2573 | 4273 | 2.34 | 4.26 | 4.06 | 8.32 | 0.21 | 31.4 | 36.5 | 58.4 | 4.4 |
| C | 6 | 12.011 | 92 | 4100 | 5100 | 2.25 | 6.24 | 4.99 | 11.3 | 1.25 | 75.2 | 80.8 | 141 | 9.55 |
| N | 7 | 14.007 | 55 | 63 | 77.4 | 0.88 | 6.97 | 7.59 | 14.6 | -0.6 | 158 | 133 | 300 | 14.8 |
| O | 8 | 15.999 | 60 | 54.8 | 90.2 | 1.15 | 7.59 | 6.14 | 13.6 | 1.46 | 290 | 161 | 564 | 14.8 |
| F | 9 | 18.998 | 71 | 59.5 | 85 | 1.51 | 10.4 | 7.07 | 17.5 | 3.33 | 460 | 119 | 916 | 3.58 |
| Ne | 10 | 20.179 | 65 | 24.5 | 27.1 | 1.20 | 10.71 | 10.92 | 21.63 | -0.31 | 645 | -2.5 | 1310 | -20 |
| Na | 11 | 22.990 | 186 | 370.9 | 1156 | 0.971 | 2.80 | 2.89 | 5.02 | 0.52 | 6.30 | 5.10 | 12 | 0.56 |
| Mg | 12 | 24.305 | 160 | 922 | 1380 | 1.738 | 2.6 | 4.99 | 7.7 | -2.4 | 13.5 | 14.4 | 25.4 | 1.70 |
| Al | 13 | 26.982 | 143 | 933 | 2792 | 2.702 | 3.22 | 2.81 | 6.03 | 0.42 | 28.4 | 32.6 | 53.0 | 3.91 |
| Si | 14 | 28.086 | 117 | 1683 | 3538 | 2.33 | 4.68 | 3.43 | 8.22 | 1.25 | 57.3 | 62.8 | 107 | 7.45 |
| P | 15 | 30.974 | 110 | 317 | 553 | 1.82 | 5.62 | 4.89 | 10.5 | 0.73 | 107 | 105 | 203 | 12.1 |
| S | 16 | 32.066 | 104 | 386 | 718 | 2.07 | 6.24 | 4.16 | 10.4 | 2.08 | 188 | 156 | 359 | 17.2 |
| Cl | 17 | 35.453 | 99 | 172.2 | 138.3 | 1.56 | 8.32 | 4.68 | 13 | 3.64 | 315 | 211 | 608 | 21.4 |
| Ar | 18 | 39.948 | 174 | 83.8 | 87.3 | 1.38 | 7.7 | 8.11 | 15.81 | -0.42 | 486 | 247 | 951 | 21.6 |
| Ti | 22 | 47.88 | 145 | 1941 | 3560 | 4.54 | 3.64 | 3.22 | 6.86 | 0.42 | 10.2 | 9.34 | 19.5 | 1.06 |

**TABLE 4.18** Continued

| Element | Z | A | R (pm) | MP (K) | BP (K) | ρ (g/cm³) | χ-FD [eV] | η-FD [eV] | EXP-IP [eV] | EXP-EA [eV] | χ-DFT [3] [eV] | η-DFT [3] [eV] | DFT [3]-IP [eV] | DFT [3]-EA [eV] |
|---|---|---|---|---|---|---|---|---|---|---|---|---|---|---|
| V | 23 | 50.942 | 131 | 2173 | 3723 | 5.8 | 3.85 | 2.91 | 6.76 | 0.94 | 11.5 | 11.2 | 21.8 | 1.29 |
| Cr | 24 | 51.996 | 125 | 2130 | 2938 | 7.19 | 3.74 | 3.12 | 6.76 | 0.62 | 12.9 | 13.4 | 24.3 | 1.57 |
| Mn | 25 | 54.938 | 137 | 1518 | 2334 | 7.43 | 3.85 | 3.64 | 7.90 | 0.31 | 14.5 | 15.8 | 27.0 | 1.88 |
| Fe | 26 | 55.847 | 125 | 1808 | 3135 | 7.87 | 4.26 | 3.64 | 7.90 | 0.62 | 16.1 | 18.6 | 30.1 | 2.24 |
| Co | 27 | 58.933 | 125 | 1768 | 3200 | 8.90 | 4.37 | 3.43 | 7.90 | 0.94 | 17.9 | 21.7 | 33.1 | 2.64 |
| Ni | 28 | 58.693 | 124 | 1726 | 3186 | 8.90 | 4.37 | 3.22 | 7.7 | 1.14 | 19.8 | 25.2 | 36.5 | 3.0 |
| Cu | 29 | 63.546 | 128 | 1356.4 | 2840 | 8.92 | 4.47 | 3.22 | 7.8 | 1.25 | 22.1 | 29.6 | 40.6 | 3.66 |
| Zn | 30 | 65.39 | 133 | 693 | 1179 | 7.14 | 4.26 | 5.2 | 9.46 | -0.9 | 24.4 | 34.0 | 44.6 | 4.24 |
| Sc | 21 | 44.956 | 161 | 1814 | 3103 | 2.99 | 3.43 | 3.22 | 6.55 | 0.21 | 9.11 | 7.70 | 17.3 | 0.85 |

**TABLE 4.19**  Testing the Second Period Periodicity by Predicting the Atomic Number Z for (Ne, Z=10) by Employing Various QSPR Equations with the Structural Factors of Table 4.18; the Best Predictions Are Further Limited by the First Three Best Correlations (r) (Putz et al., 2011)

| Model | QSPR | r | $Z^{\text{PREDICTED}}$ |
|---|---|---|---|
| *Atomic mass* | Z=–0.5010+ 0.51843*A | 0.99381 | 9.9603 |
| *Radius* | Z=11.013–0.0545*R | 0.8856 | 7.470 |
| *Melting point* | Z=6.5696–0.005*MP | 0.3279 | 6.447 |
| *Boiling point* | Z=7.0631–0.005*BP | 0.5175 | 6.927 |
| *Density* | Z=5.9556+0.02961*$\rho$ | 0.00942 | 5.991 |
| *FD-Electronegativity* | Z=1.2029+0.80124*$\chi^{FD}$ | 0.97426 | 9.7841 |
| *FD-Chemical hardness* | Z=-0.4942+1.131*$\eta^{FD}$ | 0.69613 | 11.863 |
| *EXP-Ionization potential* | Z=0.26558+0.50120*$IP^{EXP}$ | 0.95574 | 11.106 |
| *EXP-Electron affinity* | Z=5.5853+0.77004*$EA^{EXP}$ | 0.64740 | 5.346 |
| *DFT[3]-Electronegativity* | Z=4.2818+0.01167*$\chi^{DFT[3]}$ | 0.92437 | 11.808 |
| *DFT[3]-Chemical hardness* | Z=3.5234+0.03183*$\eta^{DFT[3]}$ | 0.91961 | 3.443 |
| *DFT[3]-Ionization potential* | Z=4.3246+0.0058*$IP^{DFT[3]}$ | 0.91848 | 11.922 |
| *DFT[3]-Electron affinity* | Z=4.5291+0.2111*$EA^{DFT[3]}$ | 0.59675 | -0.307 |

**TABLE 4.20**  The Same Analysis as in Table 4.19, for the Third Period (Ar, Z=18) (Putz et al., 2011)

| Model | QSPR | r | $Z^{\text{PREDICTED}}$ |
|---|---|---|---|
| *Atomic mass* | Z=0.08671+0.48489*A | 0.99268 | 19.453 |
| *Radius* | Z=22.329–0.0634*R | 0.9591 | 11.297 |
| *Melting point* | Z=14.918–0.0013*MP | 0.3311 | 14.809 |
| *Boiling point* | Z=15.043–0.0007*BP | 0.4098 | 14.981 |
| *Density* | Z=12.436+0.82985*$\rho$ | 0.21426 | 13.581 |
| *FD-Electronegativity* | Z=9.2614+0.99074*$\chi^{FD}$ | 0.96357 | 16.890 |
| *FD-Chemical hardness* | Z=9.5955+1.1071*$\eta^{FD}$ | 0.47846 | 18.574 |
| *EXP-Ionization potential* | Z=7.6803+0.72676*$IP^{EXP}$ | 0.93678 | 19.163 |
| *EXP-Electron affinity* | Z=13.183+0.91703*$EA^{EXP}$ | 0.78112 | 12.797 |
| *DFT[3]-Electronegativity* | Z=12.208+0.01754*$\chi^{DFT[3]}$ | 0.92076 | 20.732 |
| *DFT[3]-Chemical hardness* | Z=11.729+0.02710*$\eta^{DFT[3]}$ | 0.97058 | 18.422 |
| *DFT[3]-Ionization potential* | Z=12.231+0.00906*$IP^{DFT[3]}$ | 0.91791 | 20.847 |
| *DFT[3]-Electron affinity* | Z=11.561+0.26543*$EA^{DFT[3]}$ | 0.98192 | 17.294 |

**TABLE 4.21**     The Same Analysis as in Table 4.18 and Table 4.19, for the Fourth Period (Sc, Z=21) (Putz et al., 2011)

| Model | QSPR | r | $Z^{PREDICTED}$ |
|---|---|---|---|
| *Atomic mass* | Z=−0.4114+0.46777*A | 0.98813 | 20.617 |
| *Radius* | Z=49.250−0.1784*R | 0.4595 | 20.527 |
| *Melting point* | Z=34.166−0.0049*MP | 0.8052 | 25.277 |
| *Boiling point* | Z=32.774−0.0023*BP | 0.6488 | 25.637 |
| *Density* | Z=15.786+1.3784*ρ | 0.75253 | 19.907 |
| *FD-Electronegativity* | Z=-5.320+7.6577*$\chi^{FD}$ | 0.88379 | 20.945 |
| *FD-Chemical hardness* | Z=17.521+2.4148*$\eta^{FD}$ | 0.59648 | 25.296 |
| *EXP-Ionization potential* | Z=5.2487+2.7051*$IP^{EXP}$ | 0.83326 | 22.967 |
| *EXP-Electron affinity* | Z=26.480−0.8091*$EA^{EXP}$ | 0.1901 | 26.310 |
| *DFT[3]-Electronegativity* | Z=16.675+0.56175*$\chi^{DFT[3]}$ | 0.99527 | 21.792 |
| *DFT[3]-Chemical hardness* | Z=19.633+0.32043*$\eta^{DFT[3]}$ | 0.98950 | 22.100 |
| *DFT[3]-Ionization potential* | Z=16.214+0.31739*$IP^{DFT[3]}$ | 0.99539 | 21.704 |
| *DFT[3]-Electron affinity* | Z=20.020+2.4938*$EA^{DFT[3]}$ | 0.98725 | 22.139 |

affinity, being these more favorite respecting the so-called "measured" indicators as the melting and boiling points are. The results reload the idea that indices measured through a perturbation of the atomic system are somehow inappropriate for being considered as proper, intimate, structural indicators. However, this study was mainly based on mono-linear regression while further extended analysis to many-linear and higher order variations has to be as well undertake to properly model the elemental periodicity within Z=Z(structure) paradigm through the quantum indices (Putz et al., 2011).

Therefore, the statistical quantitative structure-property relationship (QSPR) methodology may be undertaken towards checking the elemental periodicity on the 2nd, 3rd, and transitional 4th chemical periods through correlating the atomic order's number Z with various physical-chemical properties, including various formulations for electronegativity and chemical hardness.

## 4.11   CONCLUSION

Aiming to hint at the solution to the current debate regarding the *physical vs. chemical definition of an atom* and as a special stage of a larger project regarding quantum chemical orthogonal spaces (Putz, 2012b), the present work addresses the challenging problem of defining and characterizing valence states with respect to the ground state within conceptual density functional theory. We are aware of the earlier warnings (Parr & Bartolotti, 1982; Bergmann & Hinze, 1987) regarding the limits of DFT and of the total energy of atomic systems combined with a Slater-based working density to provide a quadratic form in terms of system charge, as required by the general theory of chemical reactivity of atoms and molecules in terms of electronegativity and chemical hardness. Fortunately, we discovered that the Bohmian form of the total energy of such atomic systems provides, instead, the correct behavior, although it is only density-function-dependent and not a functional expression (Putz, 2012a) consistent with the recent advanced chemical orthogonal spaces approaches of chemical phenomenology (Putz, 2012b) as being at least complementary to the physical description of many-electronic systems when they are engaging in reactivity or equilibrium as the atoms-in-molecules theory prescribes (Bader, 1990). With the present Bohmian approach, the total energy is in fact identified with the quantum potential, thus inherently possessing non-locality and appropriate reactivity features, which are manifested even over long distances (Putz, 2010b; Putz & Ori, 2012); this also generalizes the previous (Boeyens, 2008) electronegativity formulation from the direct relationship between a quantum potential and its charge derivative. The double variational algorithm was implemented to discriminate the valence from the ground state charges, by using the golden ratio imbalance equation as provided by adaption of the Heisenberg type relationship to chemical reactivity for atoms. This corresponds to an analytical unfolding of the physical and chemical imbalance of the electronic charge stability of atomic systems, paralleling the deviation from the equal electron-to-proton occupancy in physical systems toward electron deficiency in the valence states of chemical systems. As a consequence, the difference between valence and ground state charge systems is naturally revealed and allows for the explanation of chemical reactivity and bonding in terms of fractional electron pairs, although driven by the golden ratio under the so-called physical-to-chemical charge difference wave function and associated normalizations, all of which represent elaborated or integral forms of the basic imbalance atomic equation. The present results are based on 10th-order polynomial fitted over 32 elements from the first 54 elements of the first four periods of Periodic Table of elements and can be further pursued by performing such systematic interpolations that preserve *the golden ratio relationships*, as advanced therein; they may also provide a comprehensive picture of how valence electrons may always be projected/equalized/transposed into ground state electrons within the perspective of further modeling chemical reactions when chemical reactivity negotiates the physical molecular stabilization of atoms in molecules.

In this context of quantum chemical atom, a unified Mulliken valence with Parr ground state for *electronegativity-chemical hardness* picture is presented. It provides an analytical useful tool on which also the absolute hardness as well *IP* and *EA* functionals are based. For all these chemical reactivity indices, within the density functional softness theory, systematic approximate density functionals are formulated and applied for atomic systems. For the absolute hardness, a special relationship with the new electronegativity ansatz and a particular atomic trend paralleling the absolute *EA* are founded, that should complement and augment the fashioned finite-difference energetic (Mulliken based) approach (Putz, 2006). In the framework of the DFT one shows that the new electronegativity functional formulation *relates* the Feynman-Kleinert path integral formalism in the Wigner/Markovian limit is proposed. The computation of the electronic density follows, in terms of partition function, the same procedure of the Levy's constrained-search for the wave function (Putz et al., 2005, 2009b). However, in all cases, the obtained electronegativity scale seems to respect the main criteria largely used for its acceptability for *chemical atomic periodicity* across the Periodic Table.

Next, in an effort to unify the chemical reactivity principles of electronegativity and chemical hardness, the *electrophilicity index* seems to offer the natural way through its definition, and it is accordingly introduced (Putz & Chattaraj, 2013). Nevertheless, due to its inverse dependence with the global chemical hardness it opens the door of being treated within the DFT of softness, where a lot of conceptual and analytical prescriptions are available. As such, the present chapter presents the route to introduce the local and kernel electrophilicities as based on long-range local and kernel softness, fulfilling the main hierarchical constraints, as successive integration rules, symmetry, and with conceptual and analytic consistency.

The obtained expressions are general, analytic and not being limited by the knowledge of the E(N) dependence and therefore not being affected by the discontinuity in energy-charge transfers (derivatives) as earlier formulations were shown. Instead, the present approach may be reduced exclusively to density and potential dependence of local and kernel electrophilicities once a working ansatz for kernel softness is provided. In such context, either atomic or molecular implementation may be worked out and applied towards modeling chemical phenomena. As an illustration, a special form of softness kernel was employed and the associated atomic scales were derived while showcasing the *periodicity reliability* and hierarchical connections among local, kernel and global forms of electrophilicity through periods and groups in the periodic table. Electronegativity trends especially for the alkali metal atoms and the halogen atoms are faithfully mimicked.

Finally, there is widely know that, although of intrinsic quantum mechanical nature, the *atomic radii has not a special operator associate*, nor a definitive working definition. Therefore, many physical-chemical approaches should be undertaken in providing reliable scales to parallel the periodic behavior of the Table of Elements. As such, the atomic scales as related with force, energy and potential are firstly reviewed in a historical context of modern physical chemistry, until the most celebrated connection with electronegativity (Putz, 2012c). However, since the proof of the quantum mechanical observable character of electronegativity, see the Chapter 3 of the present volume, there appears that the atomic radii based electronegativity computation should be considered as the main quantum chemical realization of the atomic radius itself. For physical completeness, also the atomic radii based physically observable quantum quantities like diamagnetic susceptibility and polarizability scales are considered in relation with atomic radii ones as computed by means of electronegativity information. Alongside, the intriguing problem of chemical hardness density functional computation in terms of atomic radii is approached due to the fact they are indices measuring the stability and extension of an atomic system, respectively. This way, the atomic radii concept and scales appear as the fundamental linking tools between the chemical quantum information of a system (compiled by electronegativity) and the physical observables, assuring therefore the needed convergence of the quantum natural hierarchy: *chemical information* → *atomic radius* (to be then generalized to *bonding length*) → *physical observable* (Putz, 2012c).

The present Chapter's approach is therefore valuable both from conceptual and computational views: for the first since it showcases the non-reducibility of the chemical knowledge to the physical laws, while assuring instead the chemical–to-physical knowledge transfer in modeling the observed world; and for the second instance since it constitutes through the connection with electronegativity—at its turn in relation with the so-called chemical action functional the veritable nano-roots of the chemical larger ensemble properties: the atomic radii information combined to provide the effective bonding length in molecules enters in characterization of the recently advanced bondons' existence—the quantum particle of chemical bond (Putz, 2010b), see also the Volume III of the present five-volume set; or describing the electronic delocalization length in molecules and complex nanostructures through the recent geometrical kernel realization of the chemical action functional (Putz, 2009a). It may be finally related with aromaticity or chemical reactivity through the further connection of atomic radii and/or bonding length with chemical hardness (Putz, 2010c), assuring therefore a comprehensive coverage of the fundamental chemical phenomena rooting in nano-scale information (Putz, 2012c).

## KEYWORDS

- **absolute electronegativity and chemical hardness**
- **atomic polarizability and scale**
- **Bethe rules**
- **Bohmian quantum theory**
- **charge waves**
- **chemical action**
- **chemical reactivity**
- **chemical softness**
- **density functional electronegativity and chemical harness**

- diamagnetic susceptibility and scale
- electronegativity scales (Pauling, Mulliken, Allred-Rochow, Gordy, Sanderson, Iczkowski-Margrave-Huheey, Klopman, Hinze-Witehead-Jaffe)
- electrophilicity
- local-to-global hierarchy
- quantitative property-periodicity correlations (QPPC)
- quantum electronegativity
- quantum semiempirical electronegativity
- radii scales
- radii-electronegativity relationships
- softness density functional theory
- valence state

## REFERENCES

### *AUTHOR'S MAIN REFERENCES*

Putz, M. V. (2016a). *Quantum Nanochemistry. A Fully Integrated Approach: Vol. III. Quantum Molecules and Reactivity*. Apple Academic Press & CRC Press, Toronto-New Jersey, Canada-USA.

Putz, M. V. (2016b). *Quantum Nanochemistry. A Fully Integrated Approach: Vol. V. Quantum Structure-Activity Relationships (Qu-SAR)*. Apple Academic Press & CRC Press, Toronto-New Jersey, Canada-USA.

Putz, M. V., Chattaraj P. K. (2013). Electrophilicity kernel and its hierarchy through softness in conceptual density functional theory. *Int. J. Quantum Chem.* 113, 2163–2171 (DOI: 10.1002/qua.24473).

Putz, M. V. (2012a). Valence atom with Bohmian quantum potential: the golden ratio approach. *Chemistry Central Journal*, 6, 135/16 pages (DOI: 10.1186/1752-153X-6-135).

Putz, M. V. (2012b). *Chemical Orthogonal Spaces*, Mathematical Chemistry Monographs, Vol. 14, University of Kragujevac, Kragujevac.

Putz, M. V. (2012c). Nanoroots of Quantum Chemistry: Atomic Radii, Periodic Behavior, and Bondons. In: *Nanoscience and Advancing Computational Methods in Chemistry: Research Progress*, Castro, E. A., Haghi, A. K. (Ed.) IGI Global (formerly Idea Group Inc.), Hershey (PA) Chapter 4, pp. 103–143 (DOI: 10.4018/978-1-4666-1607-3.ch004).

Putz, M. V., Ori, O. (2012). Bondonic characterization of extended nanosystems: application to graphene's nanoribbons. *Chem Phys Lett* 548, 95–100 (DOI: 10.1016/j. cplett.2012.08.019).

Matito, E., Putz, M. V. (2012). New link between conceptual density functional theory and electron delocalization. *J. Phys. Chem. A* 115(45), 12459–12462 (DOI: 10.1021/ jp200731d).

Putz, M. V. (2011a). Quantum and electrodynamic versatility of electronegativity and chemical hardness. In *Quantum Frontiers of Atoms and Molecules*, Putz, M. V. (Ed.), *Series "Chemistry Research and Applications,"* NOVA Science Publishers, Inc., New York, Chapter 11, pp. 251–270.

Putz, M. V. (2011b). Chemical action concept and principle. *MATCH Commun Math. Comput. Chem.* 66, 35–63.

Putz, M. V., Ionaşcu, C., Chiriac, A. (2011). Testing elemental periodicity by qspr. *Int. J. Chem. Model.* 3(1–2), 23–28.

Putz MV (2010a). On Heisenberg uncertainty relationship, its extension, and the quantum issue of wave-particle duality. *International Journal of Molecular Sciences* 11, 4124–4139 (DOI: 10.3390/ijms11104124).

Putz, M. V. (2010b). The bondons: the quantum particles of the chemical bond. *Int. J. Mol. Sci.* 11, 4227–4256 (DOI: 10.3390/ijms11114227).

Putz, M. V. (2010c). Compactness aromaticity of atoms in molecules. *Int. J. Mol. Sci.* 11(4), 1269–1310 (DOI: 10.3390/ijms11041269).

Tarko, L., Putz, M. V. (2010). On electronegativity and chemical hardness relationships with aromaticity. *Journal of Mathematical Chemistry*, 47(1), 487–495 (DOI: 10.1007/s10910-009-9585-6).

Putz, M. V., Ed. (2011c). *Carbon Bonding and Structures: Advances in Physics and Chemistry*, Series of Carbon Materials: Chemistry and Physics, Vol. 5, Springer Verlag, Dordrecht-London.

Putz, M. V. (2011d). Electronegativity and chemical hardness: different patterns in quantum chemistry. *Current Physical Chemistry* 1(2), 111–139 (DOI: 10.2174/1877946811101020111).

Putz, M. V. (2009a). Chemical action and chemical bonding. *J. Mol. Struct. (THEOCHEM)* 900(1–3), 64–70 (DOI: 10.1016/j.theochem.2008.12.026).

Putz, M. V. (2009b). Path integrals for electronic densities, reactivity indices, and localization functions in quantum systems. *Int. J. Mol. Sci.* 10, 4816–4940 (DOI: doi:10.3390/ ijms10114816).

Putz, M. V. (2008a). *Absolute and Chemical Electronegativity and Hardness*, Nova Publishers Inc., New York.

Putz, M. V. (2008b). Density functionals of chemical bonding. *Int. J. Mol. Sci.* 9, 1050–1095 (DOI: 10.3390/ijms9061050).

Putz, M. V. (2006). Systematic formulation for electronegativity and hardness and their atomic scales within density functional softness theory. *Int. J. Quantum Chem.* 106, 361–389 (DOI: 10.1002/qua.20787).

Putz, M. V., Russo, N., Sicilia, E. (2005). About the Mulliken electronegativity in DFT. *Theor. Chem. Acc.* 114(1–3), 38–45 (DOI: 10.1007/s00214-005-0641-4).

Putz, M. V. (2003). *Contributions within Density Functional Theory with Applications in Chemical Reactivity Theory and Electronegativity*, Dissertation.com, Parkland.

Putz, M. V., Russo, N., Sicilia, E. (2003). Atomic radii scale and related size properties from density functional electronegativity formulation. *J. Phys. Chem. A* 107(28), 5461–5465 (DOI: 10.1021/jp027492h).

## SPECIFIC REFERENCES

Allen, L. C. (1989). Electronegativity is the average one-electron energy of the valence-shell electrons in ground-state free atoms. *J. Am. Chem. Soc.* 111, 9003–9014.

Allen, L. C., Huheey, J. E. (1980). The definition of electronegativity and chemistry of the noble gases. *J. Inorg. Nucl. Chem.* 42, 1523–1524

Allred, A. L., Rochow, E. G. (1958). A scale of electronegativity based on electrostatic force. *J. Inorg. Nucl. Chem. 5,* 264–268.

Allred, A. L. (1961). Electronegativity values from thermochemical data. *J. Inorg. Nucl. Chem.* 17 (3–4), 215–221.

Ashby, J., Tennant, R. W. (1991). Definitive relationships among chemical structure, carcinogenicity and mutagenicity for 301 chemicals tested by the, U. S. NTP. *Mutat. Res.* 257(3), 229–306.

Ayers, P. W. (2000). Density per particle as a descriptor of coulombic systems. *Proc. Natl. Acad. Sci.* 97(5), 1959–1964.

Ayers, P. W. (2001). Strategies for computing chemical reactivity indices. *Theor. Chem. Acc.* 106(4), 271–279.

Ayers, P. W. (2008). The dependence on and continuity of the energy and other molecular properties with respect to the number of electrons. *J. Math. Chem.* 43, 285–303.

Ayers, P. W. Parr, R. G. (2001). Variational principles for describing chemical reactions: reactivity indices based on the external potential. *J. Am. Chem. Soc.* 123, 2007–2017.

Ayers, P. W., Parr, R. G. (2000). Variational principles for describing chemical reactions: the Fukui function and chemical hardness revisited. *J. Am. Chem. Soc.* 122, 2010–2018.

Ayers, P. W., Parr, R. G. (2008b). Beyond electronegativity and local hardness: Higher-order, *J. Chem. Phys. 129,* 054111–054117.

Ayers, P. W., Parr, R. G. (2008a). Local hardness equalization: exploiting the ambiguity. J. Chem. Phys. 128, 184108- 184115.

Bachelet, G. B., Hamann, D. R., Schülter, M. (1982). Pseudopotentials that work: from H to Pu. *Phys. Rev. B* 26, 4199–4228.

Bader, R. (1990). *Atoms in Molecules: A Quantum Theory*, Oxford University Press, New York.

Bader, R. F. W. (1990). *Atoms in Molecules—A Quantum Theory*, Oxford University Press, Oxford.

Bartolotti, L. J., Gadre, S. R., Parr, R. G. (1980). Electronegativities of the elements from simple Xα theory. *J. Am. Chem. Soc.* 102, 2945–2948.

Bergmann, D., Hinze, J. (1987). Electronegativity and charge distribution. Struct. Bond. 66, 145–190.

Berkowitz, M., Ghosh, S. K., Parr, R. G. (1985). On the concept of local hardness in Chemistry. *J. Am. Chem. Soc.* 107, 6811–6814.

Berkowitz, M., Parr, R. G. (1988). Molecular hardness and softness, local hardness and softness, hardness and softness kernels, and relations among these quantities. *J. Chem. Phys.* 88, 2554–2557.

Berski, S., Andres, J., Silvi, B., Domingo, L. R. (2003). The joint use of catastrophe theory and electron localization function to characterize molecular mechanisms. A density functional study of the Diels-Alder reaction between ethylene and 1,3-butadiene. *J. Phys. Chem. A* 107, 6014–6024.

Bethe, H. (1997). Theory of the passage of fast corpuscular rays through matter (Translated). In: *Selected Works of Hans, A. Bethe with Commentary* (World Scientific Series in twentieth century Physics), Bethe, H.(Ed), World Scientific, Singapore, Chapter 18 pp. 77–154.

Bethe, H., Jackiw, R. (1968). *Intermediate Quantum Mechanics*, 2nd Ed., Benjamin, New York, NY, USA, Chapter 11.

Boeyens, J. C., Levendis, D. C. (2012). The structure lacuna. *Int. J. Mol. Sci.* 13, 9081–9096.

Boeyens, J. C. A. (2005). *New Theories for Chemistry*, Elsevier, Amsterdam.

Boeyens, J. C. A. (2010). A molecular–structure hypothesis. *Int. J. Mol. Sci.* 11, 4267–4284.

Boeyens, J. C. A. (2011). Emergent properties in bohmian chemistry. In: *Quantum Frontiers of Atoms and Molecules*, Putz, M. V. (Ed.), Nova Publishers Inc., New York, pp. 191–215.

Boeyens, J. C. A., Levendis, D. C. (2008). *Number Theory and the Periodicity of Matter*, Springer, Heidelberg-Berlin.

Bohm, D. (1952). A suggested interpretation of the quantum theory in terms of "hidden" variables. I. *Phys Rev* 85, 166–179; 180–193.

Bohm, D., Vigier, J. P. (1954). Model of the causal interpretation of quantum theory in terms of a fluid with irregular fluctuations. *Phys. Rev.* 96, 208–216.

Bohr, N. (1921). *Abhandlungen über Atombau aus des Jahren 1913–1916*. Vieweg & Sohn, Braunschweig.

Boyd, R. J., Markus, G. E. (1981). Electronegativities of the elements from a nonempirical electrostatic model. *J. Chem Phys.* 75, 5385–5388.

Bransden, B. H., Joachain, C. J. (1983). *Physics of Atoms and Molecules*, Longman, London.

Bratsch, S. G. (1984). Electronegativity equalization with Pauling units. *J. Chem. Ed.* 61(7), 588–589.

Bratsch, S. G. (1985). A group electronegativity method with Pauling units. *J. Chem. Ed.*, 62(2), 101–103.

Bredow, T., Jug, K. (2005). Theory and range of modern semiempirical molecular orbital methods. *Theor. Chem. Acc.* 113, 1–14.

Bruss, D. (2002). Characterizing entanglement. *J. Math Phys.* 43, 4237–4251.

Burresi, E., Sironi, M. (2004). Determination of extremely localized molecular orbitals in the framework of density functional theory, *Theor. Chem. Acc.* 112, 247–253.

Cárdenas, C., Ayers, P., De Proft, F., Tozer, D. J., Geerlings, P. (2011). Should negative electron affinities be used for evaluating the chemical hardness? Phys. Chem. Chem. Phys. 13, 2285–2293.

Cárdenas, C., Echegaray, E., Chakraborty, D., Anderson, J. S. M., Ayers, P. W. J. (2009). Should negative electron affinities be used for evaluating the chemical hardness? *Chem. Phys.* 130, 244105.

Chamorro, E., Chattaraj, P. K., Fuentealba, P. (2003). Variation of the electrophilicity index along the reaction path. *J. Phys. Chem. A* 107, 7068–7072.

Chattaraj, P. K., (1996). Sengupta, S. Popular electronic structure principles in a dynamical context. *J. Phys. Chem.* 100, 16129–16130.

Chattaraj, P. K., Duley, S. (2010). Electron affinity, electronegativity, and electrophilicity of atoms and ions. *J. Chem. Eng. Data.* 55, 1882–1886.

Chattaraj, P. K., Fuentealba, P., Gomez, B., Contreras, R. (2000). woodward–hoffmann rule in the light of the principles of maximum hardness and minimum polarizability: DFT and ab initio SCF studies, *J. Am. Chem. Soc.* 122(2), 348–351.

Chattaraj, P. K., Lee, H., Parr, R. G. (1991). Principle of maximum hardness. *J. Am. Chem. Soc.* 113, 1854–1855.

Chattaraj, P. K., Liu, G. H., Parr, R. G. (1995). The maximum hardness principle in the Gyftpoulos-Hatsopoulos three-level model for an atomic or molecular species and its positive and negative ions. *Chem. Phys. Lett.* 237, 171–176.

Chattaraj, P. K., Maiti, B. (2001). Electronic structure principles and atomic Shell structure. *J. Chem. Educ.* 78(6), 811–813.

Chattaraj, P. K., Maiti, B. (2003). HSAB principle applied to the time evolution of chemical reactions. *J. Am. Chem. Soc.* 125, 2705–2710.

Chattaraj, P. K., Maiti, B., Sarkar, U. (2003). Philicity: a unified treatment of chemical reactivity and selectivity. *J. Phys. Chem. A* 107(25), 4973–4975.

Chattaraj, P. K., Parr, R. G. (1993). Density functional theory of chemical hardness. *Struct. Bond.* 80, 11–25.

Chattaraj, P. K., Sarkar, U., Roy, D. R. (2006). Electrophilicity index. *Chem. Rev.* 106, 2065–2091.

Chattaraj, P. K., Schleyer, P.v.R. (1994). An ab initio study resulting in a greater understanding of the HSAB principle. *J. Am. Chem. Soc.* 116, 1067–1071.

Chermette, H. (1999). Chemical reactivity indexes in density functional theory. *J. Comput. Chem.* 20, 129–154.

Clementi, E., Raimondi, D. L. (1963). Atomic screening constants from SCF functions. *J. Chem. Phys.* 38, 2686–2689.

Clementi, E., Raimondi, D. L., Reinhardt, W. P. (1967). Atomic screening constants from SCF functions. II. Atoms with 37 to 86 electrons. *J. Chem. Phys.* 47, 1300–1307.

Cohen, A. J., Mori-Sánchez, P., Yang, W. (2012). Challenges for density functional theory. *Chem. Rev.* 112, 289–320.

Cook, B. D. (1974). *Ab Initio Valence Calculations in Chemistry*, Butterworths, London.

Coulson, C. A. (1960). *Valence*, Oxford University Press, Oxford.

Cushing, J. T. (1994). *Quantum Mechanics—Historical Contingency and the Copenhagen Hegemony*, The University of Chicago Press, Chicago & London.

De Proft, F., Liu, S., Parr, R. G. (1997). Chemical potential, hardness and softness kernel and local hardness in the isomorphic ensemble of density functional theory. *J. Chem. Phys.* 107, 3000–3006.

Deb, B. M. (1973). The force concept in chemistry. *Rev. Mod. Phys.*, 45, 22–43.

Delone, N. B., Krainov, V. P. (1994). *Multiphoton Processes in Atoms*. Springer, Berlin, Germany.

Domingo, L. R., Asensio, A., Arroyo, P. (2002a). Density functional theory study of the Lewis acid catalyzed Diels-Alder reactions of nitroalkenes with vinyl ethers Using Aluminium Derivatives. *J. Phys. Org. Chem.* 15, 660–666.

Domingo, L. R., Aurell, M. J., Perez, P., Contreras, R. (2002b). Quantitative characterization of the global electrophilicity power of common diene dienophile pairs in Diels-Alder reactions. *Tetrahedron* 58, 4417–4423.

Eremets, M. I., Struzhkin, V. V., Mao, H., Hemley, R. J. (2001). Superconductivity in Boron. *Science* 293, 272–4.

Ferreira, R. (1968). Is one electron less than half what an electron pair is? *J. Chem. Phys.* 49, 2456–2457.

Feynman, R. P. (1939). Forces in molecules. *Phys. Rev.* 56, 340–343.

Feynman, R. P. (1948). Space-time approach to non-relativistic quantum mechanics. *Rev. Mod. Phys. 20,* 367–387.

Fuentealba, P. (1998). Reactivity indices and reponse functions in density functional theory. *J. Mol. Struct.THEOCHEM.* 433, 113–118.

Fuentealba, P., Parr, R. G. (1991). Higher-order derivatives in density-functional theory, especially the hardness derivative $\partial\eta/\partial N$. *J. Chem. Phys.* 94, 5559–5564.

Fujita, S. (2005). Graphs to chemical structures 1. Sphericity indices of cycles for stereochemical extension of Polya's theorem. *Theor. Chem. Acc.* 113, 73–79.

Gál, T., Geerlings, P., De Proft, F., Torrent-Sucarrat, M. (2011). A new approach to local hardness *Phys. Chem. Chem. Phys.* 13, 15003–15015.

Garza, J., Robles, J. (1993). Density functional theory softness kernel. *Phys. Rev. A* 47, 2680–2685.

Gazquez, J. L., Cedillo, A., Vela, A. (2007). Electrodonating and electroaccepting powers. *J. Phys. Chem. A* 111(10), 1966–1970.

Gázquez, J. L., Galván, M., Vela, A. (1990). Chemical reactivity in density functional theory: the N-differentiability problem. *J. Mol. Struct. THEOCHEM* 210, 29–38.

Geerlings, P., De Proft, F., Langenaeker, W. (2003). Conceptual density functional theory. *Chem. Rev.* 103, 1793–1874.

Geerlings, P., De Proft, F. (2008). Conceptual DFT: the chemical relevance of higher response functions *Phys. Chem. Chem. Phys.* 10(21), 3028–3042.

Ghanty, T. K., Ghosh, S. K. (1996). New scale of atomic orbital radii and its relationships with polarizability, electronegativity, other atomic properties, and bond energies of diatomic molecules. *J. Phys. Chem.* 100, 17429–17433.

Ghosh, D. C., Biswas, R. (2002). Theoretical calculation of absolute radii of atoms and ions. Part 1. The atomic radii. *Int. J. Mol. Sci.* 3, 87–113.

Ginsberg, A. P., Miller, J. M. (1958). An empirical method for estimating the electron affinities of atoms. *J. Inorg. Nucl. Chem.* 7(4), 351–367.

Gordy, W. (1946). A relation between bond force constants, bond orders, bond lengths, and the electronegativities of the bonded atoms. *J. Chem. Phys.* 14(5), 305–320.

Gordy, W., Thomas, W. J. O. (1956). Electronegativities of the elements. *J. Chem. Phys.* 24, 439–444.

Guantes, R., Sanz, A. S., Margalef-Roig, J., Miret-Artés, S. (2004). Atom–surface diffraction: a trajectory description. *Surf. Sci. Rep.* 53, 199–330.

Haasnoot, C. A. G., de Leeuw, F. A. A. M., Altona, C. (1980). The relation between proton-proton NMR coupling constants and substituent electronegativities. I. An empirical generalization of the Karplus equation. *Tetrahedron*, 36, 2783–2792.

Hartree, D. R. (1957). *The Calculation of Atomic Structures*, Wiley & Sons, New York.

Hati, S., Datta, D. (1995). Electronegativity and static electric dipole polarizability of atomic species. A semiempirical relation. *J. Phys. Chem.* 99, 10742–10746.

Hawking, S. (2001). *The Universe in a Nutshell*, Bantam Books, New York.

Hellmann, H. (1937). *Einfürung in die Quantum-chemie*, Deuticke, Leipzig.

Hinze, J. (1968). Elektronegativität der Valenzzustände. *Fortschritte der chemischen Forschung* 9, 448–485.

Hinze, J., Jaffe, H. H. (1963a). Electronegativity. IV. Orbital electronegativities of the neutral atoms of the periods three a and four a and of positive ions of periods one and two. *J. Phys. Chem.* 67(7), 1501–1506.

Hinze, J., Jaffe, H. H. (1962). Electronegativity. I. Orbital electronegativity of neutral atoms. *J. Am. Chem. Soc.* 84(4), 540–546.

Hinze, J., Jaffe, H. H. (1963b). Electronegativity: III. Orbital electronegativities and electron affinities of transition metals. *Canadian, J. Chem.* 41(5), 1315–1328.

Hinze, J., Whitehead, M. A., Jaffe, H. H. (1963). Electronegativity. II. Bond and orbital electronegativities, *J. Am. Chem. Soc.* 85 (2), 148–154.

Hohenberg, P., Kohn, W. (1964). Inhomogeneous electron gas. *Phys. Rev.* 136:B864-B871.

Horovitz, O., Sârbu, C., Pop, H. (2000). *Rational Classification of the Chemical Elements (in Romanian)*, Dacia Publishing House, Cluj-Napoca, pp.11–17, 221–247.

Hotop, W., Lineberger, W. C. (1985). Binding energies in atomic negative ions. II. *J. Phys. Chem. Ref. Data* 14, 731–750.

Huheey, J. E. (1978). *Innorganic Chemistry. Principles of Structure and Reactivity*, 2nd ed, Harper and Row, New York.

Iczkowski, R. P., Margrave, J. L. (1961). Electronegativity. *J. Am. Chem. Soc.* 83(17), 3547–3551.

Jackiw, R. (1967). Quantum mechanical sum rules. *Phys. Rev.* 157, 1220–1225.

Jordan, K. D., Burrow, P. D. (1987). Temporary anion states of polyatomic hydrocarbons. *Chem. Rev.* 87, 557–588.

Ketelaar, J. A. A. (1958). *Chemical Constitution*, Elsevier, New York.

Kleinert, H. (2004). *Path Integrals in Quantum Mechanics, Statistics, Polymer Physics, and Financial Markets*, 3rd ed., World Scientific, Singapore.

Klopman, G. (1965). Electronegativity, *J. Chem. Phys.* 43:S124-S129.

Klopman, G. (1968). Chemical reactivity and the concept of charge- and frontier-controlled reactions. *J. Am. Chem. Soc.* 90(2), 223–234.

Kohn, W., Becke, A. D., Parr, R. G. (1996). Density functional theory of electronic structure. *J. Phys. Chem.* 100, 12974–12980.

Kohout, M., Pernal, K., Wagner, F. R., Grin, Y. (2004). Electron localizability indicator for correlated wavefunctions. I. Parallel-spin pairs. *Theor. Chem. Acc.* 112, 453–459.

Komorowski, L. (1983a). Fractionally charged ions in crystal lattices of organic ion-radical salts. Chem. Phys. 76, 31–43.

Komorowski, L. (1983b). Calculation of the charge transfer in ion-radical salts. *J. Phys. Colloques* 44, 1211–1214.

Komorowski, L. (1983c). Electronegativity through the energy function. *Chem. Phys. Lett.* 103(3), 201–204.

Komorowski, L. (1987). Electronegativity and hardness in chemical approximation. *Chem. Phys.* 55, 114–130.

Komorowski, L. (1987b). Chemical hardness and Pauling's scale of electronegativities. *Z. Naturforsch. A*, 42, 767–773.

Koopmans, T. (1934). Uber die zuordnung von wellen funktionen und eigenwerter zu den einzelnen elektronen eines atom. *Physica* 1, 104–113.

Krylovetsky, A. A., Manakov, N. L., Marmo, S. I. (1997). Quadratic stark effect and dipole dynamic polarizabilities of hydrogen-like levels. *Laser Phys.* 7, 781–796.

Kuhn, W. (1925). Regarding the total strength of a condition from outgoing absorption lines. *Z. Phys.* 33, 408–412.

Lackner, K. S., Zweig, G. (1983). Introduction to the chemistry of fractionally charged atoms: electronegativity. *Phys. Rev. D* 28, 1671–1691.

Levy, M. (1982). Electron densities in search of hamiltonians. *Phys. Rev. A* 26, 1200–1208.

Manly, B. F. J. (1986). *Multivariate Statistical Methods*, Chapman and Hall, London.

March, N. H. (1991). *Electron Density Theory of Many-Electron Systems*. Academic Press, New York.

March, N. H. (1992). *Electron Density Theory of Atoms and Molecules*, Academic Press: London, Appendix 8.1.

March, N. H. (1993,) The ground-state energy of atomic and molecular ions and its variation with the number of elections. *Struct. Bond.* 80, 71–86.

Margenau, H., Murphy, G. M. (1964). *The Mathematics of Physics and Chemistry*, Van Nostrand, Princeton.

Matcha, R. L. (1983). Theory of the chemical bond. 6. Accurate relationship between bond energies and electronegativity differences, J. Am. Chem. Soc. 105(15), 4859–4862.

Mathews, J., Walker, R. L. (1970). *Mathematical Methods of Physics*. Addison-Wesley Publishing Company, Inc.

Matsunaga, Y. (1969). Orbital electronegativities in molecules and ionic character in molecular complexes. *Bull. Chem. Soc. Jap.* 42(8) 2170–2173.

McDowell, H. K. (1976). Exact static dipole polarizabilities for the excited S states of the hydrogen atom, *J. Chem. Phys.* 65, 2518–2522.

Mehra, J., Rechenberg, H. (1982). *The Historical Development of Quantum Theory: The Formulation of Matrix Mechanics and its Modifications 1925–1926*, Springer-Verlag: New York, USA, Chapter IV.

Meneses, L., Fuentealba, P., Contreras, R. (2006). On the variations of electronic chemical potential and chemical hardness induced by solvent effects. *Chem. Phys. Lett.* 433, 54–57.

Miller, J. A., Miller, E. C. (1977). *Ultimate Chemical Carcinogens as Reactive Mutagenic Electrophiles*, In: Hiatt, H. H., Watson, J. D., Winsten, J. A. Eds., *Origins of Human Cancer*, Cold Spring Harbor Laboratory Press, Cold Spring Harbor, NY, pp 605–627.

Moore, C. E. (1970). National Bureau of Standards, Washington, DC, Report No. NSRDS-NBS 34.

Morse, P. M., Feshbach, H. (1953). *Methods of Theoretical Physics*, McGraw-Hill, New York, USA.

Mullay, J. (1987a). Estimation of atomic and group electronegativities. *Struct. Bond.* 66, 1–25.

Mullay, J. (1987b). A relationship between impact sensitivity and molecular electronegativity. Propellants Explos. Pyrotech.12(2), 60–63.

Mulliken, R. S. (1934). A new electroaffinity scale: together with data on valence states and an ionization potential and electron affinities. *J. Chem. Phys.* 2, 782–793.

Murphy, L. R., Meek, T. L., Allred, A. L., Allen, L. C. (2000). Evaluation and test of Pauling's electronegativity scale. *J. Phys. Chem. A* 104, 5867–5871.

Nalewajski, R. F. (Ed.) (1996). *Density Functional Theory*, Springer-Verlag, Berlin, Volume, I.

Padmanabhan, J., Parthasarathi, R., Subramanian, V., Chattaraj, P. K. (2006b). Group philicity and electrophilicity as possible descriptors for modeling ecotoxicity applied to chlorophenols. Chem. Res. Toxicol. 19(3) 356–364.

Padmanabhan, J., Parthasarathi, R., Subramanian, V., Chattaraj, P. K. (2006a). Chemical reactivity indices for the complete series of chlorinated benzenes: solvent effect. J. Phys. Chem. A 110, 2739–2745.

Padmanabhan, J., Parthasarathi, R., Subramanian, V., Chattaraj, P. K. J. (2005). Molecular structure, reactivity, and toxicity of the complete series of chlorinated benzenes. *Phys. Chem. A* 109, 11043–11049.

Parr, R. G. (1972). *The Quantum Theory of Molecular Electronic Structure*, WA Benjamin Inc., Reading-Massachusetts.

Parr, R. G. (1983). Density functional theory. *Annu. Rev. Phys. Chem.* 34, 631–656.

Parr, R. G., Bartolotti, L. J. (1982). On the geometric mean principle of electronegativity equalization. *J. Am. Chem. Soc.* 104, 3801–3803.

Parr, R. G., Bartolotti, L. J. (1983). Some remarks on the density functional theory of few-electron systems. J. Phys. Chem. 87(15), 2810–2815.

Parr, R. G., Chattaraj, P. K. (1991). Principle of maximum hardness. *J. Am. Chem. Soc.* 113(5), 1854–1855.

Parr, R. G., Donnelly, R. A., Levy, M., Palke, W. E. (1978). Electronegativity: the density functional viewpoint. *J. Chem. Phys.* 68, 3801–3808.

Parr, R. G., Pearson, R. G. (1983). Absolute hardness: companion parameter to absolute electronegativity. *J. Am. Chem. Soc.* 105, 7512–7516.

Parr, R. G., Szentpály, L.v., Liu, S. (1999). Electrophilicity index. *J. Am. Chem. Soc.* 121(9), 1922–1924.

Parr, R. G., Yang, W. (1984). Density functional approach to the frontier-electron theory of chemical reactivity. *J. Am. Chem. Soc.* 106(14), 4049–4050.

Parr, R. G., Yang, W. (1989). *Density Functional Theory of Atoms and Molecules*, Oxford University Press, New York.

Parr, R. G., Zhou, Z. (1993). Absolute hardness: unifying concept for identifying shells and subshells in nuclei, atoms, molecules, and metallic clusters. *Acc. Chem. Res.* 26(5), 256–258.

Parthasarathi, R., Elango, M., Subramanian, V., Chattaraj, P. K. (2005). Variation of electrophilicity during molecular vibrations and internal rotations. Theor. Chem. Acc. 113, 257–266.

Parthasarathi, R., Padmanabhan, J., Subramanian, V., Maiti, B., Chattaraj, P. K. (2003). Chemical reactivity profiles of two selected polychlorinated biphenyls. J. Phys. Chem. A 107(48), 10346–10352.

Parthasarathi, R., Subramanian, V., Roy, D. R., Chattaraj, P. K. (2004). Electrophilicity index as a possible descriptor of biological activity. *Bioorg. Med. Chem.* 12(21), 5533–5543.

Pauling, L. (1932). The nature of the chemical bond IV. The energy of single bonds and the relative electronegativity of atoms. *J. Am. Chem. Soc.* 54, 3570–3582.

Pauling, L. (1960). *The Nature of the Chemical Bond*, Cornell University Press, Ithaca, NY.

Pearson R G. (1988b). Chemical hardness and bond dissociation energies. J. Am. Chem. Soc. 110(23), 7684–7690.

Pearson, R. G. (1988a). Absolute electronegativity and hardness: application to inorganic chemistry. *Iorg. Chem.* 27, 734–740.

Pearson, R. G. (1973). *Hard and Soft Acids and Bases*, Dowden, Hutchinson & Ross, Stroudsberg (PA).

Pearson, R. G. (1986). Absolute electronegativity and hardness correlated with molecular orbital theory. Proc. Natl. Acad. Sci. USA 83(22), 8440–8441.

Pearson, R. G. (1987). Recent advances in the concept of hard and soft acids and bases. J. Chem. Ed. 64(7), 561–567.

Pearson, R. G. (1989). Absolute electronegativity and hardness: applications to organic chemistry. J. Org. Chem. 54(6), 1423–1430.

Pearson, R. G. (1990). Hard and soft acids and bases—the evolution of a chemical concept. *Coord. Chem. Rev.* 100, 403–425.

Pearson, R. G. (1993). The principle of maximum hardness. *Acc. Chem. Res.* 26(5), 250–255.

Pearson, R. G. (1997). *Chemical Hardness*, Wiley-VCH, Weinheim.

Perdew, J. P., Parr, R. G., Levy, M., Balduz, J. L. (1982). Density functional theory for fractional particle number: derivative discontinuities of the energy. *Phys. Rev. Lett.* 49, 1691–1694.

Pérez, P., Chamorro, E., Ayers, P. W. (2008). Universal mathematical identities in density functional theory: Results from three different spin-resolved representations. *J. Chem. Phys.* 128(2), 204108.

Pérez, P., Toro-Labbé, A., Aizman, A., Contreras, R. (2002). Comparison between experimental and theoretical scales of electrophilicity in benzhydryl cations. J. Org. Chem. 67(14), 4747–2752.

Petrucci, R. H., Harwood, W. S., Herring, F. G., Madura, J. D. (2007). *General Chemistry: Principles & Modern Applications*, 9th Ed, Pearson Education Inc., New Jersey.

Preuss, H. (1969). *Quantenchemie fuer Chemiker*, Verlag Chemie, Weinheim, Germany.

Reiche, F., Thomas, W. (1925). Uber die Zahl der dispersionselektronen, die einem stationären Zustand zugeordnet sind. Z. Phys. 34, 510–525.

Robert Ponec (1981). Generalization of electronegativity concept. *Theor. Chim. Acta* 59(6), 629–637.

Robles, J., Bartolotti, L. J. (1984). Electronegativities, electron affinities, ionization potentials, and hardnesses of the elements within spin polarized density functional theory. *J. Am. Chem. Soc.* 106, 3723–3727.

Rong, C., Lian, S., Yin, D., Zhong, A., Zhan, R., Liu, S. (2007). Effective simulation of biological systems: Choice of density functional and basis set for heme-containing complexes. *Chem. Phys. Lett.* 434, 149–154.

Roy, D. R., Pal, N., Mitra, A., Bultinck, P., Parthasarathi, R., Subramanian, V., Chattaraj, P. K. (2007). An atom counting strategy towards analyzing the biological activity of sex hormones. *Eur. J. Med. Chem.* 42, 1365–1369.

Sablon, N., De Proft, F., Geerlings, P. (2010). The linear response kernel: inductive and resonance effects quantified. *J. Phys. Chem. Lett.* 1(28), 1228–1234.

Sanderson, R. T. (1983a). Electronegativity and bond energy. *J. Am. Chem. Soc.* 105 (8), 2259.

Sanderson, R. T. (1983b). *Polar Covalence*, Academic Press, New York.

Sanderson, R. T. (1986a). The inert-pair effect on electronegativity. *Inorg. Chem.*(25), 1856–1858.

Sanderson, R. T. (1986b). Electronegativity and bonding of transitional elements. Inorg. Chem. 25(19), 3518–3522

Sanderson, R. T. (1988a). Principle of electronegativity, I. *J. Chem. Educ.* 65, 112–118.

Sanderson, R. T. (1988b). Principles of electronegativity II. *J. Chem. Ed.* 65, 227–231.

Selwood, P. W. (1956). *Magnetochemistry*, 2nd ed. Interscience Publishers, New York.

Sen, K. D., Jørgensen, C. K. (Eds.) (1987). Electronegativity, *Structure and Bonding*, Volume 66, Springer Verlag, Berlin.

Sen, K. D., Mingos, D. M. P. (Eds.) (1993). Chemical Hardness, *Structure and Bonding*, 80, Springer Verlag, Berlin.

Senet, P. (1996). Nonlinear electronic responses, Fukui functions and hardnesses as functionals of the ground-state electronic density. *J. Chem. Phys.* 105, 6471–6490.

Senet, P. (1997). Kohn-Sham orbital formulation of the chemical electronic responses, including the hardness. *J. Chem. Phys.* 107, 2516–2525.

Simons, G., Zandler, M. E., Talaty, E. R. (1976). Nonempirical electronegativity scale. *J.Am.Chem.Soc.* 98, 7869–7870

Slater, J. C. (1930). Atomic shielding constants. *Phys. Rev.* 36, 57–64.

Slater, J. C. (1939). *Introduction to Chemical Physics*. McGraw-Hill, New York, Chapter 22.

Slater, J. C. (1964). Atomic radii in crystals. *J. Chem. Phys.*, 41(10), 3199–3204.

St. John, J., Bloch, A. N. (1974). Quantum-defect electronegativity scale for nontransition elements. *Phys. Rev. Lett.* 33, 1095–1098.

Stuttgart Pseudopotentials (2011): http://www.theochem.uni-stuttgart.de/pseudopotentials/index.en.html and the foregoing references (accessed 2001, 2011)

Szasz, L. (1985). *Pseudopotential Theory of Atoms and Molecules*, John Wiley, New York, USA.

Thom, R. (1975). Structural stability and morphology (An Outline of a General Theory of the Models), Benjamin Inc: Reading, Massachusetts.

Thomas, W. (1925). Uber die zahl der dispersionselectronen, die einem starionären zustande zugeordnet sind. *Naturwissenschaftern* 13, 510–525.

Tomasi, J. (2004). Thirty years of continuum solvation chemistry: a review, and prospects for the near future. *Theor. Chem. Acc.* 112, 184–203.

Torrent–Sucarrat, M., De Proft, F., Geerlings, P., Ayers, P. W. (2008). Do the local softness and hardness indicate the softest and hardest regions of a molecule? *Chem. Eur. J.* 14(28), 8652–8660.

Torrent-Sucarrat, M., De Proft, F., Ayers, P. W., Geerlings, P. (2010). On the applicability of local softness and hardness. *Phys. Chem. Chem. Phys.* 12(5), 1072–1080.

van Setten, M. J., Uijttewaal, M. A., de Wijs, G. A., de Groot, R. A. (2007). Thermodynamic stability of boron: The role of defects and zero point motion. *J. Am. Chem. Soc.* 129, 2458–2465.

Vishveshwara, S., Brinda, K. V., Kannany, N. (2002). Protein structure: insights from graph theory. *J. Theor. Comput. Chem.* 1, 187–211.

Von Szentpály, L. (2000). Modeling the charge dependence of total energy and its relevance to electrophilicity. *Int J Quant Chem* 76, 222–234.

Weast, R. C., Astle, M. J., Beyer, W. H. (Eds.) (1989). *CRC Handbook of Chemistry and Physics*, 69th Edition, Section E-78, CRC Press, Boca Raton.

Web Elements (2011). http://www.webelements.com/electronegativity.html (January 2011)

Wentorf Jr, R. H. (1965). Boron: another form. *Science.* 147, 49–50.

White, H. E. (1934). *Introduction to Atomic Spectra*, McGraw-Hill, New York.

Widom, M., Mihalkovic, M. (2008). Symmetry-broken crystal structure of elemental boron at low temperature. *Phys Rev B* 77, 064113.

Yang, W., Cohen, A. J., De Proft, F., Geerlings, P. (2012). Analytical evaluation of Fukui functions and real-space linear response function. J. Chem. Phys. 136, 144110.

Yang, W., Parr, R. G. (1985). Hardness, softness, and the Fukui function in the electronic theory of metals and catalysis. *Proc. Natl. Acad. Sci. U. S. A.*, 82, 6723–6726.

# CHAPTER 5

# QUANTUM ALGEBRAIC AND STOCHASTIC DYNAMICS FOR ATOMIC SYSTEMS

## CONTENTS

## ABSTRACT

This chapter reviews and advances the two related quantum ways for quantum description of valence/interacting/exchanged electrons among atoms at their turn involved in binding or molecular systems:

- by abstract formalization within quantum algebra of open systems; and
- by analytical formulation within stochastic/dissipative systems.

This way one models the electronic distribution, exchange and localization in chemical (inherently open) system, dealing therefore with that chemistry is at its ultimate description: the science of moving electrons from one state to another (either by intra- or inter-atomic framework).

## 5.1    INTRODUCTION

Algebraic characterization of many-electronic systems, and especially by formalization of chemical potential quantity, as the may mathematical-physical-chemical conceptual link and observable, offers the appropriate premises in further considering the chemical potential (aka minus electronegativity as previously introduced, consistently within density functional theory (DFT)) as the triggering quantity for chemical reactivity and electronic exchange and correlations in isolated atoms as well as for atoms-in-molecules combinations. To this aim, the Thomas-Fermi (TF) theory is firstly exposed in a way continuing the previous quantum mathematical analysis which widely assures the matter stability at the atomic level by the uniform electronic density, acceptable for many-electronic systems in general and for valence states (in atomic chemical zones) in special. The next natural step is considering the inhomogeneities specific to electronic localization in atomic shells, which will be nevertheless later solved by stochastically treatment of many-electronic systems at non-equilibrium (see Section 5.4).

In the history of chemistry, the revolutionary concepts of quantum mechanics lead with both conceptual and innovative understanding and designing of molecular structures. In this review, we would like to survey the main references in this rich and fascinating field of bonding knowledge. In this respect, the intensive level of chemical bonding such as the Schrödinger many-electronic-poly nuclei problem is firstly approached under the consecrated Hartree-Fock (*HF*), Roothaan and Kohn-Sham Self-Consistent Field (*SCF*) quantum frames. The localization problem is considered as the next level, in which context both the orbital and density localization functions are discussed. Finally, the chemical reactivity is indexed through the global density functionals of electronegativity and chemical hardness and of the associate principles. A study case of the particular series of acidic halogens in reactions with hydrogen peroxide is undertaken at each level of chemical bond characterization. It is found that the quantitative structure-property (activity) multi-linear relationships – *QSP(A)R*s – may be faithfully employed aiming to unify the levels of chemical bonding in single equation.

Very often, the famous words of Dirac, i.e., "The underlying physical laws necessary for the mathematical theory of a large part of physics and the whole of chemistry are thus completely known," are quoted by theorists in physics when they like to underline that chemistry is in principle solved by the basics of quantum mechanics so that some more interesting problems should be solved. Despite this, from 1929 nowadays, quantum physics of atoms and molecules largely turns into quantum chemistry, an interdisciplinary discipline that still struggles with the elucidation of the actual behavior of electrons in nano- and bio-systems. While the total success is still not in sight, the achievements in the arsenal of concepts, principles, and implementation was considerable and already enters goes into the arsenal of humankind hall-of-fame giving thus hope for a shining dawn in the poly electronic interaction arena (Putz & Chiriac, 2008). However, when questing for the underlying principles of the chemical bond, the first compulsory level of expertise may be called as the intensive level of analysis in which the main ingredients of a many-electronic-many-nuclear problem has to be clarified. These are subjected in the below following sections.

Despite the fact that *HF* or Kohn-*SCF* equations provide in principle the complete set of electronic orbitals that describe the multi-electronic-poly center bonds, their main drawback is that of providing the delocalized description over an entire molecular space. Such an analysis has to be accomplished with special techniques through which the localized orbitals and localized chemical bond are to be recovered (Putz & Chiriac, 2008; Putz, 2009). Only this way can quantum mechanics provide a viable rationale, i.e., quantum chemistry, in chemical bond characterization. Nevertheless, such a rationale can be achieved in two ways: one of them involves the orbital transformation producing the localized set of orbitals and indices (Daudel et al., 1983); the other one, based on electronic density, includes the electronic density, to a certain degree, into an *electronic localization (super) function –* ELF so as to generate a local, analytical indication of the electronic pair of the chemical bond (Becke & Edgecombe, 1990; Schmidera & Becke, 2002; Berski et al., 2003; Kohout et al., 2004; Nesbet, 2002; Putz, 2005).

In what follows, we are going to outline the modeling the chemical electronic behavior in atomic structure by both of these major mathematical-algebraically and stochastic-localization quantum approaches.

## 5.2   QUANTUM ALGEBRA OF ELECTRONIC SYSTEMS

### 5.2.1   INTRODUCING KUBO-MARTIN-SCHWINGER (KMS) STATES

Aiming to establish the fundaments of the chemical reactivity on quantum mathematical-physical concepts, a new way can be approached, by considering the statistical quantum phenomenology of the multi-electronic processes. In this context, the relationship between the Heisenberg and Schrödinger dynamic formalisms is constituted to be the starting point.

For a multi-particles system, the canonical coordinates associated to the operators impulse $\hat{p}_i$ and position $\hat{q}_i$, satisfy the canonical commutation relations (Davies, 1976).

$$\begin{cases} \hat{p}_i\hat{p}_j - \hat{p}_j\hat{p}_i = 0 = \hat{q}_i\hat{q}_j - \hat{q}_j\hat{q}_i \\ \hat{p}_i\hat{q}_j - \hat{q}_j\hat{p}_i = -i\hbar\delta_{ij} \end{cases} \tag{5.1}$$

and the develop equation for any operator $\hat{O}$ in time is written such as:

$$\frac{\partial\hat{O}}{\partial t} = \frac{i\left(\hat{H}\hat{O}_t - \hat{H}\hat{O}_t\right)}{\hbar} \tag{5.2}$$

equation in which $\hbar$ represents the reduced Planck constant, and $\hat{H}$ denotes the Hamiltonian operator of the system, conventionally written as a functional of the operators associated to the impulse and the position of each electron in the system:

$$\hat{H} = \sum_{i=1}^{N}\frac{\hat{p}_i^2}{2m} + V\left(\hat{q}_1,\hat{q}_2,...,\hat{q}_N\right) \tag{5.3}$$

From the physical point of view, any vector from the Hilbert space (normalized) $H = L^2(R^N)$, $H = L^2\left(R^N\right)$ corresponds to a system state, while the scalar product $\langle\psi|O_t|\psi\rangle$ is the observable $\hat{O}$ value at the moment $t$. The Heisenberg formalism is an operatorial one and the solution of Eq. (5.2) is actually the time-evolution equation for the considered operator. The passage and the equivalent with the Schrödinger formalism of the wave functions is made by identifying:

$$\hat{p}_i|\psi\left(x_1,...,x_N\right)\rangle = -i\hbar\frac{\partial}{\partial x_i}|\psi\left(x_1,...,x_N\right)\rangle \tag{5.4}$$

$$\hat{q}_i|\psi\left(x_1,...,x_N\right)\rangle = x_i|\psi\left(x_1,...,x_N\right)\rangle \tag{5.5}$$

$$\hat{H}|\psi\left(x_1,...,x_N\right)\rangle = \left[\sum_{i=1}^{N}\frac{\hat{p}_i^2}{2m} + V\left(\hat{q}_1,\hat{q}_2,...,\hat{q}_N\right)\right]|\psi\left(x_1,...,x_N\right)\rangle \tag{5.6}$$

while the complementarity between the two dynamic algorithms corresponds to the transposition law:

$$\langle \psi | O_t | \psi \rangle = \langle \psi_t | O | \psi_t \rangle \tag{5.7}$$

where $\hat{O}$ and $\psi$ correspond to the values $\hat{O}_t$ and $\psi_t$ at the moment $t = 0$.

In the '20s, '30s Stone and (von Neumann, 1929) have worked to clarify the connections between the two formalisms, in a coherent mathematical description, showing that the quantum theory is essentially unique. For example, in 1930, Stone (von Neumann, 1961) showed that if $t \mapsto \hat{U}_t$ is a continuous unitary representation along a real line, then there is a unique self-adjoint operator for $\hat{H}$ so that will be satisfied the equation:

$$\frac{d\hat{U}_t}{dt} = i\hat{U}_t \hat{H} \tag{5.8}$$

on the Hamiltonian operator field values; reciprocally, if $\hat{H}$ is self-adjunct, then the temporary equation determines a unique unitary representation continuous on $R$.

Therefore, in the early '30s the quantum mechanics was based on the following basic rules:

1.   An observable is a self-adjoint operator on the Hilbert space.
2.   A physical state (pure) is a vector from H.
3.   The estimated value of the operator $\hat{O}$ on the state $\psi$ is given by the scalar product $\langle \psi | \hat{O} | \psi \rangle$.
4.   The dynamic evolution of the system is determined by the self-adjoint properties of the Hamiltonian operator, and by Eq. (5.8) of the development operator

$$\hat{U}_t = \exp\left[ it\hat{H} \right] \tag{5.9}$$

and fulfills the operator and wave field rules:

$$\begin{cases} \hat{O} \mapsto \hat{O}_t = \hat{U}_t \hat{O} \hat{U}_{-t} \\ \psi \mapsto \psi_t = \hat{U}_{-t} \psi \end{cases} \tag{5.10}$$

by which the unity of transposition operation is respected.

5. The mixed state, $\omega$, are defined as functionals of the observables such as:

$$\frac{m}{2}\int_0^{\hbar\beta}d\tau\dot{x}^2(\tau)=\frac{m}{2}\int_0^{\hbar\beta}d\tau\left[\sum_{m=-\infty}^{+\infty}ix_m\omega_m\exp(i\omega_m\tau)\right] \qquad (5.11\text{a})$$

where

$$\lambda_i\geq0,\sum_i\lambda_i=1,\|\psi_i\|=1 \qquad (5.11\text{b})$$

and $\hat{p}$ represents the traces class operator, with the trace as unit.

The two dynamic approaches (Heisenberg and Schrödinger) can be unitary integrated to the *Stone-von Neumann* uniqueness theorem, which state that for the unitary groups,

$$\begin{cases}\hat{U}_i(t)=\exp\left[i\hat{p}_it\right]\\\hat{V}_j(t)=\exp\left[i\hat{q}_jt\right]\end{cases} \qquad (5.12)$$

associated to the self-adjoint operators of impulse and position, which satisfy the Weyl form of the commutation relationships

$$\begin{cases}\hat{U}_i(s)\hat{U}_j(t)-\hat{U}_j(t)\hat{U}_i(s)=0=\hat{V}_i(s)\hat{V}_j(t)-\hat{V}_j(t)\hat{V}_i(s)\\\hat{U}_i(s)\hat{V}_j(t)=\hat{V}_j(t)\hat{U}_i(s)\exp\left[ist\delta_{ij}\right]\end{cases} \qquad (5.13)$$

while noting that just the representations through the unitary continuous groups in the Hilbert space are written as sums of identical Schrödinger representations.

The algebra generated by the unitary groups satisfying the commutation relationships is called *algebra C\**. Therefore, appears useful and even necessary the quantum mechanics reformulation of algebraic perspective.

After 1933, Jordan suggested that the quantum observables can be characterized by their algebraic structure, introducing the Jordan algebras of operators, which satisfy the following composition axioms:

$$
\begin{cases}
\hat{A} \circ \hat{B} = \dfrac{\hat{A}\hat{B} + \hat{B}\hat{A}}{2} \\[2mm]
\hat{A} \circ \left(\hat{B} + \hat{C}\right) = \hat{A} \circ \hat{B} + \hat{A} \circ \hat{C} \\[2mm]
\left(\hat{B} + \hat{C}\right) \circ \hat{A} = \hat{B} \circ \hat{A} + \hat{C} \circ \hat{A} \\[2mm]
\lambda\left(\hat{A} \circ \hat{B}\right) = \left(\lambda \hat{A}\right) \circ \hat{B} = \hat{A} \circ \left(\lambda \hat{B}\right) \\[2mm]
\hat{A} \circ \hat{B} = \hat{B} \circ \hat{A} \\[2mm]
\left(\left(\hat{A} \circ \hat{A}\right) \circ \hat{B}\right) \circ \hat{A} = \left(\hat{A} \circ \hat{A}\right) \circ \left(\hat{B} \circ \hat{A}\right)
\end{cases}
\qquad (5.14)
$$

$\lambda \in \mathfrak{R}$. In the composition axioms set, the last relationship is a weaker form of the associative law, which satisfies the quantum operators.

Thus, there is defined by (Murray & von Neuman, 1936) the *algebra* $W^*$ in which the quantum observables are identified with the self-adjoint elements of the weak algebra of operators (in the Jordan sense) closed over the Hilbert-M space, while the states are described as mixed states. Based on these reasons, Gelfand and Naimark (1943) and Segal (1947) have argued that the observables uniform convergence has a direct physical interpretation, while the weak convergence has only an ana-lytical meaning. Therefore, the quantum observables can be identified as elements of a uniformly closed Jordan algebra. For finite electrons systems, the Stone–von Neumann uniqueness theorem indicates the fact that the distinction between $C^*$ and $W^*$ algebras is a matter of technical convenience.

Theoretically, are introduced as an idealization of the finite physical systems, the infinite systems. Therefore, becomes natural the study of the infinite systems models of electrons and the role which the $C^*$ and $W^*$ algebras properly play.

If is first considered a finite system as a subset in the electrons space, $\Lambda$, then it can be constructed the $C^*$ algebra on the vectorial space of opera-tors associated, $U_\Lambda$. Therefore, the observables corresponding to an elec-trons system however great can be determined by the algebra formed on the space consisting of the spaces $U_\Lambda$ reunion.

The algebra $U$, built without the specification (as sub-indices) of a state, or particular representations, corresponds to the $C^*$ algebra of observables for an infinite system. The algebras constructed in this way, from $U_\Lambda$ sub-algebra families, are called quasi-local algebras and the algebras $U_\Lambda$ are called local algebras.

Since Hagg (1962) had argued the importance of the quasi-local structures in the field theoretical models, they have been applied to the quantum statistical mechanics.

In this context, the algebraic structure is proving as an essential base of analysis the equilibrium states for the multi-electronic systems.

According to the arguments presented, for each operator $\widehat{O} \in U_\Lambda$ and for each set of electrons $\Lambda$, is introduced the equilibrium state of the system into the thermodynamic limit,

$$\omega_\alpha\left(\widehat{O}\right) = \lim_{\Lambda' \to \infty} \omega_{\Lambda', \alpha}\left(\widehat{O}\right) \tag{5.15}$$

where the infinite limit indicates the increase of the electrons set until the increase, which contains any compact subset of electrons for which the limit is reached.

The lower indices $\alpha$ denotes a thermodynamic parameter, for example, the temperature, the electron density or the chemical potential – and any relevant algorithm of constructing the equilibrium states depends on the parameter type selection.

The set of the thermodynamic limits for various observables associated to the electronic system represents the systemic equilibrium data, regardless of the size and the form of the considered electrons system.

Broadly, there are two ways of constructing the equilibrium state:

1. If it is started from the Hamiltonian specified for an electrons set $\widehat{H}_\Lambda$, which contains incorporated the descriptions of the interactions and the conditions of equilibrium for the electrons in a finite region, the Gibbs equilibrium states of the system – in the parameterization of the thermodynamic temperature inverse $\beta = 1/T$ – will be written as follows:

$$\omega_{\Lambda, \beta}\left(\widehat{O}\right) = \frac{\mathrm{Tr}\left(\widehat{O} \exp\left[-\beta \widehat{H}_\Lambda\right]\right)}{\mathrm{Tr}(\exp[-\beta \widehat{H}_\Lambda])} \tag{5.16}$$

From here starts the discussion of the existence or nonexistence of the thermodynamic limit applied to the equilibrium states. Unfortunately, this type of construction gives little information about the critical phenomena.

2. The second way starts from the specification of the infinite system evolution attached to the studied system, for which the simplest and also the strongest assumption is that the temporal evolution $t \mapsto \tau_t\left(\hat{O}\right)$ of the observable $\hat{O}$ is given by a continuous uni-parameter group $\tau$ for the *auto-morphisms $U$ of the algebra $C^*$, for all the observable. Therefore, the stationary criterion in time for the equilibrium states will be,

$$\omega\left(\tau_t\left(\hat{O}\right)\right) = \omega\left(\hat{O}\right), \forall \hat{O} \in U \, \& \, t \in R \qquad (5.17)$$

In 1957, Kubo, and respectively Martin and Schwinger (K-M-S) in 1959, had formulated a condition equivalent with the stationary criterion for the Gibbs states of finite volume, using the equilibrium-states form – (i) and performing the thermodynamic limit,

$$\omega_\beta\left(\tau_t\left(\hat{A}\right)\hat{B}\right) = \omega_\beta\left(\hat{B}\tau_{t+i\beta}\left(\hat{A}\right)\right), \forall \hat{A}, \hat{B} \in U \, \& \, t \in R \qquad (5.18)$$

it is obtained the so-called *KMS* condition, which was proposed as an equilibrium criterion by Haag (1962) and Hugenholtz (1967). This condition implies that the function $t \mapsto \omega\left(\hat{B}\tau_t\left(\hat{A}\right)\right)$ is analytic in the interval $0 < \text{Im}\, t < \beta$ and therefore expresses an approximate commutation of the observables $\omega$.

Then Tomita (1956) assigns to each normal state (mixed) $\omega$ over M the $W^*$ algebra, an uni-parameter canonical group of the automorphisms $\tau^\omega$. Moreover, Takesaki (1973) showed that the state $\omega$, satisfying the equilibrium condition *KMS* in relation with the group $\tau^\omega$, with the only difference of considering the evolutions $t \mapsto \tau_t^\omega\left(\hat{A}\right), \hat{A} \in M$, not necessarily has a continuous norm.

Although the second formulation of the equilibrium states is more realistic, in the way of including the temporal evolutions, however in this case, the assumption of the existence of a continuous uni-parameter group of the *automorphisms of $U$ of the algebra $C^*$, is hardly satisfied for the complex electronic systems. Therefore, the construction of the equilibrium states $\omega$, and also the temporal evolution $\tau$ must be simultaneously implemented.

The following situation "proper" for the temporal evolution compatible with the system equilibrium states, would be the group of the weak closure automorphisms, $\pi_\omega\left(U\right)"$, of the space $U$ in the cyclic representation associated to the state $\omega$.

However, Haag et al. (1970) made some suggestions concerning the equilibrium state and its physical equivalence, in order to obtain a better phenomenological interpretations. Thus, if the observables $\hat{A}_1,...,\hat{A}_n$ are measured in the equilibrium state $\omega$, are obtained the numbers

$$\omega\left(\hat{A}_1\right)=\lambda_1,...,\omega\left(\hat{A}_n\right)=\lambda_n \qquad (5.19)$$

due to the inherent imprecision in the measurement process, the observed values being included in intervals as $(\lambda_i - \varepsilon, \lambda_i + \varepsilon)$. Therefore, the state $\omega$ is equivalent to the state $\omega'$ which satisfies:

$$\left|\omega\left(\hat{A}_i\right)-\omega'\left(\hat{A}_i\right)\right|<\varepsilon \ , i=1,...,n \qquad (5.20)$$

so that the physical equivalence of the equilibrium states is determined by the neighborhoods of the *weak topology of any *weak dense representations $\pi_\omega(U)$.

The mixed states are a set of states large enough to can physically describe an electronic system, for various temperatures, densities.

Since the introduction of the electronegativity as chemical potential with changed sign is considered to be a fundamental observable for the characterization of the equilibrium states of the electronic systems in interaction, the chemical reactivity description in terms of quantum statistics and algebraic theory is considered to be a fundamental step in elucidating the tendencies of evolution to and from the equilibrium states, admitted by an electronic system (finite) and also of the afferent critical states.

### 5.2.2 BASIC ALGEBRAIC STRUCTURES AND TRANSFORMATIONS

In this section the following symbols for the affiliation to the concrete algebraic structures will be used; the present discussion follows (Bratteli & Robinson, 1987a,b):

- $U, B, C$ ... algebras $C^*$;
- $M, N, Z$ ... algebras $W^*$ (von Neumann);
- $I$ ... ideals in algebras $C^*$ and $W^*$;
- $H$ ... Hilbert space; $A, B, C$ elements in algebras $C^*$ and $W^*$;
- $P, E, F$ ... orthogonal projections; $G$ group;

- $\omega, \varphi$ ... quantum statistical states (connected) $\pi$ *morphism;
- $\xi, \eta, \psi, \Omega$ ... vectors in Hilbert space; $v$ measure;
- $\alpha, \beta$ ... coefficients from the real complex space etc.

Let U be a vectorial space with complex coefficients. The space U will be called algebra if is endowed with a multiplicative law which associates for each two elements $A, B \in U$, the product AB satisfies the associativity and distributivity properties (Bratteli & Robinson, 1987a,b; Araki & Kishimoto, 1977; Araki et al., 1977; Mebkhout, 1979; Haag et al., 1970; Hepp, 1972; Maksimov, 1974; Takesaki, 1970; Doplicher et al., 1966, 1969a-b; Connes, 1973):

$$(1) A(BC) = (AB)C$$
$$(2) A(B + C) = AB + AC$$
$$(3) \alpha\beta(AB) = (\alpha A)(\beta B) \quad \alpha, \beta \in C \tag{5.21}$$

A subspace $B$ of $U$, which is also an algebra in relation to the operations from $U$, is called a subalgebra. If the algebra $U$ is commutative or abelian, the $AB$ product is commutative,

$$AB = BA \tag{5.22}$$

The application $A \in U \rightarrow A^* \in U$ is called involution or self-adjoint operation of the $U$ algebra, if satisfies the following properties:

$$(1) A^{**} = A$$
$$(2) (AB)^* = B^* A^*$$
$$(3) (\alpha A + \beta B)^* = \bar{a} A^* + \bar{\beta} B^* \tag{5.23}$$

The algebra with the involution operation is called *algebra*, and the subset $B$ from $U$, is called auto deputy, if:

$$A \in B \Rightarrow A^* \in B \tag{5.24}$$

The algebra $U$ is called normalized algebra, if for each element $A \in U$ is associated a real number $\|A\|$, called the norm of $A$, satisfying the following requirements:

$$(1i) \|A\| \geq 0; \|A\| = 0 \Leftrightarrow A = 0$$

$$(2i)\|\alpha A\| = |\alpha|\|A\|$$

$$(3i)\|A + B\| \le \|A\| + \|B\| \quad \text{(triangle inequality)}$$

$$(4i)\|AB\| \le \|A\|\|B\| \quad \text{(product inequality)} \tag{5.25}$$

The norm defines the metric topology from $U$, which is interpreted as uniform topology, according to which the neighborhood of an element $A \in U$ is given as follows:

$$U(A; \varepsilon > 0) = \{B; \ B \in U, \|B - A\| < \varepsilon\} \tag{5.26}$$

If $U$ is a complete algebra in relation to the uniform topology, then is a Banach algebra. In other words, an normalized algebra with the operation of involution is complete in relation with the uniform topology and has the property,

$$\|A\| = \|A^*\| \tag{5.27}$$

is called the *Banach* algebra.

It will be called the algebra $C^*$, the Banach $U^*$ algebra, which has the property:

$$\|A^* A\| = \|A\|^2, \quad \forall A \in U \tag{5.28}$$

The normalized algebras are the substance of the Hilbert spaces.

Let $H$ be a Hilbert space. It will be noted by $L(H)$ the set of all bordered operators on $H$, equipped with the norm of the operators, such as:

$$\|A\| = \sup\{\|A\psi\|; \|\psi\| = 1, \psi \in H\} \tag{5.29}$$

The self-adjoint operation on the Hilbert space defines the involution operation on $L(H)$, and in relation with these operations and this norm, $L(H)$ is an algebra $C^*$:

$$\|A\|^2 = \sup\{\|\langle A\psi \,|\, A\psi \rangle\|; \|\psi\| = 1, \psi \in H\} \tag{5.30}$$

$$= \sup\{\langle \psi \,|\, A^* A\psi \rangle; \|\psi\| = 1, \psi \in H\}$$

$$\le \sup\{\|A^* A\psi\|; \|\psi\| = 1, \psi \in H\}$$

$$= \left\| A^* A \right\| \le \left\| A^* \right\| \left\| A \right\| = \left\| A \right\|^2 \qquad (5.31)$$

Note that any subalgebra $U$ form $L(H)$, which is auto deputy it is also an algebra $C^*$. In this sense the algebra of the compact operators which action on $H$, noted by $LC(H)$ is also an algebra $C^*$.

Let $U$ be an algebra $C^*$ without the identical element and $\tilde{U}$ be the algebra of the pairs like $\{(\alpha, A); \alpha \in C, A \in U\}$ with the operations:

$$(1i)(\alpha, A) + (\beta, B) = (\alpha + \beta, A + B)$$
$$(2i)(\alpha, A)(\beta, B) = (\alpha\beta, \alpha B + \beta A + AB)$$
$$(3i)(\alpha, A^*) = (\bar{\alpha}, A^*) \qquad (5.32)$$

The norm defined as follows:

$$\left\| (\alpha, A) \right\| = \sup \{ \left\| \alpha B + AB \right\|; \left\| B \right\| = 1, B \in U \} \qquad (5.33)$$

makes from $\tilde{U}$ an algebra $C^*$.

We can say that the algebra $U$ is the subalgebra $C^*$ of the $\tilde{U}$ algebra formed from the pairs $(0, A)$. Moreover, if is used the notation $\alpha 1 + A$ for the pair $(\alpha, A)$, with 1 the identical element, then it can be rewritten the relationship between the algebras $C^*$:

$$\tilde{U} = C1 + U \qquad (5.34)$$

where $C$ denotes the complex coefficients field.

A subspace $B$ of an algebra $U$ is called left ideal, if:

$$A \in U \ \& \ B \in B \ \Rightarrow \ AB \in B \qquad (5.35)$$

and respectively right ideal, if

$$A \in U \ \& \ B \in B \ \Rightarrow \ BA \in B \qquad (5.36)$$

If $B$ is both left and right ideal, is called two-sides ideal.

Any ideal is automatically an algebra.

If $U$ is an *algebra Banach and $I \subseteq U$ is a closed* two-sides ideal, then the rest space $U/I$ can be seen as an *algebra Banach. Therefore, an element $\hat{A} \in U/I$ is actually a subset (a class) of elements defined for any element $A \in U$, through:

$$\hat{A} = \{A + I; I \in I\} \tag{5.37}$$

and which satisfies the relationship of multiplication, addition, involution, such as:

$$(1i)\hat{A}\hat{B} = \hat{AB}$$

$$(2i)\hat{A} + \hat{B} = (\hat{A + B})$$

$$(3i)\hat{A}^* = \hat{A}^* \tag{5.38}$$

The remnant space $U/I$ becomes an *algebra Banach, if is introduced the norm:

$$\|\hat{A}\| = \inf\{\|A + I\|; I \in I\} \tag{5.39}$$

If $U$ is an algebra $C^*$ then the rest algebra $U/I$ endowed with the properties above is also an algebra $C^*$.

A $C^*$ *algebra* $U$ it will be called simple, if its unique two-sides ideals are only the trivial ideals $\{0\}$ and $U$. If the algebra $U$ has the identical element, then the algebra $U$ does not present any two-side ideal.

There is defined a *morphism between two algebras $U$ and $B$ through the following application:

$$A \in U \rightarrow \pi(A) \in B, \forall A \in U \tag{5.40}$$

which satisfies the following properties:

$$(1i)\pi(\alpha A + \beta B) = \alpha\pi(A) + \beta\pi(B)$$

$$(2i)\pi(AB) = \pi(A)\pi(B)$$

$$(3i)\pi(A^*) = \pi(A)^* \tag{5.41}$$

Let $U$ be an algebra with the unitary element, 1. Is defined the resolvent set $r_U(A)$ of the element $A \in U$, the of complex numbers set $\lambda \in C$, so that $\lambda 1 - A$ is irreversible. The resolvent of $A$ $in$ $\lambda$ will be given by the inverse $(\lambda 1 - A)^{-1}, \lambda \in r_U(A)$.

*The A spectrum,* $\sigma_U(A)$, is defined by the complement of the set $r_U(A)$ in $C$.

Each $^*$ morphism $\pi$ between the algebras $C^*$, $U$ and $B$, is positive,

$$\pi(A) = \pi(B^*B) = \pi(B)^* \pi(B) \geq 0 \qquad (5.42)$$

continuous, and satisfies the relation:

$$\|\pi(A)\| \leq \|A\| \qquad (5.43)$$

with the norm defined as:

$$\|\pi(A)\| = \sup\{|\lambda|; \lambda \in \sigma(\pi(A))\} \qquad (5.44)$$

Moreover, the $B_\pi = \{\pi(A); A \in U\}$ set determined by the $\pi^*$ morphism is a subalgebra $C^*$ from $B$.

One defines a representation of $C^*$ algebra $U$, by the pair $(H, \pi)$, where $H$ is the Hilbert complex space and $\pi$ is a $^*$morphism of $U$ in $L(H)$. The representation is called proper representation, if and only if $\pi$ is an $^*$iso-morphism between $U$ and $\pi(U)$, i.e., if and only if the nucleus of the $^*$morphism $\pi$ contains only the null element,

$$ket\pi = \{A \in; \pi(A) = 0\} = \{0\} \qquad (5.45)$$

From the presented arguments above, immediately results that any representation $(H, \pi)$ of $C^*$ algebra $U$ is a proper representation for the rest algebra

$$U_\pi = U / \ker \pi \qquad (5.46)$$

An automorphism $\tau$ of a $C^*$ *algebra* $U$, is defined as an $^*$isomorphism of the algebra $U$ in itself, so that $\tau$ is a $^*$morphism of $U$ of equal dimension with $U$ and the nucleus equal to zero.

Each $^*$automorphism $\tau$ of a $C^*$ algebra $U$, preserve the norm

$$\|\tau(A)\| = \|A\|, \quad \forall\, A \in U \qquad (5.47)$$

If $(H, \pi)$ is a representation of $C^*$ algebras $U$, and $H_1$ is a subspace of $H$, then it says that $H_1$ *is invariant, or stable under the action of* $\pi$, if:

$$\pi(A)H_1 \subseteq H_1, \forall\, A \in U \tag{5.48}$$

If $H_1$ is a closed subspace of $H$, and $P_{H1}$ is the orthogonal projector with the dimension $H_1$, then $H_1$ is *invariant* under the action $\pi$ if and only if:

$$\pi(A)P_{H1} = P_{H1}\pi(A) \tag{5.49}$$

There can be concluded that if $H_1$ is invariant under the $\pi$ action and if $\pi_1$ is defined by the orthogonal projector so that,

$$\pi_1(A) = P_{H1}\pi(A)P_{H1} \tag{5.50}$$

then also $(H_1, \pi_1)$ is a representation of $U$, called sub-representation.

If $H_1$ is invariant under the action $\pi$, then its orthogonal complement

$$H_2 = H_1^{\perp} \tag{5.51}$$

is also invariant.

Therefore, can be automatically defined the second sub-representation, $(H_2, \pi_2)$, by the action of specific orthogonal projectors

$$\pi_2(A) = P_{H2}\pi(A)P_{H2} \tag{5.52}$$

If one naturally writes

$$H = H_1 \oplus H_2 \tag{5.53}$$

and respectively each operator $\pi(A)$ is decomposes into the direct sum

$$\pi(A) = \pi_1(A) \oplus \pi_2(A) \tag{5.54}$$

for each $A \in U$, one can also write:

$$\pi = \pi_1 \oplus \pi_2 \tag{5.55}$$

$$(H, \pi) = (H_1, \pi_1) \oplus (H_2, \pi_2) \tag{5.56}$$

in terms of decomposition of the *morphisms and representations.

It says that a set $M$ of bounded operators acts non-degenerative on the Hilbert spaces $H$, if:

$$\{\psi; A\psi = 0, \forall \in M\} = \{0\} \tag{5.57}$$

An important class of non-degenerate representations is the class of cyclic representations.

There is firstly defined the cyclic vector $\Omega \in H$, if for a set $M$ of bounded operators, the set $\{A\Omega; A \in M\}$ is dense in $H$.

A cyclic representation of a $C^*$ *algebra* $U$ is defined by the triplet $(H, \pi, \Omega)$, where $(H, \pi)$ is a representation of $U$, and $\Omega \in H$, is a cyclic vector for $\pi$ in $H$.

If $(H, \pi)$ is a non-degenerate representation of $C^*$ algebra $U$, then $\pi$ is the direct sum of a family of cyclic sub-representations.

A set $M$ of bounded operators on Hilbert $H$ is defined as irreducible, if the single closed subspaces of $H$, which are invariant under the action of $M$, are trivial subspaces $\{0\}$ and $H$. A representation $(H, \pi)$ of a $C^*$ *algebra* $U$, is called irreducible if the set $\pi(U)$ is irreducible.

Having a representation $(H, \pi)$, is easy to construct other representations. For example, if $U$ is a unitary operator on $H$, it can be introduced the application

$$\pi_U(A) = U\pi(A)U^*, \forall\ A \in U \tag{5.58}$$

from where results the second representation $(H, \pi_U)$. This type of distinction in representation is not relevant, if the construction is based on the unitary operators.

Two representations $(H_1, \pi_1)$, $(H_2, \pi_2)$ *are called equivalent, or unitary, if there is an unitary operatory* $U$ *from* $H_1$ *to* $H_2$, *so that*

$$\pi_1(A) = U\pi_2(A)U^*, \forall\ A \in U \tag{5.59}$$

The equivalence of $\pi_1$ with $\pi_2$ will be noted by $\pi_1 \cong \pi_2$.

A linear functional $\omega$ *over* $^*$*algebra* $U$ *is defined as being positive, if:*

$$\omega(A^*A) \geq 0, \forall\ A \in U \tag{5.60}$$

A linear functional $\omega$ positive over a $C^*$ algebra $U$, with $\|\omega\| = 1$, is called the state.

If one considers a non-degenerate representation $(H,\pi)$ of $C^*$ algebra $U$ and a vector $\Omega \in H, |\Omega| = 1$, there can be defined as a state vector the linear functional over $U$:

$$\omega_\Omega(A) = \langle \Omega | \pi(A)\Omega \rangle, \ \forall \ A \in U \tag{5.61}$$

Reciprocally, giving a state over $C^*$ algebra $U$, it follows that there is an unique cyclic representation $(H_\omega, \pi_\omega, \Omega_\omega)$ of $U$, until the unitary equivalence, such that,

$$\omega(A) = \langle \Omega_\omega | \pi_\omega(A)\Omega_\omega \rangle, \ \forall \ A \in U \tag{5.62}$$

and consequently

$$\|\Omega_\omega\|^2 = \|\omega\| = 1 \tag{5.63}$$

A *state* $\omega$ over an algebra $C^*$ is defined as being *pure* if the only linear positive functionals, upper bordered by $\omega$, have the form $\lambda\omega$, $0 \le \lambda \le 1$.

The set of all the states is noted with $E_U$, and the set of the pure states with $P_U$.

As a corollary of this reciprocity, if $\omega$ is a state over $C^*$ algebra $U$, and $\tau$ is an *automorphism of $U$, letting the state $\omega$ invariant,

$$\omega(\tau(A)) = \omega(A), \ \forall \ A \in U \tag{5.64}$$

then there is a single unitary operator determined $U_\omega$, on the space of the *cyclic representation* $(H_\omega, \pi_\omega, \Omega_\omega)$ constructed on the state $\omega$, so that:

$$U_\omega \pi_\omega(A) U_\omega^{-1} = \pi_\omega(\tau(A)), \ \forall \ A \in U \tag{5.65}$$

and

$$U_\omega \Omega_\omega = \Omega_\omega \tag{5.66}$$

The cyclic representation $(H_\omega, \pi_\omega, \Omega_\omega)$ constructed on the state $\omega$ over $C^*$ algebra $U$, is called the canonical cyclic representation of $U$ *associated with* $\omega$.

The fact that the representation $(H_\omega, \pi_\omega)$ is irreducible and the associated state $\omega$ is a pure, are equivalent conditions.

Let $C^*$ be the abelian algebra $U$, and is defined by the character $\omega$ *for* $U$, the non-null linear application,

$$A \in U \to \omega(A) \in C \qquad (5.67)$$

which leads the elements from algebra $U$ in the space of complex numbers $C$, so that:

$$\omega(AB) = \omega(A)\omega(B), \ \forall \ A, B \in U \qquad (5.68)$$

*The spectrum $\sigma(U)$ of $U$*, is defined as the set of all characters on $U$.

The connection of the $\omega$ characters with the $\omega$ states defined above is very simple and consists in: the fact that $\omega$ is a pure state on $U$ is equivalent with the fact that $\omega$ is a character on $U$.

Therefore, the spectrum $\sigma(U)$ *on $U$ is a subset of the dual space $U^*$*, defined as the functionals space, $f$, linear and continuous over $U$, with the norm:

$$\|f\| = \sup\{|f(A)|; \|A\| = 1\}, \ \forall \ A \in U \qquad (5.69)$$

Also, the sets of states $E_U$, $P_U$, are subsets of the dual space $U^*$, which allows their topological organization by the restriction to any topology from $U^*$.

*The norm*, or the uniform topology, is determined by the specification of the neighborhoods of the state $\omega$ under the form:

$$U(\omega; \varepsilon > 0) = \{\omega'; \omega' \in U^*, \|\omega - \omega'\| < \varepsilon\} \qquad (5.70)$$

In the *weak topology, neighborhoods of the state $\omega$ are indexed by finite sets of elements under the form:

$$U(\omega; A_1, A_2, \ldots, A_n; \varepsilon > 0) = \{\omega'; \omega' \in U^* \|\omega'(A_i) - \omega(A_i)\| < \varepsilon, i = 1, 2, \ldots n\}$$
$$(5.71)$$

Further, are introduced the topologies specific to the algebra of operators $L(H)$:

(i)  If $\xi \in H$, then the application $A \to \|A\xi\|$ defines a seminorm on $L(H)$. The strong topology represents the local convex topology on $L(H)$ *defined with the aid of these seminorms.*

(ii)  *The strong topology $\sigma$ is* correlated to the strong topology and is obtained from it, considering all the sequences $\{\xi_n\}$ from H which satisfy the condition

$$\sum_n \|\xi_n\|^2 < \infty \qquad (5.72)$$

Therefore, one has:

$$\sum_n \|A\xi_n\|^2 \le \|A\|^2 \sum_n \|\xi_n\|^2 < \infty, \ \forall \ A \in L(H) \qquad (5.73)$$

which allows the definition of the seminorm

$$A \rightarrow \left[ \sum_n \|A\xi_n\|^2 \right]^{1/2} \qquad (5.74)$$

on $L(H)$, which in turn defines the strong topology $\sigma$.

(iii)  If $\xi, \eta \in H$, then

$$A \rightarrow \left| \langle \xi | A\eta \rangle \right| \qquad (5.75)$$

is a seminorm on $L(H)$ which defines the weak topology, where the introduced seminorm satisfies the polarization identity:

$$\langle \xi | A\eta \rangle = \frac{1}{4} \sum_{n=0}^{3} i^{-n} \left\langle \xi + i^n \eta \left| A \left( \xi + i^n \eta \right) \right. \right\rangle \qquad (5.76)$$

(iv)  Analogically to the previous case, it can be defined the weak topology $\sigma$, considering the sequences $\{\xi_n\}$, $\{\eta_n\}$ from $H$ which satisfy the condition

$$\sum_n \|\xi_n\|^2 < \infty, \ \sum_n \|\eta_n\|^2 < \infty \qquad (5.77)$$

Then the succession of next inequalities occurs:

$$\sum_n \left| \langle \xi_n | A\eta_n \rangle \right| \le \sum_n \|\xi_n\| \|A\| \|\eta_n\|$$
$$\le \|A\| \left[ \sum_n \|\xi_n\|^2 \right]^{1/2} \left[ \sum_n \|\eta_n\|^2 \right]^{1/2} < \infty, \ \forall \ A \in L(H) \qquad (5.78)$$

which allows the insertion of the weak topology $\sigma$, by the seminorm

$$A \rightarrow \sum_n \left| \langle \xi | A\eta \rangle \right| \tag{5.79}$$

which induce the local convex topology on $L(H)$.

(v)  *The \*strong topologies and $\sigma^*$ strong are introduced by considering the seminorms:*

$$A \rightarrow \|A\xi\| + \|A^*\xi\|$$

$$A \rightarrow \left[ \sum_n \|A\xi\|^2 + \sum_n \|A^*\xi\|^2 \right]^{1/2}, \forall\ A \in L(H) \tag{5.80}$$

where condition (5.72) holds.

The main distinction between the \*strong topology and the strong one arise from the fact that the application $A \rightarrow A^*$ is continuous in the first topology, but not also in the last one.

Moreover, between the topologies presented on $L(H)$, can be established the relationships:

| Uniform Topology | < | $\sigma$ \* strong Topology | < | $\sigma$ \* strong Topology | < | $\sigma$ weak Topology |
|---|---|---|---|---|---|---|
| | | ∧ | | ∧ | | ∧ |
| | | \* strong Topology | < | strong Topology | | weak Topology |

Let $H$ be a Hilbert space. For each subset $M$ from $L(H)$, it will be considered its switched, $M'$, the set of bounded operators from H which switches with each operator from M. Therefore, $M'$ is a Banach algebra of operators which includes the identical element 1.

If $M$ is an self-adjoint algebra, then $M'$ is a $C^*$ algebra of operators on $H$, which is closed for any local convex topology, previously defined.

Between the original set $M$ and its successive switches, there are the relations:

$$M \subseteq M'' = M^{(IV)} = M^{(VI)} = \dots \tag{5.81a}$$

$$M' = M''' = M^{(V)} = M^{(VII)} = \dots \tag{5.81b}$$

An algebra *von Neumann* on $H$, is defined a \*subalgebra $M$ from $L(H)$, which satisfies the equality:

$$M = M''$$ (5.82)

*The von Neumann algebra center* will be defined as follows:

$$Z(M) = M \cap M'$$ (5.83)

An von Neumann algebra is called *factor*, if it has a trivial center,

$$Z(M) = C1$$ (5.84)

Let $U$ be a non-degenerate *algebra of operators acting on Hilbert space $H$. Then, $U$ is dense in $U''$, in the way of all the topologies that are placed on $L(H)$ (the Von Neumann density theorem). Moreover, there is the relation:

$$\pi(U'') = \pi(U)''$$ (5.85)

The space with the weak $\sigma$ topology of the continuous linear functionals on $L(H)$ is called the $L(H)$ predual and is noted by $L_*(H)$.

Similarly, is obtained the *von Neumann* algebra predual, $M_*$.

A $C^*$ algebra $U$ is *isomorphic with an von Neumann algebra, if and only if $U$ is the dual Banach space (Sakai theorem).

Let $M$ be an von Neumann algebra, and $\omega$ a positive linear functional on $M$. If occurs the relation:

$$\omega \text{ (smallest bounded}_\alpha \text{operator } A_\alpha) = \text{smallest bounded }_\alpha \text{operator } \omega(A_\alpha)$$ (5.86)

for any sets $\{A_\alpha\}$ from $M_+$, with a superior border, then $\omega$ is defined as being the normal operator.

This definition allows the insertion of the following equivalent conditions:

(1) $\omega$ is a normal operator.
(2) $\omega$ is continuous weak topological $\sigma$.
(3) There is a density matrix $\rho$, with the property that the operator $\rho$ is from the positive trace class, satisfying $\text{Tr}(\rho) = 1$, so that it can be written:

$$\omega(A) = \mathrm{Tr}(\rho A), \ \forall \ A \in U \qquad (5.87)$$

If $U$ is an algebra $C^*$, the following conditions are also equivalent (*Kadison's theorem*): $U$ is *isomorphic with a von Neumann algebra.

Any set of increase or bounded operators from $U$ has the smallest bounded element, and for any element $A \in U$ positive and non-null, there is a normal state $\omega$ over $U$, so that $\omega(A) \neq 0$.

If $\pi$ is a representation of an $U$ algebra $C^*$, then o state $\omega$ from $U$ is called $\pi$-*normal* if there is a normal state $\rho$ from $\pi(U)''$ such as:

$$\omega(A) = \rho(\pi(A)), \ \forall \ \in U \qquad (5.88)$$

Using this fact, it can be introduced a new definition of equivalence in a weaker but more useful sense for physicochemical characterizations.

Two representations $\pi_1, \pi_2$, of an $U$ algebra $C^*$, are called as being quasi-equivalent and are written as $\pi_1 \approx \pi_2$, if each $\pi_1$-normal state is $\pi_2$-normal and vice versa.

A state $\omega$ of an $U$ algebra $C^*$, is called primal state or state factor, if $\pi_\omega(U)''$ is a factor, where $\pi_\omega$ is an associated cyclic representation.

Two states $\omega_1, \omega_2$, from $U$ are called quasi-equivalent, if their cyclic representations $\pi_{\omega 1}, \pi_{\omega 2}$ are quasi-equivalent.

If $(H, \pi, \Omega)$ is a cyclic representation associated with the state $\omega$, for which:

$$\omega(A) = \int A(x) dv(x), \ \forall \ A \in M, x \in X \qquad (5.89)$$

where $v(x)$ represents the probability measure, and $X$ a compact *Hausdorff space*, then it can be established the biunivocal correspondence:

(i)   $H$ is identifiable with $L^2(X, \mu)$;
(ii)  $\Omega$ is identifiable with the identical function 1;
(iii) $\pi(M)$ is identified with $L^\infty(X, \mu)$, acting as a multiplicative operator on $L^2(X, \mu)$.

One can construct the convex self-dual cone of operators with elements from $C^*$ algebra $M_+$, by the inequality:

$$\int dv \, fg \geq 0, \ \forall \ f, g \in L^2_+(X, \mu) \qquad (5.90)$$

Each element from this cone has the form $AA^*$, with $A \in M$, so that the vectors $\pi(AA^*)\Omega$ are being positive, and $L^2_+$ is being the weak closure of these vectors. Therefore, the self-duality occurs because:

$$\langle \pi(A^*A)W \mid \pi(B^*B)W \rangle = \omega(A^*AB^*B) = \omega((AB)^*AB) \geq 0 \quad (5.91)$$

where in the last stage it had been used the commutativity on $M$.

It can be constructed the convex cone of vectors $AA^*\Omega$ in $H$, but if $M$ is not abelian, this cone not necessarily has the property of self-duality. Under these conditions, the associated state

$$\omega(A) = \langle W \mid AW \rangle \quad (5.92)$$

satisfies

$$\omega(A^*AB^*B) \geq 0 \ \text{ if } \ \omega(AB) = \omega(BA), \ \forall \ A, B \in M \quad (5.93)$$

in other words, $\omega$ is a trace, fact which allows the comparison of this case with the abelian situation above.

In defined the conjugate operator $J$ on $H$, by the relation:

$$JA\Omega = A^*\Omega \quad (5.94)$$

The trace property of the state $\omega$, allows the writing:

$$\|AW\|^2 = \omega(A^*A) = \omega(AA^*) = \|A^*W\|^2 \quad (5.95)$$

through which $J$ can be enlarged to the *antiunitary operator*:

$$JAJB\Omega = JAB^*\Omega = BA^*\Omega \quad (5.96)$$

If $M$ is abelian, the calculation (4.57) shows that $J$ actually implements the *conjugation,

$$A^* = j(A), \ A^* = j(A) = JAJ \quad (5.97)$$

If the trace is performed, the action of $j$ is more complex, for example:

$$\langle B_1W \mid A_1 j(A_2) B_2W \rangle = \omega(B_1^*A_1B_2A_2^*) = \langle B_1W \mid j(A_2) A_1 B_2W \rangle \quad (5.98)$$

which demonstrates that $j(A) \in M'$, a property overlapped by the abelian case.

It has been proved that a general self-dual cone can be constructed by modifying the *conjugation on the $AA^*\Omega$ set, where $A^*$ can be replaced by the conjugate element $j(A)$ and conjugation $j$ is expected to produce an application from $M$ to $M'$.

A *von Neumann algebra* $M$, is called *$\sigma$-finite*, if a collection of orthogonal mutual projections has a cardinal at least countable.

Any von Neumann algebra on a separable Hilbert space is $\sigma$-finite.

The reciprocal is not generally valid: not all the Neumann algebras $\sigma$-finite, can be represented on a separable Hilbert space.

Let be a von Neumann algebra $M$ on a Hilbert space $H$. A subset $R \subseteq H$ is separator for $M$ if:

$$\forall\ A \in M,\ A\xi = 0,\ \forall \xi \in R \Rightarrow A = 0 \tag{5.99}$$

If $M$ is a subset from $L(H)$, and $R \subseteq H$, then is noted by $[MR]$ the linear closing of the overlapping of the elements as $A\xi$, $A \in M$, $\xi \in R$. Also $[MR]$ denotes the orthogonal projection in $[MR]$-space.

If $[MR] = H$, then the subset $R \subseteq H$ is cyclic over $M$. This is a dual relationship between the properties of cyclicity for an algebra and those of separation for the commutant. Thus, the following conditions are equivalent:

(i)   $R \subseteq H$ is cyclic over $M$;
(ii)  $R$ is separator for $M'$;

Let be an von Neumann algebra $M$ on a Hilbert space $H$. A close operator $A$ on $H$ is called affiliated with $M$, and is written as $A\eta M$, if:

$$M'D(A) \subseteq D(A) \wedge AA' \supseteq A'A,\ \forall A' \in M' \tag{5.100}$$

where $D(A)$ represents the field of linear operators $A$.

If $\Omega$ is cyclic and separator for $M$, then is also cyclic and separator for $M'$. Therefore, the anti-linear operators $S_0$ and $F_0$ defined by:

$$S_0 A\Omega = A^*\Omega,\ \forall A \in M \tag{5.101}$$

$$F_0 A'\Omega = A'^*, \ \forall \ A' \in M' \tag{5.102}$$

there are both very well defined in dense fields, respectively:

$$D(S_0) = M\Omega, \ D(F_0) = M'\Omega \tag{5.103}$$

There are defined the closing operators $S_0$, $F_0$, by

$$S = \overline{S}_0, \ F = \overline{F}_0 \tag{5.104}$$

Let $\Delta$ be the single positive self-adjoint operator and $J$ the unique antiunitary operator in the polar decomposition of $S$:

$$S = J\Delta^{1/2} \tag{5.105}$$

Then, $\Delta$ is called the modular operator associated with the pair $\{M,\Omega\}$, and $J$ is called modular conjugation.

Let $M$ be an algebra von Neumann with the cyclic and separator vector $\Omega$. Also, let $\Delta$ be the modular operator associated, and $J$ the modular conjugation. Then, the (*Tomita-Takesaki Theory*) relationships occur

$$JMJ = M' \tag{5.106}$$

$$\Delta^{-it} M \Delta^{-it} = M, \ t \in R \tag{5.107}$$

Let $M$ be an algebra von Neumann, $\omega$ a proper state, normal on $M$, $(H_\omega, \pi_\omega, \Omega_\omega)$, the corresponding cyclic representation and $\Delta$ the modular operator associated with the pair $(\pi_\omega(M), \Omega_\omega)$. The *Tomita-Takesaki Theory* actually establishes the existence of a uni-parameter group $\sigma$- weak continuous $t \to \sigma_t^\omega$ of the *automorphisms of $M$, through the definition:

$$\sigma_t^\omega(A) = \pi_\omega^{-1}\left(\Delta^{it}\pi_\omega(A)\Delta^{-it}\right) \tag{5.108}$$

The $t \to \sigma_t^\omega$ group *is called the molecular group of the automorphisms associated with the pair* $(M,\omega)$.

The *quasi-local algebras* are generated by the increasing subalgebras set, $\{U_\alpha\}_{\alpha \in I}$, which satisfies a number of structural relations which will be discussed further.

The set of indices $I$ is said that possesses the orthogonality relationship, if there is the relation $\perp$ between the elements pairs of $I$, such that:

(i) $$\alpha \in I \Rightarrow \exists \beta \in I\big|\ \alpha \perp \beta \qquad\qquad (5.109a)$$

(ii) $$\alpha \leq \beta\ \&\ \beta \perp \gamma \Rightarrow \alpha \perp \gamma \qquad\qquad (5.109b)$$

(iii) $$\alpha \perp \beta\ \&\ \alpha \perp \gamma \Rightarrow \exists \delta \in I\big|\ \alpha \perp \delta\ \&\ \delta \geq \beta, \gamma \qquad (5.109c)$$

If $I$ is a subset opened on the space of the configurations $R^{\upsilon}$, then the condition $\alpha \perp \beta$, is deduced that $\alpha, \beta$ correspond to some separated algebras.

Assuming that each pair $\alpha, \beta$ from the set of indices $I$ admits the smallest bounded index $\alpha \vee \beta \in I$, can also be introduced the relations:

(iv) $$\alpha \vee \beta \geq \alpha\ \&\ \alpha \vee \beta \geq \beta \qquad\qquad (5.109d)$$

(v) $$\gamma \geq \alpha\ \&\ \gamma \geq \beta \Rightarrow \gamma \geq \alpha \vee \beta \qquad\qquad (5.109e)$$

A quasi-local *algebra is a* $C^{*}$ algebra U and a set $\left\{U_{\alpha}\right\}_{\alpha \in I}$ of $C^{*}$ subalgebras, so that the $I$ indices set has the relation of orthogonality and are satisfied the properties:

1) $\alpha \geq \beta \Rightarrow U_{\alpha} \supseteq U_{\beta}$;
2) $U = \overline{\cup_{\alpha} U_{\alpha}}$ where the bar denotes the uniform closing;
3) the algebras $U_{\alpha}$ has an identity element 1 in common;
4) there is an automorphisms $\sigma$ so that $\sigma^2 \to \iota$ (the identical automorphisms),

$$\sigma\left(U_{\alpha}\right) = U_{\alpha} \qquad\qquad (5.110a)$$

$$\left[U_{\alpha}^{p}, U_{\beta}^{p}\right] = \{0\} \qquad\qquad (5.110b)$$

$$\left[U_{\alpha}^{p}, U_{\beta}^{i}\right] = \{0\} \qquad\qquad (5.110c)$$

$$\left[U_{\alpha}^{i}, U_{\beta}^{i}\right] = \{0\} \qquad\qquad (5.110d)$$

as long as $\alpha \perp \beta$, where $U_\alpha^i \subseteq U_\alpha$, $U_\alpha^p \subseteq U_\alpha$ are the odd elements and respectively even in relation with $\sigma$; in this case,

$$\{A,B\} = AB + BA \tag{5.111}$$

represents the notation for the anticommutator).

One should note that if $\sigma$ is an automorphisms of a $C^*$ algebra $U$ which satisfies

$$\sigma^2 = \iota, \text{ i.e., } \sigma(\sigma(A)) = A, \forall A \in U \tag{5.112}$$

so that each element $A \in U$ has an unique decomposition in its odd and even parts in relation with $\sigma$, defined as:

$$A = A^+ + A^- \quad , \quad A^\pm = \frac{A \pm \sigma(A)}{2} \tag{5.113}$$

The even elements of $U$ form a $C^*$ subalgebra $U^p$ of $U$ and the odd elements, $U^i$, of $U$ form a Banach space.

The algebra $U_\alpha$ is interpreted as the physical observables algebra for a subsystem located in the $\alpha$ region of the $R^\upsilon$ configurations space. The quasi algebra $U$ corresponds to the observables algebra enlarged to an infinite system.

A state $\omega$ over $U$ represents a physical state of a system in evolution, and the values $\omega(A)$, $\omega(B)$, ... represent the observables values $A$, $B$,... .

Therefore, the representation $(H_\omega, \pi_\omega, \Omega_\omega)$ allows a more detail description of the individual state $\omega$ and the von Neumann algebra $\pi_\omega(U)''$ is interpreted as the observables algebra for this state.

If $\omega$ is a state over an $U$ quasi local algebra, it can be defined as commutative algebra $z_\omega^c$ for the associated representation $(H_\omega, \pi_\omega, \Omega_\omega)$ through:

$$z_\omega^c = \bigcap_{\alpha \in I} \left(\pi_\omega(U_\alpha)' \cap \pi_\omega(U)\right)'' \tag{5.114}$$

and also the algebra on infinite, by:

$$z_\omega^\perp = \bigcap_{\alpha \in I} \left(\bigcup_{\beta \perp \alpha} \pi_\omega(U_\beta)\right)'' \tag{5.115}$$

Let $U = \{M_\alpha\}_{\alpha \in I}$, be a quasi-local algebra, whose generator set is formed by the Neumann $M_\alpha$ algebras. A state $\omega$ over $U$ is defined as being local normal if $\omega$ is normal in the restriction to each algebra $M_\alpha$.

In the physicochemical systems symmetries analysis in algebraic terms, the concept of dynamic system is essential.

A dynamical system $C^*$ is a triplet $\{U, G, \alpha\}$, where $U$ is an algebra $C^*$, $G$ is a local compact group, and $\alpha$ is a strongly continuous representation of $G$ in the group of the automorphisms of $U$, so that for each element $g \in G$, $\alpha_g$ is an automorphisms of $U$, with satisfying the following relations:

$$\alpha_e = \iota, \; \alpha_{g1}\alpha_{g2} = \alpha_{g1g2} \tag{5.116}$$

The application $g \to \alpha_g(A)$ is continuous in norm for each $A \in U$; ($e$ represents the identity element in $G$, $\iota$ is the application identical on $U$).

A dynamic system $W^*$ is a triplet $\{M, G, \alpha\}$ where $M$ is a von Neumann algebra, $G$ is a local compact group, and $\alpha$ is a weak continuous representation of $G$ in the automorphisms group of $M$.

*A covariant representation of a dynamical system* is a triplet $(H, \pi, U)$ where $H$ is a Hilbert space, $\pi$ is a non-degenerate representation of the algebra on $H$, (which is normal for the $W^*$ case), and $U$ is a strongly continuous unitary representation of the group $G$ on $H$, so that the relation is valid:

$$\pi(\alpha_g(A)) = U_g \pi(A) U_g^*, \; \forall \in U(M), g \in G \tag{5.117}$$

For each dynamic system $C^*$ (respectively $W^*$) is defined a new algebra $C^*$ respectively a von Neumann algebra) called the direct product between $U$ and $G$ and noted by $C^*(U, \alpha) \equiv U \otimes_\alpha G$ (respectively $W^*(M, \alpha) \equiv M \otimes_\alpha G$).

Let $\{M, G, \alpha\}$ be a dynamic system $W^*$ and the $U$ algebra action over the Hilbert space $H$. The new Hilbert space is defined $L^2(H, G, dg)$ *as a completion of the set* $R(H, G)$, where $R(H, G)$ represents the set of the continuous functions from $G$ in $H$, with compact support *endowed with the internal product:*

$$\langle \xi | \eta \rangle = \int_G \langle \xi(g) | \eta(g) \rangle dg \tag{5.118}$$

For the system $C^*$ or $W^*$ dynamic $\{U,G,\alpha\}$ where $U$ is abelian, the action $\alpha$ is *defined as ergodic, if* $U$ do not contain any nontrivial double ideal nontrivial $\alpha$-global invariant.

The action will be called free if:

$$\forall g \neq e \ \& \ A>0 \ \exists B \mid A \geq B > 0 \ \& \ \alpha_g(B) \neq B \qquad (5.119)$$

Should be noted that the ergodicity for the $C^*$ case means that all the orbitals of the action induced by the application $\alpha$ in the spectrum $\sigma(U)$ are dense in $\sigma(U)$. On the other side, the ergodicity for the $W^*$ case means that $U$ does not contain non-trivial projectors $\alpha$-invariant.

Suppose that $C^*$ is a dynamic system $\{U,G,\alpha\}$ where $U$ is abelian and separable, $G$ is a discreet group, countable and allows invariant states and the action $\alpha$ is ergodic and free. Then the direct product $C^*(U,\alpha)$ *is simple.*

Let $W^*$ be a dynamic system $\{M,G,\alpha\}$ where $M$ is abelian and $\sigma$-finite, $G$ is a countable group acting freely and *ergodic* on $M$. Then the direct product $W^*(M,\alpha)$ *is factor.*

Is defined the weight on $C^*$ *algebra* $U$ as being positive linear functional $\omega:U_+ \rightarrow [0,\infty]$ (with the convention $0 \cdot \infty = 0$) which satisfies the relations:

$$\omega(A+B) = \omega + \omega(B), \ \forall \ A, B \in U_+ \qquad (5.120a)$$

$$\omega(\alpha A) = \alpha\omega(A), \ \forall \ \alpha \in R_+, \ A \in U_+ \qquad (5.120b)$$

*The trace on* $U$ is defined as being the weight $\omega$ which satisfies the relation:

$$\omega(A^*A) = \omega(AA^*), \ \forall \ A \in U \qquad (5.121)$$

Let $\omega$ be a weight on the algebra von Neumann $M$. The following conditions are equivalent:

(i)   if $\{A_i\}$ is a sequence from $M_+$ and

$$\sum_i A_i = A \in M_+, \text{ then } \omega(A) = \sum_i \omega(A_i) \qquad (5.122)$$

(ii)  there is a set $\{\omega_\alpha\}$ of normal functionals, positive from $M$ so that

$$\omega(A) = \sup_\alpha \omega_\alpha(A), \; \forall \; A \in M_+ \qquad (5.123)$$

(iii)  there is a set $\{\omega_\alpha\}$ of normal functionals, positive from $M$ so that

$$\omega(A) = \sum_\alpha \omega_\alpha(A), \; A \in M_+ \qquad (5.124)$$

Suppose that $\omega$ is a weight on a von Neumann algebra $M$. It is called normal weight if satisfies any conditions from above. It is called proper weight if $A \in M$ & $\omega(A) = 0$ implies $A = 0$; it is called semifinite weight if $M_\omega$ is dense $\sigma$-weak in $M$.

Each algebra von Neumann allows a normal, faithful and semifinite weight.

Two projections $E$, $F$ in an algebra von Neumann $M$ are considered equivalent, writing $E \sim F$, if:

$$\exists W \in M \mid E = W^* W \; \& \; F = WW^* \qquad (5.125)$$

A projection $E$ in $M$ is considered finite, if it is not equivalent with an own sub-projection of itself. Therefore, it is called as infinite projection.

*The algebra M is considered semifinite* if any projection in $M$ contains a non-null finite projection; in other words, there is a set with increasing elements $\{E_\alpha\}$ of finite projections in $M$ so that $E_\alpha \to 1$.

*The algebra M is considered finite* if its identical element, 1, is finite. Thus, *M is infinite.*

The algebra $M$ is called proper infinite if all the non-null projections in the center $M \cap M'$ are infinite. The algebra $M$ is called *pure-infinite* if all the non-null projections in $M$ are infinite.

For any pairs of weights $\varphi$, $\psi$ proper, normal and semifinite from $M$, there is a continuous family of uni-parameter unitary applications in $M$, $(D\psi : D\varphi)_t$, called Radon-Nikodim cocycles, which have the properties *(Connes Radon-Nikodym Theorem)*:

(i)  $$\sigma_t^\psi = (D\psi : D\varphi)_t \, \sigma_t^\varphi(A) (D\psi : D\varphi)_t^* \qquad (5.126a)$$

(ii)  $$(D\psi : D\varphi)_{t+s} = (D\psi : D\varphi)_t \, \sigma_t^\varphi \big((D\psi : D\varphi)_s\big) \qquad (5.126b)$$

(iii) $\left(D\psi : D\varphi\right)_t^* = \left(D\varphi : D\psi\right)_t; \left(D\varphi : D\psi\right)_t \left(D\varphi : D\omega\right)_t = \left(D\psi : D\omega\right)_t$

$$(5.126c)$$

(iv) $\psi\left(A\right) = \varphi\left(UAU^*\right)$ with $U$ unitary in $M\left(D\psi : D\varphi\right)_t = U^*\sigma_t^\varphi\left(U\right)$

$$(5.126d)$$

A M factor is considered of type *I*, if it has a minimal projection, non-null. It is of type *II* if is semifinite, but not of type *I*.

A factor of type *II* is of type $II_1$ if is finite and of type $II_\infty$ if is infinite.

A factor $M$ is considered of type *III*, if is not semifinite, i.e., if is pure infinite.

If $M$ is a factor, is defined:

$$S\left(M\right) = \cap_\omega \sigma\left(\Delta_\omega\right) \qquad (5.127)$$

where the index $\omega$ overlaps all the semifinite normal weights from $M$, and $\sigma\left(\Delta_\omega\right)$ is the spectrum of the operator $\Delta_\omega$.

This definition allows a further classification of the von Neumann factors of type *III*:

(i)  if $S\left(M\right) = [0,\infty)$ then $M$ is a factor of type $III_1$;
(ii)  if $S\left(M\right) = \{0\} + \lambda^Z$, $0 < \lambda < 1$, then $M$ is a factor of type $III_\lambda$;
(iii)  if $S\left(M\right) = \{0,1\}$ then $M$ is a factor of type $III_0$.

Certainly, these notions, properties and algebra classifications can be much more extended. However, for the intended purpose, namely to give an algebraic characterization deeply grounded in mathematics, so with a strong universal character of the physicochemical reactivity, the algebraic structures and properties presented are legitimate for this approach.

### 5.2.3 CHEMICAL POTENTIAL AND REACTIVE EQUILIBRIUM

In terms of algebraic quantum mechanics, the complete set of observables is divided into two classes: microscopic observable and macroscopic observables. The microscopic observables are currently identified with the elements of a $C^*$ algebra $U$ quasi-local. The macroscopic observables do not necessary belong to the algebra U, but to an adequate Neumann subalgebra $\pi(U)$ of an universal representation $U^{**}$ of $U$. The central elements of the subalgebra $\pi(U)$ are called classical observables, and they have the properties physical systems in the classical theories; the present discussion follows (Bratteli & Robinson, 1987a,b; Primas & Müller-Herold, 1978; Müller-Herold, 1980, 1982):

- in a macroscopic pure state they have defined values without dispersion;
- the values specification of all the central observables determine a classical state;
- their temporal evolution leads to a classical dynamic system;
- they represent the sizes directly observable (experimental) of a system.

*The temperature and the chemical potential* are direct macroscopic observables par excellence. Since the quantum mechanics say rules that the observable size may be represented by the self-adjoint operators formed on the Hilbert space, is adding the interest to associate the operators for the temperature and the chemical potential for an adequate macroscopic characterization.

To associate a self-adjoint operator to the temperature in $U^{**}$, the KMS states separation theorem for the factors of type III for various temperatures (Takesaki Theorem) makes this construction as legitimate. The KMS states were described in the beginning of this Chapter.

For the chemical potential the studies of (Araki et al., 1977) showed that it can be formed an operator associated to the chemical potential in the middle of the representation $U^{**}$, respectively in a subalgebra suitable chosen.

For a system formed from a single chemical species (the electronic fluid evolves under the same potential generated by the system nuclei for a fixed coupling constant) the construction of the operator with chemical potential includes the following ingredients:

(i)	*a C* algebra U of the quasi-local observables*, together with a set of automorphisms located $\rho$ of $U$ which correspond to the creations and annihilation of particles located in various regions from space. They form an isomorphic commutative group, additive to the integers for which the dual group is the *Lie uni-parameter compact $\gamma_s$ (a gauge group of the first rank);*

(ii)   an algebraic field $F$ was formed as a covariant algebra where the located automorphisms are unitary implemented:

$$\rho(a) = uau^*, \; a \in U, \; u \in F \tag{5.128}$$

which contains $U$ as its invariant gauge part, so that $\gamma_s(a) = a$, $\forall \, a \in U$

A local *automorphism, located $\rho$ is called of class N,* if occurs:

$$\gamma_s(u) = u \cdot \exp[iNs], \; s \in R \tag{5.129}$$

which corresponds the creations of $N$ particles;

(iii)  *a uni-parameter group $t \to \alpha_t$,* strongly continuous, of the $F$ automorphisms, for which the restriction to $U$, represents the physical temporal automorphism – and which commutes with the gauge group:

$$\alpha_t \cdot \gamma_s = \gamma_s \cdot \alpha_t, \; t, s \in R \tag{5.130}$$

(iv)   a family of unitary cocycles $v_1 \in U$

$$v_{t+s} = v_t \cdot \alpha_t(v_s), \; t, s \in R \tag{5.131}$$

which reflects the temporal evolution perturbations due to the addition of a finite number of particles:

$$Aav_t \circ a_t = \rho^{-1} \circ a_t \circ \rho, \; t \in R \tag{5.132}$$

where by definition was considered the notation:

$$Adv_1(x) := v_t x v_t^*, \; x \in U \tag{5.133}$$

(v)    a state factor $\omega\text{-}(\alpha_t, \beta)$ KMS on $U$, $0 < \beta < \infty$ which is considered as a state of *thermodynamic equilibrium* with the parameter of the inverse temperature, $\beta$, and which is quasi-equivalent to $\omega \circ \rho$.

Since $\omega$ and $\omega \circ \rho$ can be extended to become normal states matched on the same von Neumann algebra $\pi(U)''$, the Radon–Nikodim–Connes

Theorem (Connes, 1973), allows the construction of an unique cocycle in $\pi(U)''$, defined as:

$$\{D(\omega \circ \rho) : D\omega\}_t \,, \ t \in R \tag{5.134}$$

called the derivative of the cocycle Radon-Nikodim $\omega \circ \rho$ in relation with $\omega$.

For the adequate constructions of the ingredients $\omega$, $\alpha_t$, $U$, there is a unique real parameter $\mu$, and a single extension $\Omega \, \gamma_s$-invariant of $\omega$ from $U$ in $F$ in order to be a state $\beta - KMS$ in relation to the uni-parameter group $t \rightarrow \alpha_t \gamma_\mu$. Adopting the terminology established in quantum statistics, the parameter $\mu$ is identified with the chemical potential.

In other words, the chemical potential may be equivalently characterized by a unitary uni-parameter group which appears as a rest of the cocycles from above:

$$\exp\left[ in\hat{\mu}\beta t \right] = \{D(\omega \circ \rho) : D\omega\}_t \left\{ \pi_\omega\left( v_{-\beta t} \right) \right\}^{-1} \tag{5.135}$$

where by definition was considered the notation:

$$\hat{\mu} = \mu + c \tag{5.136}$$

where $c$ represents the independent real constant of $\omega$ which reflect the choosing of the referential zero in the chemical potential, and $\rho$ is an automorphism located by class $n$.

Once introduced the chemical potential in algebraic way, it becomes interesting its characterization, taking into account the fact that the chemical potential with changed sign represents the Parr electronegativity.

Being two factor states $\omega$, $\varphi$ - $(\alpha_t, \beta)$ KMS to which are associated as above, two chemical potentials, $\mu$, $v$ which satisfy the definition relations:

$$\exp\left[ in\hat{\mu}\beta t \right] = \{D(\omega \circ \rho) : D\omega\}_t \left\{ \pi_\omega\left( v_{-\beta t} \right) \right\}^{-1} \tag{5.137a}$$

$$\exp\left[ in\hat{v}\beta t \right] = \{D(\varphi \circ \rho) : D\omega\}_t \left\{ \pi_\varphi\left( v_{-\beta t} \right) \right\}^{-1} \tag{5.137b}$$

If $\mu \neq v$ then the representations $\pi_\omega$ and $\pi_\varphi$ are separated (disjoint).

Reciprocally, if $\omega$, $\varphi$ are states $\beta$-KMS factorial and normalized relative to the same uni-parameter group $\pi \circ \alpha_t \circ \pi^{-1}$ (thus considering the same representation for both states), it follows that $\omega = \varphi$ consecutively from the last relationships, $\hat{\mu} = \hat{v}$.

Therefore, different chemical potential correspond to different representation and equilibrium states.

This conclusion is also maintained for the uni-factorial states which have different chemical potential. In this case, the two states KMS involved, if are considered continuous decompositions but countable of the involved states, each of these sub-states is disjoint with any of the other decomposition sub-states, preserving the validity of the result above.

This characteristic of the chemical potential, i.e., Parr electronegativity with changed sign (Parr et al., 1978; Par & Yang, 1989), is kept also for n different species, in which case it will be used a gauge group under the form of a n-dimensional torus, and $n$ various subgroups of the located automorphisms, responsible for the creations and the annihilations of each species, as will be further considered.

Further it will be showed under what conditions the chemical potentials obtained as now satisfy the reactive equilibrium conditions:

$$\sum_j q_{ij} \mu_j = 0, \ i = 1, 2, \ldots \qquad (5.138)$$

where $\mu_j$ represents the chemical potential of the $j$ chemical species involved, and $q_{ij}$ represent the *stoichiometric coefficients* associated to the species $j$ in reaction $i$. Therefore, the reactive equilibrium equation is an of energy conservation low.

### 5.2.4 NON-REACTIVE SYSTEMS

For the statistical mechanics associated to a non-reactive finite system, the great canonical Gibbs state $\rho$ with the temperature $1/k_B\beta$ and with the chemical potential $\mu_j$ associated to the chemical species $j$, will be given by:

$$\rho\left(\hat{A}\right) = \text{Tr}\left\{\hat{D}\right\} \qquad (5.139)$$

$$\widehat{D} = \frac{\exp[-\beta \widehat{K}]}{\mathrm{Tr}\{\exp[-\beta \widehat{K}]\}} \tag{5.140}$$

$$\widehat{K} = \widehat{H} - \sum_j \widehat{\mu}_j \widehat{N}_j \tag{5.141}$$

where $\widehat{A}$ is a random observable, $\widehat{H}$ is the Hamiltonian, and $\widehat{\mu}_j$ and $\widehat{N}_j$ are the particle number operator and respectively the chemical potential operator for the species $j$; the present discussion follows (Bratteli & Robinson, 1987a,b).

Is has to be noted that the state $\rho$ does not satisfy the condition KMS in relation with the temporal evolution generated by the Hamiltonian $\widehat{H}$, condition satisfied by the uni-parameter group generated by $\widehat{K}$:

$$Ad \exp\left[i\widehat{K}t\right] = \tau_t \gamma^{(1)}_{\mu_1 t} \gamma^{(2)}_{\mu_2 t} \dots \gamma^{(n)}_{\mu_n t} \tag{5.142}$$

where,

$$t \to \tau_t = Ad \exp\left[i\widehat{K}t\right] \tag{5.143}$$

represents the physical temporal evolution, and,

$$\phi \to \gamma^{(j)}_\phi = Ad \exp\left[-i\widehat{N}_j\phi\right], \; 0 \le \phi < 2\pi \tag{5.144}$$

represents the gauge transformation of the first rank in relation to the scalar species $j$, from the total n considered, with the gauge group $G$ associated by the torus dimension $T^n$:

$$G = T^n = T^1 \times T^1 \times \dots \times T^1 \; (n \; times) \tag{5.145}$$

If we consider the algebraic field $F$ (a quasi algebra $C^*$ generated by the creation and annihilation operators, $\widehat{a}_j, \widehat{a}^*_j, j = 1, \dots, n$, associated to the $n$ species, and each element $g \in G$ is parameterized by n angles $0 \le \varphi_j < 2\pi$, then the gauge automorphisms action will be given by:

$$\gamma_g \widehat{a}_j = \exp\left[i\varphi_j\right]\widehat{a}_j \tag{5.146a}$$

$$\gamma_g \hat{a}_j = \exp\left[-i\varphi_j\right]\hat{a}_j \qquad (5.146b)$$

where $g \to \gamma_g$, $g \in G$ is a continuous representation, appropriate to $G$ in the automorphisms group of $F$.

The observables algebra $A$, generated by the even polynomials of the creation and annihilation operators will represent the gauge- invariant part of the algebra $F$:

$$A = F^\gamma \subset F \qquad (5.147)$$

Since we have not considered the chemical reactions, in other words, because the gauge and temporal automorphisms switch, $\gamma \circ \tau = \tau \circ \gamma$, results that $A$ is global $\tau$-invariant. Since $A$ is gauge invariant, it follows that the state $\rho|A$ restriction is a condition KMS in relation with the temporal physical evolution $t \to \tau_t$ on $A$, while the state $\rho$ as a state of $F$, is simply a state $\tau_t$-invariant.

Since the algebraic approach is constructed for the infinite systems, there is no Hamiltonian and respectively no canonical great density operator from where it can be possible the directly extraction of the chemical potentials. But they can be extracted from the formalism which include the states KMS conditions.

Supposing a factor state $\omega$ - $(\tau_t, \beta)$ KMS on $A$, $0 < \beta < \infty$ as was considered in the condition ($v$) above, adapted to the new conditions (the generalization to $n$ chemical species). In even general circumstances, the $\omega$ state can be extended to a factor state $\Omega$ gauge invariant state on $F$, so that $\Omega\big|_A = \omega$ will satisfy the KMS condition on $F$, in relation to the mixed gauge group uni-parameter temporal:

$$t \to \tau_t \gamma^{(1)}_{\mu_1 t} \gamma^{(2)}_{\mu_2 t} \cdots \gamma^{(n)}_{\mu_n t} \qquad (5.148)$$

which explicitly contains the various chemical potentials.

Therefore, a system can be algebraically described, consists of a set of non-reactive chemical species through a system of covariant algebraic field $W^*(M, N, \tau, \gamma, \omega)$ explained as follows:

(1) an *algebraic field M*, representing a factor of type $III_1$ with an ergodic action matched $r \to \alpha_r$ on the group from the space of the translations $R^3$ $\sigma$-weak continuous;

(2) a proper action $\sigma$-*weak continuous* $g \rightarrow \gamma_g$, $g \in G$, of the abelian gouge group which switches with the translations space, $\alpha \circ \gamma = \gamma \circ \alpha$;

(3) a *factor of type III*, *of the observables algebra* $N = M^\gamma \subset M$, which represents the invariant part of M;

(4) a temporal physical evolution $t \rightarrow \tau_t$, i.e., the uni-parameter group $\sigma$-weak continuous of the automorphisms of M which switches with the automorphisms of the spatial translations, $\tau \circ \alpha = \alpha \circ \tau$, and with the gauge automorphisms $\tau \circ \gamma = \gamma \circ \tau$;

(5) a state of thermodynamic equilibrium $\omega$, i.e., a state $\omega$ on $N$ $\tau_1 - 1 - KMS\alpha_r$ invariant.

According to this theory, the chemical potential appears now as follow: $\omega$ is enlarged to a state $\Omega$, $\tau_t$, $\alpha_r$ – invariant on M, by:

$$\Omega = \omega \circ \eta \qquad (5.149)$$

where $\eta$ is an unique projection, normal and gouge invariant, of $M$ on $N = M^\gamma$.

Since $\Omega$ is $\tau_t$– and $\gamma_g$– invariant, the modular group automorphic associated, $\sigma_t^\Omega$, switches with the temporal and gauge automorphisms, and taking into consideration the conditions (1)-(5) previously presented, it will be written:

$$\sigma_t^\Omega = \tau_t \gamma_{\mu_1 t}^{(1)} \gamma_{\mu_2 t}^{(2)} ... \gamma_{\mu_n t}^{(n)} \qquad (5.150)$$

regaining the transformation condition on the mixed uni-parameter temporal group above, and once with it the chemical potentials associated to the chemical species at non-reactive equilibrium.

### 5.2.5  REACTIVE SYSTEMS

In treating the physicochemical reactive systems the characterization (4) and (5) in the final of the preceding Section, will be modified as following (Primas & Müller-Herold, 1978; Müller-Herold, 1980, 1982; Bratteli & Robinson, 1987a,b):

(4') the temporal evolution $t \rightarrow \tau_t$ switches with the translations space and with the gauge automorphisms of a proper subgroup $G_r$ *close and conex of* $G$;

(5') the thermodynamic equilibrium is given by a state $\omega$ which is a state $\tau_1 - 1 - KMS$ on $M^{\gamma r}$, *the algebra of the actions fixed by the elements of the subgroup $G_r$.*

Generally speaking, you cannot determine if $G_r$ is a conex subgroup of $G$ but if we will consider the previous gauge torus decomposition in the direct sum:

$$T^n = T^{n-m} \oplus T^m \tag{5.151}$$

then, what it will be actually considered will be the part of $G_r$ which is reduced to the subtorus $T^{n-m}$ with $m < n$.

Unlike the non-reactive situation, the state $\omega$ which is a state $\tau_1 - 1 - KMS$ on $N = M^{\gamma r}$ is not stable in time anymore if $\tau$ and $\gamma$ do not switch. Under these conditions $M^{\gamma r}$ is the smallest subalgebra from $M$ containing the algebras of the observables that evolve in time,

$$M^{\gamma r} = \bigvee_{t \in R} \tau_t (N) \tag{5.152}$$

For the further construction, it is important the introduction of the dual group of $G = T^n$, i.e., $G^* = T^{n*} = Z^n$. This is a free abelian group, isomorphic with the group of translations in the dimensional lattice $n$, associated to the number of the chemical species involved in reactivity.

The abelian group $Z^n$ is generated by n generators $(u_1, \ldots u_n)$, so that each element $g^* \in Z^*$ will have the representation:

$$g^* = \sum_{j=1}^{n} m_j \left(g^*\right) u_j, \; m_j \in Z \tag{5.153}$$

and the dual group itself will be written as the direct sum:

$$G^* = \bigoplus_{j=1}^{n} gp\{u_j\} \tag{5.154}$$

where $gp\{u_j\}$ corresponds to the free abelian group generated by $u_j$.

Using the duality relations for the commutative groups, results that the decomposition (5.151), $G = T^n = T^{n-m} \oplus T^m$, induces a dual decomposition associated to the dual group,

$$G = Z^n = Z^{n-m} \oplus Z^m = G_r^* \oplus G_r^\perp \tag{5.155}$$

where $G_r^*$ is the dual group of $G_r$, and $G_r^\perp \subset G^*$ is the annihilation group for $G_r$.

Taking into consideration the relations

$$\left(X^\perp\right)^\perp = X, \left(X^*\right)^* = X \tag{5.156}$$

for an arbitrary abelian group $X$, based on the construction presented, the following diagram can be formed:

$$
\begin{array}{ccccc}
T^m = G_r^{*\perp} & \cdots & \leftrightarrow & \cdots & G_r^\perp = Z^m \\
\cup & & & & \cup \\
& & \ddots \qquad \cdot\cdot & & \\
T^n = G & & \times & & G^* = Z^n \\
& & \cdot\cdot \qquad \ddots & & \\
\cap & & & & \cap \\
T^{n-m} = G_r & \cdots & \leftrightarrow & \cdots & G_r^* = Z^{n-m}
\end{array}
\tag{5.157}
$$

where the horizontal arrows are referring to the duality relationship, and the diagonal references indicate the annihilation relations.

Supposing a parameterization of the group $G$ through $n$ angles $\varphi_j$ so that $\gamma_{\phi_j}^{(j)}$ is the gauge automorphism belonging to the chemical species $j$, and assuming the set $(u_1,\ldots u_n)$ of generators for its dual group, $G^*$, so that $u_j$ is the $r$ of the dual group generator corresponding to $\varphi_j$. Then, there is another set of generators $(v_1,\ldots v_n)$, with

$$v_i = \sum_{j=1}^n q_{ij} u_j \tag{5.158}$$

so that it can be written:

$$
\begin{cases}
G_r^\perp = \bigoplus_{i=1}^m gp\{v_i\} \\
G_r^* = \bigoplus_{i>m}^n gp\{v_i\}
\end{cases}
\tag{5.159}
$$

Moreover, by the virtue of duality, there is a new parameterization, $0 \leq \zeta_i \leq 2\pi$, $i=1,\ldots n$ for $T^m$ where $\zeta_i$ is the angle which parameterizes the dual group $gp\{v_i\}$.

The bond between the two parameterizations $(\varphi_i, \zeta_i)$ will be given by:

$$\zeta_i = \sum_{j=1}^{n} q_{ij} \varphi_j \qquad (5.160)$$

noting that $\zeta_{m+1}, \ldots \zeta_n$ are right the parameterized angles for $G_r$.

Appealing to the same method of construction as for the non-reactive systems, it is introduced the extension of the state $\omega$ to the state $\Omega$, $G_r$-invariant from $M^{\gamma r}$ to $M$ by:

$$\Omega = \omega \circ \eta_r \qquad (5.161)$$

where $\eta_r : M \to M^{\gamma r}$ is the unique projection, normal and $G_r$ invariant of $M$ in its part $G_p$, invariant.

Therefore, $\Omega$ satisfies now the *KMS* condition in relation to the uni-parameter group:

$$\sigma_t^{\Omega} = \tau_t \gamma_{\zeta(t)} = \tau_t \delta_{v_{m+1}t}^{(m+1)} \ldots \delta_{v_n t}^{(n)} \qquad (5.162)$$

where

$$\delta_{\zeta_i}^{(i)} := \gamma_{q_{i1}\phi_1}^{(1)} \gamma_{q_{i2}\phi_2}^{(2)} \ldots \gamma_{q_{in}\phi_n}^{(n)} \qquad (5.163)$$

is the automorphic representation of 1-torus parameterized by $\zeta_i$, and $v_i$ are the chemical potentials formed in relation with this parameterization.

If the last form is placed under the one of the mixed uni-parameter temporal transformations, Eq. (5.150) is retrieved but with the new chemical potentials of the form:

$$\mu_j = \sum_{l>m}^{n} p_{jl} v_l, \quad j = 1, \ldots, n \qquad (5.164)$$

Using this form is easy to show the relation validity from the reactive case:

$$\sum_j q_{ij} \mu_j = \sum_j q_{ij} \sum_{l>m} p_{jl} v_l = \sum_{l>m} v_l \delta_{il}^{KRONECKER} = 0, \quad i = 1, \ldots, m \qquad (5.165)$$

Therefore, it can be concluded that for any subtorus $T^{n-m}$ of $T^n$ corresponds a set of m reactions *m reactions* and *n-m conservation laws*.

Reciprocally, any set of m chemical reactions linearly independent in a system of n chemical species corresponds to a subtorus $T^{n-m}$ of $T^n$.

*The chemical reactions set is given by* stoichiometric coefficients $q_{ij}$, $i = 1, \ldots n$ & $j = 1, \ldots m$ and is unique until the unimodular transformation in $Z^m = G_r^\perp$.

*The conservation laws set is given by* the stoichiometric coefficients $q_{ij}$, $i = 1, \ldots n$ & $j = m+1, \ldots n$, representing the linear specific combinations of the number of particles which conserve the weight of the atomic species in molecules, or more generally, representing a so-called resistant group to transformations.

For the non-reactive systems, $m = 0$, and $G_r = G$, all the chemical species are conserved in a closed thermodynamic system limited to its own volume.

In an extreme case, $m = n$, the system is necessarily open, and there are no linear conservation laws.

These results, in fact currently used in experimental applications and for more or less empirical applications, specific to the chemical thermodynamics, are dressed up in the algebraic clothes of dual transformations, of group, with the invariants and the commutations specific to the reactive equilibrium conditions, of the KMS states.

This algebraic map of the chemical reactivity interpretation, allows for further interpretation and selection of the essential results abstracted from the path integral formalism, applied quantum statistical to a molecular electrons system evolution, under a potential produced by nuclei (the same type or not), as an enharmonic potential generalized (the coupling constant, left free in variation, overlaps both cases of identical set of nuclei or not), see Putz (2013).

## 5.3   FORMALIZATION OF THOMAS-FERMI THEORY

### 5.3.1   THOMAS-FERMI  THEOREMS

The TF theory was independently developed by Thomas (1927) and Fermi (1927), and proposes for the total energy as the density functional, the expression:

$$E[\rho] = \frac{3}{5} C_F \int \rho(x)^{5/3} \, dx - \int \rho(x) V(x) dx + D[\rho, \rho] + U \quad (5.166)$$

where the electrostatic potential of a k nuclei of charges $z_1, z_2, ..., z_k > 0$ located to $R_1, R_2, ..., R_k \in R^3$ is given by

$$V(x) = \sum_{j=1}^{k} z_j |x - R_j|^{-1} \tag{5.167}$$

the repulsive potential between the nuclei is given by:

$$U = \sum_{1 \le i < j \le k} z_i z_j |R_i - R_j|^{-1} \tag{5.168}$$

the repulsive potential between the electrons has the general form:

$$D[f, g] = \frac{1}{2} \iint g(x) f(x) |x - y|^{-1} dx dy \tag{5.169}$$

and the constant $C_F > 0$ makes the connection with the quantum theory through the relation:

$$C_F = \left(6\pi^2\right)^{2/3} \hbar^2 \left(2mq^{2/3}\right)^{-1} \tag{5.170}$$

with $q$ corresponding to the spin states numbers ($q = 2$ for electrons).

The correction through the exchange term was introduced by Dirac (1930) through which the TF theory becomes the DFT theory and is amended as follows:

$$E[\rho]^{TFD} = E[\rho] - \frac{3}{4} C_X \int \rho(x)^{4/3} dx \tag{5.171}$$

with

$$C_X > 0, C_X = \sqrt[3]{6 / (q\pi)} \tag{5.172}$$

Another correction this time on the kinetic term, brought by von Weizsaecker (1935) transforms the TF theory in TFW theory and the last relation will be properly modified ; the present discussion follows (Putz, 2012):

$$E[\rho]^{TFW} = E[\rho] + C_G \int \left[\left(\nabla \rho^{1/2}\right)(x)\right]^2 dx \tag{5.173a}$$

with $C_G = a\hbar^2 / (2m)$, where $a$ is an adjustable parameter.

One can consider the cumulative case of the Thomas-Fermi-Dirac-von Weizsaecker (TFDW) theory in which the total energy as the density electronic functional will have the general expression

$$E[\rho]^{TFDW} = E[\rho] - \frac{3}{4}C_X \int \rho(x)^{4/3} dx + C_G \int \left[ \left( \nabla \rho^{1/2} \right)(x) \right]^2 dx \quad (5.173b)$$

Next it will be presented a couple of definitions, sentences and theorems in terms of the distribution space – the functionals space (Lieb & Thirring, 1975; Lieb & Simon, 1977, 1978; Lieb, 1976, 1981; Lieb & Oxford, 1981) in order to prove that the molecular processes approach in form of the density functional has a rigorous mathematical support, which confers universality and motivates deep studies in this sense (Teller, 1962; Balàzs, 1967; March, 1983, 1992).

DEFINITION *TF1*: One says that a function $f$ belongs to the space $L^p$, of the p-integrable function, if its norm

$$\left( \int \left| f(x) \right|^p dx \right)^{1/p} \equiv \| f \|_p , 1 \leq p < \infty \quad (5.174)$$

is finite. In addition, if $f \in L^p \cap L^q$, $p < q$ then $f \in L^t$, $p < t < q$ and occur the relations:

$$\| f \|_t \leq \| f \|_p^{\lambda} \| f \|_q^{1-\lambda} , \lambda p^{-1} + (1-\lambda) q^{-1} = t^{-1} \quad (5.175)$$

*SENTENCE TF1*: *If $\rho \in L^{5/3} \cap L^1$ then all the terms of E[ρ] and of $E[\rho]^{TFD}$ are finite. If $\int \rho \leq N$ (N the electrons numbers, not necessary a whole number) then E[ρ] and $E[\rho]^{TFD}$ are lower bounded by a constant C(N). Moreover, for any N for a C(N) fixed occurs the relation:*

$$E[\rho] > C > -\infty \quad (5.176)$$

*SENTENCE TF2*: *The application $\rho \rightarrow E[\rho]$ is strictly convex, so that satisfies the inequality*

$$E[\lambda \rho_1 + (1-\lambda)\rho_2] < \lambda E[\rho_1] + (1-\lambda) E[\rho_2] , 0 < \lambda < 1 , \rho_1 \neq \rho_2$$
$$(5.177)$$

One introduces the following crowd of energies provided by the Sentence TF1:

$$E(N) = \inf \left\{ E[\rho] \big| \rho \in L^{5/3} \cap L^1, \int \rho = N \right\} \tag{5.178}$$

and only the electronic contribution will be considered through:

$$e(N) = E(N) - U \tag{5.179}$$

which will allow the introduction of the following theorem which provides more than the existence of a lower bounds, and even a minimal energy.

*THEOREM TF1*: The function $e(N)$ is convex, negative if $N > 0$, *not in ascending and lower bound.* Moreover, there is the crowd:

$$E(N) = \inf \left\{ E[\rho] \big| \rho \in L^{5/3} \cap L^1, \int \rho \leq N \right\} \tag{5.180}$$

The first part of the theorem is provided by the Sentence TF2, and in the second part from the *monotonicity of e(N) and E(N)*. With these it can be proved the following theorem.

*THEOREM TF2: There is a unique density $\rho$ which minimizes E[$\rho$] under the condition*

$$\int \rho \leq N \tag{5.181}$$

Thus, it can be concluded that the function $E(N)$ is non-ascending, bounded and convex and hence continuous, which allows the following definition.

*DEFINITION TF2*: The number $N_c$ is called critical value of $N$ and represents the highest value of $N$ for which $N' < N$ implies $E(N') > E(N)$. Equivalently:

$$\text{if } E(\infty) = \lim_{N \to \infty} E(N) \text{ then } N_c = \inf \left\{ N \big| E(N) = E(\infty) \right\} \tag{5.182}$$

For $N \geq N_c$ there is the relation:

$$E(N) = E(N_c) + const.(N - N_c) \tag{5.183}$$

With this one can show that the following theorem is valid.

*THEOREM TF3*: For $N \leq N_c$ there is an unique minimization for $\rho$ satisfying $\int \rho = N$. In the interval $[0, N_c]$, $E(N)$ is strictly convex and monotonically decreasing. For $N > N_c$ there is no minimization for $\rho$ that satisfying $\int \rho = N$ and $E(N) = E(N_c)$.

Performing the variational derivative of $E[\rho]$ there is obtained:

$$\frac{\delta E}{\delta \rho} = C_F \rho^{2/3}(x) - \phi_\rho(x) \qquad (5.184)$$

with the potential:

$$\phi_\rho = V(x) - \int \rho(y)|x-y|^{-1} dy \qquad (5.185)$$

In order to ensure $\int \rho = N$, the Lagrange multiplier have to be introduced, being this $\mu$ the chemical potential. Therefore, it is expected that:

$$\frac{\delta E}{\delta \rho} + \mu = 0, \text{ if } \rho(x) > 0 \qquad (5.186)$$

and respectively:

$$\frac{\delta E}{\delta \rho} + \mu \geq 0, \text{ if } \rho(x) = 0 \qquad (5.187)$$

the situations with negative densities being prohibited.

The two cases from above can be compressed in the formal Thomas-Fermi equation, as follows:

$$C_F \rho^{2/3}(x) = \max\left[\phi_\rho(x) - \mu, 0\right] \equiv \left[\phi_\rho(x) - \mu\right] \qquad (5.188)$$

This equation allows, along with the other topological properties presented, the introduction of the following theorems.

*THEOREM TF4*: If the $\rho$ density minimizes the functional $E[\rho]$ for $\int \rho = N \leq N_c$, then $\rho$ *satisfies the* Thomas-Fermi equation (the last equation) for a given (unique) value $\mu(N)$. Reciprocally, if $\rho$ *and* $\mu$ satisfy the Thomas-Fermi equation and $\rho \in L^{5/3} \cap L^1$, *then* $\rho$ *minimize* $E[\rho]$ *for* $\int \rho = N$. *If* $N = N_c$, *then* $\mu = 0$.

*THEOREM TF5*: If $E(N)$ is continuous and differentiable and

$$\frac{dE}{dN} = -\mu, \text{ if } N \leq N_C \tag{5.189}$$

*together with:*

$$\frac{dE}{dN} = 0, \text{ if } N \geq N_C \tag{5.190}$$

*then μ(N) is the chemical potential.*

An equation in the distributions sense, equivalent with the Thomas-Fermi equation, can be introduced by taking into account the expression of the Coulomb-Poisson potential, resulting the Thomas-Fermi differential equation with $\phi(x)$ instead of $\phi_p(x)$, thus:

$$-\frac{\Delta\phi(x)}{4\pi} = \sum z_i \delta(x - R_i) - \rho(x)$$

$$= \sum z_i \delta(x - R_i) - C_F^{-3/2}[\phi(x) - \mu]^{3/2} \tag{5.191}$$

The differential equation TF promotes of the following definition introduction.

*DEFINITION TF3*: A function defined on an open crowd $M \subset R^3$ is superharmonic on M if, for almost all $x \in M$ and for almost all the spheres centered in x but contained in M, *f(x)* is at least equal to its average on the sphere, i.e.

$$f(x) \geq (4\pi)^{-1} \int_{|y|=R} f(x + y) dy \tag{5.192}$$

The condition is the same as for $\Delta f \leq 0$ (in the sense of distributions) in M. It is said that the *function f(x) is subharmonic if − f(x) is superharmonic, and f(x) is said harmonic if it is both subharmonic and superharmonic.*

One can generalizes the TF functional density, by multiplying the term $D[\rho, \rho]$ through a parameter $b > 0$. Then

$$e(N) = E(N) - U \tag{5.193}$$

is a parameters function of $C_F$, $\{z_j\}$ and *b*. It will be considered also the scaling of the critical number of electrons,

$$N_c(b) = N_c / b \qquad (5.194)$$

and will be adopted the following notations:

$$K = C_F \int \rho^{5/3} dx \qquad (5.195a)$$

$$R = bD[\rho, \rho] \qquad (5.195b)$$

$$A = \int \rho v(x) dx \qquad (5.195c)$$

with $\rho$ minimizing the $E[\rho]$ for

$$\int \rho = N \le N_c \qquad (5.195d)$$

thanks to this identity, one can introduce the following theorem of Feynman-Hellman.

*THEOREM TF6: The size* $e\left(N, C_F, \{z_j\}, b\right)$ *is a function $C^1$ in the $k + 3$ arguments. Then, the size e is convex in N and concave in the arguments* $\left(C_F, \{z_j\}, b\right).$ *Moreover, the next equalities are valid:*

$$\frac{\partial e}{\partial C_F} = \frac{K}{N} \qquad (5.196a)$$

$$\frac{\partial e}{\partial b} = \frac{R}{b} \qquad (5.196b)$$

$$\frac{\partial e}{\partial N} = -\mu \qquad (5.196c)$$

$$\frac{\partial e}{\partial z_j} = -\int \rho(x) |x - R_j|^{-1} dx \qquad (5.196d)$$

*which implies also the relation:*

$$\frac{\partial E}{\partial z_j} = \lim_{x \to R_j} \left\{ \phi(x) - z_j |x - R_j|^{-1} \right\} \qquad (5.196e)$$

Equally, the following theorems of the *virial theorem* type can be formulated:

*THEOREM TF7:*

a)
$$\frac{5K}{3} = A - 2R - \mu N \qquad (5.197a)$$

b) *for an atom (k=1), $2K = A - R$* $\qquad$ (5.197b)

The first relation is obtained from the Thomas-Fermi equation, multiplied with $\rho$ and then integrated. Alternatively, one can proof the assertion (a) by noting

$$G[\rho] = E[\rho] + \mu \int \rho \qquad (5.198)$$

which is minimized for all $\rho \in L^{5/3} \cap L^1$. Then by scaling the density

$$\rho_t(x) = t\rho(x) \qquad (5.199)$$

There follows that $G[\rho_t]$ reaches its minimum to $t = 1$, while and from the extreme condition

$$dG[\rho_t] / dt = 0 \qquad (5.200)$$

the relation TF7-(a) is obtained.

The second relation of the TF7 Theorem is obtained by scaling

$$\rho_t(x) = t^3 \rho(tx) \qquad (5.201a)$$

so that

$$\int \rho_t = N \qquad (5.201b)$$

Then, $E[\rho_t]$ reaches its minimum to $t = 1$ and from the condition

$$dE[\rho_t] / dt = 0 \qquad (5.202)$$

from where the relation TF7-(b) immediately results.

However, by further scaling the coordinates with a parameter $l > 0$, $R_j \rightarrow lR_j$ and noting with $\underline{z}$ and respectively with $\underline{R}$ the assembly of the charges and nuclear coordinates, one may obtain – as a consequence of the functional scaling properties $E[\rho_t]$, the following relations for the total energy, the chemical potential, the electronic density and the Thomas-Fermi potential, respectively:

$$E\left(\underline{z},N,l\underline{R}\right) = l^{-7}E\left(l^3\underline{z},l^3N,\underline{R}\right) \tag{5.204a}$$

$$\mu\left(\underline{z},N,l\underline{R}\right) = l^{-4}\mu\left(l^3\underline{z},l^3N,\underline{R}\right) \tag{5.204b}$$

$$\rho\left(\underline{z},N,l\underline{R};x\right) = l^{-6}\rho\left(l^3\underline{z},l^3N,\underline{R};l^{-1}x\right) \tag{5.204c}$$

$$\phi(\underline{z},N,l\underline{R};x)) = l^{-4}\phi(l^3\underline{z},l^3N,\underline{R};l^{-1}x) \tag{5.204d}$$

Further, we will consider a more general form of the TF functional:

$$E[\rho] = \int j[\rho(x)]dx - \int V(x)\rho(x)dx + D[\rho,\rho] \tag{5.205}$$

where $j$ is a convex function $C^1$ with $j(0) = j'(0) = 0$; moreover, is noted the explicit absence of the repulsive term $U$. This renunciation it comes from the fact that if the potential $V(x)$ is not necessary the Columbic potential, the term $U$ does not have a clear meaning. The equation Euler-Lagrange may be associated to the last equation when considering

$$\phi_\rho = V - |x|^{-1} * \rho \tag{5.206}$$

where

$$(f * g)(x) = \int f(x-y)g(y)dy \tag{5.207}$$

represents the convolution of $f$ with $g$; that is, the correspondent of the TF formal equation will be generated as:

$$\phi_\rho(x) - \mu \dots \begin{cases} = j'[\rho(x)] \dots AlmostEveryWhere \ \ for \ \rho(x) > 0 \\ \leq 0 \qquad \dots AlmostEveryWhere \ \ for \ \rho(x) = 0 \end{cases} \tag{5.208}$$

which for

$$j'[\rho] = C_F \rho^{2/3} \tag{5.209}$$

can be rewritten in the optimal form:

$$\phi_\rho(x) - \mu \equiv \max\left[\phi_\rho(x) - \mu, 0\right] = j'\left[\rho(x)\right] \tag{5.210}$$

The TF theory and electronic description is considered as the referential for the uniform distribution of electrons in atoms and molecules, respecting which the electronic accumulation in bonding is further described, usually as a perturbation – as in DFT when density gradient expansions are considered, or by general reformulation of the problem in terms of localization – in which case special quantum treatment as provided by stochastic Fokker-Planck modeling is needed; these issues will be in next addressed and unfolded.

### 5.3.2   ELECTRONIC LOCALIZATION PROBLEM

Let's consider the "spherical" referential electronic picture as the most useful in establishing the uniform electronic distribution by indicating the occupation of the all-possible electronic levels in a semiclassical quantum frame (without explicit exchange-correlation involvement). Actually, the Fermi sphere in a momentum space finely defines the total homogeneous kinetic energy as:

$$\tau_s(\mathbf{r}) = \frac{p_F^2}{2m_0} \tag{5.211}$$

while the quantum nature of the kinetic energy (5.211) is covered by involving the quantum (Heisenberg) uncertainty

$$\Delta p_x \Delta p_y \Delta p_z \Delta x \Delta y \Delta z \cong h^3 \tag{5.212}$$

in uniform density computation. This suggests that the density of states in the Fermi volume of the impulse $p_F$ has to be normalized to the inverse of the cube power of Planck constant $h$, while the density of electrons is reached by multiplying the density of states with the electron multiplicity $2(1/2)+1 = 2$ for every occupied state. The obtained density-Fermi impulse relationship:

$$\rho(\mathbf{r})d\mathbf{r} = 2\frac{4}{3}\pi\frac{p_F^3}{h^3}d\mathbf{r} \Rightarrow p_F(\mathbf{r}) = \left(\frac{3}{8\pi}\right)^{1/3}h\rho(\mathbf{r})^{1/3} \qquad (5.213)$$

allows the Thomas-Fermi kinetic energy unfolding as the density functional (Parr & Yang, 1989; Garcia-Gonzales et al., 1996; Chan & Handy, 1999; Bartolotti & Acharya, 1982; Dawson & March, 1984; Baltin, 1987):

$$T_{TF}[\rho] = \frac{3}{5}\int\rho(\mathbf{r})\tau_s(\mathbf{r})d\mathbf{r} = C_{TF}\int\rho(\mathbf{r})^{5/3}d\mathbf{r},$$

$$C_{TF} = \frac{3h^2}{10m_0}\left(\frac{3}{8\pi}\right)^{2/3} \cong 2.871[a.u.] \qquad (5.214)$$

with the help of which the total Thomas-Fermi energy functional takes the form (Putz, 2008):

$$E_{TF}[\rho] = C_{TF}\int\rho(\mathbf{r})^{5/3}d\mathbf{r} + \frac{1}{2}\iint\frac{\rho(\mathbf{r}_1)\rho(\mathbf{r}_2)}{|\mathbf{r}_{12}|}d\mathbf{r}_1 d\mathbf{r}_2 + \int V(\mathbf{r})\rho(\mathbf{r})d\mathbf{r}$$

$$(5.215)$$

that can be seen as the first approximation for the density functional total energy.

From physical point of view worth noted that the kinetic TF energy exactly corresponds to the total energy of the *free* electrons in a crystal, $V(\mathbf{r}) = 0$ in Eq. (5.215), equivalently with the fact that the electrons are not "feeling" the nuclei, i.e., electrostatic attractions are excluded, being as close each other to avoid reciprocal repelling. Such picture suggests that free electrons are completely non-localized leaving with the condition of complete cancellation of the electronic inter-repulsion; this feature may be putted formally as Becke (1988):

$$\frac{e^2}{|\mathbf{r}_1 - \mathbf{r}_2|} \rightarrow \frac{\lambda e^2}{|\mathbf{r}_1 - \mathbf{r}_2|}, \lambda = 0 \qquad (5.216)$$

However, the model in which the (valence) electrons are completely free and are neither "feeling" the attraction nor the repulsion is certain not properly describing the nature of the chemical bond. In fact, this limitation was also the main objection brought to Thomas-Fermi model and to the atomic or molecular approximation of the homogeneous electronic gas or helium model in solids. Nevertheless, the lesson is well served because Thomas-Fermi description may be regarded as the "inferior" extreme in quantum known structures while further exchange-correlation effects may be added in a perturbation manner.

The idea of introducing exchange and correlation effects as a perturbation of the homogeneous electronic system could be considered from the interpolation of the energetic terms for $0 \le \lambda \le 1$ in Eq. (5.215). Parameter $\lambda$ is defined as a *parameter of the electronic coupling*, with a slightly (adiabatically) scaling of the perturbation from the homogeneous electronic systems, $\lambda = 0$, until the maximal inter-electronic interaction, $\lambda = 1$ (in accordance with Pauli principle). Therefore, the overall interpolation $\int_0^1 [\bullet] d\lambda$ will be spread over the terms which contain the intermediate degree of exchange and correlation interactions; since it accounts for the electronic inter-repulsion while indexing the electronic presence/absence in a given spatial region the degree of *electronic localization* is in this way furnishes.

The coupling parameter $\lambda$ will serve as a switcher between the referential Thomas-Fermi uniform case and the full interaction through the density limit (Putz, 2008):

$$\lim_{\lambda \to 1} \rho_\lambda (\mathbf{r}) = \rho(\mathbf{r} = \mathbf{r}_1 \cap \mathbf{r}_2) \tag{5.217}$$

Actually, the density (5.217) has a major role in defining exchange-correlation functionals. To see that let's firstly consider the conditional electronic density $g(\mathbf{r}_1, \mathbf{r}_2; \lambda)$ indicating that the electronic density in $\mathbf{r}_1$ is conditioned by the presence (localization) of another electron (any from the total $N$ in the system) in $\mathbf{r}_2$. Mathematically, this is expressed by using the conditional probabilities:

$$g(\mathbf{r}_1, \mathbf{r}_2; \lambda) = \frac{\rho(\mathbf{r} = \mathbf{r}_1 \cap \mathbf{r}_2)}{\rho(\mathbf{r}_2)} \tag{5.218}$$

fulfilling the Pauli principle by means of the integration rule:

$$\int \rho(\mathbf{r}_2) g(\mathbf{r}_1,\mathbf{r}_2;\lambda) d\mathbf{r}_2 = \int \rho(\mathbf{r}_1 \cap \mathbf{r}_2) d\mathbf{r}_2 = 0 \qquad (5.219)$$

saying that the spatial average of the electronic reciprocal constraint vanishes. This behavior opens the possibility in introducing the conditional probability of electronic holes,

$$h(\mathbf{r}_1,\mathbf{r}_2;\lambda) = g(\mathbf{r}_1,\mathbf{r}_2;\lambda) - 1 \qquad (5.220)$$

providing the associate integration rule (Ayers & Levy, 2001; Koch & Holthausen, 2000):

$$\int \rho(\mathbf{r}_2) h(\mathbf{r}_1,\mathbf{r}_2;\lambda) d\mathbf{r}_2 = -1 \qquad (5.221)$$

consecrating a sort of negative normalization of the exchange and correlation density of holes:

$$\rho_{xc}(\mathbf{r}_1,\mathbf{r}_2;\lambda) = \rho(\mathbf{r}_2) h(\mathbf{r}_1,\mathbf{r}_2;\lambda) \qquad (5.222)$$

Now, once this exchange-correlation hole density is mediated over the coupling factor $\lambda$ the averaged exchange-correlation density of holes is generated:

$$\bar{\rho}_{xc}(\mathbf{r}_1,\mathbf{r}_2) = \int_0^1 \rho_{xc}(\mathbf{r}_1,\mathbf{r}_2;\lambda) d\lambda \qquad (5.223)$$

allowing the formal writing of exchange-correlation density functional from as a generalized version of the inter-electronic interaction term:

$$E_{xc}[\rho] = \frac{1}{2}\iint \frac{\rho(\mathbf{r}_1)\bar{\rho}_{xc}(\mathbf{r}_1,\mathbf{r}_2)}{r_{12}} d\mathbf{r}_1 d\mathbf{r}_2 = -\frac{1}{2}\int \frac{\rho(\mathbf{r}_1)}{\bar{R}_{xc}(\mathbf{r}_1,\rho(\mathbf{r}))} d\mathbf{r}_1 \quad (5.224)$$

with the help of the introduced radius of the *$\lambda$-mediated exchange-correlation density of holes*:

$$\bar{R}_{xc}^{-1}(\mathbf{r}_1,\rho(\mathbf{r})) := -\int \frac{\bar{\rho}_{xc}(\mathbf{r}_1,\mathbf{r}_2,\rho(\mathbf{r}))}{r_{12}} d\mathbf{r}_2 \qquad (5.225)$$

The radius $\bar{R}_{xc}(\mathbf{r})$ could be considered as a functional of density (5.217) with the leading term being defined in *the short limit of the distance*, i.e., being of the inter-particle average radius order,

$$\frac{4\pi}{3}r_0^3 = \frac{1}{\rho(\mathbf{r})} \Rightarrow r_0 = \left(\frac{3}{4\pi}\right)^{1/3}\rho(\mathbf{r})^{-1/3} \qquad (5.226)$$

known as the Wigner radius for indexing the volume of a sphere containing (*localizing*) a single electron (from the total of $N$) belonging to the density family (5.217), although, also other quantities accounting for electron localization such as the domain averaged Fermi hole (Ponec & Cooper, 2007) or the electron sharing index (also known as delocalization index) have been in the last decade proposed (Matito et al., 2007).

In these conditions, the inverse radius (5.225) could be expressed around the inverse of the Wigner radius in a gradient density expansion (Becke, 1988):

$$\bar{R}_{xc}^{-1}(\mathbf{r},\rho(\mathbf{r})) = F_0\left[\rho(\mathbf{r})\right] + F_{21}\left[\rho(\mathbf{r})\right]\nabla^2\rho(\mathbf{r})$$
$$+ F_{22}\left[\rho(\mathbf{r})\right]\sum_{i=1}^{N}\left(\nabla_i\rho(\mathbf{r})\right)\left(\nabla_i\rho(\mathbf{r})\right) + ... \qquad (5.227)$$

while, by considering it back in exchange-correlation energy (5.224) produces, after the integration by parts, the generalized gradient density functional:

$$E_{xc} = \int G_0\left[\rho(\mathbf{r})\right]\rho(\mathbf{r})d\mathbf{r} + \int G_2\left[\rho(\mathbf{r})\right]\left(\nabla\rho(\mathbf{r})\right)^2 d\mathbf{r}$$
$$+ \int \left[G_4\left[\rho(\mathbf{r})\right]\left(\nabla^2\rho(\mathbf{r})\right)^2 + ...\right]d\mathbf{r} + ... \qquad (5.228)$$

The restriction to the first term of the series (5.228) corresponds to the cases where the spatial distance of variation in electronic density highly exceeds the corresponding Wigner radius (5.226) this way producing the famous *local density approximation* (LDA) (March et al., 2003):

$$E_{xc}^{LDA}\left[\rho\right] = \int e_{xc}\left[\rho(\mathbf{r})\right]\rho(\mathbf{r})d\mathbf{r} \qquad (5.229)$$

with $e_{xc}$ being the exchange and correlation density per particle, that can be further approximated (see in the following paragraphs) as (Zhao et al., 1994; Gritsenko et al., 2000; Zhao & Parr, 1992; Lam et al., 1998; Gaspar & Nagy, 1987; Levy, 1991):

$$e_{xc}[\rho] = e_x[\rho] + e_c[\rho] = \left( -\frac{0.458}{r_0} \right) + \left( -\frac{0.44}{r_0 + 7.8} \right) \; [a.u.] \quad (5.230)$$

In fact, the LDA stands as the immediate step after TF approximation; it can be extended also for systems with un-pair spins by the so-called *local spin density approximation* (LSDA) (Guo & Whitehead, 1989, 1991; Dunlap, 1988; Dunlap et al., 1990; Dunlap & Andzelm, 1992; Harrison, 1987; Becke, 1986; Manoli & Whitehead, 1988; Filippeti, 1998; Liu, 1996):

$$E_{xc}\left[ \rho = \rho_\uparrow + \rho_\downarrow \right] = E_{xc}\left[ \rho_\uparrow \right] + E_{xc}\left[ \rho_\downarrow \right] \quad (5.231)$$

while further inclusion of the gradient terms in Eq. (5.228) establishes the *general gradient approximation* (GGA).

Worth noting that when undertaken GGA, beside the gradient terms arising in exchange-correlation energy, the gradient correction of the kinetic energy functional has to be as well considered providing terms of which the standard one takes the von Weizsäcker form (Romera & Dehesa, 1994; Murphy, 1981):

$$\tau_W(\mathbf{r}) = \frac{1}{8} \frac{|\nabla \rho(\mathbf{r})|^2}{\rho(\mathbf{r})} \quad (5.232)$$

While more analytical discussions about various approximations and density functionals are the subject of different monography (Putz, 2012), here we would like only to present the practical difference between the local and gradient density approximations for a solid-state case. For instance, Figure 5.1 presents the band structure and the density of states (DOS) for the $R\bar{3}m$ oxide of the Cobalt transitional metal (CoO) calculated with either LSDA or GGA approximations (Dufek et al., 1994). Nonetheless, at the level of bands structure of the solids and crystals an inevitable *localization*

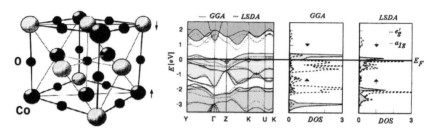

**FIGURE 5.1** Left: the anti-ferromagnetic structure CoO; right: the band structure and the density of state (DOS) in LSDA and GGA approximations, respectively; the upper and down arrows are associated with the spin orbital projections; after Dufek et al. (1994).

*paradox* emerges namely, to use the real 3D electronic densities in furnish a localization description in the reciprocal (energy) space.

Regarding the energy bands there can be noted that, around the Fermi level $E_F$, LSDA approach is less relevant in indicating the energetically gap respecting the GGA computation. The difference is even more drastic in DOS representations for employed approximations in $d$ orbital separation $\left(t_{2g} = a_{1g} + e'_g\right)$ due to the central ion of cobalt trigonal symmetry coordination. In fact, with LSDA a strong mixing of the orbitals $a_{1g}$ and $e'_g$ is recorded while in the case of GGA-DOS the bands with the symmetry $a_{1g}$ are up and down shifted for the respective down and up spin projections resulting a clear separation from states with $e'_g$ symmetry.

Recently, it was found a way to avoid the electronic localization paradox through introducing specific electronic localization functions (ELF) in real space. Nevertheless, an ELF should relay on combination of the gradient and homogeneous energetic density functionals, in accordance with Pauli principle, shaping for instance as (Becke & Edgecombe, 1990):

$$ELF = \left(1 + \left[\frac{\tau_s(\mathbf{r}) - \tau_W(\mathbf{r})}{T_{TF}[\rho(\mathbf{r})]}\right]^2\right)^{-1} \quad (5.233)$$

by emphasizing the excess of the kinetic energy difference $\tau_s(\mathbf{r}) - \tau_W(\mathbf{r})$ "normalized" to the referential kinetic TF homogeneous behavior.

Worth remarking that the localization function (5.233) acts like a sort of density, with values between 0 and 1 corresponding with maximum delocalization and localization, respectively. This heuristically proposal has the merit to give an analytical reflection of the qualitative *valence shell electron pair repulsion* (VSEPR) geometric model (Bader et al., 1988), with the immediate consequence in topological characterization of the chemical bond (Silvi & Savin, 1994). In solid-state case, the reliability of above ELF in describing the chemical bond in real space is illustrated in Figure 5.2 for the Li and Sc crystals. Atomic and molecular levels are in next section illustrated with which occasion further ELF characterization and developments are presented within the more complex formulation of quantum dynamical (open) systems.

## 5.4   FOKKER-PLANCK DYNAMICS OF QUANTUM SYSTEMS

### 5.4.1   NON-EQUILIBRIUM DYNAMIC LEVEL

The electronic states associated to the cyclic and oscillatory reactions, reactions with instabilities, etc. are obeying for the discrete state dynamics to the quantum general evolution equation (QME: quantum master equation) (see Gray & Scott, 1990; van Kampen, 1987; Gardiner, 1994; Risken, 1984; Haken, 1978, 1987, 1988):

$$\frac{\partial W_n}{\partial t_b} = \sum_m \left[ w(m \to n)W_m - w(n \to m)W_n \right] \tag{5.234}$$

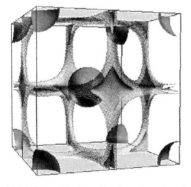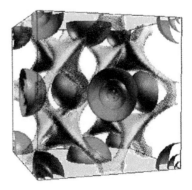

**FIGURE 5.2**   The localization domains for Li (left) and Sc (right) crystals based on the electronic localization function (ELF) (5.233); after Silvi & Gatti (2000).

and in case of continuous dynamic equation:

$$\frac{\partial W\left(x_b,t_b\right)}{\partial t_b} = \int \left[ w\left(x_b^{\#} \to x_b\right) W\left(x_b^{\#},t_b\right) - w\left(x_b \to x_b^{\#}\right) W(x_b,t_b)\right] dx_b^{\#}$$

(5.235)

where $W_n$ or $W$ represent the probability density (discreet or continuous). A special form of this general equation of quantum evolution represents the Fokker-Planck (FP), which is obtained for the transition probability $w$ given for the expression:

$$w\left(x_b^{\#} \to x_b\right) = \left[ -\frac{\partial}{\partial x_b} D^{(1)}\left(x_b\right) + \frac{\partial^2}{\partial x_b^2} D^{(2)}\left(x_b\right) \right] \delta\left(x_b - x_b^{\#}\right)$$

(5.236)

Using the filtering property of the Dirac function,

$$\int f\left(x,x'\right) \delta\left(x-x'\right) dx = f\left(x',x'\right)$$

(5.237)

along with the identities

$$\frac{\partial}{\partial x_b^{\#}} W\left(x_b,t_b\right) = \frac{\partial^2}{\partial x_b^{\#2}} W\left(x_b,t_b\right) = 0$$

(5.238)

the Fokker-Planck equation result from (5.235) under the form

$$\frac{\partial W\left(x_b,t_b\right)}{\partial t_b} = \left[ -\frac{\partial}{\partial x_b} D^{(1)}\left(x_b\right) + \frac{\partial^2}{\partial x_b^2} D^{(2)}\left(x_b\right) \right] W\left(x_b,t_b\right)$$

(5.239)

with the potential dependency of force type:

$$D^{(1)}\left(x_b\right) = -\frac{d}{dx_b} V\left(x_b\right)$$

(5.240)

This is the probability density level of the describing of the dynamic non-equilibrium of the electronic states associated to some atomic and molecular species involved in the reactive chain (cycles, oscillations).

The next level of describing roots form the introduction of the Markovian processes which will be defined next. For instance, in the dynamic reactive chain

$$\overline{A}(x_a,t_a) \xrightarrow{e^-} \overline{B}(x_b,t_b) \xrightarrow{e^-} \overline{C}(x_c,t_c) \to ... \tag{5.241}$$

the electronic evolution is equivalent with its probability density transforming so that the equivalent $(x_c, t_c)$ is correlated with the event $(x_b, t_b)$ which is correlate with the event $(x_a, t_a)$ in a probability density row:

$$W_1(x_a,t_a) \to W_2(x_b,t_b;x_a,t_a) \to W_3(x_c,t_c;x_b,t_b;x_a,t_a) \to ... \tag{5.242}$$

The Markovian processes are characterize by the fact that all the information is contained in the first two functions $W_1$ and $W_2$ correlated by the transition probability $\rho(x_b,t_b;x_a,t_a)$ (conditioned probability density):

$$W_2(x_b,t_b;x_a,t_a) = \rho(x_b,t_b;x_a,t_a)W_1(x_a,t_a) \tag{5.243}$$

the transition probability (conditioned probability density) satisfies the following properties:

$$\rho(x_b,t_b;x_a,t_a) \geq 0 \tag{5.244}$$

$$\int dx_a \rho(x_b,t_b;x_a,t_a) = 1 \tag{5.245}$$

$$W_1(x_b,t_b) = \int dx_a \rho(x_b,t_b;x_a,t_a)W_1(x_a,t_a) \tag{5.246}$$

An immediate application of those properties is constituted by the equation Chapman-Komogorov-Smoluchowski (CKS), in fact the equation on defining the Markovian processes. For example, it can be immediately write that:

$$\begin{aligned}
W_2(x_c,t_c;x_a,t_a) &= \int dx_b W_3(x_c,t_c;x_b,t_b;x_a,t_a) \\
&= \int dx_b \rho(x_c,t_c;x_b,t_b)\rho(x_b,t_b;x_a,t_a)W_1(x_a,t_a) \quad (5.247)
\end{aligned}$$

and, on the other hand,

$$W_2\left(x_c,t_c;x_a,t_a\right)=\rho\left(x_c,t_c;x_a,t_a\right)W_1\left(x_a,t_a\right) \qquad (5.248)$$

By comparing the two expressions (5.247) and (5.248) immediately results the equation CKS:

$$\rho\left(x_c,t_c;x_a,t_a\right)=\int dx_b\rho\left(x_c,t_c;x_b,t_b\right)\rho\left(x_b,t_b;x_a,t_a\right) \qquad (5.249)$$

This equation significance is immediate: the probability of transition for the event $(x_a, t_a)$ to the event $(x_c, t_c)$ can be calculated as the product of the transitions probability of the initial state $(x_a, t_a)$ to an intermediary one $(x_b, t_b)$ to the final one $(x_c, t_c)$ and summed (integrated) over the all possible intermediate values.

Using the property (5.246) form above in the Fokker-Planck equation (5.239) in probability densities, it is obtained the Fokker-Planck equation in conditioned probability density:

$$\frac{\partial\rho\left(x_b,t_b;x_a,t_a\right)}{\partial t_b}=\left[-\frac{\partial}{\partial x_b}D^{(1)}\left(x_b\right)+\frac{\partial^2}{\partial x_b^2}D^{(2)}\left(x_b\right)\right]\rho\left(x_b,t_b;x_a,t_a\right)$$
$$(5.250)$$

Now one can introduce the functional density (of probability) specific to the nonequilibrium treating of the electronic dynamic in electro-reactivity.

In essence, the Fokker-Planck equation is a continuous equation, of temporary evolution of the electronic flux, compelled to an external potential $V(x)$ with the drift factor $D^{(1)}$ and one of diffusion $D^{(2)}$ (stochastic noise). This thing can be easily notice if the Fokker-Planck equation is rewritten (5.250), as example, in a hydrodynamic form:

$$\frac{\partial\rho\left(x_b,t_b;x_a,t_a\right)}{\partial t_b}+\frac{\partial}{\partial x_b}j_{FP}\left(x_b,t_b;x_a,t_a\right)=0 \qquad (5.251)$$

in which the conditioned probability of the density flow under the functional form there was introduced:

$$j_{FP}\left(x_b,t_b;x_a,t_a\right)=D^{(1)}\left(x_b\right)\rho\left(x_b,t_b;x_a,t_a\right)-\frac{\partial}{\partial x_b}\left[D^{(2)}\left(x_b\right)\rho\left(x_b,t_b;x_a,t_a\right)\right]$$
$$(5.252)$$

If the natural condition of frontier is considered

$$\rho\left(x_b,t_b;x_a,t_a\right)\Big|_{x_b\to\pm\infty}=0 \tag{5.253}$$

automatically results also the conditioned probability flowing condition

$$j_{FP}\left(x_b,t_b;x_a,t_a\right)\Big|_{x_b\to\pm\infty}=0 \tag{5.254}$$

as well as the implication:

$$\frac{\partial}{\partial t_b}\int_{-\infty}^{+\infty}dx_b\rho\left(x_b,t_b;x_a,t_a\right)=0 \tag{5.255}$$

from where one will retrieve the normalization condition in agreement with the previous property (5.245), for the conditioned probability density.

Another important observation results from the fact that, for the initial probability density of the Dirac distribution form

$$W_1\left(x_a^{\#},t_a\right)=\delta\left(x_a-x_a^{\#}\right) \tag{5.256}$$

in conformity with the property (5.246) and with the Dirac property (5.237), results that at the moment $t_b$ the probability density will be of the form

$$W_1\left(x_b,t_b\right)=\rho\left(x_b,t_b;x_a,t_a\right) \tag{5.257}$$

This fact says that $\rho\left(x_b,t_b;x_a,t_a\right)$ is a special probability density which evolve with the increasing of tome (at $t_b$) from the local Dirac distribution (at $t_a$). When an initial Gaussian distribution is considered (since all continuous distributions can be eventually reduced to a Gaussian form, according with the central limit theorem)

$$W_1\left(x_a,t_a\right)=\frac{1}{\sqrt{2\pi}}\exp\left(-\frac{1}{2}x_a^2\right) \tag{5.258}$$

the Markovian probability density (5.246) becomes:

$$W_1\left(x_b,t_b\right)=\frac{1}{\sqrt{2\pi}}\int dx_a\rho\left(x_b,t_b;x_a,t_a\right)\exp\left(-\frac{1}{2}x_a^2\right) \tag{5.259}$$

The (5.259) distribution automatically allows also the calculation of the Markovian probability flow, abstracted from the rewritten of Fokker-Planck equation on the probability density level (5.239), in an appropriate hydrodynamic form:

$$j_{FP}(x_b,t_b) = D^{(1)}(x_b)W_1(x_b,t_b) - \frac{\partial}{\partial x_b}\left[D^{(2)}(x_b)W_1(x_b,t_b)\right] \quad (5.260)$$

The probability density flow (5.260), or the one of conditioned probability density (5.252), stay on the basis of analytically representation of the no equilibrium dynamic for the electro-reactive chains in a temporary scale which cover, but also it overcome the one of the activated chemical complex. For the complexity of this study, at least from the perspective of the involving the path integrals in the dynamic of the respective equilibrium and no equilibrium, becomes extremely instructive the solution of Fokker-Planck equation in the conditioned probability density form (5.250), from where, the calculation of the probability density as well as of the associated currents are immediate.

### 5.4.2 PROPERTIES OF FOKKER-PLANCK EQUATION

Before the effective calculation of the conditioned probability density, worth exposing several essential characteristics of the Fokker-Planck equation as well as of the connection with the Schrödinger equation. To this aim one will be starting with the effective deduction of the Fokker-Planck equation. It can be successively write for the conditioned probability density (Feynman & Hibbs, 1965; Balescu, 1975):

$$\rho(x_c,t_c;x_a,t_a) = \int_{-\infty}^{+\infty} dx \delta(x-x_b)\rho(x,t_b;x_a,t_a)$$

$$= \int_{-\infty}^{+\infty} dx \delta(x_a - x_b + x - x_a)\rho(x,t_b;x_a,t_a)$$

$$= \int_{-\infty}^{+\infty} dx \sum_{k=0}^{\infty} \frac{(x-x_a)^k}{k!}\left(\frac{\partial}{\partial z}\right)^k \delta(z)\Big|_{z=x_a-x_b} \rho(x,t_b;x_a,t_a)$$

$$= \int_{-\infty}^{+\infty} dx \sum_{k=0}^{\infty} \frac{(x - x_a)^k}{k!} \left( -\frac{\partial}{\partial x_b} \right)^k \delta(x_a - x_b) \rho(x, t_b; x_a, t_a)$$

$$= \sum_{k=0}^{\infty} \frac{1}{k!} \left( -\frac{\partial}{\partial x_b} \right)^k M_k(x_a, t_a; t_b) \delta(x_a - x_b)$$

$$(5.261)$$

where the so-called moments were induced

$$M_k(x_a, t_a; t_b) = \int_{-\infty}^{+\infty} dx_b (x_b - x_a)^k \rho(x_b, t_b; x_a, t_a) \qquad (5.262)$$

the upper expansion having the name of Moyal-Kramers series (MK). If it is used the property of the probability densities, it can be calculated the difference

$$W_1(x_b, t_b) - W_1(x_a, t_a) = \sum_{k=1}^{\infty} \frac{1}{k!} \left( -\frac{\partial}{\partial x_b} \right)^k \{ M_k(x_a, t_a; t_b) \} W_1(x_b, t_b)$$

$$(5.263)$$

which can be rewritten on its turn under the form:

$$\frac{\partial}{\partial t_b} W_1(x_b, t_b) = \hat{L}_{KM}(x_b, t_b) W_1(x_b, t_b) \qquad (5.264)$$

where the Kramers-Moyal operator (KM) was introduced

$$\hat{L}_{KM}(x_b, t_b) \bullet = \sum_{k=1}^{\infty} \left( -\frac{\partial}{\partial x_b} \right)^k \{ D^{(k)}(x_b, t_a) \bullet \} \qquad (5.265)$$

with the KM coefficients given by the expression:

$$D^{(k)}(x_b, t_a) = \lim_{\varepsilon \to 0} \frac{M_k(x_b, t_a; t_a + \varepsilon)}{\varepsilon} \qquad (5.266)$$

It is obvious that, in the light of the same property (5.246), results an equation alike and in the conditioned probability density, the direct equation KM:

$$\frac{\partial}{\partial t_b} \rho(x_b, t_b; x_a, t_a) = \hat{L}_{KM}(x_b, t_b) \rho(x_b, t_b; x_a, t_a) \qquad (5.267)$$

The Fokker-Planck equation is automatically obtained by cutting the expansion of the KM operator (5.263) in the second order (Pawula, 1967)

$$\frac{\partial \rho\left(x_b,t_b;x_a,t_a\right)}{\partial t_b} = \frac{\partial}{\partial x_b}\left\{\begin{array}{l} D^{(1)}\left(x_b\right)\rho\left(x_b,t_b;x_a,t_a\right) \\ -\frac{\partial}{\partial x_b}\left[D^{(2)}\left(x_b\right)\rho\left(x_b,t_b;x_a,t_a\right)\right]\end{array}\right\} \quad (5.268)$$

The stationary solution $\rho_{st}\left(x_b\right)$ will automatically satisfy the equation:

$$\frac{\partial}{\partial x_b}\left\{D^{(1)}\left(x_b\right)\rho_{st}\left(x_b\right)-\frac{\partial}{\partial x_b}\left[D^{(2)}\left(x_b\right)\rho_{st}\left(x_b\right)\right]\right\}=0 \quad (5.269)$$

with which help, the FP operator abstract from the KM operator (5.265), be cut off in the second order so it can be rewritten as:

$$\hat{L}_{FP}\left(x_b,t_b\right)\bullet = -\frac{\partial}{\partial x_b}\left[D^{(1)}\left(x_b\right)\bullet\right]+\frac{\partial^2}{\partial x_b^2}\left[D^{(2)}\left(x_b\right)\bullet\right]$$

$$=\frac{\partial}{\partial x_b}\left\{D^{(2)}\left(x_b\right)\rho_{st}\left(x_b\right)\frac{\partial}{\partial x_b}\left[\frac{1}{\rho_{st}\left(x_b\right)}\bullet\right]\right\}=\hat{L}_{FP}\left(x_b\right)\bullet$$

$$(5.270)$$

For revealing the adjoint Fokker-Planck equation and the associate operator one starts from the CKS equation (5.249); by taking the derivative of this equation respecting $t_b$ and with the direct KM equation (5.267) there results:

$$\int_{-\infty}^{+\infty} dx_b \rho\left(x_c,t_c;x_b,t_b\right)\hat{L}_{KM}\left(x_b,t_b\right)\rho\left(x_b,t_b;x_a,t_a\right)$$

$$+\int_{-\infty}^{+\infty} dx_b \rho\left(x_b,t_b;x_a,t_a\right)\frac{\partial}{\partial t_b}\rho\left(x_c,t_c;x_b,t_b\right)=0 \quad (5.271)$$

Next, the operational equality will be taking into consideration:

$$\int_{-\infty}^{+\infty} dx\, a(x,t)\hat{L}_{KM}b(x,t)=\int_{-\infty}^{+\infty} dx\, b(x,t)\hat{L}_{KM}^{+}a(x,t) \quad (5.272)$$

by which the last equality becomes:

$$\forall \rho\left(x_b,t_b;x_a,t_a\right):$$

$$\int_{-\infty}^{+\infty} dx_b \rho\left(x_b,t_b;x_a,t_a\right)\left\{\frac{\partial}{\partial t_b}\rho\left(x_c,t_c;x_b,t_b\right)+\hat{L}_{KM}^{+}\left(x_b,t_b\right)\rho\left(x_c,t_c;x_b,t_b\right)\right\}=0$$

(5.273)

from where, by re-labeling the events $\left(x_c,t_c\right)=\left(x_b,t_b\right)$ and $\left(x_b,t_b\right)=\left(x_a,t_a\right)$ the adjoint KM equation results:

$$\frac{\partial}{\partial t_a}\rho\left(x_b,t_b;x_a,t_a\right)=-\hat{L}_{KM}^{+}\left(x_a,t_a\right)\rho\left(x_b,t_b;x_a,t_a\right) \qquad (5.274)$$

The adjoint KM operator will result by replacing in the relation (5.272) the KM operator with its direct expression (5.265), so obtaining:

$$\hat{L}_{KM}^{+}\left(x_b,t_b\right)\bullet=\sum_{k=1}^{\infty}D^{(k)}\left(x_b,t_a\right)\left(\frac{\partial}{\partial x_b}\right)^k\bullet \qquad (5.275)$$

In case when the KM operator is limited to the adjoint FP operator, this one, on its turn, can be rewritten in function of the stationary solution (5.269) of the direct Fokker-Planck equation:

$$\hat{L}_{FP}^{+}\left(x_b\right)\bullet=D^{(1)}\left(x_b\right)\frac{\partial}{\partial x_b}\bullet+D^{(2)}\left(x_b\right)\frac{\partial^2}{\partial x_b^2}\bullet$$

$$=\frac{1}{\rho_{st}\left(x_b\right)}\frac{\partial}{\partial x_b}\left\{D^{(2)}\left(x_b\right)\rho_{st}\left(x_b\right)\frac{\partial}{\partial x_b}\bullet\right\} \qquad (5.276)$$

By comparing the operators $\hat{L}_{FP}(x_b)$ and $\hat{L}_{FP}^{+}(x_b)$ one can immediately observe that the FP operator is not an autoadjoint operator; However, in order to make the FP operator to be autoadjoint, new operators are introduced, defined in conformity with the relation between the operators from above (5.272):

$$\hat{O}(x_b)\bullet=\frac{1}{\sqrt{\rho_{st}\left(x_b\right)}}\hat{L}_{FP}\left(x_b\right)\left\{\sqrt{\rho_{st}\left(x_b\right)}\bullet\right\} \qquad (5.277)$$

$$\hat{O}^{+}(x_b)\bullet=\sqrt{\rho_{st}(x_b)}\hat{L}_{FP}^{+}(x_b)\left\{\frac{1}{\sqrt{\rho_{st}(x_b)}}\bullet\right\} \tag{5.278}$$

as based on the operational properties (5.272). Because the operator $\hat{L}_{FP}(x_b)$ do not depend on the time $t_b$, the searching of the solutions $\rho_n(x_b,t_b)$ naturally appears, which satisfy the factorization condition:

$$\rho_n(x_b,t_b)=\varphi_n(x_b)\exp(\lambda_n t_b) \tag{5.279}$$

with the help of which the direct Fokker-Planck equation (5.273) will be reduced to the eigen-values equation (the adjoint one will be similar reduced):

$$\hat{L}_{FP}(x_b)\varphi_n(x_b)=\lambda_n\varphi_n(x_b) \tag{5.280}$$

In conformity with the stationary solution case, to the proper value $\lambda_0=0$ is corresponding the eigen-function $\varphi_0(x_b)=\rho_{st}(x_b)$. If, in addition, there are considered the transformations of the eigen-functions so that:

$$\varphi_n(x_b)=\sqrt{\rho_{st}(x_b)}\psi_n(x_b) \tag{5.281}$$

$$\varphi_n^{+}(x_b)=\frac{1}{\sqrt{\rho_{st}(x_b)}}\psi_n^{+}(x_b) \tag{5.282}$$

the problem of the eigen-values of the Fokker-Planck equation (5.280) became a problem of the eigen-values for the autoadjoint operators:

$$\hat{O}(x_b)\psi_n(x_b)=\lambda_n\psi_n(x_b) \tag{5.283}$$

Observe the fact that the new the eigen-value equation (5.283) has the same eigen-values $\lambda_n$ as the original Fokker-Planck equation (5.280), having, in addition, assured the viability of the completion and ortho-normalization relations

$$\int_{-\infty}^{+\infty}dx_b\psi_n(x_b)\psi_{n'}(x_b)=\delta_{n,n'} \tag{5.284}$$

$$\sum_n \psi_n(x_b)\psi_n(x_b') = \delta(x_b - x_b') \qquad (5.285)$$

which can be rewritten also in the following way:

$$\frac{1}{\rho_{st}(x_b)} \int_{-\infty}^{+\infty} dx_b \varphi_n(x_b)\varphi_{n'}(x_b) = \delta_{n,n'} \qquad (5.286)$$

$$\frac{1}{\sqrt{\rho_{st}(x_b)\rho_{st}(x_b')}} \sum_n \varphi_n(x_b)\varphi_n(x_b') = \delta(x_b - x_b') \qquad (5.287)$$

This way the orthonormalization and completion of the FP solutions are assured, which can be developed as special non-stationary solutions:

$$\rho(x_b,t_b) = \sum_n \rho_n \varphi_n(x_b)\exp(\lambda_n,t_b) \qquad (5.288)$$

The coefficients of this expansion will be written according with the initial conditions and with the above relations under the form:

$$\rho_n = \exp(-\lambda_n t_a) \int_{-\infty}^{+\infty} dx_a \varphi_n^+(x_a)\rho(x_a,t_a) \qquad (5.289)$$

Several observations are necessary. First of all, from the way of defining the $\psi$ functions towards the $\varphi$, see Eqs. (5.281) and (5.282), one observes that for the eigen-values $\lambda_0 = 0$ be associated with $\varphi_0(x_b) = \rho_{st}(x_b)$, the condition $\varphi_0^+(x_b) = 1$ has to be fulfilled. Therefore, from the relation (5.289) one has:

$$\rho_0 = \int_{-\infty}^{+\infty} dx_a \rho(x_a,t_a) = 1 \qquad (5.290)$$

thus recovering the initial normalization condition. Then, the eigen-values $\lambda_n$ are always negative. This thing can be easily proofed by successively writing:

$$\lambda_n = \int_{-\infty}^{+\infty} dx_b \varphi_n^+(x_b)\lambda_n \varphi_n(x_b) = \int_{-\infty}^{+\infty} dx_b \varphi_n^+(x_b)\hat{L}_{FP}(x_b)\varphi_n(x_b)$$

$$= \int_{-\infty}^{+\infty} dx_b \frac{\psi_n^+(x_b)}{\sqrt{P_{st}(x_b)}} \frac{\partial}{\partial x_b} \left\{ D^{(2)}(x_b) P_{st}(x_b) \frac{\partial}{\partial x_b} \left[ \frac{1}{P_{st}(x_b)} \sqrt{P_{st}(x_b)} \psi_n(x_b) \right] \right\}$$

$$= -\int_{-\infty}^{+\infty} dx_b \left\{ \frac{\partial}{\partial x_b} \frac{\psi_n(x_b)}{\sqrt{P_{st}(x_b)}} \right\}^2 D^{(2)}(x_b) P_{st}(x_b) \leq 0$$

$$(5.291)$$

because by definition, Eqs. (5.262) and (5.266), the second moment is positive

$$D^{(2)}(x_b) = \frac{1}{2} M_2(x_b, t_a; t_b) = \frac{1}{2} \int_{-\infty}^{+\infty} dx_b (x_b - x_a)^2 p(x_b, t_b; x_a, t_a) \geq 0$$

$$(5.292)$$

due to the fact that the conditioned probability densities and the stationary solutions are positive. This property of the eigen-values of the Fokker-Planck equation gives information about the asymptotic times limit $(t \to \infty)$, when all the contributions for which $\lambda_n < 0$ decay exponentially very quick, while letting only the term for which $\lambda_0 = 0$ as surviving. This way, one will be obtain:

$$\lim_{t_b \to +\infty} p(x_b, t_b) = P_0 \varphi_0(x_b) = P_{st}(x_b) \qquad (5.293)$$

As grounded on the above properties (5.290) for $\lambda_0 = 0$. This result shows that any temporary dependent FP solution is reduced to the stationary solution $P_{st}(x_b)$ in the asymptotic limit $t \to \infty$.

As also showed in Eq. (5.257), the transition probability or the conditioned probability density $p(x_b, t_b; x_a, t_a)$ can be seen as a special probability density with the initial condition given by the delta Dirac function. In such conditions, for the conditioned probability density one can obtain the coefficients of the spectral expansion in Eq. (5.289):

$$P_n = \exp(-\lambda_n t_a) \int_{-\infty}^{+\infty} dx_a' \varphi_n^+(x_a') \delta(x_a' - x_a) = \varphi_n^+(x_a) \exp(-\lambda_n t_a)$$

$$(5.294)$$

from where the spectral decomposition of the conditioned probability density in Eq. (5.288) results as

$$p\left(x_b,t_b;x_a,t_a\right)=\sum_n \varphi_n\left(x_b\right)\exp\left[\lambda_n\left(t_b-t_a\right)\right]\varphi_n^+\left(x_a\right) \qquad (5.295)$$

wherefrom the asymptotic times' limit is recovered:

$$\lim_{t_b\to+\infty} p\left(x_b,t_b;x_a,t_a\right)=\varphi_0\left(x_b\right)\varphi_0^+\left(x_a\right)=\rho_{st}\left(x_b\right) \qquad (5.296)$$

The last result shows that the conditioned probability density is reduced in the limit $t_b \to +\infty$ to the stationary solution, in which all the initial state memory $\left(x_a,t_a\right)$ is lost. Such properties should be regained also at the end of the analytical determinations aiming to explicit the conditioned probability density by the path integral method, as follows in the next.

### 5.4.3  FOKKER-PLANCK TO SCHRLIDINGER EQUATION TRANSFORMATION

In order to obtain the transformation of Fokker-Planck equation into Schrödinger equation one starts employing the autoadjoint operator introduced in Eq. (5.277), which along (5.276) becomes:

$$\hat{O}\left(x_b\right)=\frac{1}{\sqrt{\rho_{st}\left(x_b\right)}}\frac{\partial}{\partial x_b}\left\{D^{(2)}\left(x_b\right)\rho_{st}\left(x_b\right)\frac{\partial}{\partial x_b}\left[\frac{1}{\sqrt{\rho_{st}\left(x_b\right)}}\bullet\right]\right\} \qquad (5.297)$$

Next, the diffusion factor $D^{(2)}\left(x_b\right)\equiv D$ will be considered as a constant of the nonequilibrium dynamic process. With this identification in the stationary (5.269) one yields the equation:

$$\frac{\partial}{\partial x_b}\rho_{st}\left(x_b\right)=\frac{D^{(1)}\left(x_b\right)}{D}\rho_{st}\left(x_b\right) \qquad (5.298)$$

With this, the autoadjoint operator expression form above, Eq. (5.297), can be successively developed (Gardiner, 1994; Risken, 1984):

$$\hat{O}(x_b) = \frac{D}{\sqrt{\rho_{st}(x_b)}} \frac{\partial}{\partial x_b} \left\{ \rho_{st}(x_b) \left[ \begin{array}{c} -\dfrac{1}{2\sqrt{\rho_{st}(x_b)}} \bullet \dfrac{\partial}{\partial x_b} \rho_{st}(x_b) \\[2ex] +\dfrac{1}{\sqrt{\rho_{st}(x_b)}} \dfrac{\partial}{\partial x_b} \bullet \end{array} \right] \right\}$$

$$= \frac{D}{\sqrt{\rho_{st}(x_b)}} \left\{ \begin{array}{c} \dfrac{1}{4} \dfrac{1}{\sqrt{\rho_{st}(x_b)^3}} \bullet \left( \dfrac{\partial}{\partial x_b} \rho_{st}(x_b) \right)^2 \\[3ex] -\dfrac{1}{2\sqrt{\rho_{st}(x_b)}} \bullet \dfrac{\partial^2}{\partial x_b^2} \rho_{st}(x_b) + \sqrt{\rho_{st}(x_b)} \dfrac{\partial^2}{\partial x_b^2} \bullet \end{array} \right\}$$

$$= D \frac{\partial^2}{\partial x_b^2} \bullet + \frac{1}{4} \frac{D}{\rho_{st}(x_b)^2} \frac{D^{(1)}(x_b)^2}{D^2} \bullet \rho_{st}(x_b)^2$$

$$-\frac{1}{2} \frac{D}{\rho_{st}(x_b)} \bullet \left[ \frac{\partial}{\partial x_b} \left( \frac{D^{(1)}(x_b)}{D} \rho_{st}(x_b) \right) \right]$$

$$= D \frac{\partial^2}{\partial x_b^2} \bullet - \frac{1}{2} \frac{\partial D^{(1)}(x_b)}{\partial x_b} \bullet - \frac{1}{4D} D^{(1)}(x_b)^2 \bullet$$

$$= D \frac{\partial^2}{\partial x_b^2} \bullet - U(x_b) \bullet;$$

$$\tag{5.299}$$

$$U(x_b) = \frac{1}{2} \frac{\partial D^{(1)}(x_b)}{\partial x_b} + \frac{1}{4D} D^{(1)}(x_b)^2 \tag{5.300}$$

The obtained expression permits the direct analogy with the Hamiltonian operator specific to the Schrödinger equation:

$$\hat{O}(x_b) = D \frac{\partial^2}{\partial x_b^2} \bullet - U(x_b) \bullet \equiv \frac{\hbar^2}{2M} \frac{\partial^2}{\partial x_b^2} \bullet - U(x_b) \bullet = -\hat{H}(x_b) \quad (5.301)$$

This correspondence stays at the basis of the transformation of the Fokker-Planck equation into the associated Schrödinger equation. Through the corresponding relations (5.299) and (5.301) one has in a direct way the transformation relation:

$$D = \frac{\hbar^2}{2m} \qquad (5.302a)$$

Beside, under an harmonic external potential

$$V(x) = \gamma x^2 / 2 \qquad (5.302b)$$

the drift factor automatically results as

$$D^{(1)}(x_b) = -[\partial / \partial x_b] V(x_b) = -\gamma x_b \qquad (5.302c)$$

and the effective potential characteristic to the Fokker-Planck dynamic will be given by:

$$U_{FP}(x_b) = \frac{1}{4D} \gamma^2 x_b^2 - \frac{\gamma}{2} \qquad (5.303)$$

If the relation (5.303) is placed in direct correspondence with the harmonic potential for the associated Schrödinger equation

$$U_S(x_b) = \frac{m}{2} \omega^2 x_b^2 + CT \qquad (5.304)$$

there results also the identifications:

$$CT = -\frac{\gamma}{2} \qquad (5.305)$$

$$\omega = \frac{\gamma}{\sqrt{2mD}} \qquad (5.306)$$

with $\gamma$ bearing the role of the (mesoscopic) friction coefficient.

Whit these quantities one can reveal the essential difference of the Fokker-Planck equation towards the Schrödinger one for the same type of external potential (harmonic, in here). Firstly, worth remarking the fact that, when it is consider eigen-value equation for the Hamiltonian operator specific to Schrödinger equation

$$\widehat{H}(x_b)\psi_n(x_b) = E_n\psi_n(x_b) \qquad (5.307)$$

and it is compared with that one in the autoadjoint operator Fokker-Planck (5.283), with the identification $\widehat{O}(x_b) = -\widehat{H}(x_b)$, there will automatically results also the connection between the associated eigen-values:

$$\lambda_n = -E_n \qquad (5.308)$$

This observation is fundamental, reveling the fact that the characteristic energies of the Fokker-Planck equation are reactive energies, and in the final, nonequilibrium energies. This aspect is directly correlated with the nonequilibrium character specific for the Fokker-Planck equation while modeling open systems (driven by drift diffusion and factors, stochastic noise, etc.). Moreover, if the analytical solution of the eigen-values for the Schrödinger equation with the potential $U_s(x_b)$ is considered, the consecrated expression is obtained:

$$E_n^S = \hbar\omega\left(n + \frac{1}{2}\right) + CT \qquad (5.309)$$

from which one is observing how, by employing the Eqs. (5.305), (5.306) and (5.302a), the zero level term and the introduced constant are reciprocally canceling

$$\frac{\hbar\omega}{2} = \frac{\hbar}{2}\frac{\gamma}{\sqrt{2m\dfrac{\hbar^2}{2m}}} = \frac{\gamma}{2} = -CT \qquad (5.310)$$

leading with Eq. (5.309) to the expression:

$$E_n^{FP} = \hbar\omega n = \gamma n = -\lambda_n \qquad (5.311)$$

This way $\lambda_0 = 0$ is indeed recovered yet with a fundamental different meaning respecting Schrödinger case for the harmonic oscillator: in the nonequilibrium FP modeling the zero level gives zero, since rooting in the dynamic action induced by diffusion and drift.

The transformation of Fokker-Planck equation into Schrödinger equation also serves in finding the FP solution for the transition probability density, here for the harmonic oscillator. To this aim, one firstly considers the eigen-function of the Schrödinger equation obtained for the harmonic potential (see Volume I/Section 3.3.2 of the present five-volume work):

$$\psi_n(x_b) = C_n H_n\left(\sqrt{\frac{m\omega}{\hbar}}x_b\right)\exp\left\{-\frac{m\omega}{2\hbar}x_b^2\right\} \qquad (5.312)$$

with $H_n(z)$ as the Hermite polynomials, and $m\omega/\hbar = \gamma/(2D)$, according with the above correspondence, Eqs. (5.302) and (5.305), respectively. This way, the solution can be rewritten in the form:

$$\psi_n(x_b) = C_n H_n\left(\sqrt{\frac{\gamma}{2D}}x_b\right)\exp\left\{-\frac{\gamma}{4D}x_b^2\right\} \qquad (5.313)$$

In order to determine the normalization constant $C_n$ the ortho-normalization property will be used along the Hermite functions' property:

$$\int_{-\infty}^{+\infty} dz H_n(z) H_{n'}(z)\exp\left(-z^2\right) = \sqrt{\pi}\, 2^n n!\delta_{n,n'} \qquad (5.314)$$

This way, the ortho-normalization condition (5.284) of the eigen-function $\psi_n$ (identical for Schrödinger and Fokker-Planck equations) successively becomes:

$$\delta_{n,n'} = \int_{-\infty}^{+\infty} dx\psi_n(x_b)\psi_{n'}(x_b)$$

$$= \int_{-\infty}^{+\infty} dx_b C_n^2 H_n\left(\sqrt{\frac{\gamma}{2D}}x_b\right) H_{n'}\left(\sqrt{\frac{\gamma}{2D}}x_b\right)\exp\left\{-\left(\sqrt{\frac{\gamma}{2D}}x_b\right)^2\right\}$$

$$\overset{z=\sqrt{\frac{\gamma}{2D}}x_b}{=} C_n^2\sqrt{\frac{2D}{\gamma}}\int_{-\infty}^{+\infty} dx_b H_n(z)H_{n'}(z)\exp\{-z^2\} = C_n^2\sqrt{\frac{2D}{\gamma}}\sqrt{\pi}\,2^n n!\delta_{n,n'}$$

$$(5.315)$$

leaving for the coefficients of Eq. (5.313) with the actual form:

$$C_n = \frac{1}{\sqrt{2^n n!}}\sqrt[4]{\frac{\gamma}{2\pi D}} \qquad (5.316)$$

while providing for the eigen functions (5.313) the expressions:

$$\psi_n(x_b) = \frac{1}{\sqrt{2^n n!}} \sqrt[4]{\frac{\gamma}{2\pi D}} H_n\left(\sqrt{\frac{\gamma}{2D}} x_b\right) \exp\left\{-\frac{\gamma}{4D} x_b^2\right\} \qquad (5.317)$$

In order to write the eigen-functions associated to Fokker-Planck equation the stationary solution for the harmonic external potential should be firstly evaluated: the stationary solution equation (5.298), with Eq. (5.302b), yields immediately:

$$\rho_{st}(x_b) = ct \exp\left\{-\frac{\gamma}{2D} x_b^2\right\} \qquad (5.318)$$

with the integration constant being determined from the normalization condition:

$$1 = \int_{-\infty}^{+\infty} dx_b \rho_{st}(x_b) = ct \int_{-\infty}^{+\infty} dx_b \exp\left[-\frac{\gamma}{2D} x_b^2\right] = ct \sqrt{\frac{2D\pi}{\gamma}} \Rightarrow ct = \sqrt{\frac{\gamma}{2D\pi}} \qquad (5.319)$$

thus providing the stationary solution with the form:

$$\rho_{st}(x_b) = \sqrt{\frac{\gamma}{2D\pi}} \exp\left\{-\frac{\gamma}{2D} x_b^2\right\} \qquad (5.320)$$

From now on the eigen-function solutions of the Fokker-Planck equation in the $\hat{L}_{FP}$ operator immediately results through the relations (5.281) and (5.282):

$$\varphi_n(x_b) = \frac{1}{\sqrt{2^n n!}} \sqrt{\frac{\gamma}{2\pi D}} H_n\left(\sqrt{\frac{\gamma}{2D}} x_b\right) \exp\left\{-\frac{\gamma}{2D} x_b^2\right\} \qquad (5.321a)$$

$$\varphi_n^+(x_b) = \frac{1}{\sqrt{2^n n!}} H_n\left(\sqrt{\frac{\gamma}{2D}} x_b\right) \qquad (5.321b)$$

By taking into consideration the property of the Hermite polynomials $(H_0 = 1)$, the identities

$$\varphi_0(x_b) = \rho_{st}(x_b) \qquad (5.322a)$$

$$\varphi_0^+ (x_b) = 1 \qquad (5.322b)$$

are naturally found. Finally, one can evaluate the conditioned probability density for the considered harmonic potential, by writing Eq. (5.295) in the form:

$$\rho(x_b, t_b; x_a, t_a) = \sum_n \varphi_n(x_b) \exp\left[\lambda_n(t_b - t_a)\right] \varphi_n^+(x_a)$$

$$= \sqrt{\frac{\gamma}{2D\pi}} \exp\left\{-\left(\sqrt{\frac{\gamma}{2D}} x_b\right)^2\right\} \sum_{n=0}^{\infty} \frac{\exp\left[-n\gamma(t_b - t_a)\right]}{2^n n!}$$

$$\times H_n\left(\sqrt{\frac{\gamma}{2D}} x_b\right) H_n\left(\sqrt{\frac{\gamma}{2D}} x_a\right)$$

$$(5.323)$$

By using the Mehlers formula for the Hermite polynomials, see Volume I/ Section 3.3.2 of the present five-volume set (Putz, 2016a)

$$\sum_{n=0}^{\infty} \frac{z^n}{2^n n!} H_n(x) H_n(y) = \frac{1}{\sqrt{1-z^2}} \exp\left\{\frac{2xyz - (x^2 + y^2)z^2}{1-z^2}\right\} \qquad (5.324)$$

the final expression of the conditioned probability density is obtained after elementary calculations under the form:

$$\rho_{Armonic}(x_b, t_b; x_a, t_a) = \frac{1}{\sqrt{2\pi K_2(x_a, t_a; t_b)}} \exp\left\{-\frac{\left[x_b - K_1(x_a, t_a; t_b)\right]^2}{2K_2(x_a, t_a; t_b)}\right\}$$

$$(5.325)$$

where:

$$K_1(x_a, t_a; t_b) = x_a \exp\left[-\gamma(t_b - t_a)\right] \qquad (5.326)$$

stays for the first cumulant, while the second cumulant looks like:

$$K_2(x_a, t_a; t_b) = \frac{D}{\gamma}\left\{1 - \exp\left[-2\gamma(t_b - t_a)\right]\right\} \qquad (5.327)$$

Worth remarking the Gaussian form of the conditioned probability density in the evolution dynamics under the influence of the harmonic potential: the first cumulant exponentially decreases from the value $x_a$, while the second cumulant starts its evolution from the zero value and accedes to a value constant in the asymptotic $t_b \to \infty$ limit. Practically, the transition probability as the FP solution for the external harmonic potential starts its evolution from a Dirac distribution at the initial moment $(t_b = t_a)$ and arriving, upon long times to shape as a stationary Gaussian distribution with characteristics depending by the nonequilibrium evolution parameters ($\gamma$ and $D$). The transcription of the Fokker-Planck equation into Schrödinger equation helped in finding the solution for the initial Fokker-Planck equation. Still, there were revealed specific aspects, essentially different respecting those derived from the Schrödinger equation and allied solutions, reaffirming the idea that the Fokker-Planck equation is characteristic to the nonequilibrium quantum statistical evolution and it correspond to an mesoscopic evolution picture (by the diffusion and drift factors).

The external enharmonic potential case will be also the next framework in which, by means of the path integrals formalism, the non-stationary solutions of the Fokker-Planck equation will be searched for.

### 5.4.4  NON-EQUILIBRIUM LAGRANGIAN

Here one will proceed with passage from the Schrödinger equation framework to that one characterized by (Feynman) path integrals; this, because the path integrals present the advantage of a treating by Feynman diagrams with an increase degree of intuition in treating the correlations' and interactions' dynamics. Actually, instead of solving a problem of eigenvalues and then of a spectral summation, as is the case in solving by correspondence (principle) with the Schrödinger equation (see the previous sections), when using the path integrals the conditioned probability density is directly written by the integral evaluation:

$$\rho\left(x_b t_b; x_a t_a\right) \to \left(x_b t_b; x_a t_a\right)_{FP} = \int\limits_{x(t_a)=x_a}^{x(t_b)=x_b} \mathrm{D'}x(t) L_{FP}\left(x(t), \dot{x}(t), t\right) \quad (5.328)$$

This expression represents the third level of the path integrals representation, after the classical quantum mechanical picture (see Volume I of the present five-volume work)

$$\exp\left\{\frac{i}{\hbar}\int_{t_a}^{t_b} dt \, L_{cl}\left(x(t),\dot{x}(t),t\right)\right\} \tag{5.329}$$

and after the quantum statistical of Section 2.5 of the present volume

$$\exp\left\{-\frac{1}{\hbar}\int_0^{\hbar\beta} d\tau \, L_{FP}\left(x(\tau),\dot{x}(\tau),\tau\right)\right\} \tag{5.330}$$

The formulation introduced by the Lagrangian (5.328) will be called as the Fokker-Planck picture, there where one remarks the absence of the reduced Plank constant $\hbar$, being such absence a characteristic of the Fokker-Planck formalism, emphasizing on the mesoscopic character as a next (statistical) level grounded on the microscopic $\hbar$ related world: the parametric correspondence between all these three levels of quantum dynamics was anticipated in Section 2.3.

Next, one has to specify the Lagrangian of the expression (5.328) for a nonequilibrium dynamics. To this aim one works out the parametric form (Kleinert, 2001):

$$L_{FP}\left(x(t),\dot{x}(t),t\right) = A\left[\dot{x}(t) + \frac{1}{m\gamma}\frac{d}{dx}V\left(x(t)\right)\right]^2 - B\frac{d^2}{dx^2}V\left(x(t)\right) \tag{5.331}$$

Which will be a posteriori justified as belonging to the studied Fokker-Planck equation though determining the A and B coefficients (in this section) as well as by retrieving the conditioned harmonic probability density solution (already formulated by employing the Fokker-Planck with the Schrödinger equation correspondence, see Section 5.4.3) using this Lagrangian with determined coefficients (in next sub-Section).

First of all, one will consider the Fokker-Planck conditioned probability density in the temporal coarse graining time $\varepsilon = t_n - t_{n-1}$:

$$\left(x_n,\varepsilon;x_{n-1},0\right)_{FP} = n_{FP}\exp\left\{-A\varepsilon\left[\frac{x_n-x_{n-1}}{\varepsilon} + \frac{V'(\bar{x}_n)}{m\gamma}\right]^2 + B\varepsilon V''(\bar{x}_n)\right\} \tag{5.332}$$

where for $\bar{x}_n$ an intermediate value between $x_{n-1}$ and $x_n$ will be implemented with the parametric form:

$$\bar{x}_n = \alpha x_n + (1-\alpha)x_{n-1} = x_n - (1-\alpha)(x_n - x_{n-1}) \qquad (5.333)$$

Whit such scaling, the potential dependences will be considered themselves as expanded until the second order:

$$V'(\bar{x}_n) = V'\left[x_n - (1-\alpha)(x_n - x_{n-1})\right] \cong V'(x_n) - (1-\alpha)V''(x_n)(x_n - x_{n-1}) \qquad (5.334)$$

$$V''(\bar{x}_n) \cong V''(x_n) \qquad (5.335)$$

By making use now of the relationship between the probability density and the conditioned probability density one can successively write (also by skipping the functions' indices, i.e., $W_1(x_n,t_n) = W(x_n,t_n)$, for simplicity, being no reason of confusion):

$$W(x_n,t_n) = \int_{-\infty}^{+\infty} dx_{n-1}(x_n,\varepsilon;x_{n-1},0)_{FP} W(x_{n-1},t_{n-1})$$

$$= \int_{-\infty}^{+\infty} d(\Delta x_n) n_{FP} W(x_n - \Delta x_n, t_n - \varepsilon)$$

$$\exp\left\{-A\varepsilon\left[\frac{\Delta x_n}{\varepsilon} + \frac{V'(x_n) - (1-\alpha)V''(x_n)\Delta x_n}{m\gamma}\right]^2 + B\varepsilon V''(x_n)\right\} \qquad (5.336)$$

with the notations:

$$t_{n-1} = t_n - \varepsilon, \quad x_{n-1} = x_n - \Delta x_n \qquad (5.337)$$

Next the first order in $\varepsilon$ will be employed. However, a note is compulsory here: there is so clearly that $\Delta t \propto \varepsilon$ along the dimensionality proportion $\Delta x_n \propto \sqrt{\varepsilon}$; yet the last proportionality comes from the slicing of the term $\exp\left\{-A\int_{t_a}^{t_b} d\dot{x}\right\}$, for instance, which in arbitrary units for (e.g. fixed to the unity) becomes $\exp\{-\varepsilon(\Delta x)^2 / \varepsilon^2\}$, from where the proportionality $\Delta x_n \propto \sqrt{\varepsilon}$ is immediately identified in order the exponential dimensionality to be kept.

In such conditions the next expansion is provided:

$$W\left(x_n - \Delta x_n, t_n - \varepsilon\right) \cong W\left(x_n, t_n\right) - \Delta x_n \frac{\partial W\left(x_n, t_n\right)}{\partial x_n}$$

$$+ \frac{1}{2}\left(\Delta x_n\right)^2 \frac{\partial^2 W\left(x_n, t_n\right)}{\partial x_n^2} - \varepsilon \frac{\partial W\left(x_n, t_n\right)}{\partial t_n} \quad (5.338)$$

By inserting (5.338) into the equality (5.336) it becomes successively (by retaining only the terms until the first order expansion in $\varepsilon$):

$$W\left(x_n, t_n\right) \cong \int\limits_{-\infty}^{+\infty} d\left(\Delta x_n\right) n_{FP} \exp\left[-A\frac{\left(\Delta x_n\right)^2}{\varepsilon}\right]$$

$$\times \begin{bmatrix} 1 - 2\dfrac{A}{m\gamma}V'\left(x_n\right)\Delta x_n - A\varepsilon\dfrac{V'\left(x_n\right)^2}{m^2\gamma^2} + \dfrac{2A(1-\alpha)V''\left(x_n\right)\left(\Delta x_n\right)^2}{m\gamma} \\ + B\varepsilon V''\left(x_n\right) + \dfrac{2A^2}{m^2\gamma^2}V'\left(x_n\right)^2\left(\Delta x_n\right)^2 \end{bmatrix}$$

$$\times \left[ W\left(x_n, t_n\right) - \Delta x_n \frac{\partial W\left(x_n, t_n\right)}{\partial x_n} + \frac{1}{2}\left(\Delta x_n\right)^2 \frac{\partial^2 W\left(x_n, t_n\right)}{\partial x_n^2} - \varepsilon \frac{\partial W\left(x_n, t_n\right)}{\partial t_n}\right]$$

$$= \int\limits_{-\infty}^{+\infty} d\left(\Delta x_n\right) n_{FP} \exp\left[-A\frac{\left(\Delta x_n\right)^2}{\varepsilon}\right]$$

$$\times \left[ W\left(x_n, t_n\right) - 2\frac{A}{m\gamma}V'\left(x_n\right)W\left(x_n, t_n\right)\Delta x_n - A\varepsilon\frac{V'\left(x_n\right)^2}{m^2\gamma^2}W\left(x_n, t_n\right)\right.$$

$$+ \frac{2A(1-\alpha)V''\left(x_n\right)}{m\gamma}\left(x_n\right)^2 W\left(x_n, t_n\right) + B\varepsilon V''\left(x_n\right)W\left(x_n, t_n\right)$$

$$+ \frac{2A^2}{m^2\gamma^2}V'\left(x_n\right)^2\left(\Delta x_n\right)^2 W\left(x_n, t_n\right)$$

$$- \Delta x_n \frac{\partial W\left(x_n, t_n\right)}{\partial x_n} + 2\frac{A}{m\gamma}V'\left(x_n\right)\left(\Delta x_n\right)^2\frac{\partial W\left(x_n, t_n\right)}{\partial x_n}$$

$$\left. + \frac{1}{2}\left(\Delta x_n\right)^2 \frac{\partial^2 W\left(x_n, t_n\right)}{\partial x_n^2} - \varepsilon \frac{\partial W\left(x_n, t_n\right)}{\partial t_n}\right] \quad (5.339)$$

Some of the terms in Eq. (5.339) are vanishing due to the integral property:

$$\int_{-\infty}^{+\infty} d(\Delta x_n) \Delta x_n \exp\left[-\frac{A}{\varepsilon}(\Delta x_n)^2\right] = 0 \tag{5.340}$$

while the even terms in the powers of $\Delta x_n$ under the integrals of the Eq. (5.340) type also satisfy the conditions:

$$1 = \int_{-\infty}^{+\infty} d(\Delta x_n) n_{FP} \exp\left[-\frac{A}{\varepsilon}(\Delta x_n)^2\right] = n_{FP}\sqrt{\frac{\varepsilon\pi}{A}} \Rightarrow n_{FP} = \sqrt{\frac{A}{\varepsilon\pi}} \tag{5.341}$$

and respectively:

$$\int_{-\infty}^{+\infty} d(\Delta x_n) n_{FP} (\Delta x_n)^2 \exp\left[-\frac{A}{\varepsilon}(\Delta x_n)^2\right] = \frac{\varepsilon}{2A} \tag{5.342}$$

where in the last relation the constant deduced in Eq. (5.341) was used. Finally, by considering all properties in expression (5.340)–(5.342), the expression (5.339) reduces to:

$$
\begin{aligned}
W(x_n, t_n) = &\, W(x_n, t_n) - A\varepsilon \frac{V'(x_n)^2}{m^2\gamma^2} W(x_n, t_n) \\
&+ \frac{\varepsilon}{2A} \frac{2A(1-\alpha)V''(x_n)}{m\gamma} W(x_n, t_n) \\
&+ B\varepsilon V''(x_n) W(x_n, t_n) + \frac{\varepsilon}{2A} \frac{2A^2}{m^2\gamma^2} V'(x_n)^2 W(x_n, t_n) \\
&+ 2\frac{A}{m\gamma} V'(x_n) \frac{\varepsilon}{2A} \frac{\partial W(x_n, t_n)}{\partial x_n} \\
&+ \frac{1}{2} \frac{\varepsilon}{2A} \frac{\partial^2 W(x_n, t_n)}{\partial x_n^2} - \varepsilon \frac{\partial W(x_n, t_n)}{\partial t_n}
\end{aligned} \tag{5.343}
$$

By making all possible groupings the simplified relation has the form:

$$\frac{\partial W(x_n,t_n)}{\partial t_n} = \frac{V'(x_n)}{m\gamma}\frac{\partial W(x_n,t_n)}{\partial x_n}$$

$$+\left[BV''(x_n)+\frac{(1-\alpha)V''(x_n)}{m\gamma}\right]W(x_n,t_n)+\frac{\partial^2}{\partial x_n^2}\left[\frac{1}{4A}W(x_n,t_n)\right]$$

$$(5.344)$$

This equation has explicitly the Fokker-Planck equation form (5.239), at the probability density level (with $D^{(2)}(x_n)\equiv D$):

$$\frac{\partial W(x_n,t_n)}{\partial t_n} = -\frac{\partial}{\partial x_n}\{D^{(1)}(x_n)W(x_n,t_n)\}+\frac{\partial^2}{\partial x_n^2}\{DW(x_n,t_n)\} \quad (5.345)$$

By making the correspondences and one-to-ne identifications for Eqs. (5.344) and (5.345) the next system is formed:

$$\begin{cases} \dfrac{1}{4A}=D \\[2mm] D^{(1)}(x_n)=-\dfrac{V'(x_n)}{m\gamma} \\[2mm] D^{(1)}{}'(x_n)=-BV''(x_n)-\dfrac{(1-\alpha)V''(x_n)}{m\gamma} \end{cases} \qquad (5.346)$$

with the solution:

$$\begin{cases} A=\dfrac{1}{4D} \\[2mm] B=\dfrac{\alpha}{m\gamma} \end{cases} \qquad (5.347)$$

Next the case in which $\alpha =1/2$ (midpoint calculation) will be considered, for which the nonequilibrium Euclidian Lagrangian (5.331) will have the working form:

$$L_{FP}(x(t),\dot{x}(t),t)=\frac{1}{4D}\left[\dot{x}(t)+\frac{1}{m\gamma}V'(x(t))\right]^2-\frac{1}{2m\gamma}V''(x(t))$$

$$(5.348)$$

Until now we actually proofed the fact that the Lagrangian (5.348) corresponds to a modeling (mesoscopic) dynamics which is described by the Fokker-Planck equation. Next, one will consider this nonequilibrium form of the Euclidian Lagrangian, to be characteristic to the effective electronic evolution, specific to the mesoscopic characterization.

### 5.4.5  THE NON-EQUILIBRIUM HARMONIC SOLUTION

There is very instructive to find of the nonequilibrium harmonic solution by the path integrals method as starting from the analyzed nonequilibrium Lagrangian (of the preceding Section): firstly because it regain the harmonic solution previously found by transforming the Fokker-Planck equation into the associate Schrödinger equation, being this way proved the reliability of the used Lagrangian (5.348) as well as the consistency of the path integrals method in solving the Fokker-Planck problem. The harmonic solution will be used in the following step, for evaluating the nonequilibrium solution in case of a harmonic generalized potential.

As projected, one is starting from the conditioned probability density expression in the path integral form (5.328) with (5.348) (Kleinert, 2001):

$$
\left( x_b, t_b; x_a, t_a \right)_{FP} = \int_{x(t_a)=x_a}^{x(t_b)=x_b} \mathcal{D}'x(t)\exp\left\{ \begin{array}{l} -\dfrac{1}{4D}\displaystyle\int_{t_a}^{t_b} dt\left[ \dot{x}(t)+\dfrac{1}{mg}V'\left(x(t)\right) \right]^2 \\ +\dfrac{1}{2mg}\displaystyle\int_{t_a}^{t_b} dt V''\left(x(t)\right) \end{array} \right\}
$$

$$\tag{5.349}$$

In order to consider also the spectral translation of the Fokker-P eigenvalues respecting the Schrödinger ones, the previous path integral will be reconsidered by introducing the transformation of the potential:

$$
V(x) \rightarrow m\gamma U(x) \tag{5.350}
$$

This way, the conditioned probability density becomes:

$$\left(x_b, t_b; x_a, t_a\right)_{FP} = \int_{x(t_a)=x_a}^{x(t_b)=x_b} \mathcal{D}'x(t) \exp\left\{ \begin{array}{l} -\dfrac{1}{4D}\displaystyle\int_{t_a}^{t_b} dt\left[\dot{x}(t)+U'\left(x(t)\right)\right]^2 \\[2mm] +\dfrac{1}{2}\displaystyle\int_{t_a}^{t_b} dt U''\left(x(t)\right) \end{array} \right\}$$

(5.351)

Starting from this form one will further run the Onsanger-Matchlup harmonic potential (equivalent with the Brownian movement) form:

$$U(x) = \frac{\gamma}{2} x^2$$

(5.352)

by whom the path integral, (5.351) casts as:

$$\left(x_b, t_b; x_a, t_a\right)_{FP}^{x^2} = \int_{x(t_a)=x_a}^{x(t_b)=x_b} \mathcal{D}'x(t) \exp\left\{ -\frac{1}{4D}\int_{t_a}^{t_b} dt \left[ \begin{array}{l} \dot{x}^2(t)+2\gamma\dot{x}(t)x(t) \\[1mm] +\gamma^2 x^2(t) \end{array} \right] \right\}$$

$$\times \exp\left[\frac{\gamma}{2}\left(t_b-t_a\right)\right]$$

$$= \exp\left[\frac{\gamma}{2}\left(t_b-t_a\right)\right]\exp\left[-\frac{\gamma}{4D}\left(x_b^2-x_a^2\right)\right]\int_{x(t_a)=x_a}^{x(t_b)=x_b} D'x(t)$$

$$\times \exp\left\{-\frac{1}{2D}\int_{t_a}^{t_b} dt\left[\frac{1}{2}\dot{x}^2(t)+\frac{1}{2}\gamma^2 x^2(t)\right]\right\}$$

$$= \exp\left[\frac{\gamma}{2}\left(t_b-t_a\right)\right]\exp\left[-\frac{\gamma}{4D}\left(x_b^2-x_a^2\right)\right]\left(x_b, t_b; x_a, t_a\right)_{FP}^{x^2[0]}$$

(5.353)

With the quantum statistics to Fokker-Planck correspondences (as of Section 2.3):

$$\left\{ \begin{array}{l} \gamma \leftrightarrow \omega \\[2mm] \dfrac{1}{2D} \leftrightarrow \dfrac{m}{\hbar} \end{array} \right.$$

(5.354a)

then, the path integral (5.353) is transferred into harmonic quantum mechanical path integral problem. Further consideration of the correspondence (5.354a) along those of the Section 2.3 this time in the harmonical motion path integral result (see Volume I/Section 4.3.3 of the present five-volume work), the analytical expression is obtained:

$$
\left(x_b,t_b;x_a,t_a\right)_{FP}^{x^2} = \exp\left[\frac{\gamma}{2}\left(t_b-t_a\right)\right]\exp\left[-\frac{\gamma}{4D}\left(x_b^2-x_a^2\right)\right]
$$

$$
\times\sqrt{\frac{\gamma}{4D\pi\sinh\left[\gamma\left(t_b-t_a\right)\right]}}\exp\left\{-\frac{\gamma\left[\left(x_b^2+x_a^2\right)\cosh\left[\gamma\left(t_b-t_a\right)\right]-2x_ax_b\right]}{4D\sinh\left[\gamma\left(t_b-t_a\right)\right]}\right\}
$$

(5.355)

With the algebraic elementary transformations performed (i.e., the hyperbolic functions became exponential functions, etc. as in Section 2.3 anticipated, see also Volume I/Section 4.3.3 of the present five-volume work) the path integrals' harmonic solution (5.355) will take the actual form:

$$
\left(x_b,t_b;x_a,t_a\right)_{FP}^{x^2} = \frac{1}{\sqrt{2\pi\dfrac{D}{\gamma}\left\{1-\exp\left[-2\gamma\left(t_b-t_a\right)\right]\right\}}}
$$

$$
\exp\left\{-\frac{\left[x_b-x_a\exp\left[-\gamma\left(t_b-t_a\right)\right]\right]^2}{2\dfrac{D}{\gamma}\left[1-\exp\left[-2\gamma\left(t_b-t_a\right)\right]\right]}\right\}
$$

(5.356)

equivalently with (5.325) – as found by the Fokker-Planck to Schrödinger equation transforming (having identified the respective cumulants). This way, the self-consistence of the path integrals formalism with Schrödinger equation's solution was once more revealed.

Up to now, in Chapter 2 of the present Volume the quantum mechanical (largely exposed in the Volume I/Section 4.2 (Putz, 2016a) of the present five-volume set) to quantum transcription $QM \rightarrow QS$ was implemented, while the analytical transition to Fokker-Planck (stochastic) approach was only anticipated. Instead, in the present Section explicit Fokker-Planck picture was considered, by which the Fokker-Planck equation must be satisfied by any (nonequilibrium) solution calculated by the path integrals. The two sets of identifications (QM-to-QS and QS-to-FP) are also unitary in the light of transformation

$$m = \gamma \hbar / (2D\omega) \tag{5.354b}$$

Accordingly, the conditional probability density in case of harmonic potential $(x_b, t_b; x_a, t_a)_{FP}^{x^2}$ has the analytically form given in the Eq. (5.356) and satisfy the Fokker-Planck equation under the form:

$$\frac{\partial (x_b, t_b; x_a, t_a)_{FP}^{x^2}}{\partial t_b} = \frac{\partial}{\partial x_b} \left[ (\gamma x_b)(x_b, t_b; x_a, t_a)_{FP}^{x^2} \right] + D \frac{\partial^2}{\partial x_b^2} \left[ (x_b, t_b; x_a, t_a)_{FP}^{x^2} \right]$$

$$\tag{5.357}$$

This fruitful analysis direction can be continued also for evaluating the Fokker-Planck solution in case of generalized harmonic potential.

### 5.4.6 THE NON-EQUILIBRIUM GENERALIZED ANHARMONIC SOLUTION

One considers the generalized harmonic potential

$$U(x) = \gamma \frac{x^2}{2} + g \frac{x^4}{4} \tag{5.358}$$

for which the transition probability will be calculated by neglecting all the contribution of type $g^n$, $n \geq 2$. This way, the path integral (5.328) successively becomes:

$$(x_b, t_b; x_a, t_a)_{FP}^{x^4} = \int_{x(t_a)=x_a}^{x(t_b)=x_b} \mathcal{D}'x(t) \exp \left\{ \begin{array}{l} -\dfrac{1}{4D} \displaystyle\int_{t_a}^{t_b} dt \left[ \dot{x}(t) + \gamma x(t) + gx(t)^3 \right]^2 \\[4mm] +\dfrac{1}{2} \displaystyle\int_{t_a}^{t_b} dt \left[ \gamma + 3gx(t)^2 \right] \end{array} \right\}$$

$$= \int_{x(t_a)=x_a}^{x(t_b)=x_b} \mathcal{D}'x(t) \exp\left\{-\frac{1}{4D}\int_{t_a}^{t_b} dt \begin{bmatrix} \dot{x}^2(t)+2\gamma\dot{x}(t)x(t)+\gamma^2 x^2(t) \\ +2g\dot{x}(t)x^3(t)+2g\gamma x^4(t) \end{bmatrix}\right.$$

$$\left. +\frac{1}{2}\int_{t_a}^{t_b} dt\left[3gx^2(t)\right]+\frac{1}{2}\gamma(t_b-t_a)\right\}$$

$$= \exp\left[\frac{1}{2}\gamma(t_b-t_a)\right]\exp\left[-\frac{1}{4D}\gamma(x_b^2-x_a^2)\right]\exp\left[-\frac{g}{8D}(x_b^4-x_a^4)\right]$$

$$\times \int_{x(t_a)=x_a}^{x(t_b)=x_b} \mathcal{D}'x(t)\exp\left\{-\frac{1}{4D}\int_{t_a}^{t_b} dt\left[\dot{x}(t)^2+\gamma^2 x(t)^2\right]\right\}\exp\left\{ \begin{matrix} -\dfrac{1}{2D}\int_{t_a}^{t_b} dt\, g\gamma x(t)^4 \\ \\ +\dfrac{3}{2}\int_{t_a}^{t_b} dt\, gx(t)^2 \end{matrix}\right\}$$

$$\times \int_{x(t_a)=x_a}^{x(t_b)=x_b} \mathcal{D}'x(t)\exp\left\{-\frac{1}{4D}\int_{t_a}^{t_b} dt\left[\dot{x}^2(t)+\gamma^2 x^2(t)\right]\right\}\exp\left[ \begin{matrix} -\dfrac{g^3}{2D}\int_{t_a}^{t_b} dt\, x^4(t) \\ \\ +\dfrac{3g}{2}\int_{t_a}^{t_b} dt\, x^2(t) \end{matrix}\right]$$

$$\cong \exp\left\{\frac{1}{2}\gamma(t_b-t_a)-\frac{1}{4D}\gamma(x_b^2-x_a^2)-\frac{g}{8D}(x_b^4-x_a^4)\right\}$$

$$\times\left(x_b,t_b;x_a,t_a\right)_{FP}^{x^2[0]}\left[1-\frac{\gamma g}{2D}\int_{t_a}^{t_b} dt\left\langle x^4(t)\right\rangle+\frac{3g}{2}\int_{t_a}^{t_b} dt\left\langle x^2(t)\right\rangle\right]$$

$$=\left(x_b,t_b;x_a,t_a\right)_{FP}^{x^2}\left[ \begin{matrix} 1-\dfrac{\gamma g}{2D}\int_{t_a}^{t_b} dt\left\langle x^4(t)\right\rangle \\ \\ +\dfrac{3g}{2}\int_{t_a}^{t_b} dt\left\langle x^2(t)\right\rangle \end{matrix}\right]\exp\left[-\frac{g}{8D}(x_b^4-x_a^4)\right]$$

$$=\left(x_b,t_b;x_a,t_a\right)_{FP}^{x^2}\left\{1+g\left[\frac{3}{2}\int_{t_a}^{t_b} dt\left\langle x^2(t)\right\rangle-\frac{\gamma}{2D}\int_{t_a}^{t_b} dt\left\langle x^4(t)\right\rangle-\frac{1}{8D}(x_b^4-x_a^4)\right]\right\}$$

$$\equiv\left(x_b,t_b;x_a,t_a\right)_{FP}^{x^2}+g\left(x_b,t_b;x_a,t_a\right)_{FP}^{x^2\wedge x^4}$$

$$(5.359)$$

In order to evaluate the remaining integrals the correlation functions are necessary, in relation with the statistical Wick rules. Precisely, having in mind the general connection between the action and the Lagrangian of the system, see Volume I/Section 2.3 of the present five-volume set (Putz, 2016a),

$$S_{FP}^{x^2}\left[x(t),\dot{x}(t),t\right]=\int_{t_a}^{t_b}L_{FP}^{x^2}\left[x(t),\dot{x}(t),t\right] \qquad (5.360)$$

the expected value of the paths $x(t)$ will be successively written as:

$$\langle x(t)\rangle_{FP} = \frac{\int_{x(t_a)=x_a}^{x(t_b)=x_b}\mathscr{D}'x(t)\left[x(t)\right]\exp\left\{-S_{FP}^{x^2}\left[x(t),\dot{x}(t),t\right]\right\}}{\int_{x(t_a)=x_a}^{x(t_b)=x_b}\mathscr{D}'x(t)\exp\left\{-S_{FP}^{x^2}\left[x(t),\dot{x}(t),t\right]\right\}}$$

$$= \frac{\dfrac{\delta}{\delta j}\int_{x(t_a)=x_a}^{x(t_b)=x_b}\mathscr{D}'x(t)\left[x(t)\right]\exp\left\{-\left(S_{FP}^{x^2}\left[x(t),\dot{x}(t),t\right]+\int_{t_a}^{t_b}dtx(t)j(t)\right)\right\}\Bigg|_{j=0}}{\int_{x(t_a)=x_a}^{x(t_b)=x_b}\mathscr{D}'x(t)\exp\left\{-S_{FP}^{x^2}\left[x(t),\dot{x}(t),t\right]\right\}}$$

$$= -\frac{\dfrac{\delta}{\delta j(t)}\left(x_b,t_b;x_a,t_a\right)_{FP}^{x^2[0]}\left[j(t)\right]\Bigg|_{j=0}}{\left(x_b,t_b;x_a,t_a\right)_{FP}^{x^2[0]}}$$

$$(5.361)$$

from where the general multiplied form may be advanced:

$$\langle x(t_1)x(t_2)...\rangle_{FP} = \frac{\left(-\dfrac{\delta}{\delta j(t_1)}\right)\left(-\dfrac{\delta}{\delta j(t_2)}\right)...\left(x_b,t_b;x_a,t_a\right)_{FP}^{x^2[0]}\left[j(t)\right]\Bigg|_{j=0}}{\left(x_b,t_b;x_a,t_a\right)_{FP}^{x^2[0]}}$$

$$(5.362)$$

The density matrix as a functional of the introduced interaction current $j$ can be developed under the form (Pelster & Kleinert, 1996; Kleinert et al., 1999):

$$\left(x_b,t_b;x_a,t_a\right)_{FP}^{x^2[0]}\left[j(t)\right]=\left(x_b,t_b;x_a,t_a\right)_{FP}^{x^2[0]}$$

$$\times\exp\left[\begin{array}{c}\dfrac{1}{2D}\displaystyle\int_{t_a}^{t_b}dt_1 x_{cl}(t_1)j(t_1)\\[4mm]+\dfrac{1}{4D^2}\displaystyle\int_{t_a}^{t_b}dt_1\int_{t_a}^{t_b}dt_2\,j(t_1)j(t_2)G(t_1,t_2)\end{array}\right]$$

(5.363)

where the classical paths and correlation propagators (see Section 3.3) are now specialized as (corresponding with the prescriptions of Section 2.3):

$$x_{cl-FP}(t)=\frac{x_b\sinh\left[\gamma(t-t_a)\right]+x_a\sinh\left[\gamma(t_b-t)\right]}{\sinh\left[\gamma(t_b-t_a)\right]}$$

(5.364)

$$G_{FP}(t_1,t_2)=\frac{2D}{\gamma}\frac{\sinh\left[\gamma(t_1-t_a)\right]\sinh\left[\gamma(t_b-t_2)\right]}{\sinh\left[\gamma(t_b-t_a)\right]}\Theta(t_2-t_1)$$

$$+\frac{2D}{\gamma}\frac{\sinh\left[\gamma(t_2-t_a)\right]\sinh\left[\gamma(t_b-t_1)\right]}{\sinh\left[\gamma(t_b-t_a)\right]}\Theta(t_1-t_2)\quad(5.365)$$

with the $\Theta(t_1-t_2)$ the Heaviside function.

Whit these (Feynman) rules the connected expressions are calculated following the (the modified Wick rules) prescriptions of Section 3.3.2, here particularized as:

$$\langle x(t)\rangle=x_{cl}(t)$$

(5.366)

$$\langle x(t_1)x(t_2)\rangle=x_{cl}(t_1)x_{cl}(t_2)+G(t_1,t_2)$$

(5.367)

$$\langle x(t_1)x(t_2)x(t_3)\rangle=x_{cl}(t_1)x_{cl}(t_2)x(t_3)+x(t_1)G(t_2,t_3)$$
$$+x(t_2)G(t_1,t_3)+x(t_3)G(t_1,t_2)$$

(5.368)

$$\langle x(t_1)x(t_2)x(t_3)x(t_4)\rangle = x_{cl}(t_1)x_{cl}(t_2)x(t_3)x(t_4) + x(t_1)x(t_2)G(t_3,t_4)$$
$$+x(t_1)x(t_3)G(t_2,t_4)$$
$$+x(t_1)x(t_4)G(t_2,t_3) + x(t_2)x(t_3)G(t_1,t_4)$$
$$+x(t_2)x(t_4)G(t_1,t_3)$$
$$+x(t_3)x(t_4)G(t_1,t_2) + G(t_1,t_2)G(t_3,t_4)$$
$$+G(t_1,t_3)G(t_2,t_4)$$
$$+G(t_1,t_4)G(t_2,t_3)$$

$$(5.369)$$

Next the interaction contribution will be considered. Note that the above Wick rules take into account all the classical uni-solutions combinations as well as all classical bi-solutions combinations and of the interaction propagators, as necessary to satisfy the current expanding order of interaction. However, in the present calculation, there appears the simplified situation of evaluating of the correlations functions at identical times, thus reducing the previous expressions (5.367) and (5.369) to these ones:

$$\langle x^2(t)\rangle_{FP} = x^2_{cl-FP}(t) + G_{FP}(t,t)$$

$$(5.370)$$

$$\langle x^4(t)\rangle_{FP} = x^4_{cl-FP}(t) + 6x^2_{cl-FP}(t)G_{FP}(t,t) + 3G^2_{FP}(t,t)$$

$$(5.371)$$

With expressions (5.370) and (5.371) in (5.359), the conditioned probability density for a harmonic potential can be analytically formulated, by turning the hyperbolic functions into decreasing exponentials, with results:

$$\left(x_b,t_b;x_a,t_a\right)_{FP}^{x^4}=\left(x_b,t_b;x_a,t_a\right)_{FP}^{x^2}$$

$$\times\left\{1+g\left[c_{00}\left(\tau\right)+c_{20}\left(\tau\right)x_a^2+c_{21}\left(\tau\right)x_ax_b+c_{22}\left(\tau\right)x_b^2+c_{40}\left(\tau\right)x_a^4\right.\right.$$

$$\left.\left.+c_{41}\left(\tau\right)x_a^3x_b+c_{42}\left(\tau\right)x_a^2x_b^2+c_{43}\left(\tau\right)x_ax_b^3+c_{44}\left(\tau\right)x_b^4\right]\right\}$$

$$\equiv\left(x_b,t_b;x_a,t_a\right)_{FP}^{x^2}+g\left(x_b,t_b;x_a,t_a\right)_{FP}^{x^2}\left(x_b,t_b;x_a,t_a\right)_{FP}^{gx^4}$$

(5.372)

with the calculated coefficients of expansion (Kleinert-Pelster-Putz, 2002):

$$c_{00}\left(\tau\right)=3D\frac{1+4\left(1-2\tau\right)\exp\left(-2\tau\right)-\left(5+4\tau\right)\exp\left(-4\tau\right)}{4\gamma^2\left[1-\exp\left(-2\tau\right)\right]^2}$$

$$c_{20}\left(\tau\right)=3\frac{\left(4\tau-5\right)\exp\left(-2\tau\right)+4\left(1+2\tau\right)\exp\left(-4\tau\right)+\exp\left(-6\tau\right)}{2\gamma\left[1-\exp\left(-2\tau\right)\right]^3}=c_{22}\left(\tau\right)$$

$$c_{21}\left(\tau\right)=3\frac{\left(2-\tau\right)\exp\left(-\tau\right)+2\left(1-4\tau\right)\exp\left(\left(-3\tau\right)-\left(3\tau+4\right)\exp\left(-5\tau\right)}{\gamma\left[1-\exp\left(-2\tau\right)\right]^3}$$

$$c_{40}\left(\tau\right)=\frac{2\exp\left(-2\tau\right)+3\left(1-4\tau\right)\exp\left(-4\tau\right)-6\exp\left(-6\tau\right)+\exp\left(-8\tau\right)}{4D\left[1-\exp\left(-2\tau\right)\right]^4}$$

$$c_{41}\left(\tau\right)=\frac{\left\{\begin{array}{l}-\exp\left(-\tau\right)+3\left(4\tau-3\right)\exp\left(-3\tau\right)\\+3\left(3+4\tau\right)\exp\left(-5\tau\right)+\exp\left(-7\tau\right)\end{array}\right\}}{2D\left[1-\exp\left(-2\tau\right)\right]^4}=c_{43}\left(\tau\right)$$

$$c_{42}\left(\tau\right)=3\frac{\left(3-2\tau\right)\exp\left(-2\tau\right)-8\tau\exp\left(-4\tau\right)-\left(3+2\tau\right)\exp\left(-6\tau\right)}{2D\left[1-\exp\left(-2\tau\right)\right]^4}$$

$$c_{44}\left(\tau\right)=\frac{-1+6\exp\left(-2\tau\right)-3\left(1+4\tau\right)\exp\left(-4\tau\right)-2\exp\left(-6\tau\right)}{4D\left[1-\exp\left(-2\tau\right)\right]^4}$$

(5.373)

and where the short notation $\tau=\gamma\left(t_b-t_a\right)$ was introduced, allowing for the graphical representation as given in the Figure 5.3.

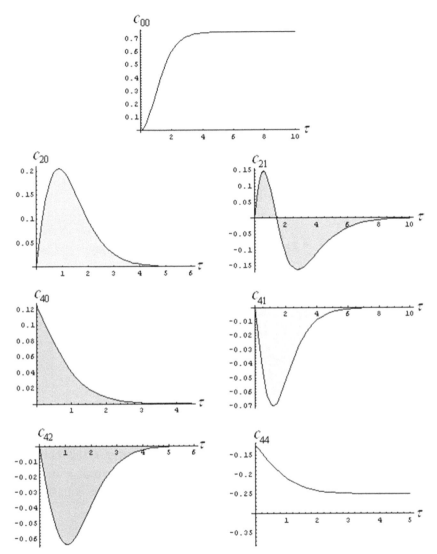

**FIGURE 5.3** Temporal behavior of the coefficients (5.373) for the Fokker-Planck enharmonic conditioned probability density (5.372) for unitary drift and diffusion constants $\gamma = D = 1$; after Voth et al. (1989) and Kleinert-Pelster-Putz (2002).

Going to cross-check the solution (5.372), it should satisfy the Fokker-Planck (5.250); by performing the required replacement one firstly gets:

$$\frac{\partial}{\partial t_b}\left[\left(x_b,t_b;x_a,t_a\right)_{FP}^{x^2} + g\left(x_b,t_b;x_a,t_a\right)_{FP}^{x^2}\left(x_b,t_b;x_a,t_a\right)_{FP}^{gx^4}\right]$$

$$= \frac{\partial}{\partial x_b}\left\{\left(\gamma x_b + gx_b^3\right)\left[\left(x_b,t_b;x_a,t_a\right)_{FP}^{x^2} + g\left(x_b,t_b;x_a,t_a\right)_{FP}^{x^2}\left(x_b,t_b;x_a,t_a\right)_{FP}^{gx^4}\right]\right\}$$

$$+ D\frac{\partial^2}{\partial x_b^2}\left[\left(x_b,t_b;x_a,t_a\right)_{FP}^{x^2} + g\left(x_b,t_b;x_a,t_a\right)_{FP}^{x^2}\left(x_b,t_b;x_a,t_a\right)_{FP}^{gx^4}\right]$$

(5.374)

from which only the first order $g$ will be retained:

$$\frac{\partial}{\partial t_b}\left(x_b,t_b;x_a,t_a\right)_{FP}^{x^2} + g\frac{\partial}{\partial t_b}\left[\left(x_b,t_b;x_a,t_a\right)_{FP}^{x^2}\left(x_b,t_b;x_a,t_a\right)_{FP}^{gx^4}\right]$$

$$= \frac{\partial}{\partial x_b}\left[\gamma x_b\left(x_b,t_b;x_a,t_a\right)_{FP}^{x^2}\right] + g\frac{\partial}{\partial x_b}\left[\gamma x_b\left(x_b,t_b;x_a,t_a\right)_{FP}^{x^2}\left(x_b,t_b;x_a,t_a\right)_{FP}^{gx^4}\right]$$

$$+ g\frac{\partial}{\partial x_b}\left[x_b^3\left(x_b,t_b;x_a,t_a\right)_{FP}^{x^2}\right] + D\frac{\partial^2}{\partial x_b^2}\left[\left(x_b,t_b;x_a,t_a\right)_{FP}^{x^2}\right]$$

$$+ Dg\frac{\partial^2}{\partial x_b^2}\left[\left(x_b,t_b;x_a,t_a\right)_{FP}^{x^2}\left(x_b,t_b;x_a,t_a\right)_{FP}^{gx^4}\right]$$

(5.375)

Now, by considering the Fokker-Planck equation for the harmonic potential (5.357) in Eq. (5.375), one notes that the terms without the coupling constant $g$ will be reduced, and the remaining equation will be shaped as:

$$\frac{\partial}{\partial t_b}\left[\left(x_b,t_b;x_a,t_a\right)_{FP}^{x^2}\left(x_b,t_b;x_a,t_a\right)_{FP}^{gx^4}\right] = \frac{\partial}{\partial x_b}\left\{\begin{bmatrix}\gamma x_b\left(x_b,t_b;x_a,t_a\right)_{FP}^{gx^4} + x_b^3\end{bmatrix} \\ \times \left(x_b,t_b;x_a,t_a\right)_{FP}^{x^2}\right\}$$

$$+ D\frac{\partial^2}{\partial x_b^2}\left[\left(x_b,t_b;x_a,t_a\right)_{FP}^{x^2}\left(x_b,t_b;x_a,t_a\right)_{FP}^{gx^4}\right]$$

(5.376)

This equation is indeed satisfied by the analytically form:

$$
\left(x_b,t_b;x_a,t_a\right)_{FP}^{gx^4} = c_{00}\left(\tau\right)+c_{20}\left(\tau\right)x_a^2+c_{21}\left(\tau\right)x_a x_b+c_{22}\left(\tau\right)x_b^2+c_{40}\left(\tau\right)x_a^4
$$
$$
+c_{41}\left(\tau\right)x_a^3 x_b+c_{42}\left(\tau\right)x_a^2 x_b^2+c_{43}\left(\tau\right)x_a x_b^3+c_{44}\left(\tau\right)x_b^4
$$

(5.377)

in which one immediately identifies the components involved in the Fokker-Planck equation, by comparing it with the expression (5.372). From the general analytical form (5.372) there is also immediately deduced the solution from asymptotic time evolution, i.e., corresponding with the stationary solution for the harmonic potential that is:

$$
\left(x_b,t_b;x_a,t_a\right)_{FP-st}^{x^4} = \sqrt{\frac{\gamma}{2\pi D}}\exp\left(-\frac{\gamma x_b^2}{2D}\right)\left[1+g\left(\frac{3}{4}\frac{D}{\gamma^2}-\frac{x_b^4}{4D}\right)\right]
$$

(5.378)

Worth to observe that by recovering the harmonic case $\left(g\to 0\right)$ there is automatically regain also the harmonic stationary solution (5.319). Besides, the general stationary solution can be cross-checked by considering the stationary condition in Fokker-Planck equation (5.268) and (5.269):

$$
0 = \frac{\partial}{\partial x_b}\left\{D^{(1)}\left(x_b\right)\left(x_b,t_b;x_a,t_a\right)_{FP-st}-\frac{\partial}{\partial x_b}\left[D^{(2)}\left(x_b\right)\left(x_b,t_b;x_a,t_a\right)_{FP-st}\right]\right\}
$$

(5.379)

under the Dirichlet (spatially limited) conditions:

$$
\lim_{x_b\to\pm\infty}\left(x_b,t_b;x_a,t_a\right)_{FP-st} = \lim_{x_b\to\pm\infty}\partial_{x_b}\left(x_b,t_b;x_a,t_a\right)_{FP-st} = 0 \qquad (5.380)
$$

From (5.379) the general solution springs as satisfying the normalization condition:

$$
\left(x_b,t_b;x_a,t_a\right)_{FP-st}^{x^4} = \sqrt{\frac{2g}{\gamma}}\frac{\exp\left(-g\dfrac{x_b^4}{4D}-\gamma\dfrac{x_b^2}{2D}-\dfrac{\gamma^2}{8Dg}\right)}{K_{1/4}\left[\dfrac{\gamma^2}{8Dg}\right]}
$$

(5.381)

with $K_v[z]$ being the Bessel modified function of the second rank. Nevertheless, this solution has to be also obtained as a perturbation of the $g \to 0$ case; therefore, the Bessel function from denominator (5.381) will be expanded in asymptotic manner, by the tabulated formula:

$$K_v[z]\overset{z\to\infty}{=}\sqrt{\frac{\pi}{2z}}\exp(-z)\left[\sum_{k=0}^{n-1}\frac{1}{(2z)^k}\frac{\Gamma\left(v+k+\frac{1}{2}\right)}{\Gamma\left(v-k+\frac{1}{2}\right)}+\theta_3\frac{\Gamma\left(v+n+\frac{1}{2}\right)}{(2z)^n n!\Gamma\left(v-n+\frac{1}{2}\right)}\right]$$

(5.382a)

where $\Gamma(x)$ stays for the Gamma Euler function and $\theta_3$ for the elliptic Theta function of the 3rd order. By keeping only the first terms for the Bessel function form in Eq. (5.381) it looks like

$$K_{1/4}\left[\frac{\gamma^2}{8Dg}\right]=\sqrt{\frac{\pi}{2}\frac{8Dg}{\gamma^2}}\exp\left(-\frac{\gamma^2}{8Dg}\right)\left[1+\frac{1}{2}\frac{8Dg}{\gamma^2}\frac{\Gamma\left(\frac{1}{4}+1+\frac{1}{2}\right)}{\Gamma\left(\frac{1}{4}-1+\frac{1}{2}\right)}\right]$$

(5.383)

with which, by considering the Gamma function property, see Appendix A.2 of the Volume I of the present five volume work (Putz, 2016a)

$$\Gamma(q+1)=q\Gamma(q)$$ (5.382b)

the solution (5.381) became finally identically with the one from (5.378).

Remarkably, as found in the previous sections, by the present approach and cross-checking we arrive to reconfirm the property by which in the stationary solution all the initial event memory $(x_a, t_a)$ is lost.

The identity verification between the asymptotical time limit abstracted from the general analytical solution (5.372) with that one derived from the asymptotical stationary solution (5.381) with (5.383), along with the fact that the general solution (5.377) satisfies the Fokker-Planck (5.376), brings sufficient arguments for adopting the present solution as a meso-scopic solution for the nonequilibrium dynamics in the first order coupling (with environment), modeling the nonequilibrium electronic evolution (aka reactivity), as marked by the influence of the generalized enharmonic potential (compatible with atoms-in-molecules description and of chemical bonding in general).

## 5.5 ELECTRONIC LOCALIZATION FUNCTIONS (ELF) BY MARKOVIAN PATH INTEGRAL

### 5.5.1 THOM'S CATASTROPHE CONCEPTS FOR CHEMICAL BONDING

The Thom's catastrophe theory (Thom, 1973) basically describes how, for a given system, a continuous action on the *control space* ($C^k$), parameterized by $c_k$'s, provides a suddenly change on its *behavior space* ($I^m$), described by variables $x_m$'s, through the stable singularities of the smooth map

$$\eta(c_k, x_m) : C^k \times I^m \to \Re \tag{5.384}$$

being $\eta(c_k, x_m)$ called the *generic potential* of the system. Therefore, catastrophes are given by the set of *critical points* $(c_k, x_m)$ for which the field gradient of the generic potential vanishes

$$M^{k \times m} = \left\{ (c_k, x_m) \in C^k \times I^m \left| \nabla_{x_m} \eta(c_k, x_m) = 0 \right. \right\} \tag{5.385}$$

or, more rigorously: a catastrophe is a singularity of the map $M^{k \times m} \to C^k$.

Next, depending on the number of parameters of space $C^k$ (named also as the *co-dimension, k*) and of the number of variables of space $I^m$ (named also as the *co-rank, m*), René Thom had classified the generic potentials (or maps) given by Eq. (5.384) as seven unfold elementary (in the sense of universally) catastrophes, i.e., providing the many-variable (with the co-rank up to two) – many-parametrical (with the co-dimension up to four) polynomials, listed in the Table 5.1. Going to the higher derivatives of the generic potential (the fields), it will be said that the control parameter $c_k^*$ for which the Laplacian of the generic potential vanishes

$$\Delta_x \eta(c_k^*, x_m) = 0 \tag{5.386}$$

**TABLE 5.1** Thom's Classification for Elementary Catastrophes, after Thom (1973)

| Name | Co-dimension | Co-rank | Universal unfolding |
|------|------|------|------|
| *Fold* | 1 | 1 | $x^3 + ux$ |
| *Cusp* | 2 | 1 | $x^4 + ux^2 + vx$ |
| *Swallow tail* | 3 | 1 | $x^5 + ux^3 + vx^2 + wx$ |
| *Hyperbolic umbilic* | 3 | 2 | $x^3 + y^3 + uxy + vx + wy$ |
| *Elliptic umbilic* | 3 | 2 | $x^3 - xy^2 + u\left(x^2 + y^2\right) + vx + wy$ |
| *Butterfly* | 4 | 1 | $x^6 + ux^4 + vx^3 + wx^2 + tx$ |
| *Parabolic umbilic* | 4 | 2 | $x^2y + y^4 + ux^2 + vy^2 + wx + ty$ |

gives the *bifurcation point*. Consequently, the set of control parameters $c^{\#}$ for which the Laplacian of a critical point is non-zero defines the *domain of stability* of the critical point. There is clear now that the small perturbations of $\eta\left(c_k^*, x_m\right)$ bring the system from a domain of stability to another; otherwise, the system is located within a *domain of structural stability*.

Remarkably, the above described cases correspond to the equilibrium limit of a dynamical (non-equilibrium) evolution for an open system

$$F\left(c_k; t; \eta\left(c_k; x_m\right); \frac{\partial \eta\left(c_k; x_m\right)}{\partial t}, \ldots\right) = 0 \qquad (5.387)$$

where the behavior space is further parameterized by the temporal paths $x_m\left(c_k, t\right)$. The connection with equilibrium is recovered through the stationary time regime imposed on the critical points. This way, the set of points giving a critical point in the stationary $t \to +\infty$ regime (the so called $\omega$-limit) corresponds to *an attractor*, and forms its *basin*, whereas the stationary regime $t \to -\infty$ (the so called $\alpha$-limit) describes *a repellor*.

These catastrophe concepts have the merit to describe the evolution of *local* properties of (in principle) any natural system. In this framework, the chemical bonds can be seen as the equilibrium part of the evolutionary binding processes. Therefore, to describe the bonds and binding, a suitable generic potential $\eta(x)$ has to be consider. In topological studies of electron localization across a chemical reaction modeled by the catastrophe approach the variable $x$ can stay also as the reaction coordinate.

With the aim of properly choosing the function $\eta(x)$, with $x$ the space-spin coordinate, the electronic wave function $\Psi$, as provided by Schrödinger or related Hartree-Fock formalisms, can be an option, but suffers from the lack in the real space significance. Next, a better choice regards the electronic density $\rho(x)$ as the real descriptor of the topological electronic distribution in space and for identifying the bonds as well. In this respect, the gradient equation of the electronic density, $\nabla\rho = 0$, provides the critical points whereas the Laplacian equation, $\nabla\rho = 0$, indicates the bifurcations and stability zones, respectively. This picture was intensively used by the Bader's theory of atoms in molecules (Bader, 1990), with partially success. An alternative electronic topological approach was performed by Mezey by considering the changes in shape of the bonding isosurfaces (Mezey, 1987), but in a form that does not allow the description of bonding in terms of Laplacian. Worth noting here that the Laplacian plays a crucial role in topological bond description, being associated with the quantum mechanical transcription of the kinetic electronic energy, $\hat{T} = \left(-\hbar^2 / 2m\right)\nabla^2$, being at its turn related with the minus of total energy of the electronic system, $E = -T$, through the virial theorem at equilibrium (Preuss, 1969).

In fact, this feature of Laplacian was extensively employed by Bader's atoms in molecules theory at the purely electronic density level in order to quantum rationalize the previous Gillespie's geometrical VSEPR (Valence Shell Electron Pair Repulsion) description of the molecular bonds (Gillespie, 1972; Bader et al., 1988).

A more elaborated choice in generic potential was proposed by Becke and Edgecombe through introducing the electron localized function (ELF), representing a density combination rather than the electronic density solely. Their approach (abbreviated as "BE") prescribes $\eta(x)$ with the form (Becke & Edgecombe, 1990)

$$\eta^{BE}(x) = \frac{1}{1 + \left[D(x)/D_h(x)\right]^2} \tag{5.388}$$

with:

$$D(x) = \frac{1}{2}\sum_i \left|\nabla\phi_i(x)\right|^2 - \frac{1}{8}\frac{\left|\nabla\rho(x)\right|^2}{\rho(x)} \tag{5.389}$$

and:

$$D_h(x) = \frac{3}{10}\left(3\pi^2\right)^{2/3}\left[\rho(\mathbf{r})\right]^{5/3} = 2.871\rho(x)^{5/3} \qquad (5.390)$$

as a combination between the kinetic electronic density terms from the Hartree-Fock (or Kohn-Sham) orbitals $\phi_i$, the Weizsäcker gradient correction, and the homogeneous (abbreviated by "$h$") Thomas-Fermi descriptions, respectively (Silvi & Savin, 1994). In Eq. (5.388) $D_h(x)$ accounts for the excess of local kinetic energy density due to Pauli repulsion, whereas $D_h(x)$ plays the role of the "renormalization" factor. However, this ELF function behaves like a density by mapping its values onto the realm [0, 1], where 1 corresponds to the perfect electronic localization, being therefore suitable for the gradient and Laplacian performances. The recent topological studies have revealed the effectiveness of the above $\eta^{BE}(x)$ function in describing both the electronic localization in bonding as well for modeling the chemical reaction pathways (Savin et al., 1992). However, despite of $\eta^{BE}(x)$ efficiency in bonding characterization, a series of aspects regarding its appearance in the context of the universal unfolding of the catastrophes (see Table 5.1) as well as within the time-dependent and the stationary ω-limit have remained unexplored. The present Markovian description of the enharmonic potentials with the help of Fokker-Planck equation and of its path integral solution aims filling this gap – as exposed next.

### 5.5.2  FOKKER-PLANCK MODELING OF THE ELECTRONIC LOCALIZATION

In order to better understand the actual ELF approach, it is worth reminding that the origin of the above $\eta^{BE}(x)$ relies in evaluation of the *conditioned pair probability* with which one electron is located at point $x_b$ with the spin $\sigma$ once the reference electron is located at point $x_a$ with the same (parallel) spin $\sigma$ (Becke, 1988)

$$P^{\sigma\sigma}\left(x_b; x_a\right) \cong A_{\sigma\sigma}x_b^2 \qquad (5.391)$$

being the coefficient $A_{\sigma\sigma}$ identified with the function $D(x)$ in Equation (5.389) within the so called "hole" function approach, whereas in the present treatment has to be re-determined. Nevertheless, $\eta^{BE}$ ELF emphasizes on the key role of the conditioned probability density – to be here considerate alternative Markovian description of natural processes.

As previously shown, the Markovian treatment of the conditioned probability density given by the general Fokker-Planck path integral (5.349) with correspondences of Table 5.1 in atomic units $(\hbar = m = 1)$

$$(x_b,t_b;x_a,t_a)_{FP} = \int\limits_{x(t_a)=x_a}^{x(t_b)=x_b} \mathcal{D}'x(t)\exp\left\{\begin{array}{l} -\dfrac{1}{2}\int\limits_{t_a}^{t_b} dt\left[\dot{x}(t) - K(x(t))\right]^2 \\[4mm] -\dfrac{1}{2}\int\limits_{t_a}^{t_b} dt K'(x(t)) \end{array}\right\}$$

$$(5.392)$$

with $K(x)$ the drift function

$$K\left(x(t)\right) = -\partial_x U(x) \qquad (5.393)$$

see its definition from (5.350) and the transformation (5.240). Within the anharmonic potential case it features the non-linear shape

$$K(x) = -hx - gx^3 \qquad (5.394)$$

assuring the connection with the electronic localization by the homogeneous and inhomogeneous (or gradient) specializations

$$h \to D_h(x_a) \qquad (5.395)$$

$$g \to D(x_a) \qquad (5.396)$$

with the help of electronic functions (5.390) and (5.389), respectively. Yet, the correspondences (5.395) and (5.396) are motivated as follows. If one considers the working effective potential (5.358) as the *bilocal dependency*

$$U(x_a,x_b) = h(x_a)\frac{x_b^2}{2} + g(x_a)\frac{x_b^4}{4} \qquad (5.397)$$

it models the field produced by the *reference electron* located at $x_a$ over its spherical neighborhood which contains the *coupled electron* at the distance (or radius) $x_b$, thus being characterized by the density (radial) equation of Poisson type

$$\rho(x_b) = \nabla^2_{x_b} U(x_a, x_b) = h(x_a) + 3g(x_a) x_b^2 \qquad (5.398)$$

while clearly revealing the role of the homogeneous and gradient related terms as being the friction $\gamma$ and perturbation factor $g$ in Eq. (5.397), respectively. Note that the form of the potential (5.397) assumes one of the most general pictures of bonding fluctuations with the enharmonic trajectories of the second electron respecting the referential one.

More, within the potential form (5.397) the $x_a$ and $x_b$ coordinates are separated and coupled, allowing the averages operations being performed firstly on the coordinates of the coupled $x_b$ electron, while replacing in the final result the referential electronic $x_a$ influences. However, the present Markovian ELF picture is summarized by the following analytical steps (Putz, 2005):

(i) Solving the path integral of Equation (5.392) for the non-linear potential (5.397) the *time-dependent (spin) conditioned probability* $\left(x_b, t_b; x_a, t_a\right)_{FP}^{x^4}$ is provided;

(ii) The $\omega$-limit $\left(t_b \to \infty\right)$ is performed on the (i) result leaving with the *stationary (spin) conditioned probability*

$$\lim_{(t_b - t_a) \to +\infty} \left(x_b, t_b; x_a, t_a\right)_{FP}^{x^4} = \left(x_b, t_b; x_a, t_a\right)_{FP/\omega}^{x^4} \qquad (5.399)$$

(iii) The result from (ii) is employed upon the specific integration rule (Parr & Yang, 1989)

$$\cdot \int_{-\infty}^{+\infty} \left(x_b, t_b; x_a, t_a\right)_{FP/\omega}^{x^4} dx_b = -1 \qquad (5.400)$$

providing the "renormalization" of the stationary spin conditioned probability (5.399) into the so called *exchange (parallel spins) conditional probability*: $\left(x_b, t_b; x_a, t_a\right)_{FP/\omega}^{x^4/\sigma\sigma}$. Note that this "unusual" normalization condition makes in fact the proper link with the Fermi hole, in close relation with Pauli exchange repulsion, telling that the $\alpha\alpha$ and $\beta\beta$ exchanged holes contain exactly minus one electron (Becke, 1988).

(iv)  Identification of the actual exchange probability $\left(x_b,t_b;x_a,t_a\right)_{FP/\omega}^{x^4/\sigma\sigma}$ with the previous general one given by Eq. (5.391) delivers the polynomial equations that can be treated either as the gradient or Laplacian equations (5.385) and (5.386), respectively, towards identifying one of the universal unfolded catastrophes given in Table 5.1. In any case, either as a gradient or Laplacian equation, the companion equation results immediately assuring therefore the necessary number of equations from which the critical solution $x_b$ as well as the bifurcation parameter $A_{\sigma\sigma}$ are evaluated in terms of $g$ and $h$.

(v)  Finally, throughout the correspondences (5.395) and (5.396) the Markovian ELF is found.

The next section is dedicated to applying the Markovian ELF algorithm for the enharmonic potential of (chemical) binding.

### 5.5.3  MARKOVIAN SHAPE FOR ELECTRONIC LOCALIZATION FUNCTIONS

The first step in the above Markovian-ELF algorithm, i.e., the analytical time-dependent conditional probability $\left(x_b,t_b;x_a,t_a\right)_{FP}^{x^4}$ for the path-integral representation (5.392) with potential (5.397) is furnished by the expression (5.378). Then, applying the $\omega$-limit on the result (5.378) one gets the stationary solution (5.381), see also the Figure 5.3 for the asymptotic behavior for the coefficients (5.373), as representing the *stationary conditioned Markovian enharmonic probability*

$$\left(x_b,t_b;x_a,t_a\right)_{FP/\omega}^{x^4} = \sqrt{\frac{h}{\pi}}\exp\left[\frac{3}{8}\frac{g}{h^2} - hx_b^2 - \frac{g}{2}x_b^4\right] \qquad (5.401)$$

where the second correspondence of Eq. (5.354a) was systematically considered in the solution (5.381) within the above specified atomic units. However, since expression (5.401) already fulfills the canonical integration

$$1 = \int_{-\infty}^{+\infty} \left( x_b, t_b; x_a, t_a \right)_{FP-st}^{x^4} dx_b \tag{5.402}$$

there is immediate that the satisfaction of "exchange hole" renormalization condition given by Eq. (5.400) requires only the changing of sign in right hand side of Equation (5.401); yet, searching for the specific catastrophe polynomials, one will consider the polynomial expansion of exponential of Eq. (5.401) up to the accustomed first order in the coupling $g$, so that the exchange conditional probability is written as

$$\left( x_b, t_b; x_a, t_a \right)_{FP/\omega}^{x^4/\sigma\sigma} = -\frac{1 - hx_b^2 - \dfrac{g}{2} x_b^4}{1 - \dfrac{3}{8}\dfrac{g}{h^2}} \tag{5.403}$$

Next, by identifying the expression (5.403) with the conditional pair probability (5.391), i.e., by putting in act the step (iv) in above formulated Markovian algorithm, it is straightforward to arrive at the polynomial equation

$$x_b^4 + \left[ \frac{2h}{g} + A_{\sigma\sigma} \left( \frac{3}{4} \frac{1}{h^2} - \frac{2}{g} \right) \right] x_b^2 - \frac{2}{g} = 0 \tag{5.404}$$

At this point, as was previously anticipated, Eq. (5.404) can be seen in two ways. Within the first case, abbreviated as "*M1: Markovian one*," Eq. (5.404) may represent the Laplacian field of Eq. (5.386) that provides the unfolded function

$$\eta^{M1}(h, g, A_{\sigma\sigma}, x_b) = \frac{1}{30} x_b^6 + \frac{1}{12} \left[ \frac{2h}{g} + A_{\sigma\sigma} \left( \frac{3}{4} \frac{1}{h^2} - \frac{2}{g} \right) \right] x_b^4 - \frac{1}{g} x_b^2 \tag{5.405}$$

which corresponds to *the butterfly elementary catastrophe* (with $v = t = 0$) in Table 5.1. Instead, when Eq. (5.404) is seen as the gradient field equation of Equation (5.385) it produces the second case, the "*M2: Markovian two*," with the unfolded function

$$\eta^{M2}\left(h,g,A_{\sigma\sigma},x_b\right)=\frac{1}{5}x_b^5+\frac{1}{3}\left[\frac{2h}{g}+A_{\sigma\sigma}\left(\frac{3}{4}\frac{1}{h^2}-\frac{2}{g}\right)\right]x_b^3-\frac{2}{g}x_b \quad (5.406)$$

associated with *the swallow tail elementary catastrophe* (with $v=0$) in Table 5.1. Up to now, it was revealed that the Markovian path-integral representation of the conditioned probability density within the non-linear drift expansion arrives to recover the unfolded elementary catas-trophes, as classified by the Thom's theory. These catastrophe forms, namely Eqs. (5.405) and (5.406), can be further transformed to shape the Markovian ELFs by eliminating, in each case, the $\left(A_{\sigma\sigma},x_b\right)$ dependence, in order to complete the (iv) step above. To do this, in each M1 and M2 cases, the Eq. (5.404) is supplemented by its topological companion (Laplacian to gradient equation and vice-versa). For instance, when Eq. (5.404) repre-sents, in the M1 case, the Laplacian equation $\Delta\eta^{M1}=0$ then, by its integra-tion the gradient equation $\nabla\eta^{M1}=0$ is also furnished

$$M1:\begin{cases}\Delta\eta^{M1}=0\\\nabla\eta^{M1}=0\end{cases}\Rightarrow\begin{cases}x_b^4+\left[\frac{2h}{g}+A_{\sigma\sigma}\left(\frac{3}{4}\frac{1}{h^2}-\frac{2}{g}\right)\right]x_b^2-\frac{2}{g}=0\\\frac{1}{5}x_b^5+\frac{1}{3}\left[\frac{2h}{g}+A_{\sigma\sigma}\left(\frac{3}{4}\frac{1}{h^2}-\frac{2}{g}\right)\right]x_b^3-\frac{2}{g}x_b=0\end{cases}$$

$$(5.407)$$

Likewise, in the M2 case, when Eq. (5.404) represents the gradient equa-tion $\nabla\eta^{M2}=0$ the correspondent Laplacian equation $\Delta\eta^{M2}=0$ is also provided through its derivation

$$M2:\begin{cases}\nabla\eta^{M1}=0\\\Delta\eta^{M1}=0\end{cases}\Rightarrow\begin{cases}x_b^4+\left[\frac{2h}{g}+A_{\sigma\sigma}\left(\frac{3}{4}\frac{1}{h^2}-\frac{2}{g}\right)\right]x_b^2-\frac{2}{g}=0\\4x_b^3+2\left[\frac{2h}{g}+A_{\sigma\sigma}\left(\frac{3}{4}\frac{1}{h^2}-\frac{2}{g}\right)\right]x_b=0\end{cases} \quad (5.408)$$

In both cases, the second equation will be solved first, in terms of bifurca-tion parameter $A_{\sigma\sigma}$, then the result is plugged in the complement equation and the critical point $x_b$ is reached out to be in each of M1 & M2 cases, respectively as

$$x_b^{M1} = \pm \left( \frac{10}{-g} \right)^{1/4} \tag{5.409}$$

$$x_b^{M2} = \pm \left( \frac{2}{-g} \right)^{1/4} \tag{5.410}$$

However, in order to form the real solutions of Equations (5.409) and (5.410), a suitable replacement has to be performed, namely

$$-g \rightarrow g*(-g,h) > 0 \tag{5.411}$$

such that the new parameter $g*$ to be strictly positive, while being a transformation of the negative of $g$, along the general homogeneous $h$ parameter dependence.

Finally, with Eqs. (5.409) and (5.410) back in the founded bifurcation parameters $A_{\sigma\sigma}$, and further into the unfolded catastrophe functions (5.405) and (5.406), the corresponding general Markovian ELF cases are displayed as (Putz, 2005):

$$\eta^{M1}(x) = \frac{1}{\left[ g*\left( -D(x), D_h(x) \right) \right]^{3/2}} \tag{5.412}$$

$$\eta^{M2}(x) = \frac{1}{\left[ g*\left( -D(x), D_h(x) \right) \right]^{5/2}} \tag{5.413}$$

Yet, while featuring the normalization between 0 and 1, in terms of $g*$ of Eq. (5.411), by close inspection of the Becke-Edgecombe (5.388) as comparing with the Markovian (5.412) and (5.413) ELFs, one may conclude that a general shape for a reliable ELF should look like (Putz, 2005)

$$\eta(x) \equiv \frac{1}{f\left( \dfrac{-g(x)}{h(x)} \right)} = \frac{1}{f\left( \dfrac{-D(x)}{D_h(x)} \right)} \tag{5.414}$$

That is, displaying an inverse function of a gradient to homogeneous ratio electronic contributions (5.389) and (5.390), respectively.

### 5.5.4 MARKOVIAN SPECIALIZATION FOR ELECTRONIC LOCALIZATION FUNCTIONS

For practical purposes the general Markovian ELFs given by Eqs. (5.412) and (5.413) are to be further specialized, according with the transformation prescribed by Eq. (5.414), in various ways. Yet, aiming to make a closer contact with previous Becke-Edgecombe ELF formulation of Eq. (5.388), a suitable choice would be

$$g*(-g,h)=1+\frac{(-g)^2}{h^2}=1+\left(\frac{g}{h}\right)^2 \rightarrow 1+\left(\frac{D(x)}{D_h(x)}\right)^2 >0 \qquad (5.415)$$

that provides the first set of Markovian ELF formulations

$$\eta^{M1}(x)=\left[\frac{1}{1+\left(D(x)/D_h(x)\right)^2}\right]^{3/2}=\left[\eta^{BE}(x)\right]^{3/2} \qquad (5.416)$$

$$\eta^{M2}(x)=\left[\frac{1}{1+\left(D(x)/D_h(x)\right)^2}\right]^{5/4}=\left[\eta^{BE}(x)\right]^{5/4} \qquad (5.417)$$

Nevertheless, the ELF cases given by Eqs. (5.416) and (5.417) correct, in a Markovian framework, the previously purely hole-pair probability approach. The question is to decide which of the above Markovian ELF' cases are more "corrective" respecting the Becke-Edgecombe one.

For better visualizing the answer the Figure 5.4 shows the BE-M1 and BE-M2 differences on the relevant homogenous ($h$ parameter) *versus* inhomogeneous ($g$ parameter) contributions to electronic localization. The analysis of Figure 5.4 clearly reveals that for a moderate inhomogeneous contribution to the electronic gas the first Markovian ELF of Equation (5.416) corrects the Becke-Edgecombe localization function up to 15%, whereas, in the same conditions, the second Markovian ELF of Eq. (5.417) improves only up to 8% the Becke-Edgecombe ELF treatment. Therefore, we can conclude that the most corrective Markovian ELF to the Becke-Edgecombe approach stays

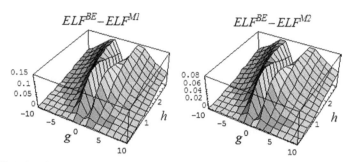

**FIGURE 5.4** The differences in electron localization functions (ELFs) between the Becke-Edgecombe (BE) and Markovian (M1) and (M2) formulations of Eqs. (5.388), (5.416), and (5.417), in left and right, respectively, versus the homogeneous ($h$-parameter) and inhomogeneous ($g$-parameter) influences on electronic distribution (Putz, 2009).

the first dependence of Eq. (5.416); this one can be further tested for prediction of the electronic localization in atoms and of bindings in molecules.

Next, we suggest another choice of the transformations (5.411)–(5.414), while maintaining the generalization of the Becke-Edgecombe ELF picture – now by the exponential form

$$g*(-g,h) = \exp\left[\left(\frac{-g}{h}\right)^2\right] \rightarrow \exp\left[\left(\frac{D(x)}{D_h(x)}\right)^2\right] > 0 \qquad (5.418)$$

with the help of which, the ELF cases of Eqs. (5.412) and (5.413) are specialized towards the new ones

$$\eta^{M1+}(x) = \exp\left[-\frac{3}{2}\left(\frac{D(x)}{D_h(x)}\right)^2\right] \qquad (5.419)$$

$$\eta^{M2+}(x) = \exp\left[-\frac{5}{4}\left(\frac{D(x)}{D_h(x)}\right)^2\right] \qquad (5.420)$$

They produce the corrections up to 30% and 20% for the Becke-Edgecombe ELF approach, respectively, in the moderate inhomogeneous electronic behavior – as the Figure 5.5 reveals. Again, the Markovian ELF

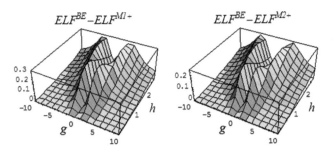

**FIGURE 5.5** The same kid of representation as in Figure 5.4, yet here for marking the Markovian (M1+) and (M2+) behaviors of Eqs. (5.419) and (5.420), in left and right, respectively (Putz, 2009).

corresponding to the first case, Eq. (5.419), is the most corrective respecting the Becke-Edgecombe treatment.

The last considered ELF particularization eventually involves the hyperbolic trigonometric function with the form:

$$g*(-g,h) = \cosh\left[\sqrt{2}\frac{(-g)}{h}\right] \to \cosh\left[\sqrt{2}\frac{D(x)}{D_h(x)}\right] > 0 \qquad (5.421)$$

producing the Markovian ELF M1++, and ELF2++ formulations, respectively, as:

$$\eta^{M1++}(x) = \left\{\operatorname{sech}\left[\sqrt{2}\frac{D(x)}{D_h(x)}\right]\right\}^{3/2} \qquad (5.422)$$

$$\eta^{M2++}(x) = \left\{\operatorname{sech}\left[\sqrt{2}\frac{D(x)}{D_h(x)}\right]\right\}^{5/4} \qquad (5.423)$$

Now, the differences to localization introduced by these Markovian ELFs as referring to the Becke-Edgecombe formulation are analyzed through the representations given in Figure 5.6; the analysis sharply indicates that the Markovian ELF++ approaches depart between 10–20% from the Becke-Edgecombe ELF, providing an intermediary situation between Markovian ELF (8–15%) and Markovian ELF+ (20–30%) predicted by

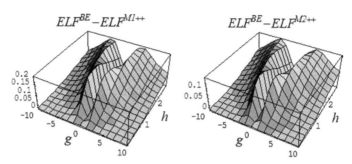

**FIGURE 5.6** The same kid of representation as in Figure 5.4, yet here for marking the Markovian (M1++) and (M2++) behaviors of Eqs. (5.422) and (5.423), in left and right, respectively (Putz, 2009).

Eqs. (5.416)–(5.417) and (5.419)–(5.420), with the representations of the Figures 3.4 and 3.5, respectively.

Overall, judging by both the analytical complexity and meaningful physical background, grounded on the Fokker-Planck approach of the non-equilibrium towards equilibrium systems, we propose that the Markovian ELF1+ of Eq. (5.419) to be adopted as the electronic localization function (ELF) for the practical topological characterization of the atomic shells and the molecular bonds (Frisch & Wasserman, 1961; Fuller, 1971; Crick, 1976; Hänggi et al., 1990; Weiss, 1993).

The combined path integral with the non-linear and electronic density aspects fully qualify our analytical results as a reliable framework within which the electronic localization targeting the bonding evolution theory to be further developed.

### 5.5.5 ELECTRONIC LOCALIZATION IN ATOM AND IN BONDING ATOMS

The definitions that are currently used in the classification of chemical bonds are often imprecise, as they are derived from approximate theories. Based on the topological analysis local, quantum-mechanical functions related to the Pauli Exclusion Principle may be formulated as "localization attractors" of bonding, non-bonding, and core types. Bonding attractors lie between nuclei core attractors and characterize shared electron interactions. The spatial arrangement of bond attractors allows for an absolute classification of ionic *versus* covalent bond to be derived from electronic density combined functions (Cioslowski, 1990).

Most modern classifications of the chemical bond are based on Lewis' theory and rely on molecular-orbital and valence-bond theories with schemes involving linear combination of atomic orbitals (LCAO). However, electron density alone does not easily reveal the consequences of the Pauli Exclusion Principle on bonding nature. While VSEPR theory indicates that the Pauli principle is important for understanding chemical structures, it has been reformulated in terms of maxima of electronic density's Laplacian $-\nabla^2 \rho(\mathbf{r})$ (Bader, 1990). Next, the exchange-correlation density functional concept was employed to achieve the coordinate-space dynamical correlation in an inhomogeneous electron gas. This way the exchange-correlation energy (5.224) further re-expresses like (Becke, 1988; Becke & Edgecombe, 1990)

$$E_{xc} = \frac{1}{2}\sum_{\alpha\alpha'}\int\int \frac{\rho(\mathbf{r}_1)\rho(\mathbf{r}_2)}{r_{12}} h_{XC}^{\alpha\alpha'} d\mathbf{r}_1 d\mathbf{r}_2 \qquad (5.424)$$

by accounting for the four $\alpha$-spin types of interactions through the "hole" functions (Levy, 1991; Liu et al., 1999)

$$h^{\alpha\alpha'}(\mathbf{r}_1,\mathbf{r}_2) = \frac{P_2^{\alpha\alpha'}(\mathbf{r}_1,\mathbf{r}_2)}{\rho_\alpha(\mathbf{r}_1)} - \rho_\alpha(\mathbf{r}_2) \qquad (5.425)$$

where $P_2(\mathbf{r}_1,\mathbf{r}_2)$ stands for the two-body or pair probability density or correlation probability of the arbitrary electrons "1 and 2," defined in terms of the $N$-body wave function $\Psi$ as follows (Parr & Yang, 1989):

$$P_2(\mathbf{r}_1,\mathbf{r}_2) = N(N-1)\int\int \cdots \int \Psi^*(\mathbf{r}_1,\mathbf{r}_2,\mathbf{r}_3,\ldots,\mathbf{r}_N)\Psi(\mathbf{r}_1,\mathbf{r}_2,\mathbf{r}_3,\ldots,\mathbf{r}_N) d\mathbf{r}_3 \ldots d\mathbf{r}_N$$
$$(5.426)$$

Within the DFT the electrons of a pair of electrons or a bond can be considered as belonging to an inhomogeneous continuum gas. In analytical terms this was translated as the ELF (5.233) index as combining the homogeneous and inhomogeneous behaviors of a many-electronic-nuclei system.

Nevertheless, as was shown in previous sections, the Markovian analytical shape of an ELF has the general qualitative form of Eq. (5.414) with the limiting constrains

$$\lim ELF = \begin{cases} 0, \nabla \rho(\mathbf{r}) >> \rho(\mathbf{r}) \\ 1, \nabla \rho(\mathbf{r}) << \rho(\mathbf{r}) \end{cases} \qquad (5.427)$$

assuring the fulfillment of the Heisenberg and Pauli principles. To clarify this (Putz, 2005), we make recourse to the Heisenberg principle, comprised in ELF. When density gradient dominates, $\nabla \rho >> \rho$ then $g >> h$ in (5.414), and $f(\infty)$ should accounted for the infinite error in assigning of momentum, therefore indicating a precisely spatial localization of electrons; thus $f(\infty) = \infty$ and $ELF \to 0$. In such, *the meaning of ELF is associated with the error in spatial localization of electrons*, being zero when the electrons are precisely located. On the contrary, when $\rho >> \nabla \rho$ then $h >> g$ in Eq. (5.414), and the resulting $f(0)$ indicates the minimum error in defining of momentum and should provide the maximum uncertain of spatial distribution; in such $f(0) = 1$ and $ELF \to 1$, where 1 stands here for 100% of coordinate localization error.

In this context, when the inverse of difference in local kinetic terms is involved, the ELF is interpreted as the *error in localization* of electrons within traps rather than where they have peaks of spatial density, as is frequently misinterpreted in literature (Santos et al., 2000; Scemama et al., 2004; Soncini & Lazzeretti, 2003; Silvi, 2003), albeit recent extensions of ELF have used the correlated (HF) wave functions, through the conditional pair probability, however not using the "kinetic energy approach" (Matito et al., 2006; Kohout et al., 2004; Jensen, 2005).

Among various classes of Markovian ELFs the most representative and efficient one was in last section proposed as having the form of eq. (5.419), with the components of Eqs. (5.389) and (5.390), being responsible for the gradient ($g$) and the homogenous ($h$) density distributions, respectively. In this frame, the *ELF* information prescribe that as it has values closer to zero as the better electronic localization is providing, according with the limits (5.427).

Establishing a proper chemical quantum index for atomic and inter-atomic shells was always a challenge because the limits of, let's say, radial density distribution $r^2 \rho(r)$ to closely follow the valence shells of heavier atoms. In this concern, as already becoming a tradition in the literature (Becke & Edgecombe, 1990; Scemama et al., 2004; Silvi, 2003), a suitable basic test of ELF will be the Ne atom system due to its property to have both quantum shells completely filled with electrons. The present analysis employs, for orbital implementation, different theoretical levels of SCF method, namely the HF orbitals (Clementi & Roetti, 1974):

$$\varphi_{1s}^{HF}(r) = 5.97998e^{-15.5659r} + 54.4121e^{-9.48486r} + 0.864275e^{-7.79242r}$$
$$+ 0.116083e^{-4.8253r} - 0.00616627e^{-2.86423r} + 0.00316777e^{-1.96184r}$$
$$(5.428)$$

$$\varphi_{2s}^{HF}(r) = -4.69086e^{-15.5659r} - 49.4426e^{-9.48486r} - 18.169e^{-7.79242r}$$
$$+ 12.2167e^{-4.8253r} + 7.17766e^{-2.86423r} + 0.775679e^{-1.96184r}$$
$$(5.429)$$

$$\varphi_{2p}^{HF}(r) = r \begin{pmatrix} 5.45138e^{-9.13464r} + 16.1988e^{-4.48489r} \\ + 5.3916e^{-2.38168r} + 0.639566e^{-1.45208r} \end{pmatrix} \qquad (5.430)$$

as well the simplified solutions of the SCF(Gombás & Szondy, 1970), both in the first approximation, providing the SCF1 spatial orbitals independently of the angular quantum number $l$:

$$\varphi_K^{SCF1}(r) = 353.397e^{-9.87r}r \qquad (5.431)$$

$$\varphi_L^{SCF1}(r) = 5.89511e^{-2.006r}r^{3/2} \qquad (5.432)$$

and in the second approximation, releasing with the SCF2 split up orbitals according to $l$ for a definite value of principal quantum number $n$:

$$\varphi_{1s}^{SCF2}(r) = 353.397e^{-9.87r}r \qquad (5.433)$$

$$\varphi_{2s}^{SCF2}(r) = -307.024e^{-9.87r}r + 2.31275e^{-2.006r}r^{3/2} \qquad (5.434)$$

$$\varphi_{2p}^{SCF2}(r) = 5.89511e^{-2.006r}r^{3/2} \qquad (5.435)$$

Note that the spatial orbitals of neon, Eqs. (5.428)–(5.435), are individually radial normalized to one such that the resulting radial density function,

$$D(r) = r^2 \rho(r) = \sum_{nl} q_{nl} \left| r \varphi_{nl}(r) \right|^2 \qquad (5.436)$$

with $q_{nl}$ the number of equivalent electrons in a subshell ($nl$), to fix by radial integration the total number of 10 electrons of the Ne structure, in each above scheme of computation.

The resulting radial density function as well the BE and actual Markovian ELFs are depicted in the Figure 5.7. (a)–(d) for all above considered levels of orbital structure of Ne. First of all, for all ELFs a clear maximum and minimum corresponding to regions within and between shells are remarked, respectively.

Nevertheless, in the spirit of interpretation of *ELF as the error in electron localization*, see the previous discussion of the (5.427) limits, for a better localized (or trapped) region of the electrons a lower ELF value has to be provided. In this respect, is evident that all the actual Markovian ELFs given by Eqs. (5.416), (5.417), (5.419), (5.420), (5.422), and (5.423) are more reliable localization indices than that of Becke-Edgecombe shaped as (5.388).

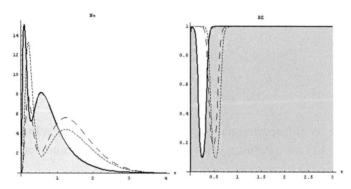

**FIGURE 5.7** (a) The comparative Ne radial density structure given by Eq. (5.436) and the Becke-Edgecombe (BE) electron localization function (ELF) given in (5.388) through different levels of self consistent field (SCF) methods: Hartree-Fock (HF) set (5.428)–(5.430) (—), simplified SCF1 as the first orbital approximation set (5.431), (5.432), independently of angular number (----), and simplified SCF2 as the second orbital approximation set (5.433)–(5.435), dependently on the angular number (....); after Putz (2005).

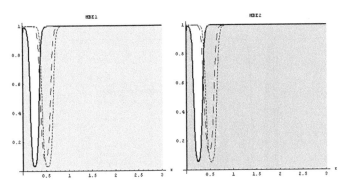

**FIGURE 5.7**    (b) The comparative Markovian ELFs, with expressions (5.416) for MBE1 and (5.417) for MBE2 cases, respectively, through the HF (—), SCF1 (----), and SCF2 (....) theoretical levels of orbital structure of Ne; after (Putz, 2005).

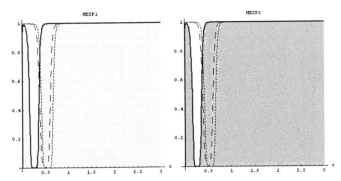

**FIGURE 5.7**    (c) The comparative Markovian ELFs, with expressions (5.419) for MEXP1 (5.420) for MEXP2, cases, respectively, through the HF (—), SCF1 (----), and SCF2 (....) theoretical levels of orbital structure of Ne; after (Putz, 2005).

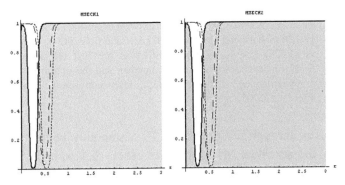

**FIGURE 5.7**    (d) The comparative Markovian ELFs, with expressions (5.422) for MSECH1 and (5.423) for MSECH2 cases, respectively, through the HF (—), SCF1 (----), and SCF2 (....) theoretical levels of orbital structure of Ne, after (Putz, 2005).

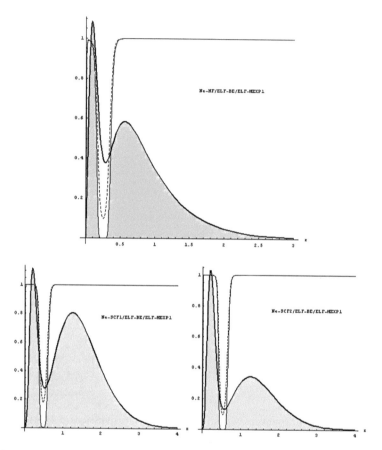

**FIGURE 5.8** (Up and down) The comparative rescaled Ne radial density structure (——) given by Eq. (5.436) with the electron localization functions (ELFs) of Becke-Edgecombe BE (- - -), given in Eq. (5.388), and with the Markovian exponential one MEXP1 (——), given by (5.419), through different levels of self consistent field (SCF) methods: Hartree-Fock (HF) set (5.428)–(5.430), simplified SCF1 as the first orbital approximation set (5.431), (5.432), independently of angular number, and the simplified SCF2 as the second orbital approximation set (5.433)–(5.435), dependently on the angular number, respectively; after Putz (2005).

Moreover, through the present ELFs a smoothly better behavior is shown by the so called M1 Markovian cases, (5.416) and (5.417) formulations, providing a closer value to zero for the error in electron localizations.

Thus, among various classes of Markovian ELFs the most representative and efficient one was advanced as having the form (5.419). In this frame, the *ELF* information prescribe that as it has values closer to zero as the better electronic localization is providing, according with above density gradient limits.

Going to illustrate a further particular application of this scheme the atomic level is firstly presented for the special case of Li atom. The main stages consist in:

- Choosing the basis of the atomic functions:

$$f_1^{Li}(r) = 8.863248 r \exp(-2.698r) \tag{5.437a}$$

$$f_2^{Li}(r) = 0.369721 r^{5/2} \exp(-0.797r) \tag{5.437b}$$

such that to fulfill the natural (radial) normalization conditions

$$\int_0^\infty \left[ f_n^{Li}(r) \right]^2 dr = 1 \,,\, n = 1,2 \tag{5.438}$$

- Generating the ortho-normalized orbital eigen-waves, here according with the Gram-Schmidt algorithm among shells and sub-shells:

$$\varphi_{1s}^{Li}(r) = f_1^{Li}(r),\ \varphi_{2p}^{Li}(r) = f_2^{Li}(r) \tag{5.439a}$$

$$\varphi_{2s}^{Li}(r) = C_{2s}^{Li} \left[ \varphi_{2p}^{Li}(r) - \alpha \varphi_{1s}^{Li}(r) \right]$$
$$= -1.2226 \exp(-2.698r) + 0.373222 r^{5/2} \exp(-0.797r) \tag{5.439b}$$

ensuring the additionally constraints:

$$\int_0^\infty \varphi_{2s}^{Li}(r) \varphi_{1s}^{Li}(r) dr = 0,\ \int_0^\infty \left[ \varphi_{2s}^{Li}(r) \right]^2 dr = 1 \tag{5.440}$$

Generating the working overall electronic density

$$\rho_{Li}(r) = 2 \left[ \varphi_{1s}^{Li}(r) \right]^2 + \left[ \varphi_{2s}^{Li}(r) \right]^2 \tag{5.441}$$

that satisfies the spatial (radial) global $N$-integral condition:

$$\int_0^\infty \rho_{Li}(r)\,dr = 3 \qquad\qquad (5.442)$$

The Li-electronic density is then used for computation of the Markovian ELF, while their comparison is in Figure 5.9 illustrated.

From the Figure 5.9, there appears that the smooth delocalization of electrons of Li represented by density structure is removed by the ELF by clearly indicating where are the regions where the electronic realm is with less uncertainty detected. This way the ELF indicates merely where the electronic transitions behave like a step-function. In this respect, ELF can be regarded as the complement of electronic density being a better indicator of the regions where the bonding may arise. For instance, in the case of Li atomic structure, the fact that the ELF does not displays localization over the second shell (due to its values approaching unity in this range) indicates a natural tendency for releasing the outermost electron to the (virtual) neighborhood atoms with uncompleted last shell(s) while preserving its delocalization feature across the bond. As such the lithium hydride (LiH)

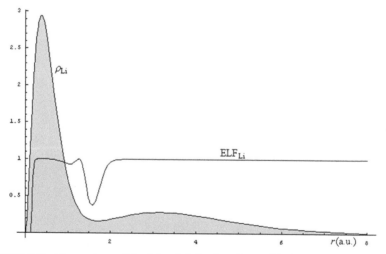

**FIGURE 5.9**  Comparison of the Li-radial density given with the electron localization function for the simplified self-consistent approximation for Li atomic structure; after Putz et al. (2006).

bond is expected to be formed with a certain degree of ionicity in resonance with its covalence: LiH $\leftrightarrow$ Li$^+$H$^-$.

The reliability of ELF to quantify the local tendency of atoms to form bonds and aggregates can be further exemplified to diatomic molecules, while the particular cases of HF, HCl, HBr and HI structures are considered in Figure 5.10. In the bonding region, i.e., in the space between the hydrogen and halogen atomic centers in H-X molecules there are represented

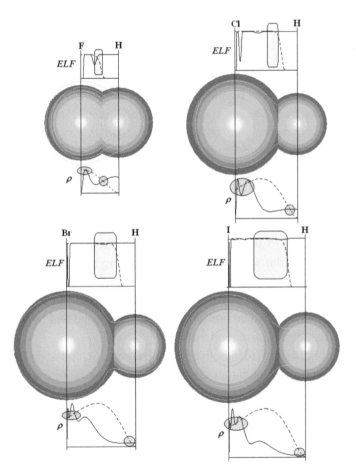

**FIGURE 5.10** Comparative analysis of the charge density contours, electronic localization functions (ELFs), and radial densities for the H (dashed lines), F, Cl, Br, and I (full lines) atoms in molecular combinations HF, HCl, HBr, and HI, respectively; after Putz & Chiriac (2008).

both the electron densities, computed upon above recipe, and the associate Markovian ELFs for the concerned atoms-in-molecules (AIM).

Figure 5.10 clearly shows that while the crossing of hydrogen and halogen radial densities does not provide the right bonding region in HCl, HBr, and HI cases, the corresponding ELFs cross-lines of AIM finely indicate the frontier of atomic basins in hydracids thus confirming the ELF reliability in identifying chemical bonds and bonding.

One can equally say that in the crossing vicinity of *AIM-ELFs* the electrons are at the same time completely localized (for bonding with $ELF_X - ELF_X \rightarrow 0$) and completely delocalized for atomic systems (with $ELF_{X,H} \rightarrow 1$), according to above the *ELF* definition and present signification.

In other words it can be alleged that *ELF application on chemical bond helps in identifying the molecular region in which the electrons undergo the transition from the complete delocalization in atoms to complete localization in molecular bonding behavior.*

Actually, it also proves that localization issue of ionic and covalent classification of bonds may be solved by a "continuous" quantum reality. Such a feature gives, nevertheless, an in-depth understanding of the quantum nature of the chemical bond by associating the mysterious pairing of electrons issue to an analytical function able to distinguish the narrow regions of molecular space where the Heisenberg and Pauli principles are jointly satisfied through the ELF's extreme values. Even more such sharp differentiation between 0 and 1 in atomic and molecular ELF values offers the future possibility in quantifying the chemical bond and bonding in the frame of quantum information theory.

This work has provided both a new interpretation and definition for the ELFas an important quantum chemical index for structure and bonding. The actual view interprets an ELF as the error in localization, on ground of the Heisenberg and Pauli quantum principles, when the inverse of difference in local kinetic terms is involved in its definition. A set of practical constraints to be fulfilled by a viable ELF definition is also given. More, employing the path integral Markovian pair conditional probability density with the basic concepts of the catastrophe theory new Markovian ELF classes are introduced, which generalize the previous Becke-Edgecombe definition. Through a concrete application on basic atomic and molecular systems there was further established that the Markovian ELF given as Eq. (5.419) with components (5.395) and (5.396) behaves almost like a quantum step-function indicator of the electronic traps. Therefore, we propose it for the further analysis of modeling the forms of molecules. Nevertheless, extensions of this work can be also done in its methodological part, if, for instance, the gradient kinetic term of Eq. (5.395) is replaced by a more complex one via Padé approximation. Finally, all these efforts have to serve for our understanding of atomic and molecular structure and their reactivity to a large degree.

## 5.6  CONCLUSION

Going to use the atomic information for further characterizing of chemical bonding, one may consider the algebraic characterization of the chemical potential either for non-reactive and reactive quantum states as the main driving source for ionic-to-covalent bonding in both ground and excited/valence atoms-in-molecules' states, respectively. Accordingly, the decisive role the kinetic and gradient energetic contributions played in bonding atoms-in-molecules is abstractly (thus universally) proven by the dual-quantum co-existence of the reactive (dynamical) equilibrium as Kubo-Martin-Schwinger (KMS) quantum states, aka ionic-to-covalent realizations, respectively.

Then, naturally, the TF theory was formally presented, continuing the abstract-mathematical analysis discourse, modeling the atomic and atoms-in-molecules stability, as the major quantum/mathematical source of the Life' stability at large. Being initially proposed for homogenous electronic distribution, further exchange and gradient terms corrections were added by Dirac and von Weizsaecker, respectively, this way representing the benchmark of the DFT chemical formulation of homogeneous electronic systems locally perturbed by gradient corrections (near the attraction centers) and at distance by exchange and correlation contributions (in boning and atomic inter/intra-shells distributions). Such possibility is sustained, for instance, by the Teller non-bonding theorem (thus related with the uniform electronic distribution as a sort of quantum vacuum from where the chemical reality springs as quantum dots/perturbations near and between the nuclei of a given system), by which the mathematical analysis prevails over the numerical one, by preceding it. Moreover, the convexity condition in the energy functional is also grounded by the parabolic shape of the energy-number of electronic dependency E(N) assumed in the conceptual DFT for modeling the chemical reactivity, see also the Volume III of the present five-volume set (Putz, 2016b). Nevertheless, TF theory should be considered there where the compromise between the local approximation for a homogeneous electronic distribution and the lack of a global energetic minimization for a density gradient included contribution does not impede a macroscopic/observable or a computational analysis whatever.

The conceptual and analytical line of non-equilibrium approaching equilibrium in electronic collective (by density) moving in atoms and in atoms-in-molecules is then continued by modeling the open systems through the already introduced path integrals, here within the Fokker-Planck picture like the main frame in which the effective densities and probability densities are computed (after the first level of quantum mechanics – see Volume I (Putz, 2016a), and of second level of quantum statistics – see Chapters 3 and 4 of this volume, in the present five-volume set). The principal advantage of this approach consists in the direct calculations that claim to solve only one effective path integral, instead of one equation of motion, the molecular Schrödinger and the Fokker-Planck equations, for instance. Another advantage states in the real time continuation that can be directly applied, after all the calculations and implementations were done. However, in such a way, the corresponding mesoscopic algorithms for the equilibrium and non-equilibrium were considered. They can lead with the specific density and probability density functionals implementations which carry out the chemical complex activities, either in the femto-second laser controlled reactions or in the stochastic ones (Putz, 2013). For the nonequilibrium mesoscopic algorithm, the perturbation solution of the probability density for the effective electron that stochastically moves in the general enharmonic potential was developed (since by the fourth order of coordinate oscillation in enharmonic potential the analysis may be further connected with the symmetry breaking of the quantum fields in generating –out of the uniformly oscillating electrons in homogeneous system – the chemical boning with a certain degree of density localization). Nevertheless, the main ingredients for the further mesoscopic goals are here grounded. Accordingly, the present study brings new insight connecting mesoscopic algorithms for the equilibrium and non-equilibrium dynamics with electronic localization by avoiding the fashioned quantum orbitals, while using the quantum amplitudes instead. The study opens new perspectives for the considerations of the basic physical-chemical processes based on the fields, currents and associated forces. All this efforts are designated to contribute to the construction of a self-consistent picture lying between the mesoscopic causal characterizations, and the macroscopic recorded phenomena; see also the Chapter 4 of the Volume I of the present five-volume set (Putz, 2016a).

However, the electronic localization complements, at the local level, the quantum information comprised in the atomic and atoms-in-molecules behavior described through the so called *localization functions*. These should express the balance between the local stability and the delocalization tendency of the involved electrons in the chemical bond in the view of the forthcoming transformations. This way, the localization functions indicate the ratio of the non-uniformly localized electronic distribution to the uniform delocalization of the electronic gas, accordingly with the Heisenberg quantum principle of delocalization and that of the Pauli indiscernibility. Accordingly, a new interpretation and definition for the ELF was presented as an important quantum chemical index for structure and bonding. The actual view interprets an ELF as the error in localization, on ground of the Heisenberg and Pauli combined quantum principles, through the involvement of the inverse of difference in local kinetic terms in its definition. A set of practical constraints to be fulfilled by a viable ELF definition is also given. Then, by employing the path integral Markovian pair conditional probability density with the basic concepts of the catastrophe theory new Markovian ELF classes are introduced, which generalize the previous Becke-Edgecombe definition. With the aid of general and specific applications on basic atomic and molecular systems there was further established that certain Markovian ELF behaves almost like a quantum step-function indicator of the electronic traps. Therefore, it was further proposed for modeling the formation and reactivity of molecules analysis; this was however proved, through a series of hydracids molecules that the atoms-in-molecule ELF in its exponential form and with error interpretation, as recently recommended by one of the authors' recipe

(Putz, 2005), stands as a viable quantum tool for identifying bonds within the bonding space.

Overall, while admitting that "a chemical reaction is a change in bonding" (Parr & Yang, 1989), we arrive to chemical bonding characterization through chemical reactivity concepts (Putz & Chiriac, 2008). They are, however, classified as reactivity indices that span the local and global indicators responsible for chemical affinity and bonding, and reactivity principles that consecrate the rules upon which the reactions can be rationalized when reactivity indices are employed. As such, the chemical reactivity indices, were introduced in the present volume for illustrating the atomic scales and elemental periodicities while they will be further employed for modeling molecular structure and chemical reactivity in the Volume III of the present five-volume book (Putz, 2016b). Finally, all these efforts have to serve for our understanding of atomic and molecular structure and of their reactivity to a large degree.

## KEYWORDS

- Becke-Edgecombe
- Chemical bonding
- electronic localization
- Fokker-Planck Theory
- Hartree-Fock
- Kubo-Martin-Schwinger (KMS) States
- localization functions
- Markovian path integral
- non-equilibrium
- non-reactive systems
- quantum algebra
- Quantum Algebras
- reactive equilibrium
- self-consistent field
- Thomas-Fermi Theory

# REFERENCES

## AUTHOR'S MAIN REFERENCES

Putz, M. V. (2016a). *Quantum Nanochemistry. A Fully Integrated Approach: Vol. I. Quantum Theory and Observability*. Apple Academic Press & CRC Press, Toronto-New Jersey, Canada-USA.

Putz, M. V. (2016b). *Quantum Nanochemistry. A Fully Integrated Approach: Vol. III. Quantum Molecules and Reactivity*. Apple Academic Press & CRC Press, Toronto-New Jersey, Canada-USA.

Putz, M. V. (2014). *Quantum and Optical Dynamics of Matter for Nanotechnology*, IGI Global, Hershey Passadena (DOI: 10.4018/978-1-4666-4687-2).

Putz, M. V. (2012). *Quantum Theory: Density, Condensation, and Bonding*, Apple Academics, Toronto.

Putz, M. V. (2009). Path integrals for electronic densities, reactivity indices, and localization functions in quantum systems. *International Journal of Molecular Sciences* 10(11), 4816–4940 (DOI: 10.3390/ijms10114816).

Putz, M. V., Chiriac, A. (2008). Quantum Perspectives on the Nature of the Chemical Bond. In: *Advances in Quantum Chemical Bonding Structures*, Putz, M. V. (Ed.), Transworld Research Network, Kerala, Chapter 1, pp. 1–43.

Putz, M. V. (2008). Density functionals of chemical bonding. *Int. J. Mol. Sci.* 9(6), 050–1095 (DOI: 10.3390/ijms9061050)

Putz, M. V., Ionaşcu, C., Hulesch, O. (2006). Global and local indicators of chemical bonding. In: *Proceedings of the VIII-th International Symposium Young People and Multidisciplinary Research*, Welding Publishing House, Timişoara, Romania, pp. 632–641.

Putz, M. V. (2005). Markovian approach of the electron localization functions. *Int. J. Quantum Chem.* 105(1), 1–11 (DOI: 10.1002/qua.20645).

Kleinert, H., Pelster, A., Putz, M. V. (2002). Variational perturbation theory for markov processes. *Physical Review E* 65, 066128/1–7 (arXiv: cond-mat/0202378).


## SPECIFIC REFERENCES

Araki, H., Kastler, D., Takesaki, M., Haag, R. (1977). Extension of KMS states and chemical potential. Comm. Math. Phys. 53(2), 97.

Araki, H., Kishimoto, A. (1977). Symmetry and equilibrium states. *Comm. Math. Phys.* 52(3), 211–232.

Ayers, P. W., Levy, M. (2001). Sum rules for exchange and correlation potentials. *J. Chem. Phys.* 115, 4438–4443.

Bader, R. F. W. (1990). *Atoms in Molecules: A Quantum Theory*, Oxford University Press, Oxford.

Bader, R. F. W., Gillespie, R. J., MacDougall, P. J. (1988). A physical basis for the VSEPR model of molecular geometry. *J. Am. Chem. Soc.* 110, 7329–7336.

Balàzs, N. (1967). Formation of stable molecules within the statistical theory of atoms. *Phys. Rev.* 156 (1), 42–47.

Balescu, R. (1975). *Equilibrium and Nonequilibrium Statistical Mechanics*, John-Wiley & Sons, New York.

Baltin, R. (1987). The three-dimensional kinetic energy density functional compatible with the exact differential equation for its associated tensor. *J. Chem. Phys.* 86, 947–952.

Bartolotti, L. J., Acharya, P. K. (1982). On the functional derivative of the kinetic energy density functional. *J. Chem. Phys.* 77, 4576–4585.

Becke, A. D. (1986). Completely numerical calculations on diatomic molecules in the local-density approximation. *Phys. Rev. A 33*, 2786–2788.

Becke, A. D. (1988). Correlation energy of an inhomogeneous electron gas: A coordinate-space model. *J. Chem. Phys.* 88, 1053–1062.

Becke, A. D., Edgecombe, K. E. (1990). A simple measure of electron localization in atomic and molecular systems. *J. Chem. Phys.* 92, 5397–5403.

Berski, S., Andres, J., Silvi, B., Domingo, L. R. (2003). The joint use of catastrophe theory and electron localization function to characterize molecular mechanisms. a density functional study of the Diels–Alder reaction between ethylene and 1,3-butadiene. *J. Phys. Chem. A* 107(31), 6014–6024.

Bratteli, O., Robinson, D. W. (1987a). *Operator Algebras and Quantum Statistical Mechanics*- Vol. I, Springer-Verlag New York, pp. 17–152.

Bratteli, O., Robinson, D. W. (1987b). *Operator Algebras and Quantum Statistical Mechanics*- Vol. II, Springer-Verlag New York, pp. 77–334.

Chan, G. K. L., Handy, N. C. (1999). Kinetic-energy systems, density scaling, and homogeneity relations in density functional theory. *Phys. Rev. A* 59, 2670–2679.

Cioslowski, J. (1990). Nonnuclear Attractors in the $LI_2$ Molecule. *J. Phys. Chem.* 94, 5496–5498.

Clementi, E., Roetti, C. (1974). Roothaan-Hartree-Fock atomic wave functions: Basis functions and their coefficients for ground and certain excited states of neutral and ionized atoms, $Z \leq 54$. *At. Data. Nucl. Data. Tables* 14, 177–478.

Connes, A. (1973). Une classification des facteurs de type III. Ann. Scient. Ec. Norm. Sup. (6), 133–252.

Crick, F. H. C. (1976). Linking numbers and nucleosomes. *Proc. Nat. Acad. Sci. USA* 73, 2639–2643.

Daudel, R., Leroy, G., Peeters, D., Sana, M. (1983). *Quantum Chemistry*, John Wiley & Sons, New York.

Davies, E. B. (1976). *Quantum Theory of Open Systems*, Academic Press, New York.

Dawson, K. A., March, N. H. (1984). Slater sum in one dimension: explicit kinetic energy functional. *Phy. Lett.* 106A, 158–160.

Dirac, P. A.M (1930). Note on the exchange phenomena in the Thomas atom. *Proc. Cambridge Philos. Soc.* 26, 376–385.

Doplicher, S.: Haag, R., Roberts, J. E. (1969a). Fields, observables and gauge transformations. I. *Comm. Math. Phys.* 13(1), 1–80.

Doplicher, S., Haag, R., Roberts, J. E. (1969b). Fields, observables and gauge transformations. II. *Comm. Math. Phys.* 15(3), 173–200.

Doplicher, S., Kastler, D., Robinson, D. W. (1966). Covariance algebras in field theory and statistical mechanics. *Comm. Math. Phys.* 3(1), 1–28.

Dufek, P., Blaha, P., Sliwko, V., Schwarz, K. (1994). Generalized-gradient-approximation description of band splitting's in transition-metal oxides and fluorides. *Phys. Rev. B* 49, 10170–10175.

Dunlap, B. I. (1988). Symmetry and spin density functional theory. *Chem. Phys.* 125, 89–97.

Dunlap, B. I., Andzelm, J. (1992). Second derivatives of the local-density-functional total energy when the local potential is fitted. *Phys. Rev. A* 45, 81–86.

Dunlap, B. I., Andzelm, J., Mintmire, J. W. (1990). Local-density-functional total energy gradients in the linear combination of Gaussian-type orbitals method. *Phys. Rev. A* 42, 6354–6358.

Fermi, E. (1927). Un Metodo Statistico per la Determinazione di alcune Prioprietà dell'Atomo. *Rend. Accad. Naz. Lincei* 6, 602–607.

Feynman, R. P., Hibbs, A. R. (1965). *Quantum Mechanics and Path Integrals*, McGraw Hill, New York.

Filippeti, A. (1998). Electron affinity in density-functional theory in the local spin-density approximation. *Phys. Rev. A* 57, 914–919.

Frisch, H. L., Wasserman, E. (1961). Chemical topology. *J. Am. Chem. Soc.* 83, 3789–3795.

Fuller, F. B. (1971). The Writhing number of a space curve. *Proc. Nat. Acad. Sci. USA* 68, 815–819.

Garcia-Gonzales, P., Alvarellos, J. E., Chacon, E. (1996). Kinetic-energy density functional: atoms and shell structure. *Phys. Rev. A* 54, 1897–1905.

Gardiner, C. (1994). *Handbook of Stochastic Methods*, Springer-Verlag, Berlin.

Gaspar, R., Nagy, A. (1987). Local-density-functional approximation for exchange-correlation potential. Application of the self-consistent and statistical exchange-correlation parameters to the calculation of the electron binding energies. *Theor. Chim. Acta* 72, 393–401.

Gelfand, I. M., Naimark, M. A. (1943). On the embedding of normed rings into the ring of operators in Hilbert space. *Mat. Sb.* 12, 197–213.

Gillespie, R. J. (1972). *Molecular Geometry*, Van Nostrand Reinhold, London.

Gombás, P., Szondy, T. (1970). *Solutions of the Simplified Self-Consistent Field for All Atoms of the Periodic System of Elements from Z=2 to Z=92*, Adam Hilger Ltd, London.·

Gray, P., Scott, K. (1990). *Chemical Oscillations and Instabilities*, Clarendon Press, Oxford.

Gritsenko, O. V., Schipper, P. R. T., Baerends, E. J. (2000). Ensuring proper short-range and asymptotic behavior of the exchange-correlation kohn-sham potential by modeling with a statistical average of different orbital model potential. *Int. J. Quantum Chem.* 76, 407–419.

Guo, Y., Whitehead, M. A. (1989). Application of generalized exchange local-spin-density-functional theory: electronegativity, hardness, ionization potential, and electron affinity. *Phys. Rev. A* 39, 2317–2323.

Guo, Y., Whitehead, M. A. (1991). Generalized local-spin-density-functional theory. *Phys. Rev. A* 43, 95–108.

Haag, R., Kadison, R. V., Kastler, D. (1970). Nets of C*-algebras and classification of states Comm. Math. Phys. 16(2), 81- 104.

Haag, R. (1962). The mathematical structure of the Bardeen-Cooper-Schrieffer model. Il Nuovo Cimento 25(2), 287–299.

Haken, H. (1987). *Advanced Synergetics*, Springer-Verlag, Heidelberg.

Haken, H. (1978). *Synergetics*, Springer-Verlag, Heidelberg.

Haken, H. (1988). *Information and Self-Organization*, Springer-Verlag, Heidelberg.

Hänggi, P., Talkner, P., Borkovec, M. (1990). Reaction-rate theory: Fifty years after Kramers. *Rev. Mod. Phys. 62*, 251–341.

Harrison, J. G. (1987). Electron affinities in the self-interactions-corrected local spin density approximation. *J. Chem. Phys.* 86, 2849–2853.

Hepp, K. (1972). Quantum theory of measurement and macroscopic observables. *Helv. Phys. Acta* 45, 237–248.

Hugenholtz, N. M. (1967). On the factor type of equilibrium states in quantum statistical mechanics Commun. Math. Phys. 6(3), 189–193.

Jensen, F. (2005). On the accuracy of numerical Hartree–Fock energies, *Theor. Chem. Acc.* 113, 187–190.

Kleinert, H., Pelster, A., Bachmann, M. (1999). Generating functionals for harmonic expectation values of paths with fixed end points: Feynman diagrams for nonpolynomial interactions. *Phys. Rev. E* 60, 2510–2527. (arXiv: quant-ph/9902051)

Kleinert, H. (2001). *Path Integrals in Quantum Mechanics, Statistics and Polymer Physics* (Third Edition), World Scientific, Singapore.

Koch, W., Holthausen, M. C. (2000). *A Chemist's Guide to Density Functional Theory*, Wiley-VCH, Weinheim.

Kohout, M., Pernal, K., Wagner, F. R., Grin, Y. (2004). Electron localizability indicator for correlated wave functions. I. Parallel-spin pairs. *Theor. Chem. Acc.* 112, 453–459.

Lam, K. C., Cruz, F. G., Burke, K. (1998). Viral exchange-correlation energy density in Hooke's atom. *Int. J. Quantum Chem. 69*, 533–540.

Levy, M. (1991). Density-functional exchange correlation through coordinate scaling in adiabatic connection and correlation hole. *Phys. Rev. A* 43, 4637–4645.

Lieb, E. H. (1976). The stability of matter. *Rev. Mod. Phys.* 48, 553–569.

Lieb, E. H. (1981). Thomas-fermi and related theories of atoms and molecules. *Rev. Mod. Phys.* 53, 603–641.

Lieb, E. H., Oxford, S. (1981a). Improved lower bound on the indirect Coulomb energy. *Int. J. Qunatum Chem.* 19, 427–439.

Lieb, E. H., Simon, B. (1977). The Thomas–Fermi theory of atoms, molecules and solids. *Adv. In Math.* 23(1), 22–116.

Lieb, E. H., Simon, B. (1978). Monotonicity of the electronic contribution to the Born-Oppenheimer energy. *J. Phys. B: At. Mol. Phys.* 11:L537–542.

Lieb, E. H., Thirring, W. (1975). Bound for the kinetic energy of fermions which proves the stability of matter. *Phys. Rev Lett.* 35(11), 687–689.

Liu, S. (1996). Local-density approximation, hierarchy of equations, functional expansion, and adiabatic connection in current-density-functional theory. *Phys. Rev. A* 54, 1328–1336.

Liu, S., Ayers, P. W., Parr, R. G. (1999). Alternative definition of exchange-correlation charge in density functional theory. *J. Chem. Phys.* 111, 6197–6203.

Maksimov, V. M. (1974). Macroscopic observables in algebraic statistical physics. Theor. Math. Phys. 20(1), 632–638.

Manoli, S. D., Whitehead, M. A. (1988). Generalized exchange local-spin-density-functional theory: one-electron energies and eigenvalues. Coll. Czechoslovak Chem. Commun. 53, 2279–2307.

March, N. H. (1983) 1. Origins – The Thomas–Fermi Theory, In: Lundqvist, S., March, N. H. Theory of The Inhomogeneous Electron Gas.

March, N. H. (1992). Electron Density Theory of Atoms and Molecules, Academic Press.

March, N. H., Howard, I. A., Van Doren, V. E. (2003). Recent progress in constructing nonlocal energy density functionals. Int. J. Quantum Chem. 92, 192–204.

Matito, E., Silvi, B., Duran, M., Solà, M. (2006). Electron localization function at the correlated level, J. Chem. Phys. 125, 024301.

Matito, E., Solà, M., Salvador, P., Duran, M. (2007). Electron sharing indexes at the correlated level. Application to aromaticity calculations. Faraday Discuss. 135, 325–345.

Mebkhout, M. (1979). Algebraic theory of the chemical potential in the case of the usual gauge group. Ann. Phys. 123(2), 317–329.

Mezey, P. G. (1987). Potential Energy Hypersurfaces, Elsevier, Amsterdam.

Müller-Herold, U. (1980). Disjointness of β-KMS states with different chemical potential. Lett. Math. Phys. 4(1), 45–48.

Müller-Herold, U. (1982). Chemisches potential, reaktionssysteme und algebraische quantenchemie. Fortschr. Physik 30(1), 1–73.

Murphy, D. R. (1981). Sixth-order term of the gradient expansion of the kinetic-energy density functional. Phys. Rev. A 24, 1682–1688.

Murray, F. J., von Neumann, J. (1936). On rings of operators Ann. Math. 37(2), 116–229.

Nesbet, R. K. (2002). Orbital functional theory of linear response and excitation. Int. J. Quantum Chem. 86(4), 342–46.

Parr, R. G., Donnelly, R. A., Levy, M., Palke, W. E. (1978,) Electronegativity: the density functional viewpoint. J. Chem. Phys. 68, 3801–3808.

Parr, R. G., Yang, W. (1989). Density Functional Theory of Atoms and Molecules, Oxford University Press, New York.

Pelster, A., Kleinert, H. (1996). Relations between Markov processes via local time and coordinate transformations. Phys. Rev. Lett. 78, 565–569. (arXiv: cond-mat/9608120v2)

Ponec, R., Cooper, D. L. (2007). Anatomy of bond formation. Bond length dependence of the extent of electron sharing in chemical bonds from the analysis of domain-averaged Fermi holes. Faraday Discuss. 135, 31–42.

Preuss, H. (1969). Quantenchemie fuer Chemiker, Verlag Chemie, Weinheim.

Primas, H., Müller-Herold, U. (1978). Quantum mechanical system theory. a unifying framework for observations in quantum mechanics. Adv. Chem. Phys. 38, 1–107.

Risken, H. (1984). The Fokker-Planck Equation, Springer-Verlag, Heidelberg.

Romera, E., Dehesa, J. S. (1994). Weizsäcker energy of many-electron systems. Phys. Rev. A 50, 256–266.

Santos, J. C., Tiznado, W., Contreras, R., Fuentealba, P. (2000). Sigma-pi Separation of the Electron Localization Function and Aromaticy, J. Chem. Phys. 120, 1670–1673.

Savin, A., Jepsen, J., Andersen, O. K., Preuss, H., von Schnering, H. G. (1992). Electron localization in the solid-state structures of the elements: The diamond structure. Angew. Chem., Int. Ed. Engl. 31, 187–188.

Scemama, A., Chaquin, P., Caffarel, M. (2004). Electron pair localization function: a practical tool to visualize electron localization in molecules from quantum Monte Carlo data. *J. Chem. Phys.* 121, 1725–1735.

Schmidera, H. L., Becke, A. D. (2002). Two functions of the density matrix and their relation to the chemical bond. *J. Chem. Phys.* 116(8), 3184.

Segal, I. E. (1947). The group algebra of a locally compact group. *Trans. Amer. Math. Soc.* 61, 69–105.

Silvi, B. (2003). The spin-pair compositions as local indicators of the nature of the bonding. *J. Phys. Chem.* 107, 3081–3085.

Silvi, B., Gatti, C. (2000). Direct space representation of the metallic bond. *J. Phys. Chem. A* 104, 947–953.

Silvi, B., Savin, A.( 1994). Classification of the chemical bonds based on topological analysis of electron localization functions. *Nature* 371, 683–686.

Soncini, A., Lazzeretti, P. (2003). Nuclear spin-spin coupling density functions and the fermi hole. *J. Chem. Phys.* (119) 1343–1349.

Takesaki, M. (1970). Disjointness of the KMS-states of different temperatures Comm. Math. Phys. 17, 33–41.

Takesaki, M. (1973). Duality for crossed products and the structure of von Neumann algebras of type III. Acta Math. 131(1), 249–310.

Teller, E. (1962). On the stability of molecules in the Thomas–Fermi theory. *Rev. Mod. Phys.* 34 (4), 627–631.

Thom, R. (1973). *Stabilitè Structurelle et Morphogènése*, Benjamin-Addison-Wesley, New York.

Thomas, L. H. (1927). The calculation of atomic fields. *Proc. Cambridge Phil. Soc.* 23(5), 542–548.·

Tomita, M. (1956). Harmonic analysis on locally compact groups. Math. J. Okayama Univ. 5, 133–193.

van Kampen, N. G. (1987). *Stochastic Processes in Physics and Chemistry*, North-Holland.

von Neumann, J. (1929). Zur algebra der funktionaloperatoren und theorie der normalen operatoren. *Math. Ann.* 102, 370–427.

von Neumann, J. (1961). *Collected Works*, Pergamon Press, New York.

Voth, G. A., Chandler, D., Miller, W. H. (1989). Rigorous formulation of quantum transition state theory and its dynamical corrections. *J. Chem. Phys.* 91, 7749–7760.

Weiss, U. (1993). *Quantum Dissipative Systems*, World Scientific, Singapore.

Weizsäcker, C. F. V. (1935). Zur Theorie der Kernmassen. *Zeitschrift für Physik* 96(7–8), 431–458.

Zhao, Q., Morrison, R. C., Parr, R. G. (1994). From electron densities to kohn-sham kinetic energies, orbital energies, exchange-correlation potentials, and exchange-correlation energies. *Phys. Rev. A* 50, 2138–2142.

Zhao, Q., Parr, R. G. (1992). Local exchange-correlation functional: numerical test for atoms and ions. *Phys. Rev. A* 46, R5320–R5323.

# APPENDIX

# THE STEEPEST DESCENT (SADDLE POINT) METHOD

## CONTENTS

One has the problem to appropriately solve the integral of the type

$$I(\alpha) = \int g(x)e^{\alpha f(x)}dx = ? \tag{A1}$$

so useful in solving the integrals specific to atomic scales determinations (see Sections 3.4.1 and 4.7.1)

Without going into details (Hassani, 1991), if one has to solve an integral of the (A1) type with $\alpha > 0$, the saddle point approximation or the stationary phase method or the method of the steepest descendent requires its expansion around the point $x_0$, the solution of the extreme equation:

$$\left.\frac{\partial f(x)}{\partial x}\right|_{x=x_0} = 0 \tag{A2}$$

As the general recipe in solving (A1) one uses first the condition (A2) in the second order truncated Taylor expansion of the phase $f(x)$ of (A1):

$$f(x) \cong f(x_0) + \frac{1}{2}(x - x_0)^2 f''(x_0) \tag{A3}$$

Then, because $f(x)-f(x_0)$ should be real and negative along the integration path in (A1) there is useful to adopt from (A3) the notation:

$$f(x) - f(x_0) \equiv -t^2 = \frac{1}{2}(x - x_0)^2 f''(x_0) \tag{A4}$$

with the help of which the integral (A1) rewrites accordingly, under the successive transformations:

$$I(\alpha) \cong \int g(x) \exp\left\{\alpha\left[f(x_0) - t^2\right]\right\} dx$$

$$= \exp\left[\alpha f(x_0)\right] \int \exp\left(-\alpha t^2\right) g(x) dx$$

$$= \exp\left[\alpha f(x_0)\right] \int \exp\left(-\alpha t^2\right) g(x(t)) \frac{dx(t)}{dt} dt \tag{A5}$$

Next, since further asymptotic form of the integral (A5) is assumed, the Mac Lauren series in $t$ can be considerate as:

$$g(x(t)) \frac{dx(t)}{dt} = \sum_{k=0}^{\infty} a_k t^k \tag{A6}$$

providing the expansion form of the integral (A5) too:

$$I(\alpha) = \exp\left[\alpha f(x_0)\right] \sum_{k=0}^{\infty} a_k \int t^k \exp\left(-\alpha t^2\right) dt \tag{A7}$$

When only the leading term of Eq. (A7) is retained, one yields:

$$I(\alpha) \cong \exp\left[\alpha f(x_0)\right] a_0 \int \exp\left(-\alpha t^2\right) dt \tag{A8}$$

From now, in the form (A8) there is straight to recognize the celebrated Poisson type integral, see Volume I/Appendices of the present five-volume set (Putz, 2016):

$$\int \exp\left(-\alpha t^2\right) dt = \sqrt{\frac{\pi}{\alpha}} \tag{A9}$$

whereas, for the $a_0$, by employing the (A6) definition and the right side identity of (A4) the following successive relations occur:

$$a_0 = g\left(x(t)\right)\frac{dx(t)}{dt}$$

$$= g\left(x_0 + t\frac{dx(t)}{dt} + ...\right)\frac{dx(t)}{dt}$$

$$\overset{t \to 0}{\cong} g(x_0)\sqrt{\frac{-2}{f''(x_0)}} \tag{A10}$$

Finally, with Eqs. (A9) and (A10) back in Eq. (A8) the approximate value of integral (A1) is obtained as:

$$I(\alpha) \cong g(x_0)\exp\left[\alpha f(x_0)\right]\sqrt{-\frac{2\pi}{\alpha f''(x_0)}} \tag{A11}$$

## REFERENCES

Hassani, S. (1991). *Foundation of Mathematical Physics*, Prentice-Hall International, Inc., Chapter 7.

Putz, M. V. (2016). *Quantum Nanochemistry. A Fully Integrated Approach: Vol. I. Quantum Theory and Observability*. Apple Academic Press & CRC Press, Toronto-New Jersey, Canada-USA.

# INDEX

T - #0798 - 101024 - C548 - 229/152/24 - PB - 9781774631003 - Gloss Lamination